GIS Applications in Agriculture

Volume Three: Invasive Species

GIS Applications in Agriculture

Series Editor
Francis J. Pierce
Washington State University, Prosser

GIS Applications in Agriculture, *edited by Francis J. Pierce and David E. Clay*

GIS Applications in Agriculture, Volume Two: Nutrient Management for Energy Efficiency, *edited by David E. Clay and John F. Shanahan*

GIS Applications in Agriculture, Volume Three: Invasive Species, *edited by Sharon A. Clay*

GIS Applications in Agriculture

Volume Three: Invasive Species

Edited by
Sharon A. Clay

GIS APPLICATIONS IN AGRICULTURE SERIES

CRC Press
Taylor & Francis Group
Boca Raton London New York

CRC Press is an imprint of the
Taylor & Francis Group, an **informa** business

CRC Press
Taylor & Francis Group
6000 Broken Sound Parkway NW, Suite 300
Boca Raton, FL 33487-2742

First issued in paperback 2019

ISBN-13: 978-1-4200-7880-0 (hbk)
ISBN-13: 978-0-367-38305-3 (pbk)

Library of Congress Cataloging-in-Publication Data

GIS applications in agriculture. Volume three, Invasive species / Sharon A. Clay [editor].
p. cm. -- (GIS applications in agriculture)
Includes bibliographical references and index.
ISBN 978-1-4200-7880-0
1. Agriculture--Remote sensing. 2. Geographic information systems. 3. Agricultural mapping. 4. Agriculture--Data processing. 5. Introduced organisms. I. Clay, Sharon A. (Sharon Ann) II. Title: Invasive species. III. Series: GIS applications in agriculture series.

S494.5.R4G572 2011
632--dc22 2010034459

Visit the Taylor & Francis Web site at
http://www.taylorandfrancis.com

and the CRC Press Web site at
http://www.crcpress.com

Contents

Series Preface ix
Preface xi
Acknowledgments xiii
Editor xv
Contributors xvii

Chapter 1 Introduction: Remote Sensing and GIS Techniques for the Detection, Surveillance, and Management of Invasive Species 1

Kevin Dalsted

Chapter 2 Obtaining Spatial Data 9

Mary O'Neill and Kevin Dalsted

Chapter 3 Population Ecology Considerations for Monitoring and Managing Biological Invasions 29

Patrick C. Tobin, Laura M. Blackburn, Shelby J. Fleischer, and E. Anderson Roberts

Chapter 4 Integrating GPS, GIS, and Remote Sensing Technologies with Disease Management Principles to Improve Plant Health 59

Forrest W. Nutter, Jr., Emmanuel Z. Byamukama, Rosalee A. Coelho-Netto, Sharon K. Eggenberger, Mark L. Gleason, Andrew Gougherty, Alison E. Robertson, and Neil van Rij

Chapter 5 Mapping Actual and Predicted Distribution of Pest Animals and Weeds in Australia 91

Peter West, Leanne Brown, Christopher Auricht, and Quentin Hart

Chapter 6 Use of GIS Applications to Combat the Threat of Emerging Virulent Wheat Stem Rust Races 129

David Hodson and Eddy DePauw

Chapter 7 Online Aerobiology Process Model 159

Joseph M. Russo and Scott A. Isard

Chapter 8 Site-Specific Management of Green Peach Aphid, *Myzus persicae* (Sulzer)........ 167

Ian MacRae, Matthew Carroll, and Min Zhu

Chapter 9 Analysis of the 2002 Equine West Nile Virus Outbreak in South Dakota Using GIS and Spatial Statistics........ 191

Michael C. Wimberly, Erik Lindquist, and Christine L. Wey

Chapter 10 Designing a Local-Scale Microsimulation of Lesser Grain Borer Population Dynamics and Movements........ 207

J.M. Shawn Hutchinson, James F. Campbell, Michael D. Toews, Thomas J. Vought, Jr., and Sonny B. Ramaswamy

Chapter 11 Geographic Information Systems in Corn Rootworm Management........ 233

B. Wade French, Kurtis D. Reitsma, Amber A. Beckler, Laurence D. Chandler, and Sharon A. Clay

Chapter 12 Improving Surveillance for Invasive Plants: A GIS Toolbox for Surveillance Decision Support........ 255

Julian C. Fox and David Pullar

Chapter 13 Tracking Invasive Weed Species in Rangeland Using Probability Functions to Identify Site-Specific Boundaries: A Case Study Using Yellow Starthistle (*Centaurea solstitialis* L.)........ 277

Lawrence W. Lass, Timothy S. Prather, Bahman Shafii, and William J. Price

Chapter 14 Using GIS to Map and Manage Weeds in Field Crops........ 301

Mary S. Gumz and Stephen C. Weller

Chapter 15 Adapting Geostatistics to Analyze Spatial and Temporal Trends in Weed Populations........ 319

Nathalie Colbach and Frank F. Forcella

Chapter 16 Using GIS to Investigate Weed Shifts after Two Cycles of a Corn/Soybean Rotation 373

Kurtis D. Reitsma and Sharon A. Clay

Chapter 17 Creating and Using Weed Maps for Site-Specific Management 405

J. Anita Dille, Jeffrey W. Vogel, Tyler W. Rider, and Robert E. Wolf

Index 419

Series Preface

GIS Applications in Agriculture: Invasive Species edited by Sharon A. Clay is the third volume in the book series GIS Applications in Agriculture. The first volume, *GIS Applications in Agriculture*, edited by Francis J. Pierce and David E. Clay was published by CRC Press in 2007. The second volume, *GIS Applications in Agriculture: Nutrient Management for Energy Efficiency*, edited by David E. Clay and John F. Shanahan will be published by CRC Press in 2011. These books each provide concepts and detailed examples of important aspects of agriculture that lend themselves to geospatial analysis and mapping provided by geographic information systems (GIS). The genesis of a book of applications of GIS in agriculture came from my good friend Max Crandall who recognized that few examples from agriculture were available for use in teaching or demonstrating GIS to the agricultural community. Max recruited Pierre C. Robert, Harold Reitz Jr., Matthew Yen, and myself to help formulate such a book, but the untimely death of Pierre diminished the effort. Through the help of John Sulzycki, CRC Press agreed to publish the first book in this series and later to commit to a book series, GIS Applications in Agriculture, for which I am the series editor.

GIS Applications in Agriculture: Invasive Species presents 17 chapters addressing various aspects of invasive species, including viruses, diseases, insects, animal pests, and weeds. Like the first two volumes, detailed applications are provided in many chapters with data sets and color figures on a separate CD for readers to use in teaching and learning GIS or in directly applying them to situations they encounter in agriculture. I am grateful to Sharon Clay for agreeing to develop this volume and for her hard work in organizing and editing it, and to the chapter authors for their excellent contributions to what I believe are interesting and useful applications of GIS in agriculture. I would also like to thank Randy Brehm, the CRC Press editor for this volume, and all those at CRC Press who made this volume possible.

As the series editor, it is my responsibility to seek new book ideas and capable editors to create additional volumes in this series on topics of importance to agriculture that provide relevant applications of GIS in agriculture. I invite those who have ideas for new volumes in the GIS Applications in Agriculture series to contact myself or CRC Press to discuss publishing opportunities.

Francis J. Pierce
Series Editor

Preface

A Web search for the key words "GIS" and "invasive species" throws over 1 million entries. One book can certainly not highlight all the different uses of GIS that have been researched when dealing with invasive species! In addition, it can be noted that there are only 17 chapters in this book; thus, not even a thumbnail of information can begin to be explored. So, why even think about undertaking this task?

This book is one in the series GIS Applications in Agriculture. The intent of the series is to provide the reader with a background as to how scientists use geographic information systems (GIS) to capture, store, analyze, manage, and present data that are linked to a specific location. The purpose of this book is to introduce the reader to an array of applications used to explore the invasive species problem. It highlights GIS uses that provide researchers, producers, land managers, and regulators information on the scope of invasion and insights into how to limit the potential rate of spread or impact of invasive species. In addition, many of the chapters provide step-by-step tutorials or case studies that allow manipulation of datasets featured on the CD-ROM to make maps, perform statistical analyses, and predict future problems. This allows the user to gain hands-on experience with a variety of software programs that create interactive queries (user-created searches), analyze spatial information, edit data and maps, and present the results of these operations in several different formats. Some of the programs are freeware, others are not, but each can be used to integrate, edit, share, and display geographic information.

An unofficial definition of *invasive species* is a species that does not naturally occur in a specific area and whose introduction does or is likely to cause economic and environmental problems or human health problems. In the unofficial definition, this may refer to native species that become established outside a typical "home range." For example, a species like the armadillo, native to the southern United States, is being seen in more northern states. In the United States, invasive species also has a legal connotation as Executive Order 13112 signed into law in 1999 defines invasive species as "an alien species whose introduction does or is likely to cause economic and environmental harm, or harm to human health." In this executive order, "alien" is the key term, as this means that the species is not native to the United States.

Many plant and animal species have been purposefully introduced into new areas such as ornamentals, livestock, crops, and even pets. These species have escaped into other areas and are threatening agricultural and native ecosystems. The use of geographic information for invasive species can be grouped into the following broad categories: (1) dispersal and transport, (2) prediction and forecasting, (3) mapping of current infestations, (4) using maps for management and control tactics, and (5) assessing the impacts of the species or of the control method. Chapter 1 (Dalsted) provides an introduction and general background of remote sensing and tools used in geographic information systems (GIS). Chapter 2 (O'Neill and Dalsted) provides information on the different types of maps and imageries available on various Web sites that can serve as a base for other information.

Population ecology of a new species, what is needed for the species to find a new niche and become successful, and how GIS can be used to monitor and manage these invasions are presented in Chapters 3 (Tobin et al.) and 4 (Nutter et al.). Chapters 5 (West et al.) and 12 (Fox and Pullar) discuss Australian examples of invasive species that were intentionally introduced, including rabbits, pigs, camels, and several plant species, that are now being monitored using GIS to predict potential habitats and success of control efforts. Other species have been introduced less intentionally but nevertheless by man's intervention such as through weed seed in contaminated grain, gypsy moths in cargo, zebra mussels in water ballasts of ships, or through other mechanisms. Man may not always be the cause of a new infestation, as winds, jet streams, and ocean currents can deposit many different types of propagules, such as fungus spores, seeds, or insects, into new areas thereby establishing and causing problems (see Chapters 6 [Hodson and DePauw], 7 [Russo and Isard], and 8 [MacRae et al.]).

Insects are often the invasive species of interest (see Chapters 10 [Hutchinson et al.] and 11 [French et al.]) and GIS have been used to predict their populations. Insects or other animals may also vector a disease (Chapters 3 [Tobin et al.], 8 [MacRae et al.], and 9 [Wimberly et al.]) or the disease (Chapters 4 [Nutter et al.] and 6 [Hodson and DePauw]) may be the primary species of interest.

GIS have been used in many aspects of weed management, from prediction in space or time (see Chapters 12 [Fox and Pullar], 13 [Lass et al.], and 16 [Reitsma and Clay]) to management maps (Chapters 14 [Gumz and Weller], 15 [Colbach and Forcella], and 17 [Dille et al.]). For color figures, please refer to the accompanying CD-ROM.

Acknowledgments

This book could not have been possible without the insight of Fran Pierce, the series editor of the GIS Applications in Agriculture series and the person who approached me with this challenge. Special thanks to Pierre Robert who relentlessly pushed for precision applications in agriculture and was an inspiration to both young and old agricultural scientists. Special thanks to David and Dan Clay, who kept me on track, to Stephanie Hansen, who read each chapter and provided insights from a different point of view, Vincent Obade and Kurtis Reitsma, who ran a lot of the scripts to debug potential problems, and the many other reviewers including Jeff Stein, Kelley Tilmon, and Steven Young, who generously provided technical expertise for review. This book, of course, relied on the chapters' authors who had the insight to use GIS in imaginative ways and who put up with my "just one more thing" requests.

Editor

Dr. Sharon A. Clay is a professor of weed science at South Dakota State University, Brookings, South Dakota, where she has research and teaching responsibilities. She received a BS in horticulture from the University of Wisconsin-Madison, Madison, Wisconsin, in 1977; an MS in plant science from the University of Idaho, Moscow, Idaho, in 1983, examining the sensitivity of barley varieties to various herbicides; and a PhD in agronomy from the University of Minnesota, Minneapolis, Minnesota, in 1987, examining weed management in wild rice production systems of northern Minnesota. She has conducted weed management studies in range and cropping systems that include corn, soybean, wheat, barley, wild rice, flax, and sunflower, as well as studies in weed physiology and site-specific weed management strategies.

Dr. Clay has published over 100 scientific articles and has served on the editorial boards for *Agronomy Journal*, *Weed Science*, and *Site-Specific Management Guidelines*. She has also served on numerous national committees and review panels and has active memberships in, and has served as president of, the South Dakota Chapters of the honorary societies Sigma Xi and Gamma Sigma Delta. She served as the chairperson of the Agricultural Systems Division in the American Society of Agronomy (ASA), participated in numerous ASA committees, was elected as ASA Fellow in 2009, and has held several positions in the Weed Science Society of America.

Contributors

Christopher Auricht
Auricht Projects
Adelaide, Australia

Amber A. Beckler
County of Nevada
Geographic Information System Division
Nevada, California

Laura M. Blackburn
U.S. Department of Agriculture
Northern Research Station
Morgantown, West Virginia

Leanne Brown
Agricultural Productivity Division
Department of Agriculture, Fisheries and Forestry
Canberra, Australia

Emmanuel Z. Byamukama
Department of Plant Pathology
Iowa State University
Ames, Iowa

James F. Campbell
U.S. Department of Agriculture
Agriculture Research Service
Center for Grain and Animal Health Research
Manhattan, Kansas

Matthew Carroll
Monsanto Co.
St. Louis, Missouri

Laurence D. Chandler
U.S. Department of Agriculture
Agriculture Research Service Mid-West Area
Peoria, Illinois

Sharon A. Clay
Plant Science Department
South Dakota State University
Brookings, South Dakota

Rosalee A. Coelho-Netto
Institute for National Research in the Amazon
Manaus, Brazil

Nathalie Colbach
Institut National de la Recherche
Dijon, France

Kevin Dalsted
Water Resources Institute
South Dakota State University
Brookings, South Dakota

Eddy DePauw
International Center for Agriculture Research in the Dry Area
Aleppo, Syrian Arab Republic

J. Anita Dille
Department of Agronomy
Kansas State University
Manhattan, Kansas

Sharon K. Eggenberger
Department of Plant Pathology
Iowa State University
Ames, Iowa

Shelby J. Fleischer
Department of Entomology
The Pennsylvania State University
University Park, Pennsylvania

Frank F. Forcella
U.S. Department of Agriculture
Agriculture Research Service
North Central Soil Conservation Research Laboratory
Morris, Minnesota

Julian C. Fox
Department of Forest and Ecosystem Science
Melbourne School of Land and Environment
The University of Melbourne
Richmond, Victoria, Australia

B. Wade French
U.S. Department of Agriculture
Agriculture Research Service
North Central Agricultural Research Laboratory
Brookings, South Dakota

Mark L. Gleason
Department of Plant Pathology
Iowa State University
Ames, Iowa

Andrew Gougherty
Department of Plant Pathology
Iowa State University
Ames, Iowa

Mary S. Gumz
Pioneer Hi-Bred International, Inc.
North Judson, Indiana

Quentin Hart
Australian Bureau of Agricultural and Resource Economics – Bureau of Rural Sciences
Department of Agriculture, Fisheries and Forestry
Canberra, Australia

David Hodson
Food and Agriculture Organization
AGP Division
Viale Terme di Caracalla
Rome, Italy

J.M. Shawn Hutchinson
Geographic Information Systems Spatial Analysis Laboratory
Department of Geography
Kansas State University
Manhattan, Kansas

Scott A. Isard
Departments of Plant Pathology and Meteorology
The Pennsylvania State University
University Park, Pennsylvania

Lawrence W. Lass
Plant, Soil and Entomological Sciences
University of Idaho
Moscow, Idaho

Erik Lindquist
Geographic Information Science Center of Excellence
South Dakota State University
Brookings, South Dakota

Ian MacRae
Department of Entomology
University of Minnesota
St. Paul, Minnesota

Forrest W. Nutter, Jr.
Department of Plant Pathology
Iowa State University
Ames, Iowa

Mary O'Neill
Water Resources Institute
South Dakota State University
Brookings, South Dakota

Timothy S. Prather
Plant, Soil and Entomological Sciences
University of Idaho
Moscow, Idaho

William J. Price
Statistical Programs
College of Agriculture and Life Sciences
University of Idaho
Moscow, Idaho

David Pullar
Geographical Sciences and Planning
The University of Queensland
St. Lucia, Queensland, Australia

Sonny B. Ramaswamy
College of Agricultural Science
Strand Agricultural Hall
Oregon State University
Corvallis, Oregon

Kurtis D. Reitsma
Plant Science Department
South Dakota State University
Brookings, South Dakota

Tyler W. Rider
Ness City, Kansas

E. Anderson Roberts
Department of Entomology
Virginia Polytechnic Institute and State University
Blacksburg, Virginia

Alison E. Robertson
Plant Pathology
Iowa State University
Ames, Iowa

Joseph M. Russo
ZedX, Inc.
Bellefonte, Pennsylvania

Bahman Shafii
Statistical Programs
College of Agriculture and Life Sciences
University of Idaho
Moscow, Idaho

Patrick C. Tobin
U.S. Department of Agriculture
Northern Research Station
Morgantown, West Virginia

Michael D. Toews
Department of Entomology
University of Georgia
Tifton, Georgia

Neil van Rij
Cedara Plant Disease Clinic
Cedara Department of Agriculture
Cedara, South Africa

Jeffrey W. Vogel
Kansas Department of Agriculture
Topeka, Kansas

Thomas J. Vought, Jr.
Geographic Information Systems Spatial Analysis Laboratory
Department of Geography
Kansas State University
Manhattan, Kansas

Stephen C. Weller
Horticulture and Landscape Architecture
Purdue University
West Lafayette, Indiana

Peter West
Invasive Animals Cooperative Research Centre
Orange, New South Wales, Australia

Christine L. Wey
Geographic Information Science Center of Excellence
South Dakota State University
Brookings, South Dakota

Michael C. Wimberly
Geographic Information Science Center of Excellence
South Dakota State University
Brookings, South Dakota

Robert E. Wolf
Department of Biological and Agricultural Engineering
Kansas State University
Manhattan, Kansas

Min Zhu
College of Life Science
China Jiliang University
Hangzhou, People's Republic of China

1 Introduction: Remote Sensing and GIS Techniques for the Detection, Surveillance, and Management of Invasive Species

Kevin Dalsted

CONTENTS

1.1 Executive Summary 1
1.2 Introduction 2
1.2.1 Remote Sensing 2
1.2.2 Geographic Information Systems 4
1.2.3 Data Synergy 7
1.3 Conclusions 7
Acknowledgments 8
References 8

1.1 EXECUTIVE SUMMARY

This chapter discusses a brief background of remote sensing and then touches on the convergence of remote sensing with geographic information systems (GIS). Modern GIS originated as scientists began to apply computerized data analysis approaches within their respective disciplines. Geographers/cartographers, computer scientists, and natural resource scientists are among those who had a hand in the rapid growth and adoption of this technology. Because of the convergence and synergism between remote sensing and GIS, scientists and managers have gained a capacity to understand natural systems in ways that have not been previously possible. These multidisciplinary collaborations have fueled the exponential growth in the knowledge of biological systems in space and time.

1.2 INTRODUCTION

When introducing a book with a series of chapters on a specific topic, it is sometimes useful to discuss what is not included. This book does not dwell on the many social and resource interactions that have led to necessity to coin the term *invasive species.* Otherwise, the intriguing mix of chapters in this book, as discussed briefly in the Preface, provides the reader with a well-documented range of information and techniques to manipulate datasets to better understand invasive species: from generalized overviews to specific research studies. Each of the chapters can stand alone, but it is anticipated that the whole will be greater than the individual parts.

The overriding theme associated with invasive species is location, not only in space but also in time. And much like the real estate cliché, location is known by practitioners to significantly influence establishment and success of an invasive species. Because invasive species also transcend man-made boundaries (field, county, state, and country), remote sensing and GIS are two associated tools and techniques that are ideal to address the location issue. These techniques incorporate hardware, software, data, and people. Similar to the constant adaptation of living organisms to changing environments, the tools and data associated with remote sensing and GIS are likewise evolving (in a good way). While the Internet and literature can provide numerous, exceptionally detailed discussions of the fundamentals of these techniques, two brief and understandable working definitions are suggested:

Remote sensing—utilization of sensors to collect spectral information about an object or phenomena from a distance (handheld to aircraft to satellite levels)

GIS—combination of digital map data with geographically referenced descriptive data (attributes) to capture, manage, analyze, and/or display user-specified output

1.2.1 Remote Sensing

Remote sensing sensors collect data from specific regions of the electromagnetic spectra (Figure 1.1) and can range from handheld units to those that are mounted on aerial platforms. Well over a hundred years ago, photography was collected from manned balloons, thus providing the first examples of aerial remote sensing. This was

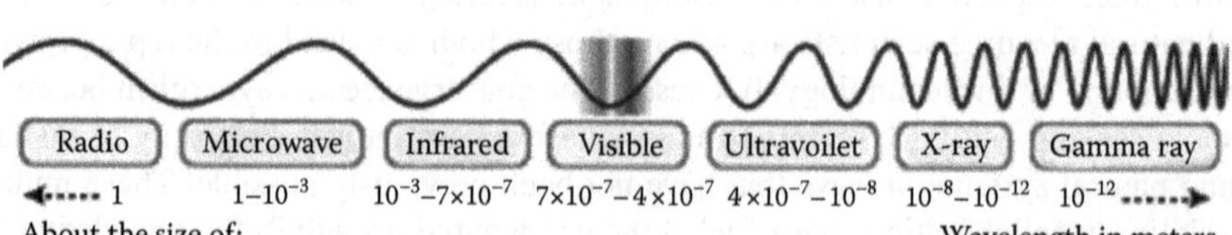

FIGURE 1.1 Electromagnetic spectrum is shown from long to short wavelength. Various sensors can be designed to capture data in any of these regions, keeping in mind that photographic film is only sensitive to a very small interval in and around the visible region. (Courtesy of http://amazing-space.stsci.edu/resources/explorations/light/ems-frames.html)

followed by airplane-mounted cameras to the more recent development of satellite-based systems. Technological advances, including multispectral sensors (nonphotographic) and digital image processing, in the 1960s and 1970s were a boon to remote sensing, and its adoption became widespread. NASA (National Aeronautics and Space Administration), which came into existence in the late 1950s, was instrumental in bringing remote sensing into the mainstream for researchers and others. An example of early research implementation is the Remote Sensing Institute (RSI), which was founded at South Dakota State University in 1969; RSI was among the early centers that helped drive the discovery of many natural resource applications of remote sensing. As is often the case, when a "new" technology is proven effective, it is eventually absorbed by mainstream scientific disciplines. Remote sensing centers at SDSU, Purdue, Texas A&M, and University of Michigan, among others, are assimilated into mainstream disciplines and associated university departments.

A noteworthy historic event happened in 1970 when congressional approval was secured to locate the Earth Resources Observation and Science (EROS) Data Center (EDC) in South Dakota near Sioux Falls. The initial building, which housed EDC, was completed in 1973. The South Dakota location was chosen because of the centralized position within the lower 48 states and, therefore, its effective communication visibility. The EDC, which currently employs over 600 employees, is overseen by the United States Geological Survey (USGS). EDC is an internationally recognized resource center, and its researchers also work directly with many U.S. and international agencies, including USGS, NASA, NOAA, U.S. Army Corps of Engineers, USAID, and the United Nations Environment Program, among others. The USGS EDC Web site (http://eros.usgs.gov/) has considerably more details about its functions and history.

EDC serves as the national archive and depository for land remote sensing data and various digital orthophoto products for the United States and contains over 2.3 TB of data.[1] Recently, the Landsat data have been made available to users at no cost, and numerous end users have been taking advantage of this great resource. As remote sensing has been adopted across the United States and the world, a new model of retrieving, processing, and using remote sensed information encompasses researchers, teachers, and end users and is often represented by various consortia, such as the Space Grant Consortiums (NASA, every state), America View (USGS, over half of the states), and multistate consortia such as Upper Midwest Aerospace Consortium (UMAC, NASA funded; includes ND, SD, WY, MT, and ID).

Numerous books and publications have provided detailed discussions of the wide range of sensors and analyses that have been or are being used in remote sensing (e.g., *Manual of Remote Sensing*, currently volumes 1–6).[2] For our purposes, standard remote sensing techniques will be emphasized, that is, applications involving visible and near-infrared bands. The visible and near-infrared spectral wavelengths have proven to be the most popular and accessible bands used (Figure 1.1). Whether in analog (film) or digital form, these data have been utilized in countless research studies and experiments, most often to collect, store, manipulate, and analyze natural resource and land cover/land use data. In most cases, the remote sensing data can be considered to be an objective and impartial data source. It is generally up to the end user to be aware of various restrictions that may impact the analyses, namely, the factors of resolution: spatial (size of pixels), spectral, temporal (frequency of collection), radiometric (number of bits), and geometric (see Chapter 2). Interpretations and analyses must be carefully documented to avoid the implication of introduced bias. For example, during war, various types of cut vegetation were used to disguise artillery, etc.; standard photography could not differentiate the differences because it all looked like green vegetation. Color infrared film, however, emerged as a solution to this problem as it can distinguish the difference between growing vegetation and nongrowing vegetation due to reflectance differences of growing/nongrowing vegetation in the near-infrared band (non-visible), hence the term camouflage detection film.

The inclusion of standard remote sensing techniques, namely, manual (human) interpretation, machine (digital image) processing, and/or a combination, is amply illustrated throughout this book. These various techniques are appropriate for many applications, much like the selection of a tool for woodworking project: with enough time and skill, a hand saw can do the same job as a power saw.

Some might say that remote sensing technology was oversold during its development as a technique to investigate natural resource information. To paraphrase any number of comic book action heroes, the earth will always need to be saved, but remote sensing will likely only be part of any potential solution. In fact, remote sensing data can most often best serve to objectively document the nature and extent of a problem or problems. While remote sensing may not offer solutions per se, understanding where the problem lies and its extent and dynamic nature can help target the solution, that is, using a patch on a flat tire rather than replacing the entire car. Remote sensing information can also be used to give a before and after view to determine if targeted techniques have accomplished the desired outcome(s), namely, monitoring and validation.

1.2.2 Geographic Information Systems

"Geospatial data," which simply means any data that have an identifiable location, is now a common term and includes remote sensing data and other GIS data. This progression makes sense as acquisition of remote sensing data and associated processing technologies have become mature. Remote sensing may now be thought of as another source of data to plug into the GIS technology.

The conceptual basis for GIS has been around for many years.[3] The use of topographic knowledge for navigation is among the best early illustrations of GIS applications, albeit without the "digital" aspect. For example, humankind probably made migration routing decisions based on several landscape and other spatial information types that most likely included ease of passage, weather situations, food and fuel sources in transit, and potential for danger from predators and/or rivals. Only in the past 50 years or so has high-speed digital computing, combined with relational databases, led to an explosion of potential uses of this powerful set of software tools.

We, as a society, are bombarded by virtually thousands of illustrations of geospatial data on a weekly basis: hour-by-hour weather forecasts, road construction and detour maps, earthquake epicenter information and tsunami predictions, realtors' open houses, and census data, to name just a few examples. In the past,

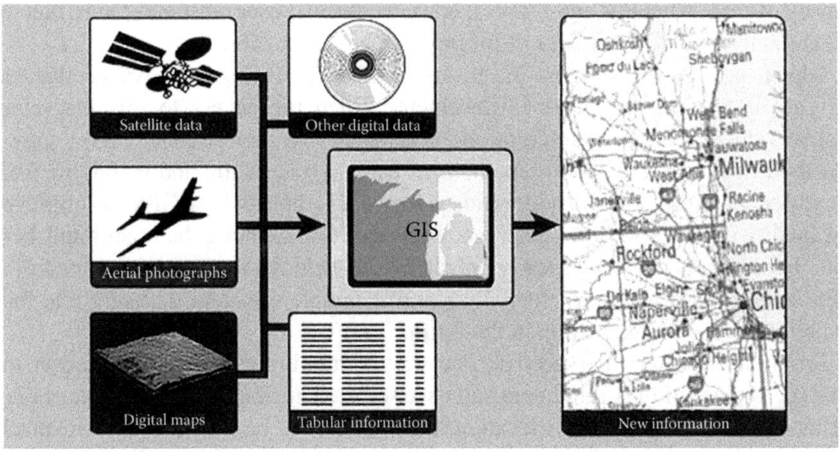

FIGURE 1.2 Data integration is the linking of information in different forms through a GIS. Many additional databases can be linked to locational elements of digital maps considering map projection, resolution, geometry, etc. (Courtesy of http://egsc.usgs.gov/isb/pubs/gis_poster/#how)

processing and/or combining geospatial data/information was a tedious process involving cumbersome equipment, manual labor, and lots of time. The computer and its associate GIS software have fundamentally changed how geospatial data are developed, stored, combined, updated, analyzed, and displayed. Users can now manipulate thousands (or hundreds of thousands) of data points simultaneously (Figure 1.2).

As more and more map data have been converted into digital form, GIS has expanded even beyond the computer science, geography, and natural resource user communities. Numerous Internet-based tools are freely available for users; anyone can now easily see their home on an aerial photo, monitor a radar map for impending rain, or map a route from home to a distant location (along with written directions).

The key element, or building block, of GIS is the geo-referenced, digital map. This element is the backbone to which we can integrate any data that we can collect or even imagine, as long as it has a spatial identifier. A variety of technical issues are tied to the digital map: map projection, a geo-rectification process to develop the digital map, and related issues that are nominally discussed in the metadata, which are data that thoroughly define/describe the aforementioned data. Another relevant background piece of knowledge is the nature of the data that comprise a GIS; these data can be points (e.g., well locations), lines (e.g., stream networks), and/or polygons (e.g., soil map units). Two distinct choices for the GIS data structures are raster (pixel based or rows and columns) or vector (coordinate based or lines that join to define a map unit). Landsat and other satellite remote sensing data are raster based, while a good example of vector data is the elevation isohyets that comprise a topographic map. With the state-of-art computing now embracing computer networks and Internet, fast and large memory, very large hard drives, and speedy graphics, data structure is not the technical factor that it once was in terms of efficiency and computing resource requirements. Most software packages can easily deal with a variety of data structures and sources; however, the end user must make sure that the data among the different layers of information are compatible.

Global positioning systems (GPS) are another technological advance that has driven GIS development and advancement. When the access to this marvelous tool was opened, a new paradigm was created in that conventional survey crews, and their time-intensive, costly methods of using a rod, chain, and theodolite were greatly supplemented. Rural addressing is a great example of how GPS has improved our quality of life, if only for emergency services (911), among the other many benefits. Researchers can now greatly enhance their field data collection by recording GPS location information to spatially identify sample sites and to insure that they can return to the same location in the future.

From an end user's perspective, one way of looking at GIS is to consider the "staying power" of the digital map and its associated attributes. For example, a digital topographic map with its elevation information can be thought to be relatively stable (earthquakes, mudslides, and man's activities, not withstanding); depth to aquifer, vegetation climate zone, and soil survey maps are a few other examples. Intermediate "stability" might be illustrated by annual cropping land cover maps, periodic census counts (human or wildlife), and annual fertilizer usage. Short-term examples could include daily traffic counts on a stretch of highway, hospital admissions, rain or snow accumulations, wind speed, etc. The reason for discussing these three levels of "staying power" is simply to illustrate the need for awareness and sensitivity to data collection frequency. As some data expire quickly, they need to be refreshed more than the so-called stable data. These are valid considerations when evaluating or designing an application.

An exciting prospect for exploiting the power of GIS is to consider analyses that connect resources and attributes from disparate sources in ways that have not been previously considered, or even possible, for that matter. An example of connecting different sources of information would be to examine environmental factors that may contribute to the occurrence of specific type of cancer in a location by combining data that include the time variable, demographics (age, occupation, housing, ethnic background, and income), climate, land use/land cover, air and water pollution sources (chemical agents and exposure duration/intensity), drinking water sources, vehicle traffic, use of ethanol versus standard gasoline versus diesel, etc. The caveat on this exciting and innovative GIS approach has to do with both the quality of data and the equivalent resolution/compatibility of the various data and their sources. We always have to be on guard against incorrectly drawing conclusions from data combinations that may just point out coincidence, whereas the "real" reason for the given phenomenon may be tied to data that were not included. Statisticians will confirm that this is a real concern.

A quick analogy of the aforementioned example could be a comparison of the early settlers and their food supply compared to modern-day society. The early settlers grew or hunted for their own food, much of the foodstuffs had limited storage, and the resultant diet, while it was often sufficient, was very limited. The recipes utilized by these individuals were probably very few in number as were their choices. Today, we have large grocery stores, refrigeration, and access to whatever foodstuffs we might desire. The recipes that we use can be very complex and include exotic ingredients that either cannot be grown in the area or may be locally grown but are "out-of-season." We still have the same need to eat, but our choices are significantly greater than the past. Bread is still bread, but today's grocery shelves hold a broad variety of types. Likewise, results from GIS analyses do not guarantee validity, just like combining a variety of spices together will not lead to a tasty dish. We do, however, have incredibly more choices today, and our scientific basis for managing our resources is much more complex than it once was. We are presented both challenges and opportunities while using GIS and remote sensing to better manage our natural resources after testing and establishing proven "recipes."

1.2.3 Data Synergy

Both remote sensing and GIS can be considered among the geospatial techniques and methods: one for data collection and other for data merging and manipulation. Remote sensing data are part of the suite of geospatial data that are available to be joined with other geo-registered data and their attributes. For the purposes of this book and its following chapters, the application of geospatial technology to the evaluation of invasive species is best left to the specialists in the pertinent scientific fields. The merger of data sources from field work, existing and new data sources, and archived data (e.g., Landsat was launched in the early 1970s) can be accomplished through the technology described above. The technical considerations, project design, field data collection, and validation leading to conclusions are all best left to the appropriate scientific experts and knowledgeable land managers. Awareness of the myriad of potential data sources and the technology that can be used to merge and analyze these data is an exciting prospect for these experts and/or end users. The technical know-how and software exist to couple any number of data sources that may lead to new knowledge, including time-dependent evaluations.

1.3 CONCLUSIONS

Invasive species are similar to any issue related to dynamic phenomena. The nature of each situation is varied and not necessarily related to other comparable issues. We can say that invasive species are opportunistic and react to their environment in somewhat predictable ways. This statement does not in any way diminish the magnitude of the problem; rather, it should serve to further drive the data collection and design/approach to monitor and document the extent of various invasive species. Establishing the range of factors that are important to these species is the beginning step to documenting the extent and ebb/flow of the problem.

The answers to controlling, slowing the rate of the extent, or destabilizing the population so that it will no longer function well in its new geographic zone will require more than mere geospatial data and analyses. However, developing and deriving models that build on geospatial data and techniques offer a means of evaluating scenarios to develop more effective detection, surveillance, and potential management options for invasive species. Improving decisions and the decision-making process is a worthwhile end goal of utilizing these technologies.

ACKNOWLEDGMENTS

Remote sensing has a very interesting history, and many of the early pioneers have moved on to higher levels but are nonetheless acknowledged en masse for their dedication to this technology and its development (http://www.geog.ucsb.edu/~jeff/115a/remotesensinghistory.html). This brief chapter came together because of Dr. Sharon Clay and her many suggestions for improvement. Mary O'Neill's review and guidance are also much appreciated. Funding support from the S.D. Space Grant Consortium and the Upper Midwest Aerospace Consortium is gratefully acknowledged.

REFERENCES

1. USGS Earth Resources Observation and Science Center (EROS). 2009. Records management report. Quarter 4, 2009. http://eros.usgs.gov/archive/nslrsda/Records_Management_Report_FY09_Q4.pdf
2. *Manual of Remote Sensing: Vol. 1* (*Earth Observing Platforms & Sensors*), *Vol. 2* (*Principles & Applications of Imaging Radar*), *Vol. 3* (*Remote Sensing for Earth Sciences*), *Vol. 4* (*Remote Sensing for Natural Resource Management & Environmental Monitoring*), *Vol. 5* (*Remote Sensing of Human Settlements*), and *Vol. 6* (*Remote Sensing of the Marine Environment*). John Wiley & Sons Publishers, Hoboken, NJ. http://www.asprs.org/publications/2009PubsCatalog.pdf
3. M. Madden (ed.). 2009. *Manual of Geographic Information Systems*. ASPRS, Bethesda, MD, 1330 pp. ISBN: 1-57083-086-X.

2 Obtaining Spatial Data

Mary O'Neill and Kevin Dalsted

CONTENTS

2.1 Executive Summary 9
2.2 Definitions 9
2.3 Factors to Consider When Acquiring and Using Spatial Data 12
2.4 Data Types: Raster and Vector 13
2.5 Raster Data Sources and Examples 13
  2.5.1 Digital Raster Graphic 13
  2.5.2 Satellite and Aerial Imagery 17
  2.5.3 Digital Elevation Data 23
2.6 Vector Data Sources and Examples 25
2.7 Software for Spatial Data Visualization and Analysis 26
2.8 Conclusion 27
Acknowledgments 27
References 27

2.1 EXECUTIVE SUMMARY

Typical spatial data for agricultural and invasive pest management include satellite images; aerial photos; soil types; elevation; soil nutrient testing results; crop type and yield; pest infestation location, type, and severity; and pesticide type, application locations, and amounts. The spatial extent of a dataset for invasive pest management ranges from a single field to the entire world. The purpose of this chapter is to

- Define the common terms associated with spatial data
- Discuss factors to consider when acquiring and using spatial data
- Show how to obtain various types of spatial data
- Contrast various data visualization and analysis possibilities
- Illustrate various types of spatial data

2.2 DEFINITIONS

An initial step in the process of acquiring spatial data is to define the areal, or spatial, **extent** of the area of interest. The **extent** can range from a single production field to the entire earth. If the spatial extent is a single field or a group of fields within a relatively small area, a single dataset (e.g., a single satellite image) may cover the entire area of interest. If, however, multiple locations are scattered over a large area (e.g., within several counties, regions, or countries), multiple datasets may be necessary.

Using multiple datasets has implications for data processing, as well as cost, if the data are not free of charge.

The **data format** is another important consideration in data acquisition and use. The importance of format is due mainly to the software used to process the data and when comparisons among datasets that use different formats are desired. As an example, many large image files are compressed in MrSID (***.sid***) format. Not all software packages have the capability to input this type of file. Thus, additional software is required to convert from the ***.sid*** format to another compatible format before processing. Although the conversion software may be available for downloading on the Internet free of charge, there is still an investment in time for doing the conversion and in computer space for storing redundant datasets.

Resolution is a term often used in conjunction with spatial data. In general, resolution refers to the quality of detail within an image or other spatial dataset. In the spatial world, four types of resolution are generally defined. **Spatial resolution** is likely the first type of resolution that comes to mind for most users of geospatial datasets. Spatial resolution refers to the fineness of the spatial detail visible in an image[1] or the land surface area contained within one picture element (pixel). Using Landsat imagery as an example, data from the Enhanced Thematic Mapper sensor aboard Landsat have a spatial resolution of 30 m, meaning that each pixel in the image is an integrated value of reflectance from a 30 × 30 m area on the Earth's surface. Spatial resolution can also refer to the scale of the original data from which a dataset is derived. For example, water features delineated on 1:10,000 scale imagery (6.33 in./mi) (10 cm/km) will have a better spatial resolution or accuracy than if delineated on 1:100,000 scale imagery (0.63 in./mi) (1 cm/km).

Temporal resolution refers to the frequency of dataset acquisition. The temporal resolution of Landsat imagery, for example, is 16 days. This means that Landsat imagery of the same area on the Earth's surface is collected every 16 days. It is important to note that the image is taken regardless of the atmospheric conditions and that not all images are usable. The image usability is best under cloud-free conditions. Another example is yield monitor datasets, which, if collected on an annual basis, have a temporal resolution of 1 year.

Spectral resolution, a third type of resolution, refers to the ability of a sensor to define fine wavelength intervals.[1] The wavelength intervals are often referred to as bands or channels. Land observation satellite sensors typically collect data in four to eight relatively broad wavelength intervals (typically ranging from 0.4 to 10 μm) (see Table 2.1) in the visible and infrared portions of the electromagnetic spectrum. Hyperspectral sensors, on the other hand, have much greater spectral resolutions, collecting data in 200 or more ultraviolet, visible, and infrared narrow wavelength intervals (typically ranging from 0.4 to 2.5 μm).

Radiometric resolution is the fourth type of resolution. It can be defined as the ability of an imaging system to record many levels of brightness.[1] Landsat sensors, for example, are 8 bit sensors, meaning that they record 256 (2^8) levels of brightness. Other sensors are capable of recording more levels of brightness, such as 1024 (10 bit) or 2048 (11 bit).

Directly from the sensor, remotely sensed imagery is not geometrically accurate enough to be used as a map. **Geometric accuracy** is especially important when one

TABLE 2.1
Landsat Spectral Bands: Wavelengths and Applications

Landsat 5 TM and Landsat 7 ETM+ Spectral Bands	Wavelength	Useful for Mapping
Band 1—blue	0.45–0.52	Bathymetric mapping, distinguishing soil from vegetation and deciduous from coniferous vegetation
Band 2—green	0.52–0.60	Emphasizes peak vegetation, which is useful for assessing plant vigor
Band 3—red	0.63–0.69	Discriminates vegetation slopes
Band 4—near infrared	0.77–0.90	Emphasizes biomass content and shorelines
Band 5—short-wave infrared	1.55–1.75	Discriminates moisture content of soil and vegetation; penetrates thin clouds
Band 6—thermal infrared	10.40–12.50	Thermal mapping and estimated soil moisture
Band 7—short-wave infrared	2.09–2.35	Hydrothermally altered rocks associated with mineral deposits
Band 8—panchromatic (Landsat 7 only)	0.52–0.90	15-m resolution, sharper image definition

Source: Courtesy of http://landsat.usgs.gov/best_spectral_bands_to_use.php
Note: Each band on Landsat 4–5 TM and Landsat 7 ETM+ is useful for capturing different land cover aspects.

spatial dataset is used in conjunction with another, for example, a yield map over a satellite image. When maps and images for the same area are viewed and processed together in a geographic information system (GIS), analysis results will be compromised if one or both are geometrically inaccurate. Preprocessing an image is required to place each pixel in its proper planimetric *x*, *y* location.[2] The **geometric error** present in an uncorrected image can be classified as **systematic** or **nonsystematic**. **Systematic error** is predictable and can be modeled for making corrections. Examples of systematic error are image skew caused by Earth rotation effects and displacement caused by topographic relief. **Nonsystematic error** is random in nature and often caused by factors such as sensor altitude changes and attitude changes (roll, pitch, and yaw).[2]

Two methods are commonly used for geometric correction: image-to-map rectification and image-to-image rectification. The first method results in an image that is planimetric, that is, features are placed in their correct horizontal positions and should be used whenever accurate area, direction, and distance measurements are required. The second method is acceptable when it is not necessary to have each pixel in its correct planimetric location.

When working with two or more layers of geospatial data, it is common to encounter problems related to the **projection** of the datasets. Projection refers to the flat-map (or two-dimensional) representation of spherical (or three-dimensional) coordinates, that is, "projecting" latitude and longitude coordinates to a two-dimensional (row and column) rectangular coordinate system. Obviously, it is impossible to create a projection that is without error. All projections are a compromise in terms of preserving angles, areas, distance, and/or direction.[3] Hundreds of projections have been defined over the years. Some commonly used projections include the Lambert conformal conic projection, the Albers equal-area conic projection, and the transverse Mercator projection.[4] It is important to know the projection of each dataset with which you are working. This information is usually included in the metadata (data about the data) that accompany geospatial datasets. Projection information is used in software packages to properly register or overlay geospatial datasets for display and analysis.

2.3 FACTORS TO CONSIDER WHEN ACQUIRING AND USING SPATIAL DATA

The cost of geospatial datasets varies from free to thousands of dollars. An increasing number of datasets can now be downloaded at no cost over the Internet. This is especially true of datasets from federal agencies such as the U.S. Geological Survey (USGS). Data available from USGS include Landsat satellite imagery, aerial imagery, digital line graphs (DLGs), digital raster graphics (DRGs), and digital elevation models (DEMs). On the other hand, geospatial datasets acquired from the private sector usually have cost associated with them. Costs in the vicinity of $18/km^2$ are typical for previously acquired high-resolution (≤4 m) satellite imagery from commercial vendors. Rush orders and/or requests for image acquisition over a designated location on a specific date may increase the cost to as much as $50/km^2$. A minimum order of, for example, 25 km^2 is usually required for commercially available products.

Geospatial datasets tend to be very large files. A single eight-band scene of Landsat 7 satellite imagery covering an area of 185 km × 170 km is nearly 600 MB (or 0.6 GB) in size. Storing and processing datasets of this size make it desirable (maybe even necessary) to have a computer with a significant amount of random-access memory (RAM) and disk space and a fast central processing unit (CPU). As an example of the hardware requirements for geospatial data processing, consider these requirements/recommendations for Arc GIS 9.3, an often-used GIS software product developed by the Environmental Systems Research Institute (ESRI) (wikis.esri.com/wiki/display/ag93bsr/ArcGIS + Desktop#ArcGISDesktop-Hardware Requirements):

CPU speed—1.6 GHz or higher
Processor—Intel Core Duo, Pentium 4, or Xeon processors
Memory/RAM—1 GB minimum, 2 GB, or higher recommended
Disk space—2.4 GB

A monitor and printer are other pieces of hardware that are used for displaying geospatial data and analysis results. For the ArcGIS 9.3 software referenced above, ESRI recommends a display with 24 bit color depth and a screen resolution of 1024 × 768 or higher at normal size (96 dots per inch [dpi]). A 24 bit capable graphics accelerator is also required, as is an Open GL 1.3 or higher compliant video card with at least 32 MB of video memory (≥64 MB recommended). Printers used for creating hardcopy versions of geospatial data and analysis results vary in size and cost from $200 desktop printers to $25,000 large-format plotters (up to 60" in width).

2.4 DATA TYPES: RASTER AND VECTOR

Geospatial datasets can be classified as either **raster** or **vector** data. **Raster data** consist of cell-like units that are often called pixels (a contraction for picture elements). A raster dataset divides the geographic area of interest into a matrix (or array) of cells (or pixels) that are of uniform size and shape. Each cell is encoded with a single value or category, which in some cases is referred to as the attribute for that cell.[1] A satellite image is an example of raster data. The image consists of many rows of data. Within each row are many pixels with values that represent the amount of reflected light recorded by the sensor for the geographic extent of each pixel.

Vector datasets use one or more *x*, *y* coordinates to represent spatial features. A **point**, the simplest type of spatial feature, consists of a single *x*, *y* coordinate. Well locations, soil-sampling sites, and elevation benchmarks are examples of point datasets. A **line**, the second type of spatial feature, consists of at least two *x*, *y* coordinates marking the end points of the line. If a line has a shape other than straight, additional *x*, *y* coordinates between the end points are required to define the shape of the line. Roads, streams, and elevation contours are examples of line features. **Polygons** (or areas) are a third type of spatial feature. Polygons are represented as a series of *x*, *y* coordinates, with the first and last coordinates being the same, thus ensuring that the coordinates define an enclosed area. Field boundaries, lakes, and soils mapping units are examples of polygon features.

2.5 RASTER DATA SOURCES AND EXAMPLES

Three types of raster data are discussed in this section: **DRGs, satellite and aerial imagery**, and **digital elevation data**. Sources of each type of data are identified and examples are shown. Although many of the examples are South Dakota datasets and sources, equivalent data are available from and for most states.

2.5.1 Digital Raster Graphic

A DRG is an image version of a USGS 1:24,000 scale 7.5 min quadrangle map. A DRG is made by scanning a published map on a high-resolution scanner. The resulting raster image is georeferenced and fit to the Universal Transverse Mercator (UTM) projection. Colors are usually standardized to duplicate the line-drawing character of the published map. The average dataset size of a 7.5 min DRG is about 8 MB in ***TIFF*** format with ***PackBits*** compression (http://topomaps.usgs.gov/drg/drg_overview.html).

DRG data for South Dakota, as well as several other types of data that will be discussed later, can be acquired via the South Dakota View (SDView) Web site at http://sdview.sdstate.edu (Figure 2.1). Note that SDView is used in this chapter's examples; however, AmericaView (http://www.americaview.org/currentmembers.htm) contains similar information for 36 U.S. states. Another Web site from which you can find links to GIS data for most U.S. states is http://web.mit.edu/dtfg/www/data/data_gis_us_state.htm. To acquire data from these databases, use similar processes. From the SDView homepage,

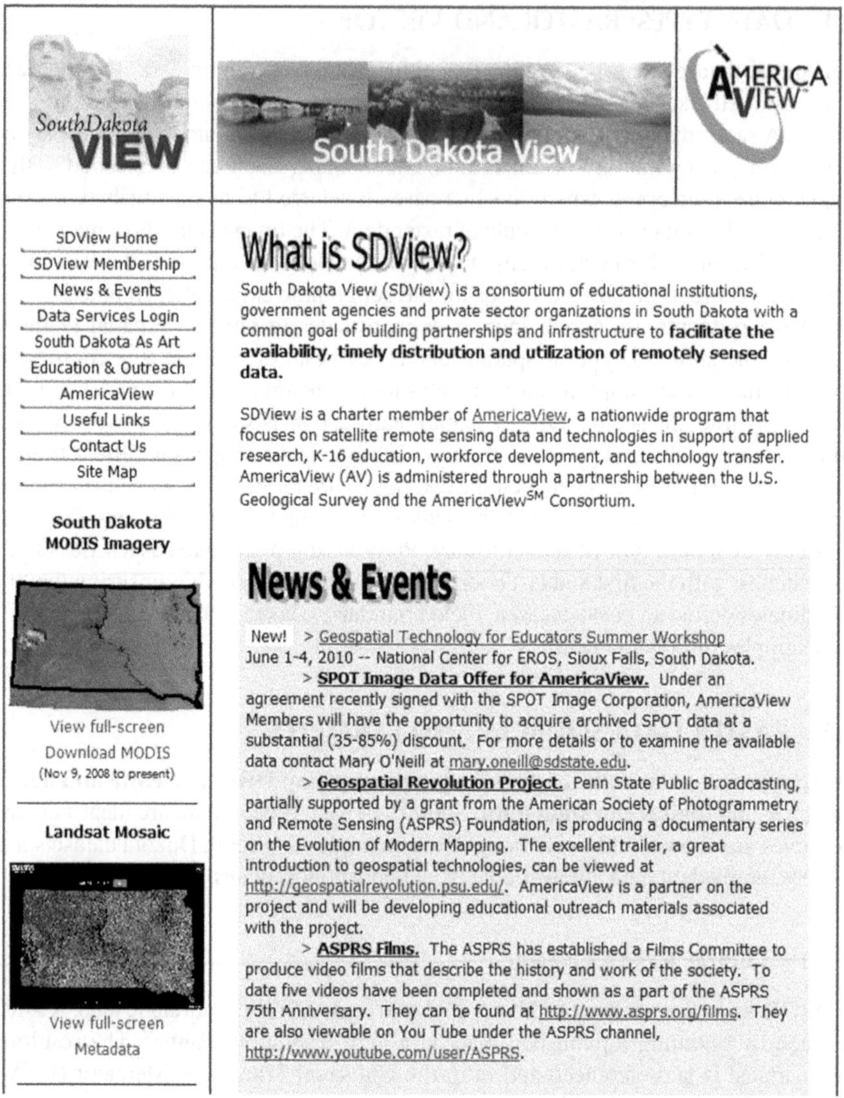

SouthDakota VIEW
South Dakota View
AMERICA VIEW
SDView Home
SDView Membership
News & Events
Data Services Login
South Dakota As Art
Education & Outreach
AmericaView
Useful Links
Contact Us
Site Map
South Dakota MODIS Imagery
View full-screen
Download MODIS
(Nov 9, 2008 to present)
Landsat Mosaic
View full-screen
Metadata
What is SDView?
South Dakota View (SDView) is a consortium of educational institutions, government agencies and private sector organizations in South Dakota with a common goal of building partnerships and infrastructure to facilitate the availability, timely distribution and utilization of remotely sensed data.
SDView is a charter member of AmericaView, a nationwide program that focuses on satellite remote sensing data and technologies in support of applied research, K-16 education, workforce development, and technology transfer. AmericaView (AV) is administered through a partnership between the U.S. Geological Survey and the AmericaView SM Consortium.
News & Events
New! > Geospatial Technology for Educators Summer Workshop
June 1-4, 2010 -- National Center for EROS, Sioux Falls, South Dakota.
> SPOT Image Data Offer for AmericaView. Under an agreement recently signed with the SPOT Image Corporation, AmericaView Members will have the opportunity to acquire archived SPOT data at a substantial (35-85%) discount. For more details or to examine the available data contact Mary O'Neill at mary.oneill@sdstate.edu.
> Geospatial Revolution Project. Penn State Public Broadcasting, partially supported by a grant from the American Society of Photogrammetry and Remote Sensing (ASPRS) Foundation, is producing a documentary series on the Evolution of Modern Mapping. The excellent trailer, a great introduction to geospatial technologies, can be viewed at http://geospatialrevolution.psu.edu/. AmericaView is a partner on the project and will be developing educational outreach materials associated with the project.
> ASPRS Films. The ASPRS has established a Films Committee to produce video films that describe the history and work of the society. To date five videos have been completed and shown as a part of the ASPRS 75th Anniversary. They can be found at http://www.asprs.org/films. They are also viewable on You Tube under the ASPRS channel, http://www.youtube.com/user/ASPRS.

FIGURE 2.1 South Dakota View (SDView) Web site. (Courtesy of http://sdview.sdstate.edu/)

1. Click on **Data Services Login** on the top-left area of the page to get to the *Data Services Login* page (Figure 2.2). You can get to the same login page by clicking on the indicated location within the *Free Download of Geospatial Data* section on the homepage.
2. On the **Login** page, you can sign up to be a new member or, if already a member, enter your e-mail address and password to proceed to the **Data Services** (Figure 2.3) page.
3. On the **Data Services** page, click on **Digital Base Data** to get to the Web site from which the DRG data can be downloaded (Figure 2.4).

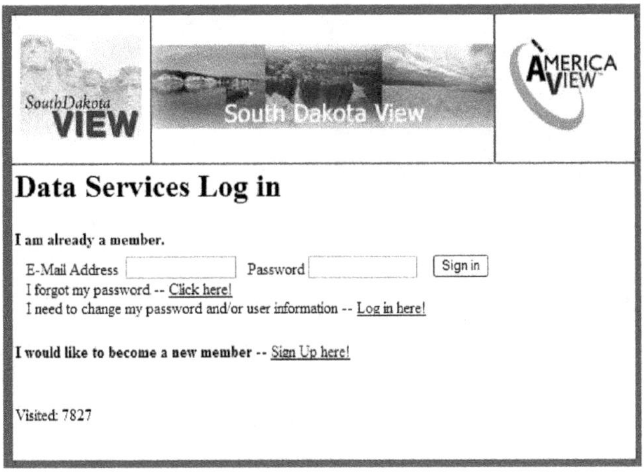

FIGURE 2.2 SDView data services login page.

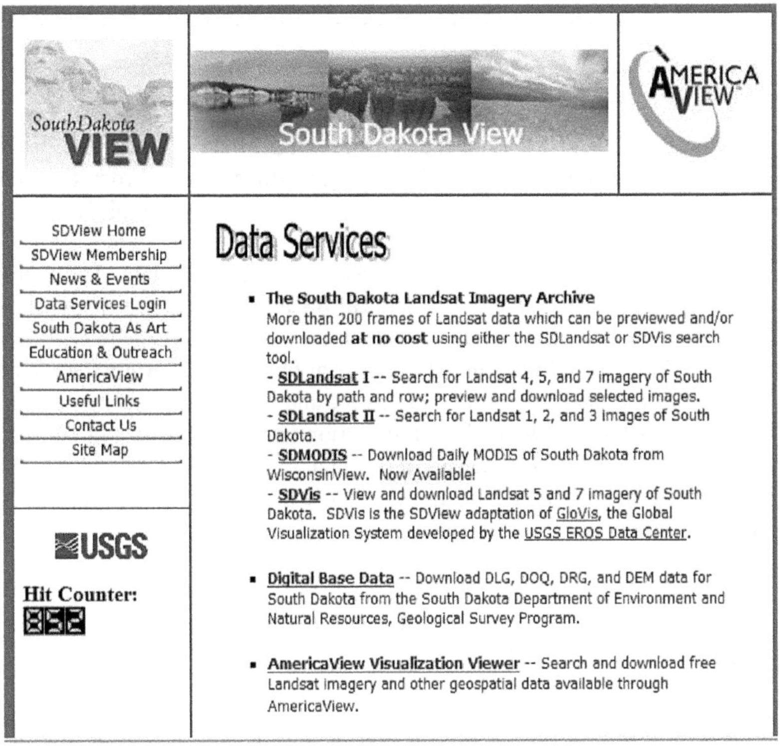

FIGURE 2.3 Data services provided by SDView.

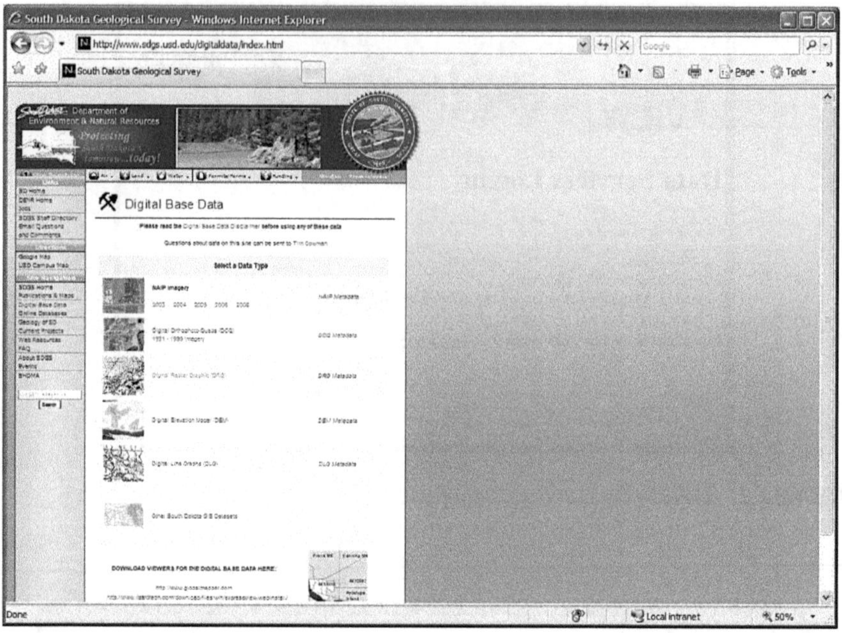

FIGURE 2.4 Webpage from which digital raster graphics and other digital base data for South Dakota can be downloaded.

You can also bypass the SDView route to the **Digital Base Data** and get there directly by going to http://sdgs.usd.edu/digitaldata/index.html. From the **Digital Base Data** Web site, DRG data are one of several types of geospatial data available for download. Before downloading DRG, or any type, data, it is advisable to read the metadata, which usually (but not always) accompany the data. The metadata files for the data types available on the **Digital Base Data** page are easily accessed by clicking on the link on the right-hand side of the page. Metadata are simply "data about the data," that is, it documents the source of the data, the original scale of the data, the date of the data, the processing steps, and other pertinent information of which the user should be aware. The information in the metadata can provide the user with clues on possible applications of the data, for example, Landsat data should not be used for extremely detailed surveys. Figure 2.5 shows the top portion of the metadata file for the DRG.

To download the DRG data,

1. Click on the **DRGs** link on the **Digital Base Data** page (see Figure 2.4). This results in a page showing the state of South Dakota and its county boundaries (Figure 2.6).
2. Click on a county of interest. This will result in a page showing all of the 7.5 min quadrangles within the county (see Figure 2.7).
3. Click on a quadrangle of interest. This will open a **File Download** window that asks whether you want to open or save the zipped file containing the DRG data for the selected quadrangle. If **Open** is selected, the software on

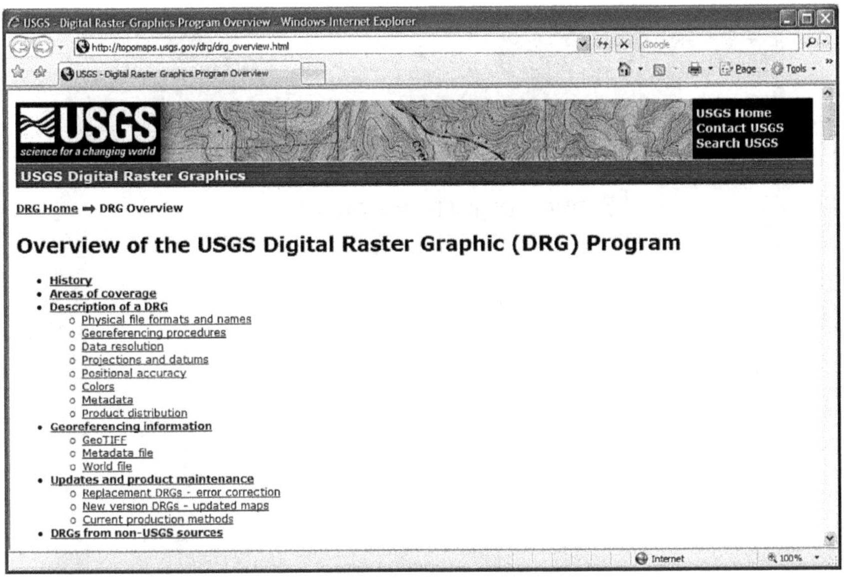

FIGURE 2.5 USGS digital raster graphics metadata.

your computer associated with ***.zip*** files will automatically open and allow you to extract the files to a folder or directory of your choice. If **Save** is selected, a **Save As** window will open, allowing you to save the zipped file to your desired location. You will then need to go to that location and unzip the file. Regardless of whether you choose the **Open** or **Save** (with subsequent unzipping) option, the result will be two files—the **image file** with a *.tif* extension and a **world file** with a ***.twf*** extension (Figure 2.8). A **world file** is a simple text file used by GIS and image processing software to georeference raster images such as DRGs. It describes the location, scale, and rotation of the image such that the image can be spatially registered with other datasets associated with the same geographic area. Figure 2.9 shows a DRG for the Bruce quadrangle within Brookings County in South Dakota and Figure 2.10 is a subset of the Bruce quadrangle showing in more detail the information available from the DRG image.

2.5.2 Satellite and Aerial Imagery

Another type of raster data is imagery acquired from sensors aboard satellites and aircraft. Numerous sources exist for both satellite and aerial imagery. Two sources of aerial imagery and two sources of satellite imagery are described in this section.

A **Digital Orthophoto Quadrangle** (DOQ) is a scanned, or digitized, aerial photograph that has been combined with a DEM to correct for image displacement caused by terrain relief and camera tilt. A DOQ, therefore, has the image characteristics of a photograph and the geometric qualities of a map (http://www.usgsquads.com/downloads/factsheets/usgs_doq.pdf). The standard USGS DOQ product is black and white, natural color, or color-infrared (CIR) images with 1 m ground resolution

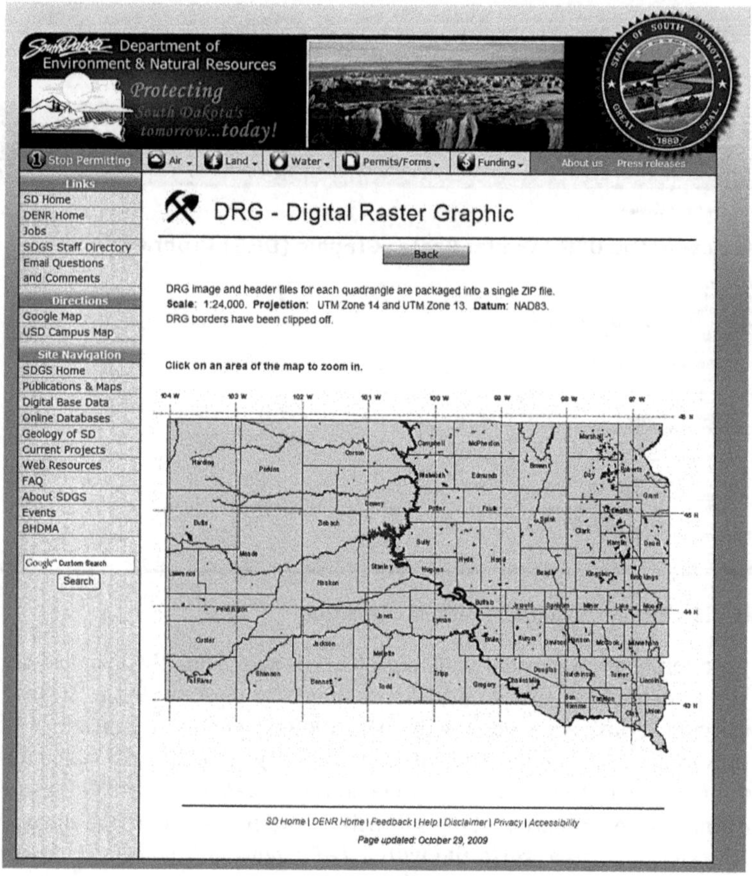

FIGURE 2.6 Webpage from which the county containing the digital raster graphic quadrangle of interest is selected.

and UTM projection. The aerial photography used to create the DOQs was predominately acquired during "leaf off" time periods. DOQs for South Dakota can be acquired from the **Digital Base Data** site described in the DRG section above. The dates of the DOQs for South Dakota vary from 1991 to 1999, making the imagery a good source of historical information. The procedure for downloading DOQ data is similar to the procedure described in the preceding text for the DRG data:

1. Click on **Digital Orthophoto Quads** on the **Digital Base Data** page.
2. Click on a county on the resulting DOQ page.
3. Click on the quadrangle of interest on the County page. The zip file that is opened or saved contains several files that when unzipped include four image files that are often called **Digital Orthophoto Quarter Quads** or **DOQQs**. Each DOQQ is a 3.75 × 3.75 min quarter of a 7.5 × 7.5 min quadrangle. Other files contained within the zipped file are a world file, header files, and a general header information file. The DOQQ image files are in

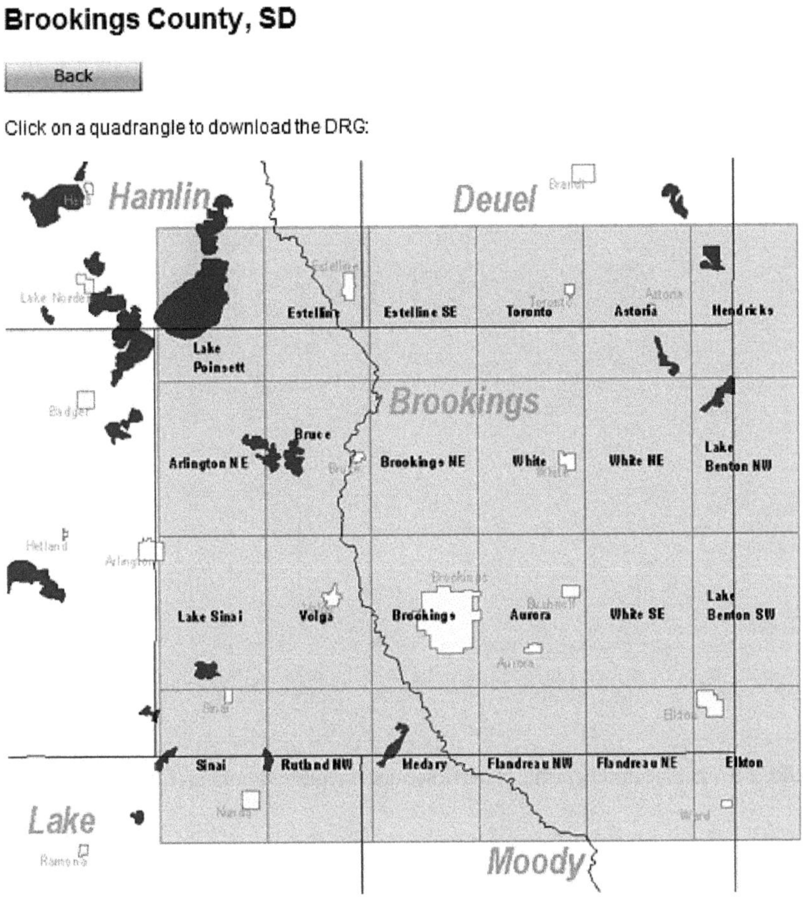

FIGURE 2.7 Webpage from which the digital raster graphic quadrangle is selected.

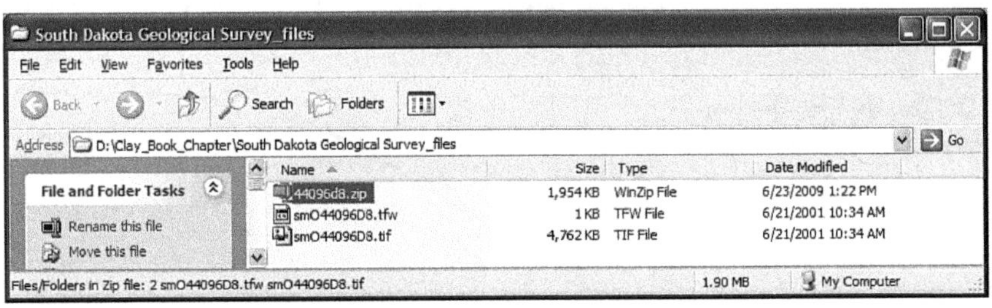

FIGURE 2.8 Files resulting from digital raster graphic quadrangle download and unzip processes.

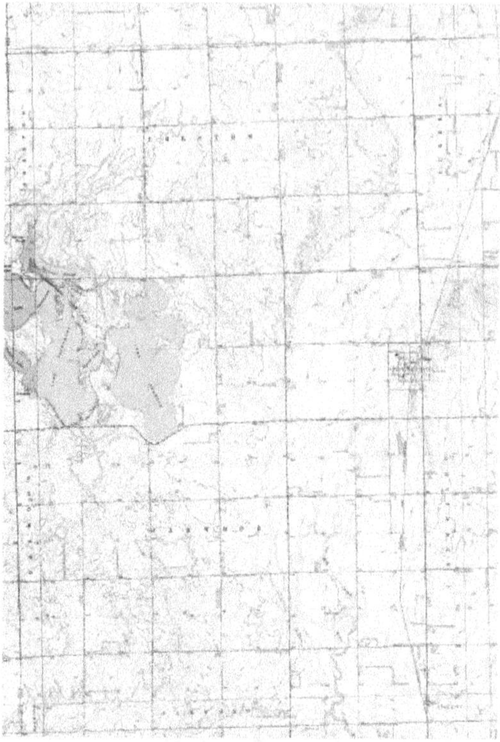

FIGURE 2.9 Digital raster graphic of the Bruce quadrangle in Brookings County, South Dakota.

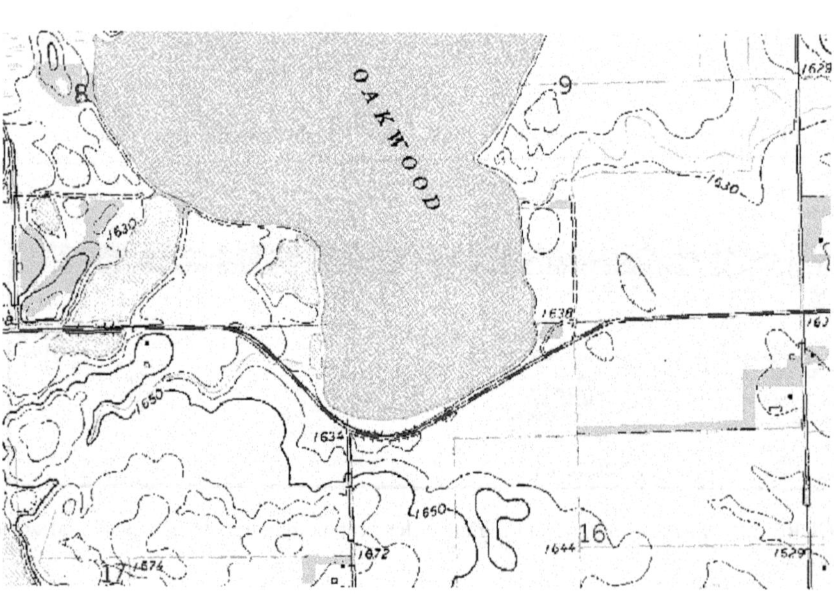

FIGURE 2.10 Subset of the digital raster graphic of the Bruce quadrangle in Brookings County, South Dakota.

FIGURE 2.11 DOQ imagery for the Bruce quadrangle in Brookings County, South Dakota.

MrSID compressed format, a format that is compatible with most GIS and image processing software. A plug-in is also available for viewing MrSID images with an Internet browser (see http://www.sdgs.usd.edu/digitaldata/doq.html). The four DOQQs comprising the Bruce quadrangle within Brookings County in South Dakota are shown in Figure 2.11.

A second and more recent data source is the imagery acquired for the **National Agricultural Imagery Program (NAIP)**, a U.S. Department of Agriculture program to acquire peak growing season "leaf-on" imagery. This natural color (three-band) imagery has 1 or 2 m spatial resolution, depending on the year it was acquired (http://www.fsa.usda.gov/Internet/FSA_File/naip_2007_infosheetpdf.pdf). NAIP imagery for South Dakota is available for 2003 through 2008, with the exception of 2007. It can be downloaded from the **Digital Base Data** Web site described in Section 2.5.1. To download an image,

1. Click on the year for which the imagery is desired.
2. Click on the county desired. As with the DRG and DOQ data, you can then either **open** or **save** the ***.zip*** file that will be downloaded. The file will be very large (typically more than 120 MB) because it will contain the imagery for the entire county rather than just a quadrangle, as is the case for the DRG and DOQ data. Figure 2.12 shows 2008 NAIP imagery for the NW quadrant of the Bruce quadrangle shown in Figure 2.11.

FIGURE 2.12 NAIP image of the NW quadrant of the Bruce quadrangle within Brookings County, South Dakota.

Many types of public and private satellite imagery are currently available. A report authored by W.E. Stoney of Noblis, Inc. for the American Society of Photogrammetry and Remote Sensing indicates that at the end of 2007, there were 13 countries that had 31 satellites with optical sensors acquiring imagery of the Earth at a resolution of 56 m or better, and 6 countries had a total of 10 radar satellites in operation. This same report indicates that, according to information available at the end of 2007, 51 optical-sensor satellites from 21 countries were expected to be collecting land remote sensing imagery, and 18 satellites from 10 countries would be collecting radar imagery. The report, *ASPRS Guide to Land Imaging Satellites*, is available online at http://www.asprs.org/news/satellites/satellites.html.

The Landsat series of satellites have been collecting moderate-resolution imagery of the Earth since 1972. Two Landsat satellites, launched by NASA and operated by USGS, are currently operational—Landsat 5 and Landsat 7. A follow-on Landsat is scheduled for launch in 2012. Information about the Landsat program and products is available at http://landsat.usgs.gov. Since October 2008, Landsat images, as well as imagery from several other USGS and NASA satellites, have been available for download at no charge via two online services—EarthExplorer and GloVis (http://edcsns17.cr.usgs.gov/EarthExplorer and http://glovis.usgs.gov). The size of each download file is approximately 160 MB. The compressed file

FIGURE 2.13 Landsat images showing. Bruce quadrangle within Brookings County, South Dakota.

(***tar.gz*** extension) consists of seven or eight bands of visible and infrared data. Software such as WinZip can be used to "unzip" the file after it has been saved. Each individual-band image file is approximately 56MB in size when "unzipped." The images are in ***geotiff*** format and have been processed to Standard Terrain Correction (Level 1T). Figure 2.13 shows the black and white Landsat image acquired on May 30, 2009, for the Bruce quadrangle in Brookings County, South Dakota (see Figure 2.11). When the website for the Landsat image in accessed, both a natural color (bands 3, 2, and 1 as red, green, and blue) and a false color (bands 4, 3, and 2 as red, green, and blue) image can be obtained.

Satellite imagery available from federal agencies such as USGS, NASA, and NOAA is generally of low to moderate resolution. The commercial sector, however, has carved out its niche in the high-resolution market. In the United States, commercial satellite imagery vendors include GeoEye (http://www.geoeye.com) and DigitalGlobe (http://www.digitalglobe.com). Satellites owned and operated by GeoEye include IKONOS with 0.82 m panchromatic resolution and 4 m multispectral resolution and GeoEye-1 with 0.41 m panchromatic and 1.65 m multispectral resolution. QuickBird imagery from DigitalGlobe has 0.61 and 2.44 m resolution in panchromatic and multispectral modes, respectively. WorldView-1, another DigitalGlobe satellite, collects panchromatic imagery only at 0.5 m resolution. The recently launched WorldView-2 will provide 0.5 m panchromatic resolution and 1.8 m multispectral resolution. Information on purchasing imagery from commercial vendors and product prices is available from their Web sites. License agreements that restrict the use and sharing of the imagery are usually required by commercial vendors when imagery is purchased.

2.5.3 Digital Elevation Data

Another commonly used type of raster data is **digital elevation data**. These data are used to create three-dimensional visualizations of land surfaces. Satellite and aerial imagery are often draped over the three-dimensional visualizations to add land cover information or, conversely, to add a third dimension to the imagery. Digital elevation data are also used to remove distortions in imagery due to terrain relief displacement. This correction, along with spacecraft or aircraft attitude corrections (roll, pitch, and yaw), produces an orthorectified image, that is, a planimetric image with uniform scale.

For the conterminous United States, Alaska, Hawaii, and territorial islands, the USGS National Elevation Dataset (NED) is a primary source of digital elevation data. NED data are available (except for Alaska) at resolutions of 1 arcsec (about 30 m),

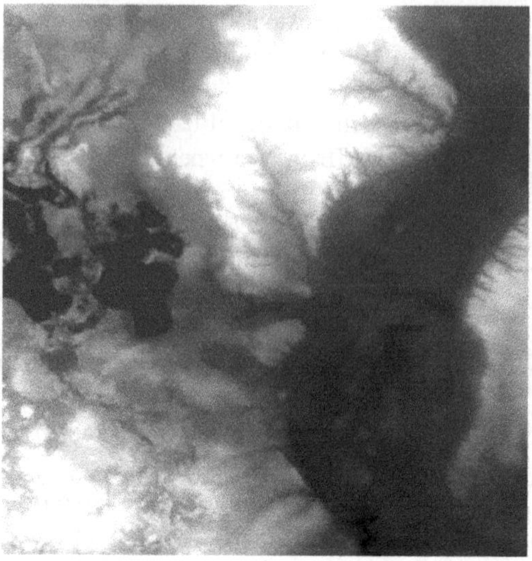

FIGURE 2.14 NED data for the Bruce quadrangle in Brookings County, South Dakota.

1/3 arcsec (about 10 m), and 1/9 arcsec (about 3 m) (http://ned.usgs.gov). NED data, along with many other types of spatial data, are available via the USGS Seamless Data Distribution System (http://seamless.usgs.gov). Figure 2.14 shows 1/3 arcsec NED data for the Bruce quadrangle in Brookings County, South Dakota.

Another source of digital elevation data for South Dakota is the DEM data found on the Digital Base Data Web site discussed earlier in this chapter. The spatial extent of each DEM file is a 7.5 min quadrangle. The horizontal resolution of each DEM file is either 10 or 30 m.

Shuttle Radar Topography Mission (SRTM) elevation data are available for a large portion of the Earth. These data were obtained by radar instruments aboard the NASA Space Shuttle Endeavor in 2000. The horizontal resolution of the SRTM data for the United States and its territories and possessions is 1 arcsec (30 m postings); the resolution for regions between 60°N and 56°S latitude is 3 arcsec (90 m postings) (http://eros.usgs.gov/#/Find_Data/Products_and_Data_available/Elevation_Products). SRTM data are available for download from the USGS EarthExplorer Web site (http://edcsns17.cr.usgs.gov/EarthExplorer) and from the Seamless Data Distribution System referenced earlier.

LiDAR (Light Detection and Ranging) is another remote sensing system that can be used to collect topographic data. The vertical accuracy of the data can be as good as 10 cm. LiDAR equipment is generally mounted on a small aircraft (http://vulcan.wr.usgs.gov/Monitoring/LIDAR/description_LIDAR.html). The data are often collected for private sector customers and, thus, not generally publicly available. The USGS, however, maintains a Web site named CLICK (Center for LIDAR Information Coordination and Knowledge) with a goal of facilitating LiDAR data access, user coordination, and education (http://lidar.cr.usgs.gov). Maps showing publicly available LiDAR data can be downloaded from this site. Much of this publicly available data have been incorporated into NED as 1/9 arcsec data.

2.6 VECTOR DATA SOURCES AND EXAMPLES

An abundance of vector datasets are available for download from the Internet. A few of the datasets commonly used for base data and for agricultural applications are briefly described below.

DLGs are digital representations of features found on USGS maps. The two basic types of DLGs are large-scale (1:24,000) and intermediate-scale (1:100,000). Depending on the scale, various layers of DLG feature types are available: Public Land Survey System, Boundaries, Transportation, Hydrography, Hypsography, Non-Vegetative Features, Survey Control and Markers, Man-Made Features, and Vegetative Surface Cover. DLG datasets can be downloaded from the USGS GeoData Web site (http://edc2.usgs.gov/geodata). Land use/land cover datasets at scales of 1:100,000 and 1:250,000 can be downloaded from this same Web site. DLG data for South Dakota at both the 1:24,000 and 1:100,000 scales can be downloaded from the Digital Base Data Web site mentioned earlier in this chapter (http://www/sdgs.usd.edu/dlgdata/dlg.html). Information for other states can also be found at various Web sites such as the GeoData Web site; state specific searches can be done to find these information types.

A clearinghouse of GIS datasets for South Dakota is available at http://arcgis.sd.gov/IMS/sdgis/Data.aspx. A few of the many datasets at this site are airports, ambulance services, bridges, cemeteries, census information, cities, hydrology, geology, flood zones, hunting areas, and zip code boundaries. A similar clearinghouse of spatial datasets is maintained by each state. Google (or similar search engine) can be used to locate a GIS clearinghouse for a particular state.

A rich source of map data for the United States is the National Atlas (http://www.nationalatlas.gov). From this Web site, a user can customize maps for printing or viewing, print preformatted maps, play with interactive maps, and download the raw map data used to generate the National Atlas maps. Geostatistical data and imagery, in addition to vector data, are available for download. The following categories of data for the United States can be downloaded from http://www.nationalatlas.gov/atlasftp.html: agriculture, biology, boundaries, climate, environment, geology, history, map reference, people, transportation, and water.

Wetland data are often of interest to agricultural producers and other natural resource managers. National Wetlands Inventory digital data are available for free download from http://www.fws.gov/wetlands/Data/DataDownload.html.

Soils data are likewise of interest to agricultural producers as well as to engineers and others. Digital soils data can be downloaded from the Geospatial Data Gateway Web site maintained by the U.S. Department of Agriculture—http://datagateway.nrcs.usd.gov. At this same site, many other layers of natural resource and environmental data may be downloaded including watersheds, quadrangle map indexes, elevation, orthoimagery, geographic names, land use/land cover, and climate (precipitation and temperature).

Although many spatial datasets are available online, some of the datasets necessary for agricultural applications may need to be acquired elsewhere or created by the user. These datasets include weather data, field boundaries, land ownership maps, and current and historical information for individual fields. The field information for each year may include crop type, variety, yield, and tillage; weed or other pest infestation areas; amounts and type of fertilizer and pesticide applications by location; and grazing intensity (animal units and duration) by location. Weather data may be obtained from state climate offices or from weather stations set up in the field. Weather data from sites some distance from the field of interest may require interpolation using geospatial data processing software. Field boundary and land ownership map data may be available from government offices such as the Farm Service Agency or the County Assessor's office. Field boundaries can also be digitized using imagery as a background layer in GIS software or by collecting GPS data along the field boundary and transferring the GPS coordinates to GIS software. GPS data can similarly be used to delineate areas within a field with variable fertilizer and herbicide applications or grazing areas. Once the field and within-field boundaries have been created, the attribute table can be used to record information such as crop type, variety, planting date, and fertilizer type and amounts.

2.7 SOFTWARE FOR SPATIAL DATA VISUALIZATION AND ANALYSIS

Several software products are available for viewing and processing spatial data. The cost of the software ranges from free to thousands of dollars. The functionality of the software likewise ranges from having basic data viewing capability to high-powered analytical capability. This section describes some of the more popular packages, but it does not present a complete listing of free or commercially available software.

One of the software products available at no cost is MultiSpec, a processing system developed at Purdue University for interactively analyzing multispectral and hyperspectral image data. The software, documentation, and tutorials can be downloaded from http://cobweb.ecn.purdue.edu/~biehl/MultiSpec.

Google Earth is another no-cost product that enables users to view satellite imagery, maps, and terrain anywhere on Earth. Users can additionally add their own vector and raster datasets for viewing. The Google Earth download site is http://earth.google.com.

The ExpressView Browser Plug-in gives browsers the ability to view MrSID and JPEG2000 images as well as magnify, print, and save the images. Its download Web site is http://www.lizardtech.com/download/files/win/expressview/webinstall. GeoExpressView is a companion product with additional functionality available from http://www.lizardtech.com/products/geo/pretrial.php for a 30 day trial.

GlobalMapper is another software package for which a free trial is available. This software features an easy-to-use viewer/editor capable of displaying the most popular raster, elevation, and vector datasets. It is also capable of converting, editing, mosaicking, reprojecting, and printing these datasets. Its download site is http://www.globalmapper.com.

Software packages with full image-processing functionality include ERDAS, ENVI, PCI-Geomatica, and PG-STEAMER. Information regarding ERDAS software is found at http://www.erdas.com. Another ERDAS product with basic capability for displaying and reprojecting various types of georeferenced data is ERDAS ViewFinder. It can be downloaded from http://www.erdas.com/Products/ERDASProductInformation/tabid/84/currentid/2537/default.aspx.

Web sites with information about ENVI, PC-Geomatica, and PG-STEAMER are, respectively, http://www.ittvis.com/ProductServices/ENVI.aspx, http://www.pcigeomatics.com, and http://www.pixoneer.com/en/PG-STEAMER.php.

GIS software is also used to display and analyze all types of spatial data. A complete line of GIS software products has been developed by ESRI. ESRI's current family of ArcGIS products ranges from the free ArcReader to its high-end program, ArcInfo. Information about all of the ESRI GIS software products can be found at http://www.esri.com/products.

2.8 CONCLUSION

Spatial information can be accessed from a number of different U.S. government agency sites at many different resolutions and formats and generally are free of charge. Spatial information is also available from many private sector organizations, but usually at a cost. Many non-U.S. public and private sector sources of spatial data also exist. Spatial information in the form of aerial and satellite imagery is available from archives that contain historical and near real-time images for land areas around the world. Additionally, image acquisition over a designated location on a specific future date can also be requested, typically only from private sector vendors. Once the information is obtained, software packages, ranging in cost from free to expensive, are available for data manipulation. Several factors, including the different types of resolution (spectral, spatial, temporal, and radiometric), should be considered when using the information. When combining layers of spatial data, projection information and geometric accuracy are also important considerations.

ACKNOWLEDGMENTS

Funding was provided by NASA and USGS.

REFERENCES

1. Campbell, J.B., *Introduction to Remote Sensing*, The Guildford Press, New York, p. 625, 2007.
2. Jensen, J.R., *Introductory Digital Image Processing: A Remote Sensing Perspective*, Pearson Prentice Hall, Upper Saddle River, NJ, p. 526, 2005.
3. Harvey, F., *A Primer of GIS Fundamental Geographic and Cartographic Concepts*, The Guilford Press, New York, p. 310, 2008.
4. Chang, K.-T., *Introduction to Geographic Information Systems*, McGraw-Hill, New York, p. 450, 2008.

3 Population Ecology Considerations for Monitoring and Managing Biological Invasions

Patrick C. Tobin, Laura M. Blackburn, Shelby J. Fleischer, and E. Anderson Roberts

CONTENTS

3.1 Executive Summary.......30
3.2 Introduction.......30
3.3 Arrival.......31
3.3.1 Invasion Pathways.......32
3.3.2 Monitoring the Arrival of Biological Invaders.......33
3.4 Establishment.......34
3.4.1 Factors That Influence Establishment Success.......34
3.4.2 Monitoring the Establishment of Nonnative Species: Space–Time Population Persistence.......35
3.5 Spread.......37
3.5.1 Types of Spread.......37
3.5.2 Estimating Invasive Species Spread.......38
3.6 Managing Biological Invasions.......40
3.7 GIS Tutorial: Estimating Spread Rates of Nonnative Species.......42
3.7.1 Introduction.......42
3.7.2 Calculating Distance to an Initial Outbreak Location.......43
3.7.3 Performing OLS Regression Analysis.......43
3.7.4 Understanding Residuals.......46
3.7.5 Testing for Spatial Autocorrelation.......46
3.7.6 Calculating Temporal Spread Rates.......47
3.7.7 Calculating Regional Spread Rates.......49
3.8 Conclusions.......50
Questions.......50
Answers to Questions.......50
Acknowledgments.......51
References.......51

3.1 EXECUTIVE SUMMARY

Biological invasions constitute a major threat to native- and agro-ecosystems and comprise three processes: arrival, establishment, and spread. Following successful establishment is an evaluation of the potential impact and management options of a nonnative species. Applications based on GIS are valuable tools that allow managers to monitor the arrival, determine successful establishment, and estimate the rate of spread of an invasive species. In this chapter, we describe the population ecology of biological invasions in a general context, focusing mostly on nonnative insects, and address conceptually the use of geospatial tools in facilitating our understanding and management of invasive species. We also include a tutorial demonstrating the utility of GIS tools in estimating invasion speed and understanding the spread dynamics of an introduced nonnative species across a landscape.

3.2 INTRODUCTION

Biological invasions constitute a major threat to native- and agro-ecosystems[1–4] and comprise three processes: arrival, establishment, and spread (Figure 3.1).[5–7] Following successful establishment is an evaluation of the potential impact and management options of a nonnative species.[8,9] The arrival process refers to a movement of individuals from a source population to a destination habitat and is facilitated through global trade and travel and atmospheric, hydrologic, or other natural transport mechanisms. This process is generally defined by invasion pathways that represent the movement of individuals from one distinct, and often but not necessarily native, habitat to a nonnative habitat, such as through the transport of infested nursery stock or solid wood packaging materials. However, the annual reinvasion of migratory species can be considered as conceptually equivalent, and a recent review considers aerobiological processes within the context of integrated pest management.[10]

Following its arrival, an invasive species becomes either established (i.e., populations grow to sufficient levels such that extinction in the absence of management activity is unlikely) or not. There are many biological and ecological factors that influence establishment success, and, due to Allee effects and stochastic forces that act upon low-density founder populations,[11–16] it often becomes a question of the size of the initial arriving population. If establishment is successful, the species will then start to spread and expand its range. The spread of biological invasions often proceeds through stratified dispersal,[5,6,17,18] in which local population growth and movement[19–21] are coupled with long-range dispersal, which can be facilitated by anthropogenic and atmospheric transport mechanisms.[7,22]

Because the population ecology of each of these three processes is unique though not necessarily independent, they each can influence the monitoring program and particularly the management guidelines and policy (Figure 3.1).[23] Applications based upon GIS are valuable tools that allow managers to monitor the arrival, determine successful establishment, and estimate the rate of spread of an invasive species. In this chapter, we describe the population ecology of biological invasions and address the use of geospatial tools in facilitating our understanding and management of invasive species. We largely but not exclusively

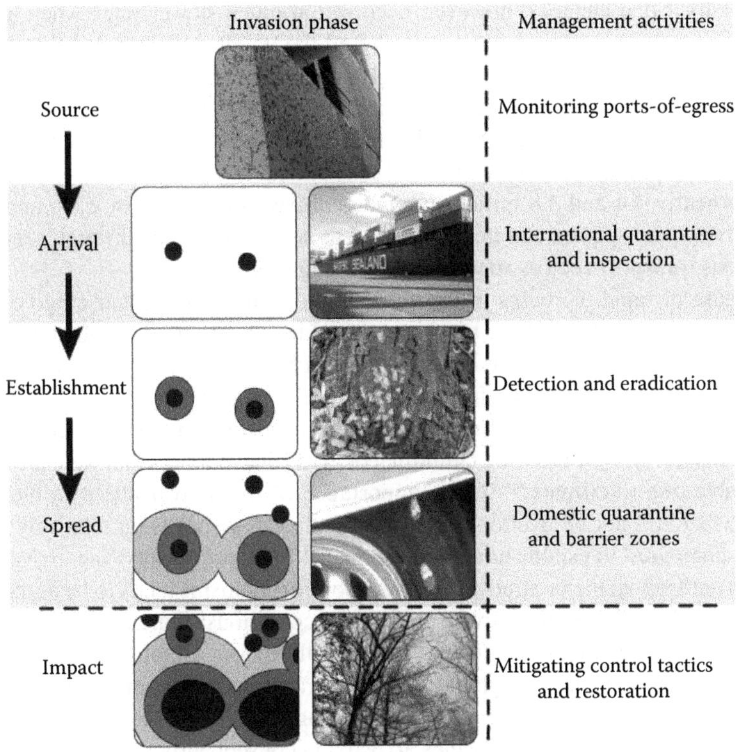

FIGURE 3.1 Example of the biological invasion process, which occurs through the arrival of propagules from source populations, after which populations may establish. If establishment is successful, the next phase is spread through diffusion[21] or more likely through stratified dispersal.[5] Impacts can occur anytime following successful establishment. The management strategy can differ depending upon the invasion phase targeted. Photos (top to bottom): Gypsy moths aggregating on a shipping vessel in Russia. (Photo courtesy of Tamara Freyman, Russian Research Institute of Plant Quarantine.) Cargo entering a port-of-entry. (Photo courtesy of U.S. Department of Agriculture, Washington, DC.) Gypsy moth egg masses at the base of a tree. (Photo courtesy of U.S. Department of Agriculture, Washington, DC.) Movement of egg masses through anthropogenic transport mechanisms. (Photo courtesy of U.S. Department of Agriculture, Washington, DC.) A recent gypsy moth outbreak at Rocky Arbor State Park, Wisconsin. (Photo courtesy of P.C.T., U.S. Department of Agriculture, Washington, DC.)

concentrate on nonnative invasive insect species given their importance as pests in agro-ecosystems, agroforestry systems, and forested ecosystems.

3.3 ARRIVAL

Throughout the world's history, species distributional changes have constantly occurred due to the shifting of land masses, receding of waters, and changing of climates. Consequently, species have been continuously introduced into new habitats, and, in some cases, they have dramatically altered ecosystem structure and biodiversity. Such changes, however, occurred at much slower rates when humans first began to wander the planet[24] or mastered the winds and explored the world in sailing ships[25] than today in an increasingly global community.[4,7,26] The world's busiest maritime ports now each handle more than 500,000 tonnes of cargo each day, and, occasionally, the movement of this cargo involves unwanted passengers in the form of pests, diseases, and weeds.[27–31] International travel continues to increase as well, as nearly 4.4 and 4.8 billion international passengers flew in 2006 and 2007, respectively, and species can be transported to new destinations through seemingly innocuous transport vectors such as personal baggage.[32]

Because of rapid increases in the movement of humans and their cargo over the last ca. 500 years, species have been introduced into novel areas in which they have not previously evolved, either intentionally, such as many North American food crops,[33] or unintentionally, such as many agricultural and forest pests.[34–37] Species may also arrive to new areas through natural transport mechanisms, such as migratory behavior,[38,39] or by crossing latitudinal or altitudinal barriers that were previously impassable due to climate.[40–42] The importance of the arrival phase in biological invasions should not be overlooked. However, this stage has been relatively poorly studied compared to establishment and spread,[43,44] in part because the arrival stage through anthropogenic or atmospheric transport mechanisms tends to be a stochastic process that is influenced by trade or climatic conditions; thus, it does not readily lend itself to population ecological questions unlike both establishment and spread dynamics. However, rates of global trade and travel pathways, and climatic conditions under which atmospheric transport mechanisms are enabled, may be used to estimate the risk of nonnative species arrival and consequently to optimize the allocation of sampling resources to detect the arrival of new species.[22,45,46]

The management of any invader is less infeasible when the target population is newly established and hence at low population density and distributed across smaller spatial scales;[47–50] thus, early detection of invading species through monitoring programs is of paramount importance. In this section, we address the arrival stage of biological invasions by considering pathways through which a species is transported from an area in which they are present to an area in which they are not. In doing so, we demonstrate the utility of a GIS in visualizing the time of arrival, the locations at which the arrival is occurring, and possibly the mechanisms facilitating their arrival.

3.3.1 Invasion Pathways

In its simplest terminology, an invasion pathway is a route through which an invasive species moves or is moved from a source population to a recipient location.[7] Pathways can be considered as anthropogenically derived, such as shipping and international travel routes, or those that are derived naturally, such as through atmospheric or hydrologic transport mechanisms. Given a particular pathway, several conditions need to be addressed to recognize its potential of transporting an invasive species.[43] First, a source population must be in a life stage that enables transport. In the aerobiology literature, this is referred to as "preconditioning," and for many invading organisms, phenology becomes relevant because the life stage must be able to survive the trek.[10,51] Second, this life stage needs to come into contact with the origin of the pathway, such as a port-of-egress located near an ongoing pest outbreak or synoptic weather conditions that would facilitate the movement of biota through the atmosphere.[22,52] When atmospheric transport pathways are involved, the species may display distinct behaviors that increase the probability of being captured in the pathway.[38] Alternatively, the cargo itself may serve as a conduit; examples include infested nursery stock containing plant pests and diseases or wood products used as packing materials and dunnage-harboring woodboring beetles.[27,35] Aquatic organisms, most notoriously zebra mussels, can be transported in the ballast water of shipping vessels.[9,53,54]

Third, a sufficient number of invading individuals must successfully survive the trek and arrive under conditions conducive to its survival. Although it is possible to identify potential invasion pathways by examining shipping routes, international flight patterns, and weather conditions through which atmospheric transport mechanisms are potentially enabled,[55] the sheer number of pathways makes it challenging to assign levels of risk. Regardless, monitoring the initial arrival of an invading species in a GIS can provide important information facilitating a timely and appropriate management response. Predicting arrival based upon existing climatic conditions can enhance integrated pest management programs against invaders transported through atmospheric transport mechanisms.[51,56]

3.3.2 Monitoring the Arrival of Biological Invaders

It can be challenging if not nearly impossible to predict the pathway through which the next detrimental invasive species will arrive to a new destination, which makes monitoring for them a daunting task. Not all pathways can be monitored at all times, though the probability of the arrival of an invasive species can often be a question of risk. For species transported through global trade routes, the monitoring of major ports-of-entry and ports-of-egress is a logical, though still nontrivial, management decision. For example, although the gypsy moth, *Lymantria dispar* (L.) (Lepidoptera: Lymantriidae), is now present over much of eastern North America,[57] it is not yet established outside of the Northeastern, Mid-Atlantic, and Midwestern United States. Also not yet established is the Asian strain of gypsy moth, whose females, unlike females from the European strain currently present in North America, are capable of sustained flight.[58] Major ports-of-entry on the west coast, such as the combined Californian ports of Los Angeles and Long Beach, as well as other locations outside of the gypsy moth established area, are monitored routinely. Many detections have been recorded, some of these triggering eradication treatments, while others, including a single Asian gypsy moth male detected at the port of Long Beach in 2005, failed to establish upon arrival.[59]

But it is not always port locations that detect new species. Cargo arriving into the United States eventually is shipped throughout the country. Again, the concept of risk could play a role in prioritizing management programs as cargo likely follows a pathway in itself: products and goods are shipped from ports-of-entry to distribution centers and then to various markets with larger markets (i.e., major cities) likely receiving more goods. However, there are certainly exceptions; for example, one male Asian gypsy moth was trapped near Hauser, Idaho, population 819, in 2004.[60]

Nevertheless, given the movement of nonnative species through global trade and travel routes, there is likely a tendency for the arrival of a new invader to be positively correlated to human population density, their movement, and the movement of their goods.[54,61–64]

3.4 ESTABLISHMENT

Upon the arrival of an invading species to a new habitat, the newly arrived species either establishes itself or not. Fortunately, most new invasions are believed to fail[65–67] or when they do establish, only pose minimal economical and environmental impacts.[2] Nevertheless, some invasions do pose severe ecological and economical costs.[1–3,68] In the United States, for example, invasions by nonnative species are thought to cause environmental and economical damages and losses adding up to nearly $122 billion annually.[69] Because many, though not all, new invaders arrive in low numbers, the establishment phase is a critical one. During this period, populations grow and expand their distribution, and forced extinction, such as through eradication management strategies, becomes increasingly less likely.[49] Because many invasions do fail, it is logical to address briefly the factors that influence the invasiveness of a nonnative species.

3.4.1 Factors That Influence Establishment Success

The probability of successful establishment is influenced by many factors, such as the size of the founder population, the susceptibility of the habitat to invasion, the presence of competitors and regulators, environmental and demographic stochasticity, and Allee effects.[7] Several previous studies on the establishment phase have sought to understand and identify components that increase or decrease the ability of a newly arrived invader to successfully invade a particular ecosystem.[16,70] For example, plant invasions tend to be more successful in habitats with an elevated availability of unused resources and disturbance.[70–73] Habitat invasibility, or the degree to which a particular habitat is susceptible to invasions, could also be influenced by climate and the presence or absence of competitors, mutualists, and regulators.[12,74–76] Establishment of plant pathogens often can be especially dependent on abiotic conditions.[77,78] Details of species natural history also play a role in species invasiveness. For example, tree-killing bark beetles that depend on mass attacking mechanisms to colonize live host trees and overcome host plant defense could be regarded as a poor invader because of the required initial population size.[79] A consequence is that certain habitats may be more invasible than others, and certain species may be better at invading novel habitats than others.[16,70,80]

For many biological invasions, it often becomes a question of numbers. The bigger the founding population, the more likely it will successfully establish.[47,81–84] Much of our knowledge of the population biology of low-density populations is due to prior work on the management of rare and endangered species. Many biological invasions are often initiated as low-density founder populations, which, like rare and endangered species,[85] are often subject to Allee effects and inimical stochastic events.

The Allee effect describes a phenomenon in which per capita population growth rates decline with declining density.[11,13,86–88] Many species depend upon cooperative or gregarious behaviors, or larger population sizes, in reproduction,[89–92] when foraging for resources,[79,93,94] when evading or satiating natural enemies,[95,96] and when avoiding the inhibitory consequences of inbreeding depression.[97] In addition to Allee effects, founder populations can also be affected by environmental and demographic stochasticity, both of which are exacerbated in low-density populations.[47,98] Thus, one primary reason why many new invasions fail is because arriving populations are small. This phenomenon is conceptually illustrated by intentional introductions of natural enemies as part of biological control programs in which the probability of establishment was higher from releases of larger numbers of individuals.[90,99–101] The primary ramification of Allee effects and stochasticity is that extinction of low-density populations of newly arriving invasive species can occur without any management intervention.

3.4.2 Monitoring the Establishment of Nonnative Species: Space–Time Population Persistence

It is an analytical challenge to quantify the spatial and temporal signature associated with an emerging invasion as opposed to a situation where a species arrived but failed to establish itself, particularly given that most newly arrived populations fail to establish. This challenge is furthermore compounded by our general lack of knowledge of the species for which we should be looking and the sheer number of invasion pathways through which new species are constantly arriving. A primary signature of an established invader is undoubtedly the spatial persistence of a population through time. Such information can be displayed as spatial layers through time in a GIS, which could shed light on whether the population is persisting at the same spatial locations through time, which in turn would suggest an established population (Figure 3.2).

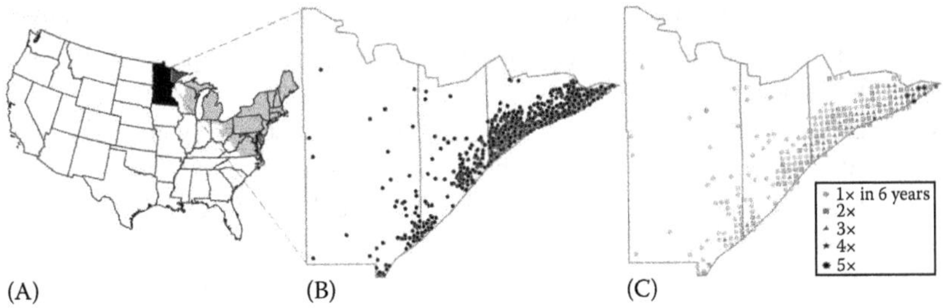

FIGURE 3.2 Using GIS to visualize arrival and establishment: (A) map of the gypsy moth infested area (gray) according to the 2005 quarantine. Newly arriving populations in northern Minnesota (St. Louis, Cook, and Lake Counties) were detected from 2000 to 2005, and all trapping locations recording at least one male moth are shown in (B) as dots. In (C), the legend relates the number of years in which a trapping location (condensed into 5 × 5 km grid cells) recorded at least one male moth; for example, 5× are locations where male moths were trapped in 5 of 6 years, which tends to argue for gypsy moth establishment as opposed to those areas where males were trapped only once in 6 years.

For several insect species, there are cost-effective, sensitive monitoring tools in the form of pheromone-baited traps generally using sexual or aggregation pheromones as bait. In monitoring programs for the gypsy moth, for example, many traps are deployed far from both the generally infested area (i.e., the area under federal quarantine) and a transition zone along the leading population front (as managed under the gypsy moth Slow-the-Spread program[102]). It is not uncommon to detect new gypsy moth colonies outside of the infested area, whether these new arrivals originated from the infested area and are transported through interstate commerce and household moves or from international trade routes.[59] A common operating practice after the initial detection of gypsy moth arrival is to deploy a more extensive trapping grid in the following year to determine establishment success, especially given that the detection of an arrival does not always equate to establishment. In Washington State, for example, Liebhold and Bascompte[47] examined 192 gypsy moth colonies that were detected using pheromone-baited traps deployed from 1974 to 1996, of which 162 (84%) went extinct without any management intervention, with 123 (64%) of these going extinct within 1 year following initial detection. Moreover, if space–time persistence of the colony is observed and eradication treatments are warranted, deploying a more extensive trapping grid can be used to determine the spatial extent of the population, so that treatment tactics can be site-specifically implemented.

The importance of the space–time persistence of a newly arrived and detected colony should not be overlooked, and GIS tools facilitate our understanding of this phenomenon. Monitoring programs that use georeferenced trapping protocols are already adapted for use in a GIS and immediately lend themselves to space–time correspondence analysis. Part of the challenge, though, is to distinguish between repeated and potentially unrelated arrivals of a nonnative species and the successful establishment of the species.[103,104] If repeated initial detections over time are the result of independent stochastic introductions that never successfully establish, then implementing an eradication program and deploying chemical pesticides or biopesticides could be both unnecessary and costly. Visualizing time layers of spatial persistence and potential colony growth in a GIS, thus, provides for a more informed decision that a management intervention is appropriate and also optimizes the spatial extent of the treated area.

Although the importance of space–time colony persistence of a newly arrived species in determining successful establishment is somewhat intuitive, it can still be a considerable ecological challenge to ascertain if establishment has actually occurred. A classic example is the presence of the Mediterranean fruit fly, *Ceratitis capitata* (Wiedemann) (Diptera: Tephritidae), in California in which debates among the scientific community, federal, state and local governments, and the public as to its establishment have existed now for decades. While no one doubts that *C. capitata* is consistently detected,[32,105,106] it remains contentious if it is an established invader[107] or an invader for which arrivals occur consistently but either do not result in successful establishment or are successfully eradicated through management intervention.[108] Since monitoring programs of many invasive species, and particularly insects, involve the trapping of adults, conclusive evidence of successful establishment can be obtained through the detection of other life stages. Space–time persistence patterns displayed in a GIS can be used to optimize sampling programs to maximize, but not guarantee, the chance that alternate life stages will be detected if populations are indeed established.

3.5 SPREAD

Spread is a process by which an invasive species moves from a habitat that it occupies to one in which it does not.[109] The rate of spread is most often estimated as the increase in distributional range over time. Rates of spread furthermore tend to vary by species[5,109,110] and vary across spatial and temporal scales within a species.[88,111,112] From an extreme perspective, the initial movement of a species from one continent to another, such as through global trade and travel, could be considered as a type of spread. Also, it is not always practical to disregard the interrelationship among arrival, establishment, and spread, especially when spread proceeds in part due to founder colonies that arrive and successfully establish outside of the organism's current distribution. Consequently, it can be ambiguous between what comprises the spread phase and what can be considered as part of the arrival and successful establishment phase. For this chapter, we address invasive species spread in reference to its expansion of distributional boundaries following its initial arrival and establishment in a new geographical area.

3.5.1 Types of Spread

In nearly all biological invasions, spread at least includes the coupling of dispersal with population growth. This process of movement was modeled in the ecological literature by Skellam,[21] who combined exponential population growth with diffusive (i.e., random) movement, which was slightly different than Fisher's model[19] that adopted logistic growth. Both models, though, result in similar predictions and patterns of spread in that the radial increase in distributional range increases linearly through time.[5] A classic example of this type of phenomenon can be seen in the muskrat, *Ondatra zibethicus* (L.) (Rodentia: Cricetidae), invasion of Europe.[110,113]

However, more often than not the spread of an invading species does not occur exclusively through simple diffusion; rather, long-distance "jumps"—such as through anthropogenic and atmospheric transport mechanisms—may occur with considerable frequency. For annually reinvasive migrants, this process is strongly tied to both their life cycle and the climate patterns, but for other invasive species, long-distance transport can be highly unpredictable.[4] The combined processes of short- and long-range dispersal is referred to as stratified dispersal.[5,6,18] Stratified dispersal can be a major driver of the spread process that in turn can be influenced by Allee effects that act upon colonies initiated through long-distance dispersal.[88,111,112,114–116]

In many biological invasions, the anthropogenic movement of life stages can be a dominant mode of long-range displacement though not an exclusive one.[4] For example, long-distance dispersal by the hemlock woolly adelgid, *Adelges tsugae* Annand (Hemiptera: Adelgidae), in North America is thought to be facilitated by migratory birds.[117] Atmospheric transport mechanisms are thought to be the primary mechanism of soybean aphid, *Aphis glycines* Matsumura (Hemiptera: Aphididae), long-range dispersal,[118] and of many plant pathogens[119,120] such as the causal agent of soybean rust, *Phakopsora pachyrhizi* Sydow and Sydow (Basidiomycotina: Urediniomycetes).[51] Both atmospheric and hydrological transport mechanisms are thought to influence the spread of *Phytophthora ramorum* Werres, de Cock, and Man in't Veld (Pythiales: Pythiaceae), the causal agent of sudden oak death.[121,122] Regardless of the mechanism of stratified dispersal, its existence has a fundamental effect on spread. The occasional long-range dispersal events initiate colonies ahead of the distributional range (essentially akin to the arrival stage). If these colonies successfully establish, then they could grow and eventually coalesce with the established range of the organism, resulting in a more rapid overall rate of spread than what would be expected under exclusively diffusive spread.[5,6,18] Empirically, past work has highlighted the ramifications of stratified dispersal in the gypsy moth,[123] Africanized honeybee, *Apis mellifera scutellata* Lepeletier (Hymenoptera: Apidae),[124] Argentine ant, *Linepithema humile* (Mayr) (Hymenoptera: Formicidae),[125] emerald ash borer, *Agrilus planipennis* Fairmaire (Coleoptera: Buprestidae)[126] invasions of North America, and the horse-chestnut leaf miner, *Cameraria ohridella* (Deshka and Dimic) (Lepidoptera: Gracillariidae), invasion in Europe.[62]

3.5.2 Estimating Invasive Species Spread

There are several methods to estimate the rate of spread of invasive species, and many have been greatly facilitated by a GIS. The ability to estimate spread is a crucial step in the development of management strategies. For example, prior to the spreading of an invader to a new area, there are several management implications. These include determining susceptible habitats that are most vulnerable to invasion, estimating the time before a new invader spreads to these susceptible areas, and predicting the eventual economic and ecological impacts. In some cases, management tactics could be to mitigate impacts prior to arrival.[127] It thus is not too terribly surprising that much past work has focused on understanding the spread of invasive species,[5,17,18,128] including historical studies that were published long before the global realization of the importance of biological invasions.[110,113,129]

The models of both Fisher[19] and Skellam[21] conceptually derive the radial rate of a species range expansion, *V*, according to

$$V = 2\sqrt{rD}, \tag{3.1}$$

where

- D is a diffusion coefficient that can be estimated as the standard deviation of dispersal distances[130,131]
- r is a measure of the intrinsic rate of population increase

There are several case studies in which Equation 3.1 has provided spread estimates consistent with empirical observations, most notably the muskrat invasion of Europe,[113] although there are many cases where it fails because it does not include long-distance dispersal.[5] Another potential problem with using Equation 4.1 in an analysis of spread is that in many cases, there simply is not enough existing data that can be used to parameterize *r* and *D*. Many nonnative species cause considerable damage in nonnative habitats where they may be free of competition and natural enemies and coevolved host resistance but often pose little economic concern in their native environment; thus, the literature could be too limited to parameterize *r* and *D* without resorting to purely theoretical estimates that may or may not have practical utility.

Alternatively, we could estimate the rate of invasive species spread by examining its historical spread based upon available empirical data.[128] In this approach, the use of GIS can facilitate the space–time reconstruction of past distributional ranges to estimate its rate of spread and to predict future range boundaries (Figure 3.3). There are, however, two primary limitations to this approach. First, for biological invasions that have only recently been detected, historical data may be limited. Second, even if data were available, spread rates estimated—especially from limited spatial and temporal scales—may not always translate into accurate future spread projections, due to, for example, spread latency. This refers to a lag between arrival and

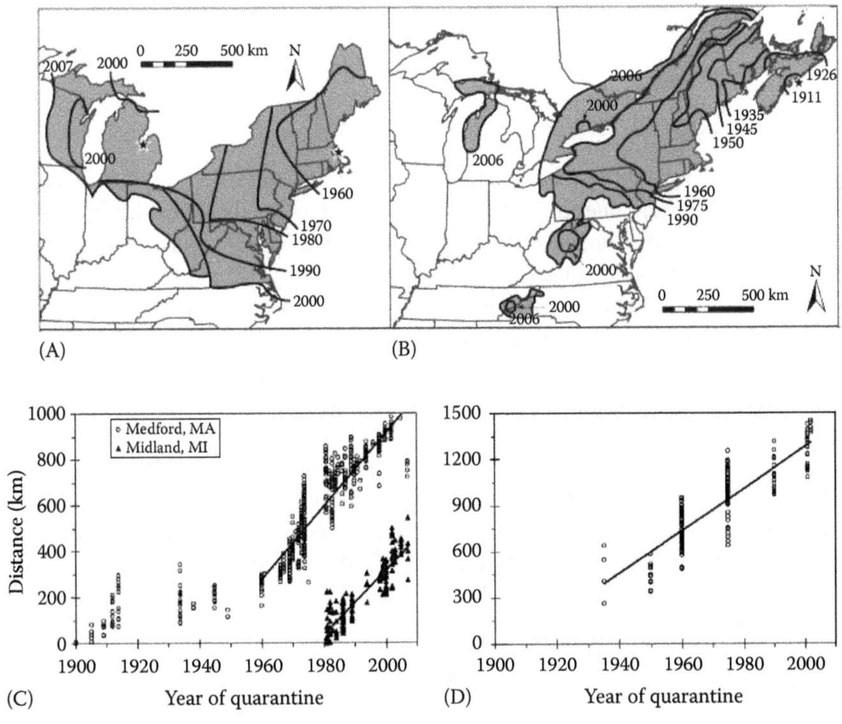

FIGURE 3.3 Using GIS to estimate the spread of the gypsy moth (A, Tobin et al.[57]) and beech bark disease (B, Morin et al.[136]). Initial introduction points for each are noted by stars (Medford, Massachusetts, 1869, and Midland, Michigan, 1981, for the gypsy moth, and Halifax, Nova Scotia [ca. 1890] for beech bark disease). The annual rate of spread, using the method of Liebhold et al.,[121] was estimated by the slope from least squares regression of the minimum distance between presence (on a county level) and the site of introduction (Medford, Midland, or Halifax) as a function of the first year of county quarantine. Estimated spread rates for the (C) gypsy moth are 15.9 km per year (from Medford) and 14.1 km per year (from Midland), and (D) 13.7 km per year for beech bark disease.

successful establishment, and the point when populations grow to sufficient levels to provide a source of dispersing (short- and long-range) individuals. This latency has been observed in several past biological invasions, perhaps arising from different mechanisms, in which initial spread occurs very slowly or that the range initially retracts prior to more rapid increases in spread.[5,48,114,132,133]

Regardless of the limitations, using past observations often can be the only available approach to estimating spread, especially when nonsynthetic estimates of r and D cannot be obtained. Several different methods have been used to estimate the rate of spread based upon historical data.[57,123,134,135] Conceptually, one approach is to use the recording of the spatial presence of an invasive species through reports, maps, or aerial photographs from successive intervals.[74,123,136–138] When spread proceeds as the circular expansion around an initial point of infestation (at time t), the radius can be estimated as the square root of the area infested at time $t + 1$ divided by π. Then, to determine the radial rate of expansion between time t and $t + 1$,

$$\frac{\left(\sqrt{\frac{A_{t+1}}{\pi}} - \sqrt{\frac{A_t}{\pi}}\right)}{T} \tag{3.2}$$

where T is the length of time between t and $t + 1$. When spread does not emanate uniformly across all radii, due to, for example, bodies of water (in-terrestrial invaders) or other unsuitable habitats, an alternative is to estimate the colony radius as the average of radii emanating from the point of infestation after excluding biologically irrelevant radii.

In the rare event that detailed population density information is available through time, a robust approach to estimating spread involves the construction of population boundaries in space, and then measuring the spatial displacement in population boundaries through time. Such data are available for the gypsy moth invasion of North America[57] in which populations are monitored across the network of pheromone-baited traps deployed each year.[102] Even prior to extensive deployment of monitoring tools such as pheromone-baited traps, Liebhold et al.[123] quantified the spread of the gypsy moth using county-level quarantine records (1900–1989) maintained by the U.S. Department of Agriculture (U.S. Code of Federal Regulations, Title 7, Chapter III, Section 301.45-3). They measured the minimum distance between a quarantined county and the site of the initial gypsy moth introduction (Medford, Massachusetts). The average yearly radial rate of spread was then estimated as the slope from least squares regression of the distance as a function of the first year of county quarantine. Interestingly, this approach, simply using spatially crude records of presence/absence, provided overall rates of spread that did not differ substantially from those obtained from the more costly deployment of extensive trapping grids.[57]

3.6 MANAGING BIOLOGICAL INVASIONS

Managing biological invasions can be a costly endeavor, both economically and ecologically, though the costs can be difficult to quantify precisely. It is furthermore not trivial to predict when a new invader will arrive, from where it will arrive, and along which invasion pathway it will be moved. Still, it tends to be far less expensive to proactively monitor ports-of-egress prior to arrival and ports-of-entry during arrival than to mitigate reactively the damage due to invasive species populations after establishment. In ports-of-egress, monitoring native outbreaks and the timing of the outbreak can be a cost-effective strategy to reduce the potential propagule pressure of a nonnative invader. In an analogous approach, control strategies that reduce source population abundance could influence population pressure in distal areas.[139] The outbreak spatial distribution of many species is currently mapped using aerial surveys and geospatial technologies, and for many species, such as insects, outbreaks tend to be temporally constrained. Monitoring areas adjacent to ports-of-egress and examining the spatial and temporal characteristics of outbreaks could reduce the risk that an invasive species "hitchhikes" through global trade routes. For example, the U.S. Department of Agriculture has implemented an overseas monitoring program in high-risk ports-of-egress (i.e., high risk due to the volume of freight processed and the degree to which the species composition in the country of origin would find U.S. habitats suitable) to limit their arrival to U.S. soil.

Because of the sheer volume of invasion pathways, new nonnative species will continue to arrive, and, thus, the need for the monitoring of ports-of-entry and inbound cargo. However, the question of successful establishment after arrival is still not always clear. Species can repeatedly arrive but never succeed in establishing, such as the tree-killing spruce bark beetle, *Ips typographus* (L.) (Coleoptera: Curculionidae), which has been detected at ports-of-entry outside of its native range but has yet to establish,[27] likely due to its host-seeking mass dispersal and mass-attacking colonization behaviors.[140] Thus, immediately implementing an eradication program each time *I. typographus* is detected could be unnecessary and costly.

However, in other invasive species, the threshold for successful establishment could be lower, or the risks due to successful establishment much greater, both of which could warrant more aggressive eradication measures. In the former, we can consider factors that affect establishment success, such as the importance of mate-finding failures at low insect population densities, relevant to many founder populations.[90,141] Thus, species that reproduce asexually, such as the hemlock woolly adelgid, could have a much lower density threshold for establishment because of the inconsequence of finding mates. The latter (i.e., greater risk due to successful establishment) often depends on the country and the ecosystem and agro-ecosystem resources potentially at risk due to an invader. Many islands, for example, tend to be more prone to invasion, and the inimical effects of invasive species could be more pronounced than those on larger land masses.[81,142] For example, New Zealand, following the detection of one male gypsy moth in a pheromone-baited trap, aggressively implemented an eradication program through the use of eight weekly aerial applications of the biopesticide *Bacillus thuringiensis var. kurstaki* over 1253 ha around the trap location.[143]

Indeed, the monitoring of ports-of-entry provides useful information regarding species arrival, but this information is most useful when placed within the context of successful establishment. Consistent space–time persistent of an invader, a process that can be visually displayed in a GIS, would argue in favor of establishment, while repeated introductions through time but without any spatiotemporal correspondence could suggest an arrival risk but the lack of an establishment risk. Since many invasions fail due to Allee effects and stochasticity, and given the general financial constraints associated with eradication tactics, it would be optimal to allow invasions to fail on their own whenever possible. A better understanding of the factors and processes that drive establishment success,[16,70,71,73] whether due to habitat invasibility, species invasiveness, or a combination of both, could facilitate the interpretation of spatial and temporal data visualized in a GIS.[84,144]

The management of spread, whether dominated by short- and/or long-range dispersal, should incorporate an understanding of the factors that drive establishment.[88,114,145] Although local population growth and local dispersal account for the spread of invasive species, propagules that disperse more distally from locally established populations, regardless of the mechanism, tend to contribute a proportionately greater amount to overall rates of spread.[5,18,146] By monitoring establishment and targeting those populations whose densities exceed those no longer subject to Allee effects or stochastic events, at least to the point of colony extinction, spread can be greatly reduced. A management program, Slow-the-Spread, exists for the gypsy moth in which isolated colonies that form ahead of the population front are targeted for eradication to limit their influence on its spread,[102] and recent spread rates under this program[57] have been greatly reduced from those rates observed prior to the implementation of the Slow-the-Spread program.[123] In some cases, range boundaries may in fact retract as a result of management programs[147] including the use of natural enemies in classical biological control.[96,148] Management programs that aim to limit spread are generally done so in a spatially explicit manner that immediately lends itself to GIS applications. The use of GIS tools and spatial modeling also allows for spatial and temporal prediction of the future boundaries of an invasive species,[138,149] allowing for land managers to prepare and in some cases preemptively mitigate the negative impacts of invasive species.[127,145] In the next section, we present a tutorial to illustrate the utility of GIS in estimating the rate of spread of an invasive species across a landscape.

3.7 GIS TUTORIAL: ESTIMATING SPREAD RATES OF NONNATIVE SPECIES

3.7.1 INTRODUCTION

The ability to estimate and predict the rate of spread in a nonnative species is a critical step in the development of management strategies and policy. When a new invader becomes established in a new area, key questions often include (1) which areas are susceptible to infestation and (2) how long before populations become established in these susceptible areas. Often, governments will impose a quarantine zone around the area of a new invasion, and the expected rate of spread can be a critical ingredient in determining the appropriate perimeter around an infestation. For established invaders not fully occupying all susceptible habitats, predicting the rates and extent of spread aids in the development of management guidelines. For example, management interventions can sometimes minimize the impacts of a nonnative species if such tactics can be implemented prior to invasion.[127,145] The gypsy moth invasion of the United States is one of the longest space–time series of a biological invasion.[57] In this exercise, you will use county-level distributional data from 1900 to 2008 to explore and estimate the rate of gypsy moth spread in a GIS.

The gypsy moth, *L. dispar* (L.), invasion of the United States is an example of the importance of both short- and long-range dispersal.[146] County-level quarantine records for gypsy moth are maintained by the U.S. Department of Agriculture (U.S. Code of Federal Regulations, Title 7, Chapter III, Section 301.45-3) and exist back to 1912, while distributional records available from various sources exist to 1900. This makes the gypsy moth invasion of the United States one of the longest space–time series of a biological invasion. In this exercise, you will use county-level distributional data from 1900 to 2008 to explore and estimate the rate of gypsy moth spread in a GIS format.

To complete this exercise, you must have **ArcGIS 9.3.1** or higher installed. Launch **ArcMap** and begin using an existing map. Highlight the option to **Browse for maps** and click **OK**. Open ***SpreadRate.mxd*** in Chapter 3 folder included on the CD that came with this book. You will see five layers listed in the table of contents located to the left of the map. If you would like more information on any of these layers, consult their associated metadata files via ArcCatalog.

3.7.2 Calculating Distance to an Initial Outbreak Location

To calculate the rate of spread in the gypsy moth, we must first know the distance from a reference point (i.e., the site of initial introduction) to each quarantined county. Compute this value by performing a spatial join on the Gypsy Moth Quarantine layer. Right click on the **Gypsy Moth Quarantine** layer, choose **Joins and Relates**, and then select **Join**. When the **Join Data** dialog box opens up, you will choose to **Join data from another layer based on spatial location** (see Figure 3.4). Because the gypsy moth was originally introduced in Medford, Massachusetts, we will select that layer as the layer to join to the Gypsy Moth Quarantine layer. Select the option to give all point attributes and distances to each polygon. Next, choose a location for the output shapefile to be saved and name the file ***QuarantineDistance***. When you have filled in your dialogue box, it should look like Figure 3.4. Click **OK**.

A new layer, QuarantineDistance, is added to the **Layers** list that contains the year of gypsy moth quarantine and the distance from the original introduction point (Medford, Massachusetts). You will use this layer to perform an **Ordinary Least Squares (OLS)** regression analysis, which will allow you to predict the rate of spread and explore various spatial and temporal relationships in the variability of spread rates. We will use OLS regression to fit the linear relationship between the year of county quarantine (explanatory variable) and the distance of each county from Medford, Massachusetts (dependent variable).

3.7.3 Performing OLS Regression Analysis

To perform an OLS, you will use **ArcToolbox** in ArcMap. To open **ArcToolbox**, click on the icon that looks like a red toolbox () found in the **Standard** toolbar typically located at the top of the GUI. Once **ArcToolbox** is open, you will find the

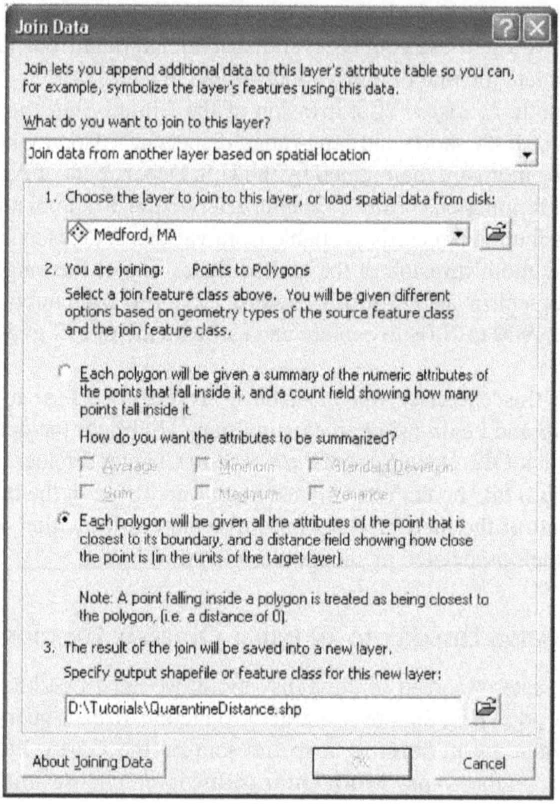

FIGURE 3.4 Join data dialogue box. (The ArcGIS® graphical user interface is the intellectual property of ESRI and is reproduced herein by permission, copyright © 1999–2009 ESRI, All rights reserved.)

Spatial Statistics Tools near the bottom of the list; either click the plus sign next to it or double-click on the text. This will reveal all of the Spatial Statistics Tools. Open the **Modeling Spatial Relationships** group and you will find the **Ordinary Least Squares** tool; double-click on it to open the OLS dialogue box (Figure 3.5). Next,

1. Choose the ***QuarantineDistance*** layer that you created as the **Input Feature Class.**
2. Select ***ID*** for the **Unique ID Field.**
3. Choose a location for the **Output Feature Class** and name it ***QuarantineDistance_OLS***.
4. Select ***Distance*** for the **Dependent Variable.**
5. Choose ***Quar_Year*** for the **Explanatory Variables.**
6. Click on the **Output Options** text to display optional output tables and fill in a location for the **Coefficient Output Table**, name it ***GM_OLScoefficient***.
7. Leave the **Diagnostic Output Table** blank. Once complete, your OLS dialogue box should mirror Figure 3.5. Click **OK**.

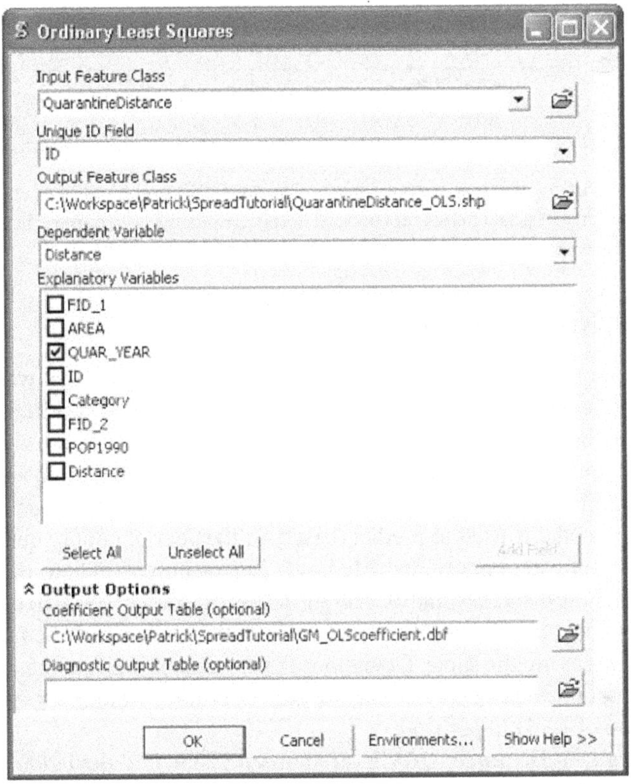

FIGURE 3.5 Ordinary least squares dialogue box. (The ArcGIS® graphical user interface is the intellectual property of ESRI and is reproduced herein by permission, copyright © 1999–2009 ESRI, All rights reserved.)

A second dialogue box will open during the OLS process with a statistical report, including a summary of OLS results, OLS diagnostics, notes on interpretation, and the location of your output table. For greater detail on interpreting these results, consult **ArcGIS Desktop Help** and search for **Interpreting OLS results**. Click **Close** to exit this report. You will see a new layer, ***QuarantineDistance_OLS***, in your table of contents; we will look at that later in the exercise. First, let us review the coefficient output table. Click on the **Add Data** icon () located to the left of the ArcToolbox icon and browse to the folder containing the coefficient output table, *GM_OLScoefficient.dbf*. Click on the **Add** button to add it to your **Layers** list. Select the coefficient data table and right click on it. When the menu appears, select **Open** (Figure 3.6). You will notice the second row has the explanatory variable, ***Quar_Year***. The coefficient **(Coef)** for that variable is the estimated slope from the OLS of the minimum distance as a function of the first year of county quarantine; in other words, it provides an estimate of the year-to-year average rate of radial spread. Because the data used in this analysis are projected in meters, the coefficient results are also in meters. Converting 12,509.96 m to kilometers, we get an overall predicted rate of gypsy moth spread of ≈12.5 km per year for the years 1900–2008.

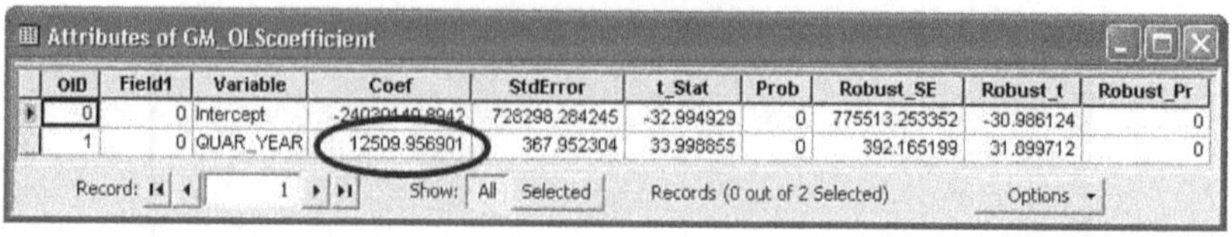

OID	Field1	Variable	Coef	StdError	t_Stat	Prob	Robust_SE	Robust_t	Robust_Pr
0	0	Intercept	-24030140.8942	728298.284245	-32.994929	0	775513.253352	-30.986124	0
1	0	QUAR_YEAR	12509.956901	367.952304	33.998855	0	392.165199	31.899712	0

FIGURE 3.6 OLS coefficient output table. (The ArcGIS® graphical user interface is the intellectual property of ESRI and is reproduced herein by permission, copyright © 1999–2009 ESRI, All rights reserved.)

3.7.4 Understanding Residuals

Let us take a closer look at the ***QuarantineDistance_OLS*** layer that was just added to your table of contents. This layer also depicts the OLS model **residuals**. Residuals are important diagnostic tools that, in this case, can be used to explore areas in which the observed rate of spread deviated from the overall model prediction and to explore spatial patterns in the rate of spread. When fitting the regression model to the data, the regression model is used to predict, based on the year of county quarantine, the distance between each county and Medford, Massachusetts. Then, the difference between predicted distances and observed distances (which you calculated earlier in the exercise) is the residual. A residual value of 0 indicates that the predicted and observed distances are the same. Deviations in the residual indicate those counties in which gypsy moth invaded sooner than or later than what would be expected under a constant rate of 12.5 km per year.

The ***QuarantineDistance_OLS*** layer shows the standard deviation of residuals for each county. In Figure 3.7, notice how some counties have a residual value that is <–2.5 standard deviation units from the predicted value, while others have a residual value that is more than 2.5 standard deviation units from the predicted value.

In the former (<–2.5 standard deviation units), we can deduce that the gypsy moth invaded these counties much later than what our constant rate of 12.5 km per year predicted. In contrast, the gypsy moth invaded those counties with a residual value >2.5 standard deviation units much sooner than what our model predicted. Notice the spatial pattern of these deviations. We will formally explore this spatial pattern next using the **Spatial Autocorrelation** tool on the regression residuals.

3.7.5 Testing for Spatial Autocorrelation

The **Spatial Autocorrelation** tool is found in the **Spatial Statistics Tools, Analyzing Patterns** group. Double-click on the **Spatial Autocorrelation (Morans I)** tool to open the dialogue box (Figure 3.8). Select the ***QuarantineDistance_OLS*** layer as the **Input Feature Class**. Choose ***Residual*** for the **Input Field** and check the box to **Display Output Graphically**. Accept the default settings for the rest of the features. Your dialogue box should look like Figure 3.8. Click **OK**.

The test for spatial autocorrelation reveals significant spatial clustering, a fact also evident in Figure 3.7. Your results panel should look like the one in Figure 3.9. Once you have had a good look at the results, click **Close**. The spatial autocorrelation tool will then complete its process, and you can **close** the report window.

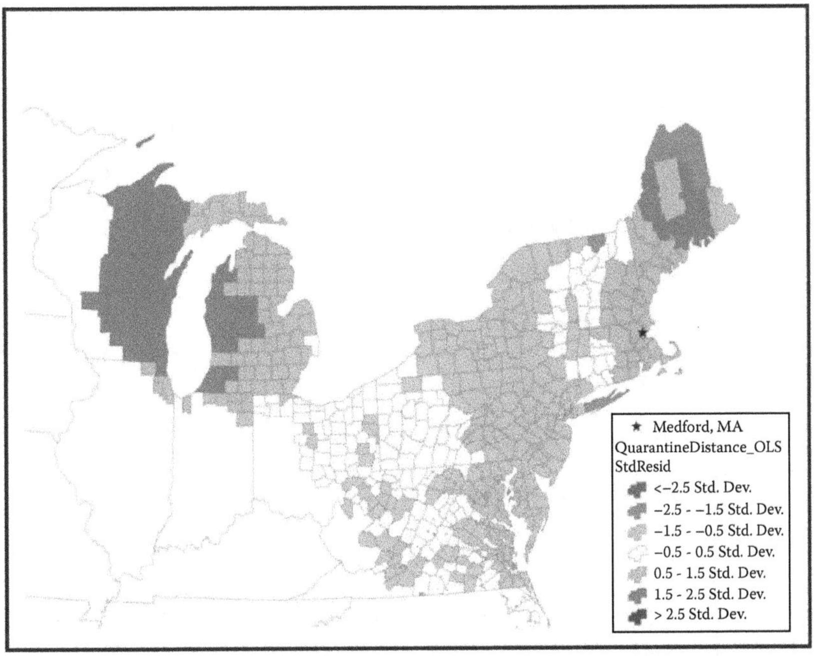

FIGURE 3.7 Map layout showing the standard deviation of residuals. (The ArcGIS® graphical user interface is the intellectual property of ESRI and is reproduced herein by permission, copyright © 1999–2009 ESRI, All rights reserved.)

The significant spatial autocorrelation indicates a clustered spatial pattern in the rate of gypsy moth spread, such as contiguous areas in which gypsy moth invaded slower or faster than expected. The results thus suggest that the rate of spread was not constant through time, and that different rates of spread were observed depending on the region that was being invaded. Let us explore this next.

3.7.6 Calculating Temporal Spread Rates

The Gypsy Moth Quarantine layer is displayed in four time periods or categories: 1900–1915 (category = 1), 1916–1965 (category = 2), 1966–1989 (category = 3), and 1990–2008 (category = 4). Let us take a look at the spread rate for each of these categories. Open the attribute table for the ***QuarantineDistance*** layer you made earlier. You will notice a field called ***Category***. Click on the **Options** button at the bottom of the attribute table and choose **Select by Attributes**. You will **Create a new selection**. Scroll down and double-click on ***Category*** in the list of fields. Click on the equal button. Then click on the button to **Get Unique Values** and double-click on the number ***1***. Your dialogue box should look like Figure 3.10. Click **Apply** and close the **Select by Attributes** window.

You will see that all of the appropriate records are selected in your table, and their associated features are highlighted in the map. Now you will complete the OLS process detailed in **Section 3.7.3** to determine the spread rate for these selected records.

What are the spread rates for each of the four time period categories? Do they differ from the overall spread rate for the entire area and time period?

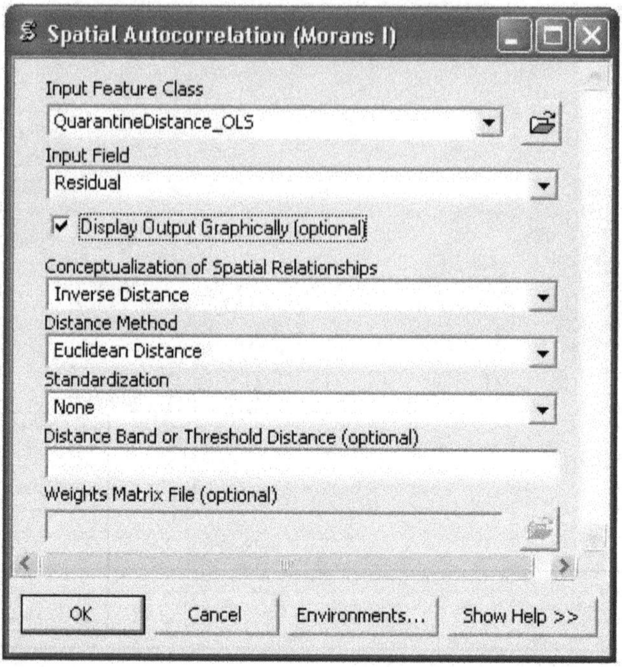

FIGURE 3.8 Spatial Autocorrelation dialogue box. (The ArcGIS® graphical user interface is the intellectual property of ESRI and is reproduced herein by permission, copyright © 1999–2009 ESRI, All rights reserved.)

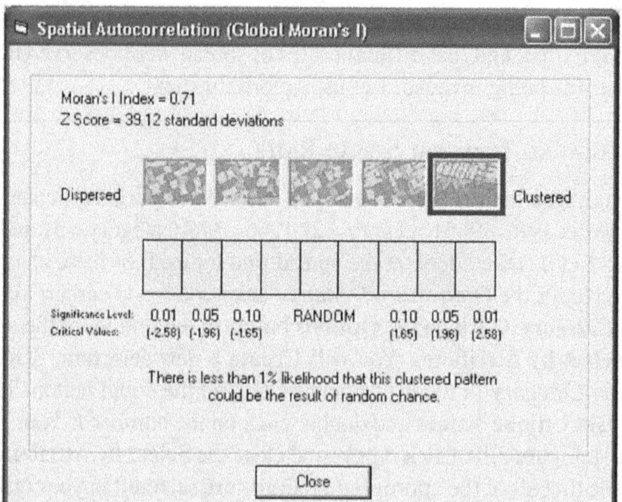

FIGURE 3.9 Spatial Autocorrelation graphic output. (The ArcGIS® graphical user interface is the intellectual property of ESRI and is reproduced herein by permission, copyright © 1999–2009 ESRI, All rights reserved.)

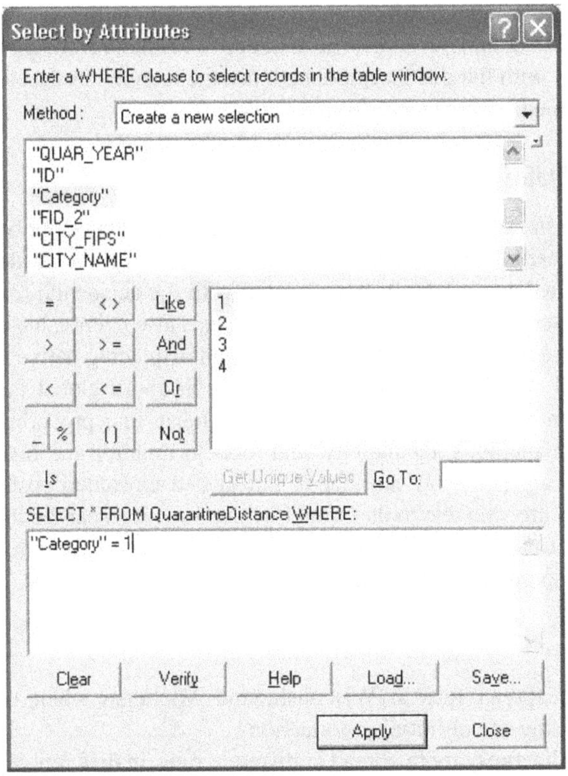

FIGURE 3.10 Select by Attributes dialogue box. (The ArcGIS® graphical user interface is the intellectual property of ESRI and is reproduced herein by permission, copyright © 1999–2009 ESRI, All rights reserved.)

3.7.7 Calculating Regional Spread Rates

Next, let us consider rates of spread across different regions. As the residual map shows, gypsy moth invaded Michigan and Wisconsin much sooner than expected when assuming a constant rate of invasion from Medford, Massachusetts of 12.5 km per year. However, it is important to note that Medford, Massachusetts, represents the initial site of gypsy moth introduction in the United States in 1869. A second introduction point is considered to be Midland, Michigan, in 1981. Let us explore the rate of gypsy moth spread in Michigan and Wisconsin again, but this time, let us use Midland, Michigan, as our initial point instead of Medford, Massachusetts.

First, open the attribute table for the ***Gypsy Moth Quarantine*** layer, select **Options** and open the **Select by Attributes** dialogue box that you used above. Create a new selection where **STATE_NAME =** ***Michiga'*** **OR STATE_NAME =** ***Wisconsin*** and click **Apply**. Right click on the ***Gypsy Moth Quarantine*** layer and choose **Data**, **Export Data**. Accept the default settings to export selected features using the same coordinate system as the layer's source data. Provide a location for the output shapefile and call it ***GMQ_MIWI.shp***. Click **Save** and then click **OK**. When asked if you want to add the exported data to the map as a layer, select **Yes**.

Now using the techniques outlined in Section 3.7.2, you will join this layer using to the Midland, Michigan, layer to get the distance to Midland from each county. After you have a layer with the distance attached, use the techniques described in Section 3.7.3 to determine the spread rate for this region.

3.8 CONCLUSIONS

Biological invasions are unfortunately a reality of today's global community. Although there are surely economic gains from increased global trade, their benefits should be weighed against the cost associated with the accidental transport of non-native invasive species. Several introduced species cause irreparable harm to native- and agro-ecosystems by damaging cash crops and displacing native species. Global trade and travel will likely continue to increase as more global markets become established. In many respects, invasion biologists must run a proverbial Red Queen's race, so that our methods, technology, and tools to monitor and manage invasions keep pace with an increasing number of introduced unwanted pests. Applications based upon GIS are valuable tools through which invasions can be monitored, from source habitats to arrival, establishment, and spread, and from which appropriate management responses can be developed and implemented.

QUESTIONS

1. What is the spread rate in Wisconsin and Michigan when using Midland, Michigan, as the site of initial introduction?
2. Why would the gypsy moth spread at different rates in different regions?
3. Let us assume a constant rate of gypsy moth spread of 12.5 km per year. Using Medford, Massachusetts, as the initial site of introduction and 1869 as the initial outbreak year, predict the year in which gypsy moth is expected to invade the following cities:
   (a) St. Louis, Missouri; (b) Atlanta, Georgia; (c) Lexington, Kentucky.
   Are there cases where the model predictions are unreasonable?
4. Based upon the spread rate you estimated in exercise #1, when do you predict the gypsy moth to invade Minneapolis, Minnesota? How does this differ from the predicted rate of spread using 12.5 km and Medford, Massachusetts as the initial site?

ANSWERS TO QUESTIONS

1. 16.05 km per year.
2. Differences in the availability of preferred host trees, different climates resulting in different rates of gypsy moth survivorship, and different degrees of human-assisted movement.
3. (a) Distance from Medford to St. Louis = 1653 km. At a constant spread rate of 12.5 km per year, it would take ≈132 years (1653 ÷ 12.5) from 1869 for the gypsy moth to invade St. Louis; thus, the predicted year is 1869 + 132 = 2001.
   (b) 1989.
   (c) 1967.

Yes—the spread rate model does not take human-assisted, long-range movement into the equation. Also, the model does not consider any management activities against the gypsy moth that could slow its rate of spread.

4. Using a spread rate of 16.05 km per year and 1981 as the initial year of establishment in Midland, Michigan, the gypsy moth is predicted to arrive in Minneapolis in 2026. Using a spread rate of 12.5 km per year and 1869 as the initial year of establishment in Medford, Massachusetts, the gypsy moth is predicted to arrive in Minneapolis in 2013. The differences in the expected time to invasion is influenced by management activities that have been implemented since the initial introduction of gypsy moth in 1869.

ACKNOWLEDGMENTS

We would like to thank John Juracko and George Racin (USDA Forest Service) for their assistance in developing the tutorial.

REFERENCES

1. Parker, I. M., Simberloff, D., Lonsdale, W. M., Goodell, K., Wonham, M., Kareiva, P. M., Williamson, M. H., Von Holle, B., Moyle, P. B., Byers, J. E., and Goldwasser, L. Impact: Toward a framework for understanding the ecological effects of invaders. *Biol. Invasions*, 1, 3, 1999.
2. Mack, R. N., Simberloff, D., Lonsdale, W. M., Evans, H., Clout, M., and Bazzaz, F. A. Biotic invasions: Causes, epidemiology, global consequences, and control. *Ecol. Appl.*, 10, 689, 2000.
3. Mooney, H. A. and Cleland, E. E. The evolutionary impact of invasive species. *Proc. Natl Acad. Sci. USA*, 98, 5446, 2001.
4. National Research Council. *Predicting Invasions of Nonindigenous Plants and Plant Pests*. National Academy Press, Washington, DC, 2002.
5. Shigesada, N. and Kawasaki, K. *Biological Invasions: Theory and Practice*. Oxford University Press, New York, 1997.
6. Shigesada, N., Kawasaki, K., and Takeda, Y. Modeling stratified diffusion in biological invasions. *Am. Nat.*, 146, 229, 1995.
7. Lockwood, J. L., Hoopes, M., and Marchetti, M. *Invasion Ecology*. Blackwell Publishing Ltd., Malden, MA, 2007.
8. Kolar, C. S. and Lodge, D. M. Progress in invasion biology: Predicting invaders. *Trends Ecol. Evol.*, 16, 199, 2001.
9. Kolar, C. S. and Lodge, D. M. Ecological predictions and risk assessment for alien fishes in North America. *Science*, 298, 1233, 2002.
10. Isard, S. A., Mortensen, D. A., Fleischer, S. J., and DeWolf, E. D. Application of aerobiology to IPM. In: *Integrated Pest Management. Concepts, Tactics, Strategies and Case Studies*. Ratcliff, E. B., Hutchison, W. D., and Cancelado, R. E., Eds. Cambridge University Press, Cambridge, U.K., 2009, p. 90.
11. Allee, W. C. *The Social Life of Animals*. W. W. Norton and Company, Inc., New York, 1938.
12. Crawley, M. J. What makes a community invasible? In: *Colonization, Succession and Stability*. Gray, A. J., Crawley, M. J., and Edwards, P. F., Eds. Blackwell, Oxford, U.K., 1987, p. 429.
13. Dennis, B. Allee effects: Population growth, critical density, and the chance of extinction. *Nat. Resour. Model.*, 3, 481, 1989.

14. Drake, J. A. and Lodge, D. M. Allee effects, propagule pressure and the probability of establishment: Risk analysis for biological invasions. *Biol. Inv.*, 8, 365, 2005.
15. Leung, B., Drake, J. M., and Lodge, D. M. Predicting invasions: Propagule pressure and the gravity of Allee effects. *Ecology*, 85, 1651, 2004.
16. Lonsdale, W. M. Global patterns of plant invasions and the concept of invasibility. *Ecology*, 80, 1522, 1999.
17. Andow, D. A., Kareiva, P. M., Levin, S. A., and Okubo, A. Spread of invading organisms. *Landsc. Ecol.*, 4, 177, 1990.
18. Hengeveld, R. *Dynamics of Biological Invasions*. Chapman and Hall, London, U.K., 1989.
19. Fisher, R. A. The wave of advance of advantageous genes. *Ann. Eugen.*, 7, 355, 1937.
20. Okubo, A. *Diffusion and Ecological Problems: Mathematical Models*. Springer, Berlin, Germany, 1980.
21. Skellam, J. G. Random dispersal in theoretical populations. *Biometrika*, 38, 196, 1951.
22. Isard, S. A. and Gage, S. H. *Flow of Life in the Atmosphere*. Michigan State University Press, East Lansing, MI, 2001.
23. Hulme, P. E. Beyond control: Wider implications for the management of biological invasions. *J. Appl. Ecol.*, 43, 835, 2006.
24. di Castri, F. History of biological invasions with special emphasis on the old world. In: *Biological Invasions: A Global Perspective*. Drake, J. A., Mooney, H. A., di Castri, F., Groves, R. H., Kruger, F. J., Rejmánek, M., and Williamson, M., Eds. John Wiley & Sons, Ltd., New York, 1989, p. 1.
25. Crosby, A. W. *Ecological Imperialism: The Biological Expansion of Europe, 900–1900*. Cambridge University Press, Cambridge, U.K., 1986.
26. Levine, J. A. Biological invasions. *Curr. Biol.*, 18, R57, 2008.
27. Brockerhoff, E. G., Bain, J., Kimberley, M., and Knížek, M. Interception frequency of exotic bark and ambrosia beetles (Coleoptera: Scolytinae) and relationship with establishment in New Zealand and worldwide. *Can. J. For. Res.*, 36, 289, 2006.
28. Carlton, J. T. Patterns of transoceanic marine biological invasions in the Pacific Ocean. *Bull. Mar. Sci.*, 41, 452, 1987.
29. McCullough, D. G., Work, T. T., Cavey, J. F., Liebhold, A. M., and Marshall, D. Interceptions of nonindigenous plant pests at U.S. ports of entry and border crossings over a 17 year period. *Biol. Invasions*, 8, 611, 2006.
30. Reichard, S. H. and White, P. Horticulture as a pathway of invasive plant introductions in the United States. *BioScience*, 51, 103, 2001.
31. Work, T. T., McCullough, D. G., Cavey, J. F., and Komsa, R. Arrival rate of nonindigenous insect species into the United States through foreign trade. *Biol. Invasions*, 7, 323, 2005.
32. Liebhold, A. M., Work, T. T., McCullough, D. G., and Cavey, J. F. Airline baggage as a pathway for alien insect species invading the United States. *Am. Entomol.*, 53, 48, 2006.
33. Kiple, K. F. and Ornelas, K. C., Eds. *The Cambridge World History of Food*. Cambridge University Press, Cambridge, U.K., 2000.
34. Campbell, F. T. The science of risk assessment for phytosanitary regulation and the impact of changing trade regulations. *BioScience*, 51, 148, 2001.
35. Liebhold, A. M., MacDonald, W. L., Bergdahl, D., and Mastro, V. C. Invasion by exotic forest pests: A threat to forest ecosystems. *For. Sci. Monogr.*, 30, 1, 1995.
36. Niemelä, P. and Mattson, W. J. Invasion of North American forests by European phytophagous insects. *BioScience*, 46, 741, 1996.
37. Pimentel, D., Eds. *Biological Invasions. Economic and Environmental Costs of Alien Plant, Animal, and Microbe Species*. CRC Press, Boca Raton, FL, 2002.
38. Dingle, H. *Migration. The Biology of Life on the Move*. Oxford University Press, New York, 1996.

39. Drake, V. A., Gatehouse, A. G., and Farrow, R. A. Insect migration: A holistic conceptual model. In: *Insect Migration*. Drake, V. A. and Gatehouse, A. G., Eds. Cambridge University Press, Cambridge, U.K., 1995, p. 427.
40. Battisti, A., Statsmy, M., Schopf, A., Roques, A., Robinet, C., and Larsson, A. Expansion of geographical range in the pine processionary moth caused by increased winter temperature. *Ecol. Appl.*, 15, 2084, 2005.
41. Carroll, A. L., Taylor, S. W., Régnière, J., and Safranyik, L. Effects of climate change on range expansion by the mountain pine beetle in British Columbia. In: *Mountain Pine Beetle Symposium: Challenges and Solutions*. T. L. Shore, J. E. Brooks, and J. E. Stone, Eds. Natural Resources Canada, Canadian Forest Service, Victoria, Canada, 2004, p. 223.
42. Logan, J. A., Régnière, J., and Powell, J. A. Assessing the impacts of global warming on forest pest dynamics. *Front. Ecol. Environ.*, 1, 130, 2003.
43. Jerde, C. L. and Lewis, M. A. Waiting for invasions: A framework for the arrival of non-indigenous species. *Am. Nat.*, 170, 1, 2007.
44. Puth, L. M. and Post, D. M. Studying invasion: Have we missed the boat? *Ecol. Lett.*, 8, 715, 2005.
45. Andersen, M. C., Adams, H., Hope, B., and Powell, M. Risk assessment for invasive species. *Risk Anal.*, 24, 787, 2004.
46. Government Accountability Office. *Invasive Forest Pests. Lessons Learned from Three Recent Infestations May Aid in Managing Future Efforts*, Report No. Report to the Chairman, Committee on Resources, House of Representatives, Washington, DC, GAO-06-353, 2006.
47. Liebhold, A. M. and Bascompte, J. The Allee effect, stochastic dynamics and the eradication of alien species. *Ecol. Lett.*, 6, 133, 2003.
48. Liebhold, A. M. and Tobin, P. C. Growth of newly established alien populations: Comparison of North American gypsy moth colonies with invasion theory. *Popul. Ecol.*, 48, 253, 2006.
49. Rejmánek, M. and Pitcairn, M. J. When is eradication of exotic pest plants a realistic goal? In: *Turning the Tide: The Eradication of Invasive Species*. Veitch, C. R. and Clout, M. N., Eds. IUCN, Gland, Switzerland and Cambridge, U.K., 2002, p. 94.
50. Simberloff, D. Eradication—Preventing invasions at the outset. *Weed Sci.*, 51, 247, 2003.
51. Isard, S. A., Gage, S. H., Comtois, P., and Russo, J. M. Principles of the atmospheric pathway for invasive species applied to soybean rust. *Bioscience*, 55, 851, 2005.
52. Showers, W. B., Whitford, F., Smelser, R. B., Keaster, A. J., Robinson, J. F., Lopez, J. D., and Taylor, S. E. Direct evidence for meteorologically driven long-range dispersal of an economically important moth. *Ecology*, 70, 987, 1989.
53. Roberts, L. Zebra mussel invasion threatens U. S. waters. *Science*, 249, 1370, 1990.
54. Drake, J. A. and Lodge, D. M. Global hotspots of biological invasions: Evaluating options for ballast-water management. *Proc. R. Soc. Lond. B. Biol. Sci.*, 271, 575, 2004.
55. Westbrook, J. K. and Isard, S. A. Atmospheric scales of biotic dispersal. *Agric. For. Meteor.*, 97, 263, 1999.
56. Pan, Z., Yang, X. B., Pivonia, S., Xue, L., Pasken, R., and Roads, J. Long-term prediction of soybean rust entry into the continental United States. *Plant Dis.*, 90, 840, 2006.
57. Tobin, P. C., Liebhold, A. M., and Roberts, E. A. Comparison of methods for estimating the spread of a non-indigenous species. *J. Biogeogr.*, 34, 305, 2007.
58. Keena, M. A., Grinberg, P. S., and Wallner, W. E. Inheritance of female flight in *Lymantria dispar* (Lepidoptera: Lymantriidae). *Environ. Entomol.*, 36, 484, 2007.
59. Hajek, A. E. and Tobin, P. C. North American eradications of Asian and European gypsy moth. In: *Use of Microbes for Control and Eradication of Invasive Athropods*. A. E. Hajek, T. R. Glare, and M. O'Callaghan, Eds. Springer, New York, 2009, p. 71.

60. Lech, G. Gypsy moth state report for Idaho. In: *Proceedings of the Annual Gypsy Moth Review*. Philadelphia, PA, 2005.
61. Bossenbroek, J. M., Kraft, C. E., and Nekola, J. C. Prediction of long-distance dispersal using gravity models: Zebra mussel invasion of inland lakes. *Ecol. Appl.*, 11, 1778, 2001.
62. Gilbert, M., Gregoire, J.-C., Freise, J. F., and Heitland, W. Long-distance dispersal and human population density allow the prediction of invasive patterns in the horse chestnut leafminer *Cameraria ohridella*. *J. Anim. Ecol.*, 73, 459, 2004.
63. Leung, B., Bossenbroek, J. M., and Lodge, D. M. Boats, pathways, and aquatic biological invasions: Estimating dispersal potential with gravity models. *Biol. Invasions*, 8, 241, 2006.
64. McFadden, M. W. and McManus, M. E. An Insect out of control? The potential for spread and establishment of the gypsy moth in new forest areas in the United States. In: *Forest Insect Guilds: Patterns of Interaction with Host Trees*. Baranchikov, Y. N., Mattson, W. J., Hain, F. P., and Payne, T. L., Eds. USDA Forest Service, Washington, DC, 1991, p. 172.
65. Ludsin, S. A. and Wolfe, A. D. Biological invasion theory: Darwin's contributions from The Origin of Species. *Bioscience*, 51, 780, 2001.
66. Simberloff, D. and Gibbons, L. Now you see them, now you don't!—Population crashes of established introduced species. *Biol. Invasions*, 6, 161, 2004.
67. Williamson, M. and Fitter, A. The varying success of invaders. *Ecology*, 77, 1661, 1996.
68. Pimentel, D., Lach, L., Zuniga, R., and Morrison, D. Environmental and economic costs of nonindigenous species in the United States. *BioScience*, 50, 53, 2000.
69. Pimentel, D., Zuniga, R., and Morrison, D. Update on the environmental and economic costs associated with alien invasive species in the United States. *Ecol. Econ.*, 52, 273, 2005.
70. Davis, M. A., Grime, J. P., and Thompson, K. Fluctuating resources in plant communities: A general theory of invasibility. *J. Ecol.*, 88, 528, 2000.
71. Burke, M. J. and Grime, J. P. An experimental study of plant community invasibility. *Ecology*, 77, 776, 1996.
72. Huebner, C. D. and Tobin, P. C. Invasibility of mature and 15-year-old deciduous forests by exotic plants. *Plant Ecol.*, 186, 57, 2006.
73. Tilman, D. Community invasibility, recruitment limitation, and grassland biodiversity. *Ecology*, 78, 81, 1997.
74. D'Antonio, C. M. Mechanisms controlling invasion of coastal plant communities by the alien succulent *Carpobrotus edulis*. *Ecology*, 74, 83, 1993.
75. Marler, M. J., Zabinski, C. A., and Callaway, R. M. Mycorrhizae indirectly enhance competitive effects of an invasive forb on a native bunchgrass. *Ecology*, 80, 1180, 1999.
76. Ohlemüller, R., Walker, S., and Bastow, W. J. Local vs. regional factors as determinants of the invasibility of indigenous forest fragments by alien plant species. *Oikos*, 112, 493, 2006.
77. Agrios, G. N. General overview of plant pathogenic organisms. In: *Handbook of Pest Management*. Ruberson, J. R., Eds. Marcel Dekker, Inc., New York, 1999, p. 263.
78. Venette, R. C. and Cohen, S. D. Potential climatic suitability for establishment of *Phytophthora ramorum* within the contiguous United States. *For. Ecol. Manag.*, 231, 18, 2006.
79. Raffa, K. F. and Berryman, A. A. The role of host plant resistance in the colonization behaviour and ecology of bark beetles. *Ecol. Monogr.*, 53, 27, 1983.
80. Rejmánek, M. and Richardson, D. M. What attributes make some plant species more invasive? *Ecology*, 77, 1655, 1996.
81. MacArthur, R. H. and Wilson, E. O. *The Theory of Island Biogeography*. Princeton University Press, Princeton, NJ, 1967.

82. Lockwood, J. L., Cassey, P., and Blackburn, T. The role of propagule pressure in explaining species invasions. *Trends Ecol. Evol.*, 20, 223, 2005.
83. Mollison, D. Modeling biological invasions: Chance, explanation, prediction. *Philos. Trans. R. Soc. Lond. B. Biol. Sci.*, 314, 675, 1986.
84. Whitmire, S. L. and Tobin, P. C. Persistence of invading gypsy moth populations in the United States. *Oecologia*, 147, 230, 2006.
85. Courchamp, F., Berec, L., and Gascoigne, J. *Allee Effects in Ecology and Conservation.* Oxford University Press, Oxford, 2008.
86. Courchamp, F., Clutton-Brock, T., and Grenfell, B. Inverse density dependence and the Allee effect. *Trends Ecol. Evol.*, 14, 405, 1999.
87. Stephens, P. A. and Sutherland, W. J. Consequences of the Allee effect for behaviour, ecology and conservation. *Trends Ecol Evol.*, 14, 401, 1999.
88. Taylor, C. M. and Hastings, A. Allee effects in biological invasions. *Ecol. Lett.*, 8, 895, 2005.
89. Courchamp, F. and Macdonald, D. W. Crucial importance of pack size in the African wild dog *Lycaon pictus*. *Anim. Conserv.*, 4, 169, 2001.
90. Hopper, K. R. and Roush, R. T. Mate finding, dispersal, number released, and the success of biological control introductions. *Ecol. Entomol.*, 18, 321, 1993.
91. Robinet, C., Lance, D. R., Thorpe, K. W., Tcheslavskaia, K. S., Tobin, P. C., and Liebhold, A. M. Dispersion in time and space affect mating success and Allee effects in invading gypsy moth populations. *J. Anim. Ecol.*, 77, 966, 2008.
92. Sharov, A. A., Liebhold, A. M., and Ravlin, F. W. Prediction of gypsy moth (Lepidoptera: Lymantriidae) mating success from pheromone trap counts. *Environ. Entomol.*, 24, 1239, 1995.
93. Christiansen, E., Waring, R. H., and Berryman, A. A. Resistance of conifers to bark beetle attack: Searching for general relationships. *For. Ecol. Manag.*, 22, 89, 1987.
94. Raffa, K. F. Mixed messages across multiple trophic levels: The ecology of bark beetle chemical communication systems. *Chemoecology*, 11, 49, 2001.
95. Wittmer, H. U., Sinclair, A. R. E., and McLellan, B. N. The role of predation in the decline and extirpation of woodland caribou. *Oecologia*, 144, 257, 2005.
96. Elkinton, J. S., Parry, D., and Boettner, G. H. Implicating an introduced generalist parasitoid in the invasive browntail moth's enigmatic demise. *Ecology*, 87, 2664, 2006.
97. Kramer, A. and Sarnelle, O. Limits to genetic bottlenecks and founder events imposed by the Allee effect. *Oecologia*, 157, 561, 2008.
98. Lande, R. Anthropogenic, ecological and genetic factors in extinction and conservation. *Res. Popul. Ecol.*, 40, 259, 1998.
99. Beirne, B. P. Biological control attempts by introductions against pest insects in the field in Canada. *Can. Entomol.*, 107, 225, 1975.
100. Fagan, W. F., Lewis, M. A., Neubert, M. G., and van den Driessche, P. Invasion theory and biological control. *Ecol. Lett.*, 5, 148, 2002.
101. Stiling, P. Calculating the establishment rates of parasitoids in classical biological control. *Am. Entomol.*, 36, 225, 1990.
102. Tobin, P. C. and Blackburn, L. M., Eds. *Slow the Spread: A National Program to Manage the Gypsy Moth.* USDA Forest Service, Newtown Square, PA, 2007.
103. Mostashari, F., Kulldorff, M., Hartman, J. J., Miller, J. R., and Kulasekera, V. Dead bird clusters as an early warning system for West Nile virus activity. *Emerg. Infect. Dis.*, 9, 641, 2003.
104. Tobin, P. C. Space-time patterns during the establishment of a nonindigenous species. *Popul. Ecol.*, 49, 257, 2007.
105. Barinaga, M. Entomologists in the Medfly maelstrom. *Science*, 247, 1168, 1990.
106. Headrick, D. H. and Goeden, R. D. Issues concerning the eradication or establishment and biological control of the Mediterranean fruit fly, *Ceratitis capitata* (Wiedemann) (Diptera: Tephritidae), in California. *Biol. Control*, 6, 412, 1996.

107. Carey, J. R. Establishment of the Mediterranean fruit fly in California. *Science*, 253, 1369, 1991.
108. California Department of Food and Agriculture. *Eradication of Medfly from California 1994–1995 Workplan*, Sacramento, CA, 1994.
109. Liebhold, A. M. and Tobin, P. C. Population ecology of insect invasions and their management. *Ann. Rev. Entomol.*, 53, 387, 2008.
110. Elton, C. S. *The Ecology of Invasions by Animals and Plants*. Methuen and Co., London, U.K., 1958.
111. Kot, M., Lewis, M. A., and van den Driessche, P. Dispersal data and the spread of invading organisms. *Ecology*, 77, 2027, 1996.
112. Tobin, P. C., Whitmire, S. L., Johnson, D. M., Bjørnstad, O. N., and Liebhold, A. M. Invasion speed is affected by geographic variation in the strength of Allee effects. *Ecol. Lett.*, 10, 36, 2007.
113. Ulbrich, J. *Die Bisamratte: Lebensweise Gang Ihrer Ausbreitung in Europa, Wirtschaftliche Bedeutung und Bekämpfung*. Verlag und Druck von C. Heinrich, Dresden, Germany, 1930.
114. Johnson, D. M., Liebhold, A. M., Tobin, P. C., and Bjørnstad, O. N. Pulsed invasions of the gypsy moth. *Nature*, 444, 361, 2006.
115. Keitt, T. H., Lewis, M. A., and Holt, R. D. Allee effects, invasion pinning, and species' borders. *Am. Nat.*, 157, 203, 2001.
116. Lewis, M. A. and Kareiva, P. Allee dynamics and the spread of invading organisms. *Theor. Popul. Biol.*, 43, 141, 1993.
117. McClure, M. S. Role of wind, birds, deer, and humans in the dispersal of hemlock woolly adelgid (Homoptera: Adelgidae). *Environ. Entomol.*, 19, 36, 1990.
118. Venette, R. C. and Ragsdale, D. W. Assessing the invasion by soybean aphid (Homoptera: Aphididae): Where will it end? *Ann. Entomol. Soc. Am.*, 97, 219, 2004.
119. Aylor, D. E. Spread of plant disease on a continental scale: Role of aerial dispersal of pathogens. *Ecology*, 84, 1989, 2003.
120. Viljanen-Rollinson, S. L. H., Parr, E. L., and Marroni, M. V. Monitoring long-distance spore dispersal by wind—A review. *N. Z. Plant. Prot.*, 60, 291, 2007.
121. Davidson, J., Rizzo, D., and Garbelotto, M. Transmission of phytophthora associated with sudden oak death in California. *Phytopathology*, 91(6 Suppl.): S108, 2001.
122. Davidson, J., Wickland, A. C., Patterson, H. A., Falk, K. R., and Rizzo, D. M. Transmission of *Phytophthora ramorum* in mixed-evergreen forest in California. *Phytopathology*, 95, 587, 2005.
123. Liebhold, A. M., Halverson, J. A., and Elmes, G. A. Gypsy moth invasion in North America: A quantitative analysis. *J. Biogeogr.*, 19, 513, 1992.
124. Winston, M. L. The biology and management of Africanized honey bees. *Ann. Rev. Entomol.*, 37, 173, 1992.
125. Suarez, A. V., Holway, D. A., and Case, T. J. Patterns of spread in biological invasions dominated by long-distance jump dispersal: Insights from Argentine ants. *Proc. Natl Acad. Sci. USA*, 98, 1095, 2001.
126. Muirhead, J. R., Leung, B., Overdijk, C., Kelly, D. W., Nandakumar, K., Marchant, K., and Maclsaac, H. J. Modelling local and long-distance dispersal of invasive emerald ash borer *Agrilus planipennis* (Coleoptera) in North America. *Divers. Distrib.*, 12, 71, 2006.
127. Waring, K. M. and O'Hara, K. L. Silvicultural strategies in forest ecosystems affected by introduced pests. *For. Ecol. Manag.*, 209, 27, 2005.
128. Hastings, A., Cuddington, K., Davies, K. F., Dugaw, C. J., Elmendorf, S., Freestone, A., Harrison, S., Holland, M., Lambrinos, J., Malvadkar, U., Melbourne, B., Moore, K., Taylor, C., and Thompson, D. The spatial spread of invasions: New developments in theory and evidence. *Ecol. Lett.*, 8, 91, 2005.
129. Cooke, M. T. The spread of the European starling in North America. U.S. Dept. Agric., Circ. 40, 1, 1928.

130. Corbett, A. and Plant, R. E. Role of movement in the response of natural enemies to agroecosystem diversification: A theoretical evaluation. *Environ. Entomol.*, 22, 519, 1993.
131. Rudd, W. G. and Gandour, R. W. Diffusion model for insect dispersal. *J. Econ. Entomol.*, 78, 295, 1985.
132. Memmott, J., P., Craze, G., Harman, H. M., Syrett, P., and Fowler, S. V. The effect of propagule size on the invasion of an alien insect. *J. Anim. Ecol.*, 74, 50, 2005.
133. Rilov, G., Benayahu, Y., and Gasith, A. Prolonged lag in population outbreak of an invasive mussel: A shifting-habitat model. *Biol. Invasions*, 6, 347, 2004.
134. Evans, A. M. and Gregoire, T. G. A geographically variable model of hemlock woolly adelgid spread. *Biol. Inv.*, 9, 369, 2007.
135. Sharov, A. A., Roberts, E. A., Liebhold, A. M., and Ravlin, F. W. Gypsy moth (Lepidoptera: Lymantriidae) spread in the central Appalachians: Three methods for species boundary estimation. *Environ. Entomol.*, 24, 1529, 1995.
136. Lonsdale, W. M. Rates of spread of an invading species—*Mimosa pigra* in northern Australia. *J. Ecol.*, 81, 513, 1993.
137. Mast, J. N., Hodgson, M. E., and Veblen, T. T. Tree invasion within the pine/grassland ecotone: An approach with historic aerial photography and GIS modeling. *For. Ecol. Manag.*, 93, 187, 1997.
138. Morin, R. S., Liebhold, A. M., Tobin, P. C., Gottschalk, K. W., and Luzader, E. Spread of beech bark disease in the eastern United States and its relationship to regional forest composition. *Can. J. For. Res.*, 37, 726, 2007.
139. Wu, K.-M., Lu, Y.-H., Feng, H.-Q., Jiang, Y.-Y., and Zhao, J.-Z. Suppression of cotton bollworm in multiple crops in China in areas with Bt toxin-containing cotton. *Science*, 321, 16, 2008.
140. Grégoire, J.-C., Piel, F., Franklin, A., and Gilbert, M. Mass-foraging for unpredictable resources: A possible explanation for Allee effects in *Ips typographus*. In: *Proceedings of the North American Forest Pest Workshop*. Hain, F. P., Coulson, R. N., Klepzip, K. D., and Rhea, J. North Carolina State University, Asheville, NC, 2006, p. 43.
141. Tobin, P. C., Robinet, C., Johnson, D. M., Whitmire, S. L., Bjørnstad, O. N., and Liebhold, A. M. The role of Allee effects in gypsy moth, *Lymantria dispar* (L.), invasions. *Popul. Ecol.*, 51, 373, 2009.
142. Veitch, C. R. and Clout, M. N. Eds. *Turning the Tide: The Eradication of Invasive Species. Proceedings of the International Conference on Eradication of Island Invasives*. Hollands Printing Ltd., Auckland, New Zealand, 2002.
143. Ross, M. Waikato residents support moth trapping programme. *Biosecurity*, 49, 7, 2004.
144. Dark, S. J. The biogeography of invasive alien plants in California: An application of GIS and spatial regression analysis. *Divers. Distrib.*, 10, 1, 2004.
145. Taylor, C. M. and Hastings, A. Finding optimal control strategies for invasive species: A density-structured model for *Spartina alterniflora*. *J. Appl. Ecol.*, 41, 1049, 2004.
146. Sharov, A. A. and Liebhold, A. M. Model of slowing the spread of gypsy moth (Lepidoptera: Lymantriidae) with a barrier zone. *Ecol. Appl.*, 8, 1170, 1998.
147. Hardee, D. D. and Harris, F. A. Eradicating the boll weevil (Coleoptera: Curculionidae): A clash between a highly successful insect, good scientific achievement, and differing agricultural philosophies. *Am. Entomol.*, 49, 82, 2003.
148. Owen, M. R. and Lewis, M. A. How predation can slow, stop or reverse a prey invasion. *Bull. Math. Biol.*, 63, 655, 2001.
149. Taylor, C. M., Davis, H. G., Civille, J. C., Grevstad, F. S., and Hastings, A. Consequences of an Allee effect on the invasion of a Pacific estuary by *Spartina alterniflora*. *Ecology*, 85, 3254, 2004.

4 Integrating GPS, GIS, and Remote Sensing Technologies with Disease Management Principles to Improve Plant Health

Forrest W. Nutter, Jr., Emmanuel Z. Byamukama, Rosalee A. Coelho-Netto, Sharon K. Eggenberger, Mark L. Gleason, Andrew Gougherty, Alison E. Robertson, and Neil van Rij

CONTENTS

4.1 Executive Summary ... 60
4.2 Introduction ... 60
4.3 Disease Management Principles ... 61
4.3.1 Disease Management Principle 1: Exclusion ... 61
4.3.1.1 Quarantine (y_0) ... 62
4.3.1.2 Seed/Plant Certification Programs (y_0) ... 62
4.3.2 Disease Management Principle 2: Avoidance (y_0 and/or t) ... 62
4.3.2.1 Avoidance of Disease Risk in Space (t) ... 62
4.3.2.2 Avoidance of Disease Risk in Time (t) ... 62
4.3.3 Disease Management Principle 3: Eradication (y_0) ... 63
4.3.3.1 Roguing of Diseased Plants (y_0) ... 63
4.3.3.2 Removal and Burial of Crop Residues/Debris (y_0) ... 64
4.3.3.3 Soil Fumigation (y_0) ... 64
4.3.4 Disease Management Principle 4: Protection (y_0 and/or r) ... 64
4.3.4.1 Use of Chemical Barriers to Protect Crops (y_0 and r) ... 64
4.3.5 Disease Management Principle 5: Host Resistance ... 65
4.3.5.1 Resistance That Reduces Initial Inoculum (y_0) ... 65
4.3.6 Disease Management Principle 6: Therapy (y_0 and Sometimes r) ... 65
4.4 Case Study: Asian Soybean Rust ... 66

4.5 Case Study: Ash Yellows Disease of Green Ash 68
4.6 Case Study: *Plum Pox Virus* of *Prunus* spp. 71
4.7 Case Study: Moko Disease of Banana 73
4.8 Case Study: Stewart's Disease of Corn 76
4.9 Case Study: Gray Leaf Spot of Corn 78
4.10 Case Study: Bean Pod Mottle Virus of Soybean 79
4.11 GIS Tutorial: Moko Disease in Amazon Region of Brazil 84
4.11.1 Saving Chapter 4 Files to Your Computer 84
4.11.2 Opening Data in ArcMap 84
4.11.3 Changing Map Symbology 84
4.11.4 Creating and Printing Map Layouts 85
4.12 Conclusions 87
Acknowledgments 88
References 88

4.1 EXECUTIVE SUMMARY

Disease and pest populations, host (crop) populations, climate, differences in management tactics, and local economic conditions all cause disease risks to vary, not only temporally, but also spatially. Adding a spatial component to disease risk assessment will help farmers and disease managers make more informed management decisions, by geospatially defining "prescription" management zones that coincide with geospatially defined disease risk zones. Global Positioning Systems (GPS) and Geographic Information Systems (GIS) technologies can be utilized to geospatially reference information from disease forecasting models, disease surveys, and maps of abiotic (e.g., climate, weather, soil type, elevation) and biotic factors (presence/survival of initial inoculum, successful overwintering of insect vectors, disease status in previous years, etc.) that influence disease risk, and then used to accurately define prescription management zones (both temporally and spatially).[1–4] Disease intensity survey data, when coupled with GPS, GIS, and geostatistical tools can be used to generate maps that more precisely delineate relative levels of disease risk (low, moderate, high) across numerous spatial scales.[1,2,5–8] This chapter illustrates, through specific pathosystem case studies, how GPS, GIS, and remote sensing technologies can be integrated to support principles of plant disease management.

4.2 INTRODUCTION

Successful plant disease management programs are underpinned by six disease management principles.[9–11] These are (1) exclusion, (2) avoidance, (3) eradication, (4) protection, (5) resistance, and (6) therapy (Figure 4.1). These disease management principles are used to achieve one or more of the following management strategies: (1) reduce the level of initial inoculum (y_0), (2) reduce the rate of pathogen development (r), and/or (3) reduce the time that host and pathogen populations intersect in time and space. These can be accomplished by excluding, eliminating, reducing, and/or avoiding potential sources of pathogen inoculum (y_0).[11–13] Sanitation is the process that eliminates, reduces, or avoids sources of initial inoculum (y_0) from which plant disease

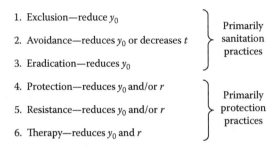

Principle	Category
1. Exclusion—reduce y_0	Primarily sanitation practices
2. Avoidance—reduces y_0 or decreases t	Primarily sanitation practices
3. Eradication—reduces y_0	Primarily sanitation practices
4. Protection—reduces y_0 and/or r	Primarily protection practices
5. Resistance—reduces y_0 and/or r	Primarily protection practices
6. Therapy—reduces y_0 and r	Primarily protection practices

FIGURE 4.1 The six principles of plant disease management and their epidemiological effects on plant disease epidemics, where y_0 represents initial inoculum, r represents the rate of pathogen development (infection), and t represents the period of time that host and pathogen populations interact to affect final disease intensity (y_{final}).

epidemics start.[11,14] Sanitation practices usually fall under one of three disease management principles: exclusion, avoidance, and eradication (Figure 4.1). For sanitation to be effective, the rate of an epidemic must be low, or sufficiently slowed to a rate below which sanitation practices will effectively delay epidemic onset.[9–11]

The infusion of spatially referenced information into processes of both disease risk assessment and disease risk management can help to optimize a farmer's return on investment. This is because disease management tactics will be deployed not only **when** needed (i.e., on a timely basis), but tactics will also be deployed only **where** they are needed (i.e., to geospatially defined management zones). The integration of GPS, GIS, and remote sensing technologies has tremendous potential to improve disease management across numerous spatial scales (leaf, plant, field, farm, county, area-wide, whole production regions, etc.).[2,7,15,16] The capture and visual display (mapping) of geospatially referenced biotic and abiotic factors that influence disease risk can help to elucidate stimulus–response relationships at multiple spatial scales, some of which have never before been conceptualized.[2] In this chapter, we will present several case studies that exemplify how GPS, GIS, and remote sensing technologies can be integrated with principles of disease management.

4.3 DISEASE MANAGEMENT PRINCIPLES

4.3.1 Disease Management Principle 1: Exclusion

Exclusion is considered to be the first line of defense in integrated disease management programs.[9,10] The concept is quite simple: keep the pathogen from entering crop production areas, so that initial inoculum (y_0) is kept at zero. In order to be effective, this management principle requires prior information about where (geographically) the pathogen is currently present, and where it is not.[1,7,15,16] In short, if the pathogen is not present in a crop production area, there can be no epidemic. Crop production areas may still be at risk, however, even when the pathogen is absent, if (1) the crop is susceptible to exotic (non-endemic) pathogens found in other areas of the globe and (2) if the local environment is favorable for both pathogen infection and pathogen dissemination. Two exclusion tactics will be illustrated: quarantine and seed/plant certification programs.

4.3.1.1 Quarantine (y_0)

The prevention of both the introduction and the interstate, intercountry, and intercontinental movement of plant pathogens and pests is the responsibility of state and national regulatory agencies. Exclusion of pathogen inoculum ($y_0=0$) is often attempted by establishing phytosanitary restriction zones (quarantines) that prohibit the importation (and exportation) of agricultural commodities known to serve as potential routes of entry for new disease threats. If successful, quarantines will keep initial inoculum at zero, and no epidemic will develop. Quarantines for specific pathogen threats may totally prohibit the importation of specific agricultural products. Quarantines may be imposed for entire countries, regions, states, or provinces, or even at the county or field scale.[9,13,17] The use of GPS and GIS technologies to display, in near real time, where plant pathogens are present throughout the globe has greatly enhanced the effectiveness of quarantine measures worldwide.[15,16]

4.3.1.2 Seed/Plant Certification Programs (y_0)

The importation of agricultural seeds and plants may be allowed if the biological material is inspected prior to export and/or entry. Seed and plant certification, inspection, and testing programs involve the sampling and testing of plant material (seed, tubers, bulbs, corms, root stocks, seedling transplants, shoots, budwood, etc.), before infected (or infested) plants or seeds enter production fields or orchards. Planting media and/or containers are also often subject to inspection and testing for the presence of threatening plant pathogens and pests. In the future, one of the primary applications of GPS and GIS technologies will be to map the locations of new disease outbreaks and to use these technologies to backtrack (and determine) the initial sources of pathogen spores or infected (infested) plant materials (i.e., nurseries, production areas, contaminated shipments, country of origin, etc.).[15]

4.3.2 Disease Management Principle 2: Avoidance (y_0 and/or t)

4.3.2.1 Avoidance of Disease Risk in Space (t)

If a plant pathogen threat cannot be kept out of a crop production area by exclusion, then the next line of defense is to attempt to avoid disease risk in either time and/or space. Examples of the "avoidance" disease management principle include the use of management tactics to avoid high disease risk planting sites (i.e., fields, regions, etc.) and exposure to local climate (environmental) conditions that favor disease risk. Geographic areas where pathogen inoculum is absent (or present at low levels) or where climate is unfavorable for disease development can be delineated using GPS and GIS tools, thereby helping producers select planting sites where initial inoculum (or the effectiveness of initial inoculum) approaches zero ($y_0 \rightarrow$ zero).[1,2,6,7,15]

4.3.2.2 Avoidance of Disease Risk in Time (t)

Avoidance of pathogen populations in time, e.g., through choice of planting date, can be a highly effective management strategy for some plant pathosystems.[9,10] The theory behind this strategy is to limit or avoid the period of time (t) when pathogen and host populations interact, thereby reducing injury and crop loss.[18,19] In some pathosystems, early planting gives the crop more time to develop before pathogen inoculum is active or present.[10] In other pathosystems, delayed planting may substantially decrease disease risk by avoiding early-season sources of pathogen inoculum. A primary factor limiting the deployment of this management principle has been the lack of geospatial information regarding where (geographically) early or delayed planting would effectively avoid pathogen inoculum.

4.3.3 Disease Management Principle 3: Eradication (y_0)

Eradication can be defined as any practice that reduces initial inoculum (y_0) at its source.[10,11] Therefore, management practices that involve the burial, burning, or removal of pathogen-infested crop debris at the source (e.g., within a field) are examples of eradication.[9,20] Removal of alternative weed hosts or alternate hosts that can serve as potential sources of initial inoculum also falls under the principle of eradication. Other practices that reduce initial inoculum (y_0) at the source include crop rotation, introduction of biological control agents, soil fumigation, steam sterilization/pasteurization, soil solarization, green manure crops, trap crops, and the incorporation of crop residues of other crop species (sometimes referred to as biopesticides).[9,12,13,17,20,21] All reduce (to differing degrees) the amount of initial inoculum. Initial inoculum (y_0) also can be reduced by a number of physical, chemical, and/or biological practices. Soil amendments, fertilizer regimens, green manure crops, and trap crops can further increase the eradicative benefits of crop rotation by increasing biological activity of soil microbiota during intercrop periods.[17,20] However, the use and success of specific eradication tactics depends upon the acquisition of accurate, precise spatial information indicating where specific (and often costly) eradication tactics are most needed, and, just as important commercially, where they are not needed.

4.3.3.1 Roguing of Diseased Plants (y_0)

Roguing is the practice of removing infected or diseased plants in crops, orchards, or plantations.[2,3,11,14,22] Because roguing removes within-field or local sources of initial inoculum, this practice epidemiologically reduces y_0 (see the case studies on Moko disease of banana and *Plum pox virus* of *Prunus* spp. in this chapter). The roguing of volunteer plants, alternate hosts, and alternative hosts is often epidemiologically important, because hosts can serve as intercrop (or overwintering) bridges that will then place ensuing crops at greater disease risk.[11,22] These sources of inoculum can be geospatially referenced for removal before spring planting. For example, barberry (*Berberis* spp.), the alternate host for stem rust of wheat (caused by *Puccinia graminis* f. sp. *tritici*), is an epidemiologically important risk factor that not only provides a local source of inoculum, but also plays a critical role in increasing genetic diversity of the pathogen population. Peterson et al.[22] reported that the diversity of races detected in spore collections of *P. graminis* f. sp. *tritici* from wheat fields in Minnesota declined sharply from 1912 to 1930 as a result of an extensive, national barberry eradication program. Pathogen diversity has remained relatively low.[22]

4.3.3.2 Removal and Burial of Crop Residues/Debris (y_0)

The removal and/or burning of infested crop residue is particularly effective in reducing initial inoculum (y_0) in pathosystems where the major source of inoculum is located in the above-ground parts of infested crop debris (residue).[4,9] Conversely, the recommended practice of leaving crop residue on the soil surface (e.g., in conservation tillage, minimum tillage, and reduced tillage systems) can provide an epidemiologically important source of initial inoculum that can significantly increase disease risk.[2,4,7] (See Case Studies 4.7 and 4.9.)

4.3.3.3 Soil Fumigation (y_0)

Soil fumigation is the practice of injecting chemical fumigants into the soil to reduce initial inoculum (at the source). This practice is primarily used in the production of high-value crops, such as strawberries (and other small fruits), ornamental nurseries, tree nurseries, and turfgrass. Because of the high costs for soil fumigants, these crops are good candidate pathosystems for the use of spatial maps that delineate within-field areas where soilborne pathogens are limiting crop yield potentials, and for the use of variable rate technologies that deploy prescribed amounts of fumigant that correspond to specific levels of disease risk within a field.

4.3.4 Disease Management Principle 4: Protection (y_0 and/or r)

When plant pathogens cannot be excluded, avoided, or eradicated, the next line of defense against plant pathogens is crop protection. Protection can be defined as any tactic that provides a physical or chemical barrier to prevent the pathogen from entering host tissue.[9,10] Such tactics generally reduce initial inoculum at the point of crop infection (i.e., protection of potential infection courts), whereas eradication tactics reduce initial inoculum at the source. In many cases, protection tactics greatly reduce the rate of infection (r), because alloinfection (pathogen dissemination from plant-to-plant) is greatly reduced.[11] The use of GPS, GIS, and geostatistical tools has greatly increased our knowledge concerning the spatial and temporal dynamics of within-field pathogen spread.[5,8,17]

4.3.4.1 Use of Chemical Barriers to Protect Crops (y_0 and r)

Chemical barriers may be deployed to protect a crop at many stages of crop development. This involves the application of chemical (or biological) agents that protect the seed and/or developing plant from both foliar and soilborne pathogens.[9] Fungicide seed treatments are also employed to reduce the effectiveness of seedborne and/or soilborne infection units (mycelium, conidia, sclerotia, etc.), thereby reducing the effectiveness of initial inoculum at the site of infection. Insecticide seed treatments may be deployed, not only to reduce injury from insect pests, but also as a tactic to disrupt the acquisition and/or transmission of plant pathogens by insect vectors.[23,24] The coupling of GPS and GIS tools with geospatially-referenced weather and site risk data (elevation, aspect, landscape features) offers great promise to generate preplant disease risk maps, which can advise farmers where fungicide/insecticide seed treatments should be deployed (in high disease risk management zones) or not deployed (in low-risk management zones) (see Case Studies 4.8 and 4.10).

4.3.5 Disease Management Principle 5: Host Resistance

While some authors categorize host resistance within the principle of "protection", host resistance deserves recognition as a separate disease management principle. Host resistance affects epidemics in two ways, either by reducing initial inoculum (by reducing y_0 to zero, or nearly to zero) or by slowing the rate (r) of an epidemic.[9,10,14] When host resistance reduces y_0 to zero, no epidemic can develop. For example, in pathosystems for which disease (or pathogen) progress over time is best described by the logistic model (which states that the absolute rate of disease progress $dy/dt = ry(1 - y)$, where r is the rate of pathogen infection and y is the initial, or present, level of disease, expressed as a proportion), if disease resistance effectively reduces y_0 to zero, then $dy/dt = 0$, irrespective of the value of r. The same concept applies for those polycyclic pathogens that are best described by other population growth models (such as the exponential and Gompertz models).[14,25]

4.3.5.1 Resistance That Reduces Initial Inoculum (y_0)

When host resistance at the cultivar (or variety) level is effective in reducing initial inoculum to zero for some pathogen strains (races), but not for others, then such resistance is said to be "strain specific" or "race specific." In these pathosystems, y_0 is not often reduced to zero. The epidemiological impact of this type of resistance is to reduce initial inoculum, and possibly delay the epidemic, but strain- or race-specific resistance will not reduce the rates of epidemics that develop in susceptible host cultivars. Since $y_0 > 0$, race-specific (or strain-specific) resistance will behave epidemiologically in a manner similar to sanitation, resulting in a delay in time (t) of the epidemic. However, since r is not affected by this type of host resistance, little benefit is achieved in plant pathosystems in which r is high.[8,10,26] To increase the effectiveness of this resistance strategy, gene rotation in time, multilines, gene pyramiding (stacking), and gene deployment[10,11,14] have been utilized. Geospatial information concerning the geographical extent of cultivar-specific pathogen races and strains (that can overcome specific host resistance genes) can potentially enhance the effectiveness and durability of host resistance, by enabling growers to deploy host resistance genes where they will be most effective against the local, prevailing races.[23,27] Prescriptive combinations of host resistance genes, selected for use in specific, geospatially defined areas, have great promise for improving area-wide disease management programs.[10]

4.3.6 Disease Management Principle 6: Therapy (y_0 and Sometimes r)

The "Therapy" disease management principle primarily reduces y_0, and, in some cases, alloinfection (r). Therapy stands alone from the other disease management principles in that this principle comes into play only after a plant becomes infected; therapy is an attempt to "cure" or increase the survival time[3] of an infected plant. This is somewhat analogous to the use of chemotherapy in humans to cure or increase the survival time of cancer patients.[28] Antibiotics and chemicals have been injected into trees infected by phytoplasmas or fungi, primarily as attempts to increase survival time, rather than to "cure" infected plants. Chemical injection of systemic fungicides has been used to increase survival times of elm trees infected by the fungus that causes Dutch elm disease and of ash trees infected with the ash yellows phytoplasma. Employing GPS and GIS tools as part of disease surveys provides geospatially referenced data concerning locations of trees that should receive therapeutic antibiotic or chemical injections.

A second application of the therapy principle involves the physical removal of infected plant parts from an infected host plant.[10,11] Blister rust, caused by the fungus *Cronartium ribicola*, is an invasive species that was introduced into Western North America in 1910. Since then, the pathogen has spread south. Blister rust currently threatens sugar pine trees (*Pinus lambertiana*) in California and Oregon. In a recent study, limbs of sugar pine trees were pruned, up to a height of 8 ft above ground, to reduce the number of subsequent blister rust infections, and to increase host longevity and productivity.[29]

4.4 CASE STUDY: ASIAN SOYBEAN RUST

One of the basic tenets of the "exclusion" management principle states that the geographic locations and extent of pathogen invasion must be known, so that the risk/probability of pathogen introduction (via a natural event, by accidental introduction via ports of entry, and/or by deliberate introduction) can be accurately assessed.[15,30]

The integration of GPS, GIS, and remote sensing technologies to extract pathogen-specific temporal and spatial signatures is a new approach for detecting and accurately identifying plant pathogens.[3] To test this approach, IKONOS satellite imagery was obtained for a soybean field in Cedara, South Africa, that was naturally infected with Asian soybean rust (Figure 4.2A). The majority of the field was sprayed once with a protectant fungicide early in the growing season, but a small area of the northwest part of the soybean field (the area within the circle) was not sprayed. As a result, the unsprayed area of the field developed a severe epidemic of Asian soybean rust, which functioned as an area source of rust spores (inoculum) for the rest of the field. As the protectant fungicide lost effectiveness, a dispersal gradient developed—more rust spores were dispersed and deposited close to the area source, and fewer spores were dispersed and deposited as distance from the area source increased. This large dispersal/deposition gradient resulted in a disease gradient that could be detected and quantified using IKONOS satellite imagery (1 m^2 resolution) (Figure 4.2A).

Estimates of soybean rust severity were assessed visually at 12 randomly selected, geospatially referenced points within the soybean field.[31] Soybean canopy reflectance (800 nm) was measured at these 12 points using a handheld multispectral radiometer. Image intensity (pixel) values in the NIR band were extracted from areas of the satellite image that corresponded to the 12 visually assessed locations within the soybean field. Satellite image intensity values were regressed against the corresponding ground-based reflectance measurements (Figure 4.2B) and visual disease severity assessments (Figure 4.2C).

The darker areas in the near-infrared satellite image (Figure 4.2A) represent low image intensity values (i.e., where the soybean canopy is less healthy due to soybean rust) and the lighter shades of gray indicate areas where image intensities are

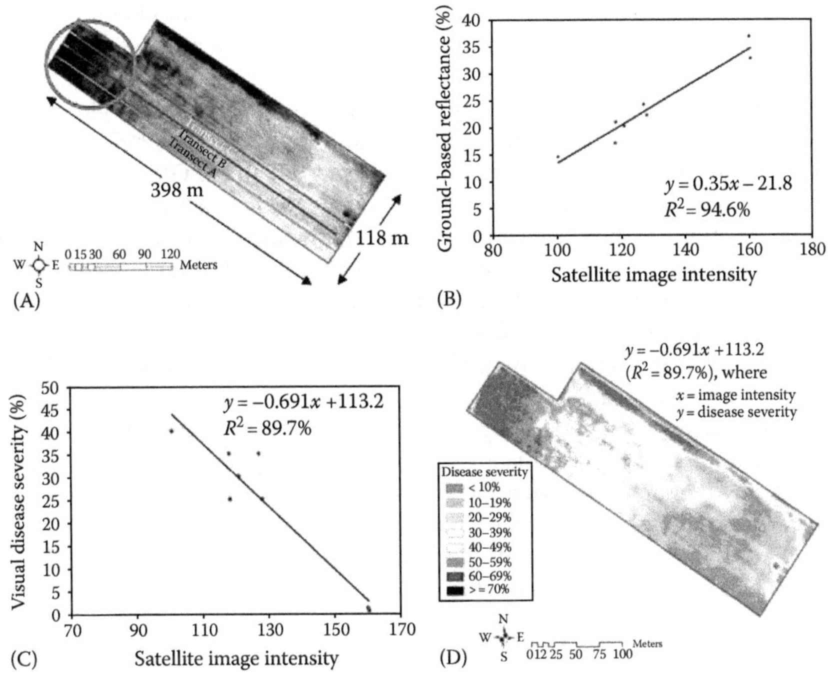

FIGURE 4.2 IKONOS satellite image (near-infrared band) of a soybean field in Cedara, South Africa, infected with Asian soybean rust in 2006, showing (A) the non-sprayed portion of the field (circled) and transects used for conducting ground-based disease assessment sampling. Regressions of satellite image intensity pixel values against (B) GPS-referenced ground-based radiometer reflectance values (%) and (C) visual estimates of disease severity (%) are shown. Satellite image intensity pixel values were transformed using the equation in C and remapped to show predicted disease severity (D).

higher (i.e., the soybean canopy is relatively more healthy and rust severity is lower). Percentage reflectance values (NIR) obtained using the handheld multispectral radiometer were in very close agreement with NIR image intensity values obtained from IKONOS satellite images ($R^2 = 94.6\%$, $P < .0001$) (Figure 4.2B). Such close agreement indicates that remotely sensed data obtained from the IKONOS satellite platform (positioned miles above the earth) provided essentially the same information as did the remotely sensed data obtained using a handheld multispectral radiometer platform (2 m above the ground). Thus, the effect of miles of atmosphere in this study (and others we have conducted) may be minute. Moreover, visual assessments for disease severity on the ground were also highly related to satellite image intensity pixel values in the NIR band ($R^2 = 89.7\%$, $P < .0001$) (Figure 4.2C). The equation in Figure 4.2C was subsequently used to interpret (predict) percent disease severity from satellite image intensity values. Predicted, geospatially referenced disease severity values were used to generate a false-color disease severity (%) map of the soybean field (Figure 4.2D). This map clearly shows the presence of a classic disease severity gradient, in which disease severity decreases as distance from the inoculum source increases.

Quantitative spatial and temporal "signatures" obtained from satellite imagery of plant disease epidemics can be used not only to detect, but also possibly to identify, plant diseases that are negatively impacting crop health.[32] Satellite imagery can also be used to identify the presence of specific diseases in areas of the globe where it is difficult or cost-prohibitive to obtain ground-truth assessments.[30] One potential application for the use of integrated remote sensing, GPS, and GIS technologies is the monitoring of disease development in sentinel locations throughout the globe. This information can then be incorporated into risk prediction models. For example, Asian soybean rust is now established in the southern United States, where the year-round existence of kudzu (an alternative host for the Asian soybean rust pathogen) provides a source of initial inoculum to infect soybean crops early in the growing season. Soybean rust severity is being monitored visually throughout the United States. County scale maps are posted to a Web site (in near real-time) to indicate areas where the soybean rust pathogen is present and can serve as a potential source of inoculum for short, intermediate, and long-distance dissemination.[27] These postings are closely watched by extension personnel, researchers, and soybean growers, who are poised to apply fungicides if and when Asian soybean rust threatens yields in specific regions. In the not too distant future, remote sensing, GPS, and GIS technologies will be integrated and used to replace ground-based visual assessments of disease severity.[27]

4.5 CASE STUDY: ASH YELLOWS DISEASE OF GREEN ASH

Ash yellows is a recently discovered disease of white and green ash (*Fraxinus americana* and *Fraxinus pennsylvanica*) that causes slow growth and dieback.[33,34] Ash yellows is caused by a specialized group of bacteria called phytoplasmas. Insect vectors, particularly leafhoppers (which are known vectors of other phytoplasma diseases in other pathosystems), have been implicated; however, mystery still surrounds how the ash yellows phytoplasma is spread from tree to tree.[33–35]

Populations of healthy and diseased green ash trees within municipal landscapes can be considered as a system of random variables. However, it is quite probable that individual green ash trees infected with the ash yellows pathogen are having an impact on the disease status of neighboring green ash trees. Thus, diseased ash trees within a municipality may be dependent upon (and interact with) neighboring green ash trees in time and space.

Stands of green ash trees within municipalities were surveyed and tested for ash yellows.[36] Disease survey data for ash yellows were coupled with GPS and GIS technologies to map the spatial patterns of diseased and healthy ash trees. Survey results were mapped and analyzed, using geospatial and statistical tools, to test hypotheses concerning the presence or absence of spatial dependence (clustering).[37–39] Clustering would indicate that ash trees infected with ash yellows were having an impact on the health status of other ash trees with respect to distance from ash yellows-positive trees. Such spatial information may provide statistical evidence for the presence of insect vectors. Moreover, spatial information that defines the size of the eradication zone needed to achieve successful eradication could lead to better management of ash yellows disease, specifically by indicating where to rogue all ash trees (diseased or otherwise) that are growing within a defined distance from ash yellows-positive trees. The current management recommendation for ash yellows is to remove only symptomatic trees.[33] Recommendations for establishing defined eradication zones that extend beyond symptomatic trees have not been proposed.

Surveys were conducted in three municipalities: Iowa City, IA; Milwaukee, WI; and St. Paul, MN. All green ash trees found within municipal right-of-ways (usually along city streets and within city parks) that possessed high densities of green ash trees were tested. Green ash roots were tested for the presence of the ash yellows pathogen using DAPI staining.[36] The area surveyed within each municipality was approximately 1 mi^2 (2.59 km^2) in size. All surveyed green ash trees were geospatially referenced at the time of sampling, using a Geotracker GPS unit (Trimble Corp., Sunnyvale, CA). Locations of healthy and diseased trees were mapped using Universal Transverse Mercator (UTM) coordinates.

To test for spatial dependence, a modification of the marked point process[39] was used in this analysis: the "points" are the positions of green ash trees (stated using a Cartesian coordinate system), and the "marks" are the disease status (ash yellows-positive or negative) of individual trees in the survey population.[38,39] Ripley's *k*-function analysis was used to test for spatial dependence (clustering) of ash yellows-positive trees.[37] This function was used for exploratory analysis of the interaction among diseased (ash yellows-positive) and healthy populations of green ash trees within geographically defined municipalities.

Ripley's *k*-function analysis involves overlaying circles about all marked points (green ash trees) to look at changes in the *k*-function with respect to increasing the radius (distance) of circles around individual ash yellows-positive trees and around individual ash yellows-negative (healthy) trees (Figure 4.3A,C,E).[37,39] This method generalizes the nearest neighbor analysis and uses information from points beyond the closest neighboring point. Mean distances to the first, second, third, ..., *n*, nearest neighbors were calculated and compared to expected distances (values that would be observed if the marked population was under complete spatial randomness). The pair-correlation function $G(r)$ characterizes relationships between diseased and healthy ash green trees, conditional upon distance.[37] The *k*-function for mapped data from this analysis captures the spatial dependence between different regions of the marked point process.[39] Its estimation is based on an empirical average replacing the expectation operator. Because the estimator counts the number of marked (diseased) green ash trees within a specified range of distances (radii), and because the density of trees per unit area may change with respect to distance, the "difference in *k*-function is plotted with respect to distance (m)."[37,39] For all estimates of k with respect to distance, an estimate of λ is needed. The parameter λ is the intensity of marked trees within the plot estimated by $\lambda = n/A$ for n trees on a plot with area A. In this study, A increases with respect to the radius of the circle about each marked point. Thus, the *k*-function takes the form:

$$k(h) = \lambda^{-1}E\,(\text{number of extra marked points within distance } h \text{ of an arbitrary marked point}),\ h \geq 0.$$

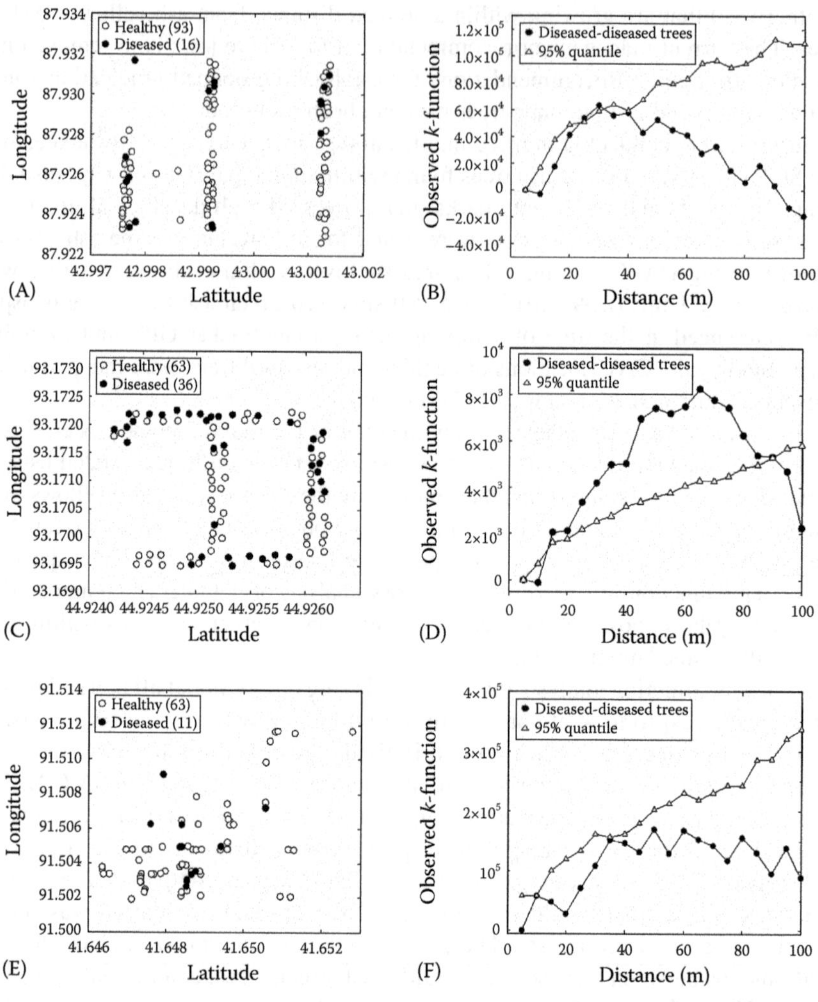

FIGURE 4.3 GPS maps depicting (A, C, E) the location of ash yellows-positive ash trees and healthy ash trees and (B, D, F) the observed differences in *k*-function values for areas surveyed in 1994 in Milwaukee, WI (A, B); St. Paul, MN (C, D); and Iowa City, IA (E, F).

Under the condition of regularity, $k(h)$ tends to be less than πh^2, whereas under the condition of clustering, $k(h)$ tends to be greater than πh^2. The *k*-function has obvious advantages because it presents spatial information at all scales of spatial patterns that are tested, and precise spatial locations of marked events are used in its estimation.[37]

The spatial patterns for ash yellows-positive (•) and negative (○) trees are shown in Figure 4.3A,C,E. The observed difference in the *k*-function, with respect to distance from each diseased tree and with respect to all other diseased trees, showed that there was strong clustering (above the 95% confidence envelope) in the spatial pattern of diseased ash trees at distances ranging from 5 to 10 m in all three municipalities (Figure 4.3B,D,F). Significant clustering with respect to distance is indicated when the observed difference in the *k*-function (lines with data points shown as filled circles) lies above the 95% confidence envelopes (lines with triangle-shape data symbols).

Our preliminary findings, which detected the presence of significant spatial clustering within 5–10m of infected trees, indicate that all ash trees within 5–10m of an ash yellows-positive tree should be rogued (eradication principle). The study also exemplifies how GPS and GIS technologies can be used to support specific disease management principles (i.e., eradication) to more effectively manage ash yellows in green ash. Significant spatial dependence was also detected at distances of 30 (Milwaukee, WI) and 90m (Iowa City, IA) from diseased ash trees. These distances may be related to the dispersal gradient of an insect vector, and/or the requirement of a latent period within the insect vector (time from acquisition of the phytoplasma by a vector to the time that a vector can successfully transmit the pathogen), in which case the vector would likely disperse 30–90m before an ash yellows-positive insect could transmit the phytoplasma to healthy ash trees. Beyond this distance–time capsule, vectors may lose the ability to transmit the pathogen. As with many spatial studies, this study answered some questions with regard to spatial dependence in this pathosystem, but also uncovered new questions that need to be addressed in future studies.

4.6 CASE STUDY: *PLUM POX VIRUS* OF *PRUNUS* SPP.

The introduction of exotic plant pathogens and pests by accidental or natural (weather-related) events is continually on the increase, due to increased global trade. Conceivably, the deliberate introduction of exotic plant pathogens to attack a nation's agricultural and economic security further emphasizes the need for taking appropriate and timely response measures to minimize the impacts of such threats.[30]

When a nonindigenous plant pathogen is first detected and confirmed in a new geographic area (state, region, country, continent), a critical question immediately arises: Can the new threat be successfully eradicated? The answer depends on the geospatial extent of the infestation, pathogen dispersal mechanisms (wind, insect vectors, waterways, seed, infected plant material, etc.), biological characteristics of the pathogen, and host characteristics (e.g., susceptibility and density). One of the key biological risk factors is the distance that dispersal units (spores, pathogen-infested insect vectors, waterborne pathogens, etc. originating from infected plants or fields) can be disseminated to adversely affect the health status of other plants or fields. The spatial patterns of infected plants (and/or fields) can be mapped using the marked point process (described in the ash yellows case study), and then analyzed using geostatistical tools to test for spatial dependence (i.e., over what distance can an infected plant or field impact the health of other susceptible hosts?).[37]

To test for the presence of significant spatial dependence of *Plum pox virus* (PPV) infected blocks of *Prunus* spp., we again used Ripley's *k*-function analysis.[37] In this analysis, the "points"[38,39] are the positions of blocks of *Prunus* spp. (peaches, plums, apricots) and the "marks" are the disease status of blocks (PPV-positive or

PPV-negative). In 2000, approximately 1200 *Prunus* orchard blocks were surveyed by the Pennsylvania Department of Agriculture and tested, using ELISA, for the presence of PPV. The point map (shown without displaying county lines, land marks, or other geographical landmarks) is shown in Figure 4.4A. The spatial pattern for PPV-infected *Prunus* blocks was analyzed using a nearest neighbor type analysis (*k*-function analysis, as described in the previous case study)[37]. Mean distances to the first, second, third, …, *n*th nearest neighbors of PPV-positive blocks were calculated and compared

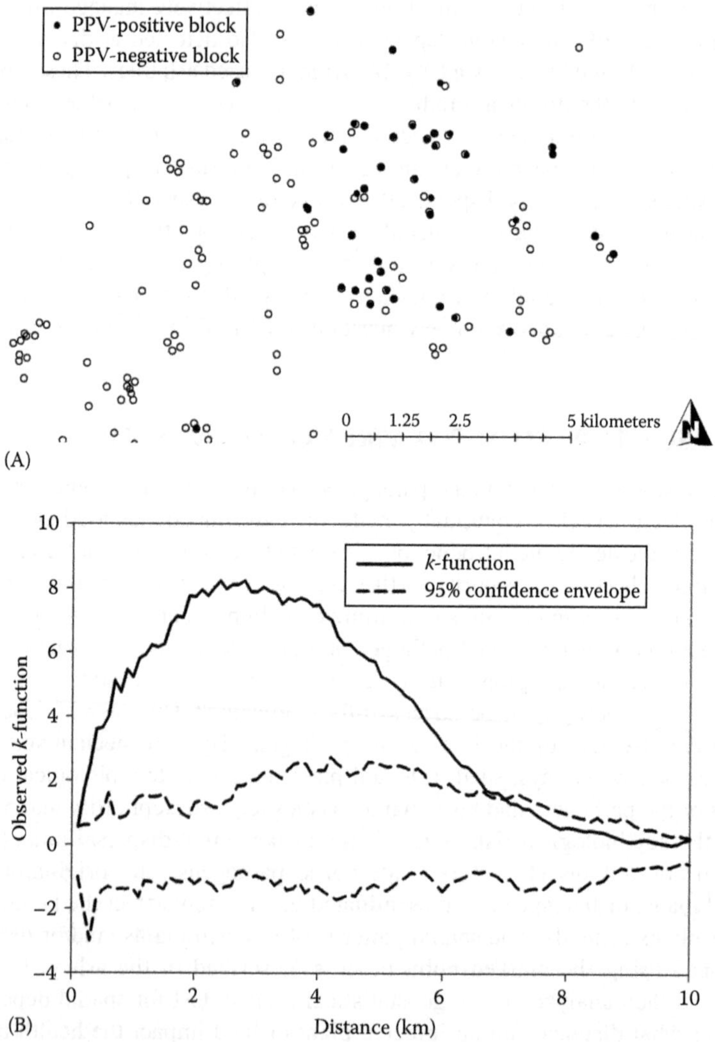

FIGURE 4.4 Map depicting (A) the relative GPS locations of *Prunus* spp. blocks in Pennsylvania identified as having *Plum pox virus* (PPV) in 2000 and (B) the observed *k*-function with respect to distance from PPV-positive blocks of *Prunus* spp. For distances (x) where the observed *k*-function is above the 95% confidence envelope, events or conditions are said to exhibit spatial dependence (i.e., significant clustering).

to the expected nearest neighbor distances that would occur by random chance (i.e., no spatial dependence), if the marked points were distributed randomly.[37,38]

The observed difference in the *k*-function, plotted with respect to distances from each PPV-diseased block to all other PPV-diseased blocks, shows that there was a strong spatial dependence up to a distance of 7.1 km (Figure 4.4B). Spatial dependence is present when the observed *k*-function value (solid line) is outside the 95% confidence envelope for a given distance. As in the previous case study, when the observed *k*-function is above the upper boundary of the confidence envelope, the spatial pattern of the marked points (blocks) at a distance *x* is said to be clustered. When the observed *k*-function is below the lower boundary, the pattern is dispersed. If the observed *k*-function for any distance (*x*) is within the 95% confidence envelope, the spatial pattern is random.

Such point maps and tests for spatial dependence can provide critical quantitative data and can be used by scientists and policy-makers to make sound, science-based decisions regarding whether eradication efforts should be undertaken. In addition, spatial analyses can be used to determine if the efforts will be cost-effective, and to determine the extent to which eradication and quarantine zones should be extended beyond each infected, marked point. Pennsylvania Department of Agriculture and USDA APHIS government officials successfully eradicated *Plum pox virus* from Pennsylvania by adopting a 500 m eradication zone around each PPV-positive block (all *Prunus* blocks within 500 m were removed regardless of disease status), and then surveying and testing each *Prunus* tree within all blocks up to a distance of 5 mi from PPV-positive blocks. *Plum pox virus* was declared successfully eradicated from Pennsylvania in 2009, after no PPV-positive trees had been found within *Prunus* blocks in Pennsylvania for a period of 3 years.

4.7 CASE STUDY: MOKO DISEASE OF BANANA

Moko disease of banana, caused by the bacterium *Ralstonia solanacearum* race 2, causes a lethal disease of banana in many parts of the world, and is especially devastating to banana growers in the Amazon River Basin in Brazil. In this region, Moko disease is the major production constraint limiting banana yields and explains why the Amazon region imports, rather than exports, bananas. Although movement of bacteria-infested insects, farm implements, and Moko-infected vegetative material (ratoons) has been identified as the primary mechanism of dissemination for this pathogen,[40] there is little quantitative information concerning the prevalence and incidence of Moko disease in the Amazonas region.[2] Moreover, how the bacterium is disseminated among subsistence farms has not been elucidated.

One of the basic tenets of integrated disease management is that the presence, distribution, and intensity of any yield-reducing factor must be known.[10,30] The integration of GPS and GIS technologies with disease prevalence and incidence data has tremendous potential to derive new information and hypotheses concerning factors contributing to disease risk.[41] In an attempt to identify large-scale mechanisms for pathogen dissemination within the Amazon River Basin, we used GPS and GIS to determine if individual banana subsistence farms infected with *R. solanacearum* affected the disease status of banana in neighboring banana subsistence farms.[2]

In 2001, 107 subsistence farms (with banana) in the Amazonas region were arbitrarily selected for assessment to quantify the prevalence and incidence of Moko disease. Prevalence was defined as the number of subsistence farms with Moko-infected banana plants ÷ total number of subsistence farms assessed × 100. Incidence was defined as the number of Moko-infected banana plants ÷ total number of banana plants assessed within a subsistence farm × 100. In this region, most subsistence farms with banana are owned by small land holders who grow approximately 50–300 banana plants. In our study, the number of banana plants assessed per farm ranged from 45 to 1025. Each subsistence farm was geospatially referenced and the locations were mapped using ArcGIS (ESRI, Redlands, CA).

The GPS locations of subsistence farms that were subject to periodic flooding ($n = 52$), and those not subject to periodic flooding ($n = 55$), were mapped using ArcGIS (Figure 4.5A). The locations of subsistence farms found to have Moko disease versus farms not found to have Moko disease were likewise mapped (Figure 4.5B). When the two maps were superimposed, it was apparent that the presence of Moko disease was associated with banana subsistence farms that were subject to periodic flooding by river water. Moko disease occurred in 30 of 52 (57.7%) banana farms subject to periodic flooding, but in only 1 of 55 (1.8%) farms not subject to periodic flooding. Chi-square analysis indicated that the null hypothesis that Moko disease would occur equally in both flooded and non-flooded banana farms was strongly rejected ($\chi^2 = 40.55$, $P < .0001$), indicating that subsistence farms (with banana) in the Amazonas River Basin that are subject to periodic flooding are at a much higher risk for Moko disease, compared to farms that are not subject to periodic flooding. Therefore, it is quite probable that individual banana farms subject to periodic flooding are having an impact on the disease status of neighboring subsistence farms with bananas. Thus, in addition to other well-documented dispersal mechanisms, such as the transport of Moko-infected vegetative planting material, dispersal by pathogen-infested farm implements, and pathogen-infested insects,[42,43] the present case study strongly implicates river water as a mechanism for long-distance pathogen dispersal of *R. solanacearum*.

This new information has important ramifications with regard to the management of Moko disease. In the Amazon region, it is recommended that areas with Moko-infected banana plants be cordoned off (with bamboo and twine) to prevent access, as well as to prevent Moko-infected plants (ratoons) from being removed and transported to other farms. These small-scale quarantines are examples of the exclusion principle. Second, it is recommended that Moko-infected banana plants be rogued as soon as plants are symptomatic (eradication principle). Unfortunately, rogued plants are often left at the field site after being cut down, and bacteria-infested rogued plants are not destroyed. The discovery of a new long-distance dispersal mechanism (periodic flooding via river water) may explain why severe Moko disease pandemics occurred in the late 1990s and early 2000s, even though rigorous roguing of Moko-infected plants was practiced. Spatial analyses using *k*-function analysis (described in the ash yellows and *Plum pox virus* case studies) also supported the hypothesis that Moko-infected banana subsistence farms were strongly influencing the health status of other banana subsistence farms, up to a distance of 200 km. Therefore, it

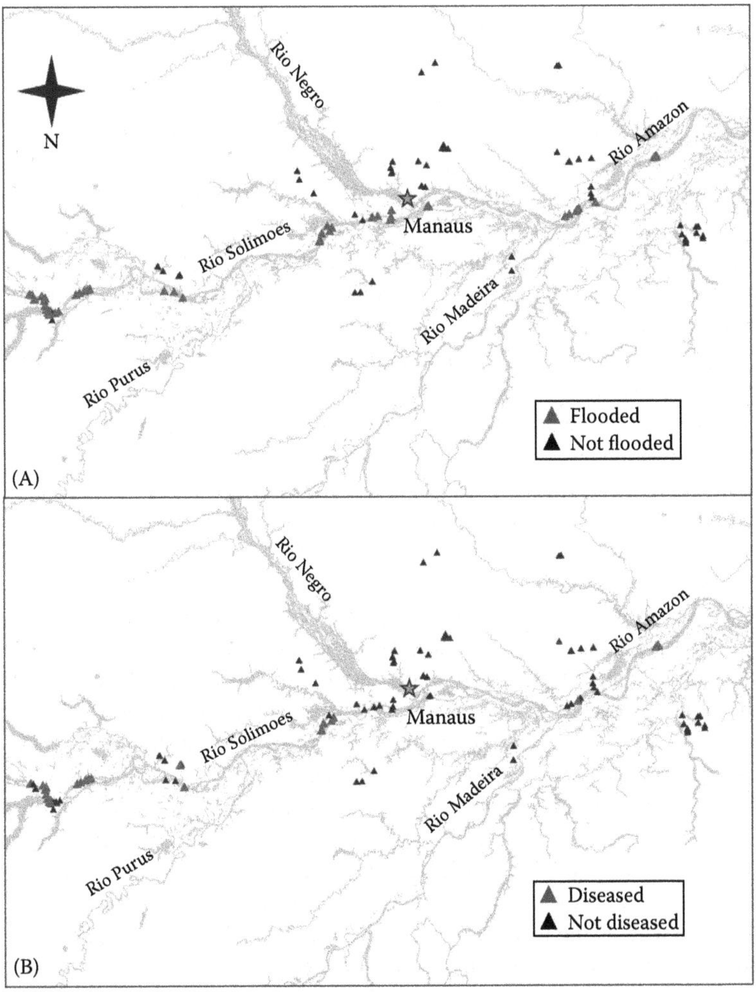

FIGURE 4.5 Geospatial maps depicting the locations of banana subsistence farms in the Amazonas River Basin of Brazil that were assessed in 2001, showing (A) locations of subsistence farms that are subject to periodic river flooding and (B) locations of subsistence farms found to have banana plants infected with Moko disease, caused by the bacterial pathogen *R. solanacearum*).

is now recommended that Moko-infected banana plants be rogued and destroyed by burning before seasonal floods arise.

This study illustrates how the discovery of a yield-reducing risk factor, and quantification of the presence, distribution, and intensity of this risk factor, can improve integrated disease management. This case study also demonstrates how geographic information systems can be used to delineate common geographic management zones (individual subsistence farms), where integrated disease management strategies and tactics can more effectively underpin disease management at the regional scale as well as the local scale.

4.8 CASE STUDY: STEWART'S DISEASE OF CORN

Pest and disease populations, climate, and economic conditions cause agricultural production patterns to be very regionalized. Hence, impacts of pests and diseases can differ widely from year to year, from region to region, and even from farm to farm, or field to field.[16,23] In hybrid seed corn production, sound planning requires accurate and reliable information concerning the seasonal and geographical risks associated with corn and corn-based cropping systems. Before making a decision to deploy a pest or disease management tactic, a critical level of information is usually required.[44] This is especially important for Stewart's disease of corn, a troubling disease for the seed corn industry. Disease forecasting is an integral component for many integrated disease and pest management programs.[10] Forecasting comes under the disease management principle known as protection. The coupling of GPS and GIS technologies with disease forecasting models can tell crop producers when and where specific fungicides, bactericides, insecticides, biological controls are needed to protect crops from plant pathogens and pathogen vectors.

Stewart's disease of corn, or "Stewart's wilt," caused by *Pantoea stewartii*, is economically important because its presence within seed corn fields prevents the export of hybrid seed corn to countries with phytosanitary (quarantine) restrictions.[45] The corn flea beetle (*Chaetocnema pulicaria*) plays an important role in this pathosystem for two reasons: (1) the bacterium survives the winter period in adult corn flea beetles, which overwinter in grassy areas surrounding fields, and (2) the corn flea beetle is the primary means for dissemination of the bacterium from plant to plant.[6,41] The development of an accurate and precise preplant forecasting system, that would identify high-risk seasons and high-risk geographical locations within those seasons, would be of tremendous economic benefit to hybrid corn growers and companies operating within the U.S. Corn Belt. The selection of low-risk planting sites should greatly reduce or even eliminate the need for insecticides. Site-specific information regarding the risk of Stewart's disease would also help seed corn production managers determine where (and to what degree) foliar insecticides are needed to reduce the occurrence of Stewart's disease in high-risk seasons.[41]

The prevalence of Stewart's disease in Iowa seed corn fields for the years 1972–2003 was determined by analyzing seed corn inspection reports obtained from the Iowa Department of Agriculture and Land Stewardship. Yearly prevalence (the number of fields where Stewart's disease was observed divided by the total number of fields inspected) was determined for all counties inspected, and also for the state of Iowa as a whole. The number of fields inspected each year ranged from 500 to more than 1300.[6] Geospatial maps were made using ArcGIS (ESRI, Redlands, CA) and compared with predicted risks for each growing season.

Mean monthly temperatures (average of all mean daily temperatures for a specific month) were generated for December, January, and February, for the years 1972 to 1998, using weather data from the National Oceanic and Atmospheric Administration (NOAA). Following the Iowa State Method, the number of months with mean monthly temperatures ≥−4.4°C (24°F) was used as a predictor of the potential risk for Stewart's disease in the upcoming growing season.[41] Geospatial maps representing the predicted risk for Stewart's disease risk in each Iowa county were generated using ArcGIS software. By coupling weather data and actual prevalence data for the 30 year period, we found that the risk of Stewart's disease was zero to low (<2% prevalence) in counties that experienced mean monthly temperatures ≥ -4.4°C (24°F) in none or one of the three winter months of December, January, and February risk was moderate to high in counties that experienced mean monthly temperatures ≥ -4.4°C during two or three of the winter months. Using individual monthly mean temperatures for December, January, and February (Iowa State Method), we accurately predicted that 1998 (Figure 4.6) and 1999 (not shown) would be high-risk years for Stewart's disease in Iowa.[41]

Information from this study provides a base for accurately assessing, several months prior to planting the seasonal and site-specific risks associated with the occurrence of Stewart's disease. Using GIS software, risk maps are generated prior to planting to graphically depict seasonal, regional, and county-level risks for Stewart's disease. This advance warning helps seed corn producers make more informed disease management decisions and choose low-risk planting sites (which will minimize or eliminate the use of insecticides). Just as important, advanced warnings for high-risk planting sites alert seed corn producers that insecticides will be required to minimize the risk of Stewart's disease in high-risk seasons and sites.

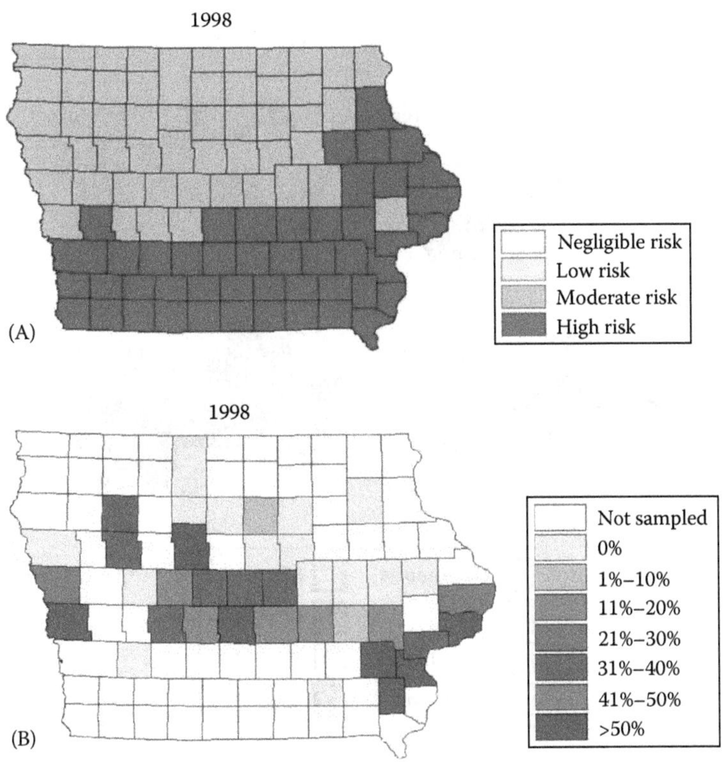

FIGURE 4.6 Geospatial maps showing (A) the predicted risk of Stewart's disease of corn in 1998, generated using the Iowa State model; and (B) the actual prevalence of Stewart's disease in Iowa in 1998.

4.9 CASE STUDY: GRAY LEAF SPOT OF CORN

The coupling of GPS and GIS technologies with historical disease survey data provides a unique opportunity to look back in time to analyze the spatial dynamics of plant disease pandemics. Using the same seed corn inspection database described in the case study concerning Stewart's disease, seed corn inspection reports for the period from 1972 to 2001 were analyzed to assess the presence/absence (prevalence) of gray leaf spot of corn (caused by the fungal pathogen *Cercospora zeae*).[7] Prevalence (the number of seed corn fields found to have gray leaf spot ÷ the total number of seed corn fields inspected × 100) was calculated for each Iowa county for each inspection year. Prevalence data at the county scale was mapped using ArcGIS to depict the spatial change in gray leaf spot prevalence (%) over time (1981–2001).

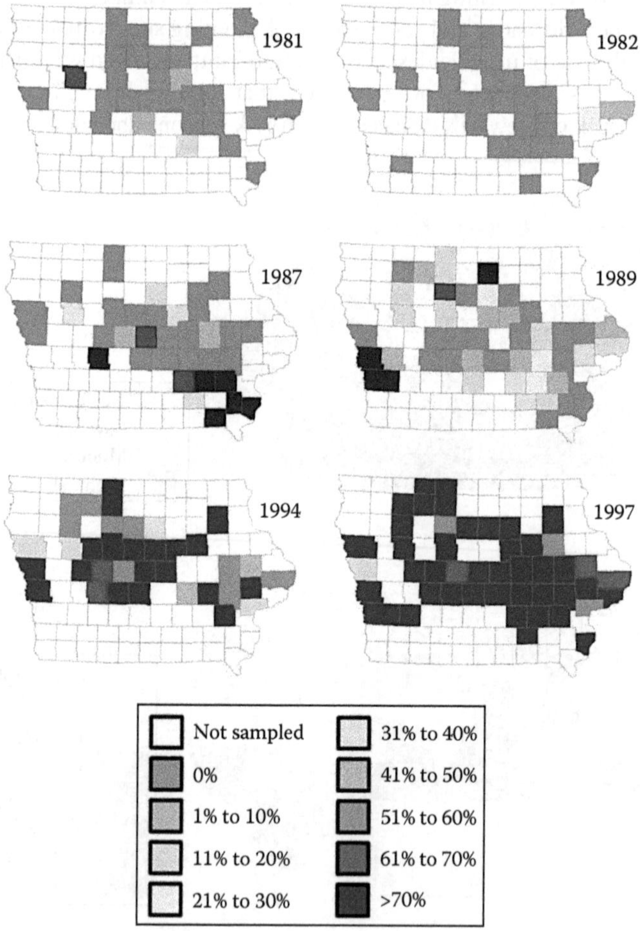

FIGURE 4.7 Mean prevalence of gray leaf spot in state-inspected seed corn fields in Iowa counties in six years (1981, 1982, 1987, 1989, 1994, 1997), determined from seed corn field inspection records obtained from the Iowa Department of Agriculture and Land Stewardship.

Prevalence maps for seed corn inspection years 1981, 1982, 1987, 1989, 1994, and 1997 were selected to depict key historical benchmarks during the pandemic. Gray leaf spot was detected in Iowa in 1981 in four Iowa counties (Figure 4.7), but these sources were eradicated by deep plowing. In year 2 of the pandemic (1982), gray leaf spot was detected for the first time in the easternmost Iowa county (Clinton). The temporal and spatial spread of gray leaf spot increased dramatically in the 1980s and 1990s. During this period, more and more corn producers adopted conservation tillage practices (and discontinued deep plowing of infested residue) because of incentives provided by the federal government. The adoption of conservation tillage practices (minimum or no till), which require that a minimum amount of corn stubble and crop debris be left on the soil surface to reduce erosion, led to a substantial increase in initial inoculum, and loss of the eradicant benefits of deep plowing. By 1988, 55% of Iowa corn producers were practicing some form of conservation tillage, and by 1997, this number had increased to 78%. Thus, government incentives resulted in large amounts of pathogen inoculum (pathogen-infested corn debris) being left on the soil surface instead of being eradicated by deep plowing.

By 1987, the gray leaf spot pandemic had spread to the southern Iowa border. Two years later, in 1989, gray leaf spot reached the northern and western regions of the state, and the disease was found from border to border, stretching from north to south and from east to west. In 1997, gray leaf spot prevalence was <70% in only 5 of the 45 Iowa counties that were inspected. As a result of the change in tillage practices, management of this disease changed from using the eradication principle (using deep plowing to reduce y_0) to the protection principle (using fungicides and/or disease resistance to reduce r).[4]

4.10 CASE STUDY: BEAN POD MOTTLE VIRUS OF SOYBEAN

Bean pod mottle virus (BPMV) has been reported to be on the increase in soybean production areas in the United States.[46] There is little information, however, concerning the present levels of BPMV prevalence and incidence in Iowa. Results from a statewide soybean disease survey conducted in Iowa were coupled with GPS and GIS technologies to elucidate where BPMV is present, and at what level of disease intensity. During the 3 year survey, more than 1000 soybean fields (per year) were surveyed and tested for the presence/absence of BPMV (prevalence data), and the percentage of plant subsamples testing positive for BPMV within each soybean field was determined (incidence data).[1] GPS locations of all surveyed soybean fields were recorded using the same marked point process as described for the ash yellows case study, but at a much larger spatial scale. Risk factors evaluated at a county-level scale included weather variables (winter temperature, snow depth, snow days, etc.), latitude/longitude, elevation, and the number of alfalfa acres within a county (alfalfa serves as an early-season alternative host for the bean leaf beetle, which is the primary insect vector of BPMV). Examples showing the application of GPS and GIS tools to evaluate temperature and latitude/longitude as risk factors are discussed here.

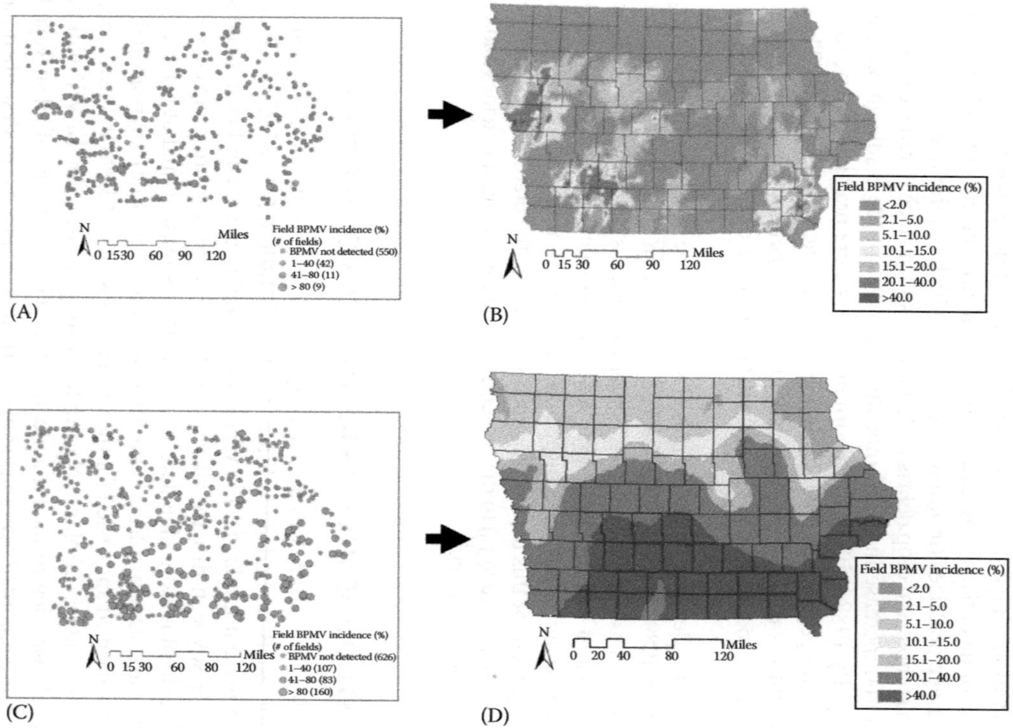

FIGURE 4.8 Incidence (%) of BPMV in individual soybean fields surveyed in Iowa in 2005 (A, C) and 2006 (B, D). Light gray dots indicate geographical (GPS) locations of surveyed fields where BPMV was not detected (shown without county boundaries to protect confidentiality). Darker gray dots indicate locations of BPMV-infected fields; the sizes of dark gray dots indicate levels of BPMV incidence. Kriged maps for 2005 (B) and 2006 (D) display interpolated values for BPMV incidence that were generated (using ArcGIS) from the field-level point data.

The incidence of BPMV in individual fields is shown in point maps for 2005 and 2006. Light gray dots represent locations of sampled soybean fields where BPMV was not detected (Figure 4.8A and C). Dark gray dots indicate the level of BPMV incidence in BPMV-positive fields—the larger the dot, the higher the incidence. These point maps are shown without county lines or topographic features because some growers have legitimate concerns about confidentiality and do not wish to have their fields identified spatially. We show the point maps without county lines or landmarks to illustrate how point data information from individual growers' fields can be analyzed using the marked point process. Using GIS software, field-level information can be displayed in publications as kriged maps without revealing specific field locations (Figure 4.8B and D). The kriging process employs a group of geospatial techniques to interpolate between point values of a random variable (in this case, the point data for incidence of BPMV within individual soybean fields).

Maps depicting the mean incidence of BPMV at the county scale in 2005 and 2006 suggest that counties with higher levels of BPMV incidence tend to be neighbored by counties that also have higher BPMV incidence (data not shown). Conversely, counties with zero or low levels of BPMV incidence tend to be neighbored by counties with similar low levels. These patterns suggest the presence of spatial dependence (clustering) of counties with regard to risk for BPMV incidence.

A tremendously useful geospatial analysis known as Moran's Index[47] can be used to test for spatial dependence. Moran's Index (also referred to as Moran's I) provides a measure of the global, spatial autocorrelation of the overall clustering of data. Moran's I values range from −1 (indicating perfect dispersion) to +1 (indicating perfect correlation, i.e., strong clustering). Values close to zero indicate a random spatial pattern. Based on this analysis, there was weak, but significant, spatial dependence for BPMV incidence among Iowa counties, indicating that counties with higher incidence levels were indeed clustered with other counties with high BPMV incidence. Conversely, counties with zero to low BPMV incidence also tended to be clustered with other counties that had zero to low BPMV incidence. This new information has important implications with regard to area-wide BPMV management in that management recommendations for BPMV should not be generalized for the entire state. Disease risk is not random within the state, and therefore, recommendations for area-wide BPMV management should be made at a spatial scale with a "neighborhood" structure of county clusters. Moreover, the neighborhood structure for BPMV risk varies from year to year, indicating that there is an opportunity to identify and incorporate county- and site-specific biotic and abiotic risk factors that affect the seasonal and neighborhood (geospatial) structure for BPMV risk.

Using logistic regression, we discovered that winter temperature is one of the best abiotic risk factors for predicting BPMV risk. When a county-level map showing BPMV incidence for 2006 (Figure 4.9A) was overlaid with a map of the number of days with average daily temperatures <32°F (0°C), (Figure 4.9B), the two maps were remarkably similar, suggesting a cause-and-effect relationship. When the number of days with average daily temperatures <32°F (x-axis) was regressed against mean BPMV incidence at the county scale (y-axis), there was a significant linear relationship: number of days with mean temperature <32°F explained 46.3% of the variation in BPMV county incidence (Figure 4.9C). Because disease risk decreased

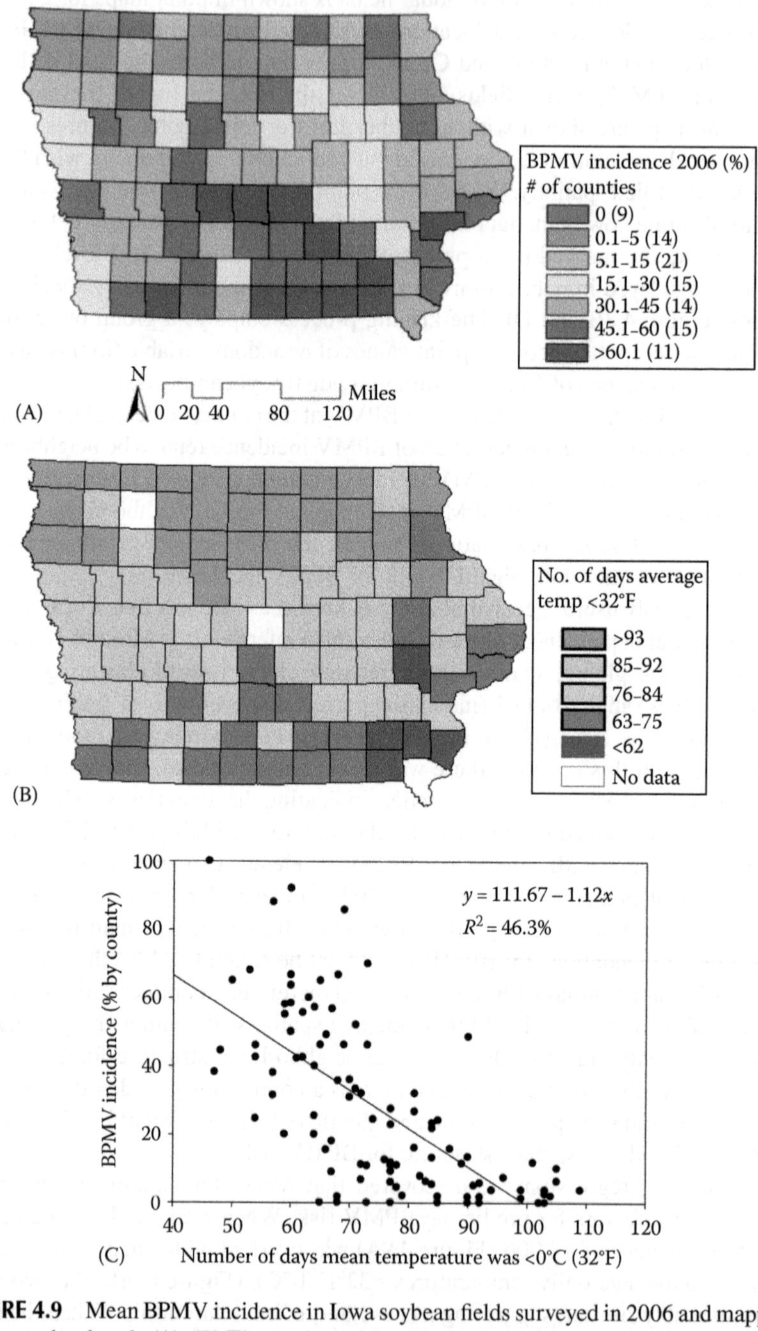

FIGURE 4.9 Mean BPMV incidence in Iowa soybean fields surveyed in 2006 and mapped at the county-level scale (A). (B) The total number of days with a daily mean temperature below freezing (<32°F or 0°C) during the period from October 1, 2005 through April 15, 2006. (C) Regression of county means for BPMV incidence (y) versus the number of days with average temperatures below freezing (x).

as the number of days <32°F increased, it can be hypothesized that colder winters lower the survival rate of the bean leaf beetle (the primary insect vector for BPMV). With regard to disease management, a forecasting model for bean leaf beetle survival could use GIS maps to delineate management zones (clusters of counties) where there is a high risk for BPMV, thereby warning soybean growers in those management zones to deploy protective management tactics (e.g., insecticide seed treatment and/or the application of foliar insecticide) to reduce the rate of virus spread by bean leaf beetles (protection principle). Using GIS software, risk maps depicting BPMV risk at the county level could also be used to alert seed production company personnel where, geospatially, the risk for BPMV incidence is lowest, thereby lowering the probability of plant-to-seed transmission. This practice should further reduce BPMV risk in subsequent growing seasons by lowering BPMV seed infection statewide.

Iowa consists of counties arranged in 9 horizontal tiers (north to south) and 12 vertical tiers (east to west). To determine if BPMV incidence gradients existed in north–south or east–west directions, geographic centroids for counties were averaged across all counties in horizontal tiers (latitude coordinates) and vertical tiers (longitudinal coordinates). Mean horizontal and vertical coordinates for each tier of counties were plotted on the y-axis, and the corresponding average BPMV incidence for each tier of counties was plotted on the x-axis. BPMV incidence data from all three years of the survey (2005, 2006, and 2007) were used.

When mean longitudinal coordinates for the vertical county tiers were plotted against mean BPMV incidence for vertical county tiers, no BPMV-incidence gradients were detected, indicating that the risk for BPMV east to west was random.

However, the presence of a north–south gradient for BPMV incidence was detected by regression analysis, and is clearly shown in the horizontal bar graph in Figure 4.10. The highest risk for BPMV incidence occurs in the southern tier of

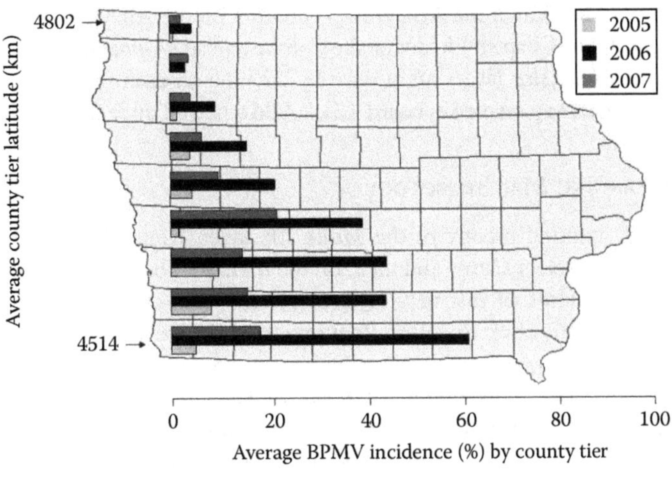

FIGURE 4.10 Incidence of BPMV in Iowa soybean fields, averaged across horizontal tiers of Iowa counties surveyed in 2005, 2006, and 2007. In general, BPMV incidence was lowest in the northern tiers of Iowa counties and was highest in the southern tiers of Iowa counties, with a strong gradient of low to high BPMV incidence from north to south present in all three years.

counties in all three years, and the lowest risk for BPMV incidence occurs in the northernmost tier of counties. This latitude gradient for BPVM incidence is likely to be related to winter temperature gradients that affect the survival of bean leaf beetles. However, additional research is needed to understand why soybean producers in the southern part of the state are at greater risk for BPMV incidence, and why the level of risk varies from season to season. Greater risk in the southern tiers may eventually translate into area-wide recommendations for increased use of crop protection tactics (protection principle).

4.11 GIS TUTORIAL: MOKO DISEASE IN AMAZON REGION OF BRAZIL

The goal of this tutorial is to generate and overlay two GIS maps to determine whether Moko disease of banana is present within subsistence farms located in the Amazon region in Brazil, and whether or not each subsistence farm is subject to periodic flooding. By completing this exercise, participants will generate and overlay GIS maps (to create a third map), and create legends that visually depict whether or not farms subject to periodic flooding are at greater disease risk than subsistence farms not subjected to periodic flooding by the Amazon River or its tributaries.

4.11.1 Saving Chapter 4 Files to Your Computer

1. Open the Chapter 4 files from the CD provided and copy all files into a folder on your computer.

4.11.2 Opening Data in ArcMap

1. Open ArcMap. Click the **Add Data** (+) button on the ArcMap toolbar.
2. Navigate to the Chapter 4 folder on the C drive and add ***Amazonas.shp, Rivers.shp***, and ***Moko.shp*** layers all at once by clicking on each file while holding down the **Ctrl** key on the keyboard. Click **Add** when all three files are selected.

4.11.3 Changing Map Symbology

1. Begin by making a copy of the ***Moko*** file layer. Right-click on the layer name ***Moko***, select **Copy**, and then, on the main tool bar, select **Edit/Paste**. Change the name of this new layer to ***Flood***: right-click on the duplicate layer name, left-click to select **Properties**, and type ***Flood*** in the **Layer Name** box.
2. To change the legend symbols for the ***Moko*** layer, double click on this layer in the **Table of Contents** and click the **Symbology** tab.
3. On the left side of the screen, in the list of options displayed under **Show**: Select **Categories**, and then select **Unique values**.
4. From the drop-down **Value Field** menu, choose ***Moko***. Click **Add All Values** at the bottom of the box. Two values will appear, **0** and **1**. In this file, **0** indicates the absence of Moko disease and **1** indicates the presence of the disease.

5. Change the symbology of these values. First, double click the **1** and in the **Symbol selector** choose **Circle 1**, change the color to **40% Gray**, and change the **Size** to **16**. Click **OK**.
6. Change the **0** category symbol: double-click the **0**, choose a new symbol (e.g., **Circle 1**), a different color (e.g., 10% gray) and a symbol **Size** (e.g., **16**).
7. When the symbols for both categories are changed, click **OK** in the **Layer Properties** window.
8. Repeat Steps 2–7 above to change the symbology of the ***Flood*** layer. In the **Value Field** drop-down menu choose ***Flood***. In this example, the **1** category was symbolized as **Cross 1**, **Black**, **Size 14**, and the **0** category was symbolized as **X**, **Black**, **Size 14**.
9. In the Table of Contents, rename the **0** category for the ***Moko*** layer to ***Absent***. Double-left-click on the **0**, and type ***Absent*** in the name box. Change the name of the **1** category to ***Present***. For the ***Flood*** layer, change the name of the **0** category to ***Not periodically flooded*** and the **1** category to ***Periodically flooded***.
10. Once completed, the map will show locations with both Moko and periodic flooding as 40% Gray + (⊕), locations with Moko but no periodic flooding as 40% Gray **x** (⊗), locations with no Moko but periodic flooding as 10% Gray + (⊕), and finally locations with no Moko and no periodic flooding as 10% Gray **x** (⊗).

4.11.4 Creating and Printing Map Layouts

1. The first map we will make will show the presence/absence of Moko disease. Begin by turning off the ***Flood*** layer by unchecking the box to the left of the layer name. Right click the ***Moko*** layer and click **Zoom to Layer**. On the main menu toolbar click **View** and then choose and click **Layout View**. Because the data is spread in an east–west direction, it is appropriate to use a landscape orientation. To change to landscape orientation, click **File** on the main menu toolbar, select **Page and Print Setup**, and in the **Paper** box, change the orientation to **Landscape**. Click **OK**.
2. In the **Layout View**, adjust the size of the data frame to fit the size of the paper. Add a legend by clicking **Insert** on the main menu toolbar, then **Legend**. Remove all the Legend Items except ***Moko*** and ***Flood***. Click **Next** twice. Change the background color from **hollow** to **white**. Click **Next** twice more, then **Finish**. Move the Legend to the lower left corner of the layout.
3. Export the map that has only the ***Moko*** layer checked by clicking **File** on the main menu and then **Export Map** (Figure 4.11).
4. Now, turn off the ***Moko*** layer by unchecking the box to the left of the layer, and turn on the ***Flood*** layer by checking the box to the left of the layer. Export this map, as above (Figure 4.12).
5. Turn on both the ***Moko*** and ***Flood*** layers. Export this map (Figure 4.13).

The spatial information gained by using geospatial ground-truth data, known information about the area's environment, and geospatial software allows for a better

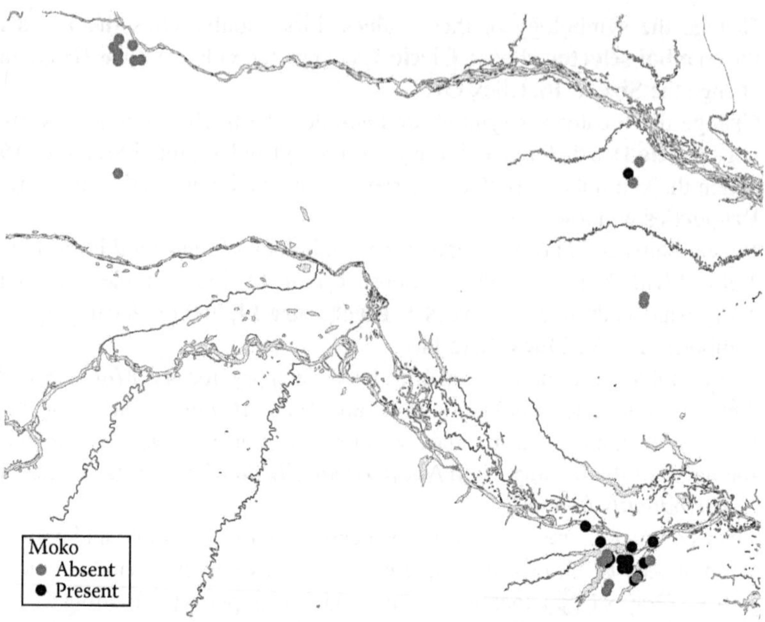

FIGURE 4.11 Geospatial map, created in ArcMap using the Chapter 4 GIS tutorial, that depicts the locations of banana subsistence farms in the Amazonas River Basin of Brazil and shows the absence (lighter color) or presence (darker color) of Moko disease.

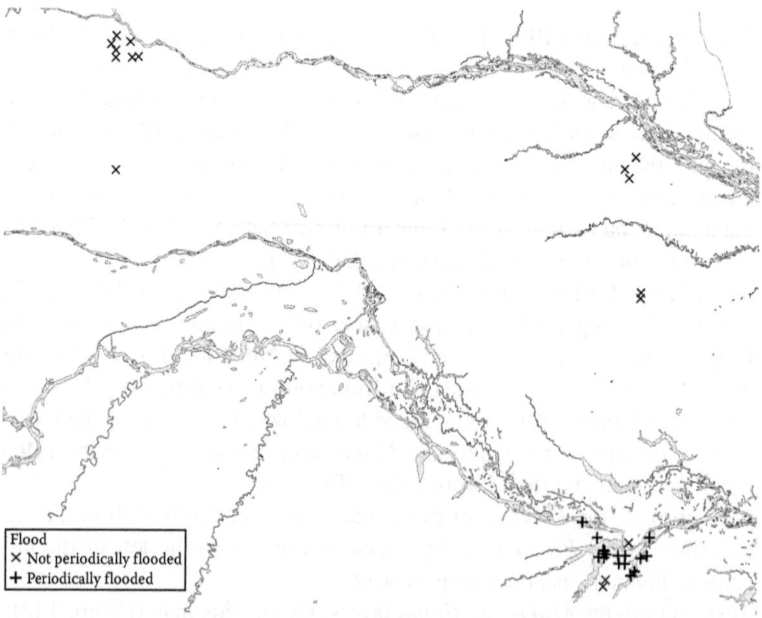

FIGURE 4.12 Geospatial map, created in ArcMap using the Chapter 4 GIS tutorial, that depicts the locations of banana subsistence farms in the Amazonas River Basin of Brazil in areas that are periodically flooded (+) or not periodically flooded (×).

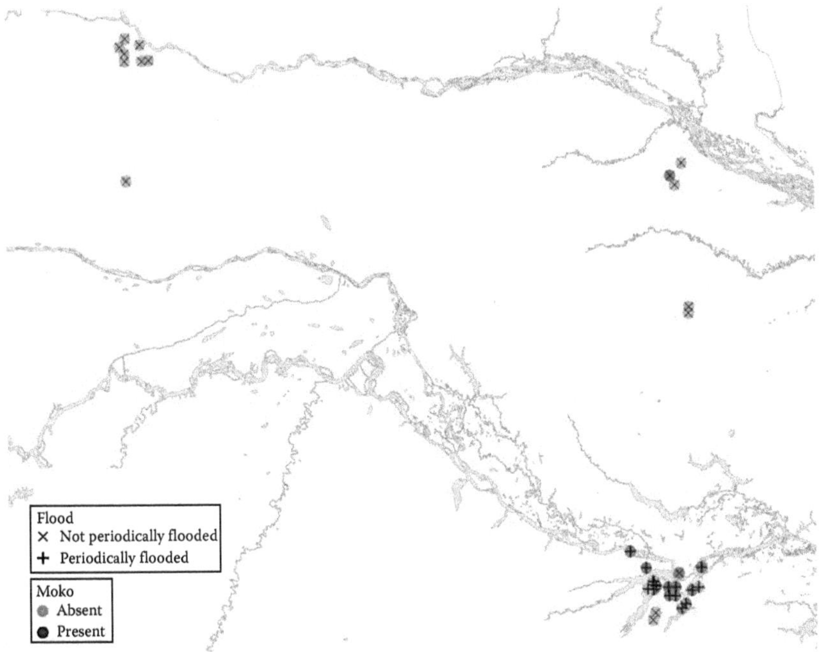

FIGURE 4.13 Geospatial map, created in ArcMap using the Chapter 4 GIS tutorial, that depicts the locations of banana subsistence farms in the Amazonas River Basin of Brazil showing (1) areas without Moko disease in areas with no flooding (light circle with ×), (2) areas without Moko disease in areas with flooding (light circle with +), (3) areas with Moko disease in areas with no flooding (dark circle with ×), and (4) areas with Moko disease in areas with flooding (dark circle with +).

understanding of this disease and its movement. As discussed in Case Study 4.7, these maps illustrate that subsistence farms (with banana) in the Amazonas River Basin that are subject to periodic flooding are at a much higher risk for Moko disease compared to farms not subject to periodic flooding. These data, in turn, can be used to delineate treatment areas and aid in developing effective treatment zones and strategies to better limit disease spread.

4.12 CONCLUSIONS

Integrated disease management programs are underpinned by strategic objectives concerning the need to reduce the rate of temporal and spatial spread (r), the level of initial inoculum (y_0), and/or the time that host and pathogen populations interact in space (t). Once a strategy is adopted, the appropriate disease management principles (exclusion, avoidance, eradication, protection, resistance, therapy) can be selected and integrated to develop cost-effective disease management programs that limit epidemic risk. The term "plant disease epidemic" can be defined as the interaction of host and pathogen populations in **time and space**, as affected by the environment. Thus, the spatial component of plant disease epidemics is of utmost importance with regard to epidemic risk, yet this component is less well understood relative to the temporal dynamics of plant disease epidemics. The purpose of this chapter was to demonstrate how GPS and GIS technologies could be used to support specific plant disease management principles to develop improved integrated disease management programs. Through case studies, we have shown that GPS and GIS tools can greatly improve our ability to detect and better understand the mechanisms of pathogen dissemination at various spatial scales. In addition, the case studies in our chapter also provide excellent examples of the types of spatial analyses available in GIS software packages.

ACKNOWLEDGMENTS

We would like to thank Dr. Khalil Ahmad and Dr. Noha Holah for helping to prepare figures for several of the Asian soybean rust examples used in this chapter. We also want to acknowledge grant support from the NRI-USDA Plant Biosecurity Grants Program and from the Department of Homeland Security, Office of Special Programs, to conduct several of the case studies used in this chapter.

REFERENCES

1. Byamukama, E., Robertson, A., and Nutter, F. W. Jr., Prevalence, incidence, and spatial analysis of the distribution of *Bean pod mottle virus* on soybean in Iowa, *Phytopathology*, 100, 931, 2010.
2. Coelho-Netto, R. A. and Nutter, F. W. Jr., Use of GPS and GIS technologies to map the prevalence of Moko disease of banana in the Amazonas Region of Brazil, in: *Bacterial Wilt Disease and the Ralstonia solanacearum Species Complex*, Allen, C., Prior, P., and Hayward, A. C., eds, p. 431, St. Paul, MN: APS Press, 2005.
3. Esker, P. D., Gibb, K. S., Padovan, A., Dixon, P. M., and Nutter, F. W. Jr., Use of survival analysis to determine the post-incubation time-to-death of papaya due to yellow crinkle disease in Australia, *Plant Disease*, 90,102, 2006.
4. Ward, J. M. J., Stromberg, E. L., Nowell, D. C., and Nutter, F. W. Jr., Gray leaf spot: A disease of global importance in maize production, *Plant Disease*, 83, 884, 1999.
5. Byamukama, E., Robertson, A. E., and Nutter, F. W. Jr., Quantifying the within-field temporal and spatial dynamics of *Bean pod mottle virus* in soybean, *Phytopathology*, 100, Submitted, 2010.
6. Esker, P. D., Harri, J., Dixon, P. M., and Nutter, F. W. Jr., Comparison of Models for forecasting of Stewart's disease of corn in Iowa, *Plant Disease*, 90, 1353, 2006.
7. Nutter, F. W. Jr., Esker, P. D., and Rubsam, R., Mapping the temporal and spatial spread of gray leaf spot of corn in Iowa using GIS, *Phytopathology*, 91, S179, 2001.
8. Steinlage, T. A., Hill, J. H., and Nutter, F. W. Jr., Temporal and spatial spread of *Soybean mosaic virus* (SMV) in soybeans transformed with the coat protein gene of SMV, *Phytopathology*, 92, 478, 2002.
9. Fry, W. E., *Principles of Plant Disease Management*, London, U.K.: Academic Press, Inc., 1982.
10. Nutter, F. W. Jr., The role of plant disease epidemiology in developing successful integrated disease management programs, in: *General Concepts in Integrated Pest and Disease Management*, Ciancio, A. and Mukerji, K. G., eds, p. 45, Amsterdam, The Netherlands: Springer, 2007.
11. Zadoks, J. C. and Schein, R. D., *Epidemiology and Plant Disease Management*, New York: Oxford University Press, 1979.
12. Sharvell, E. G., *Plant Disease Control*, Westport, CT: The AVI Publishing Company, 1979.

13. Sill, W. H. Jr., *Plant Protection: An Integrated Interdisciplinary Approach*, Ames, IA: The Iowa State University Press, 1982.
14. Van der Plank, J. E., *Plant Diseases: Epidemics and Control*, New York: Academic Press, 1963.
15. Isard, S. A., Russo, J. M., and DeWolf, E. D., The establishment of a national pest information platform for extension and education, Online. *Plant Health Progress*. doi: 10.1094/PHP-2006-0915-01-RV, 2006. Available at http://www.plantmanagementnetwork.org/pub/php/review/2006/platform/ (accessed September 18, 2010).
16. Nelson, M. R., Felix-Gastelum, R., Orum, T. V., Stowell, L. J., and Myers, D. E., Geographic information systems and geostatistics in the design and validation of regional plant virus management programs, *Phytopathology*, 84, 898, 1994.
17. Palti, J., *Cultural Practices and Infectious Crop Diseases*, New York: Springer-Verlag, 1981.
18. Nutter, F. W. Jr. and Guan, J., Disease losses, in: *Encyclopedia of Plant Pathology*, O. C. Maloy and T. D. Murray, eds, p. 340, New York: John Wiley & Sons, Inc., 2001.
19. Savary, S., Teng, P. S., Willocquet, L., and Nutter, F. W. Jr., Quantification and modeling of crop losses: A review of purposes, *Annual Review of Phytopathology*, 44, 89, 2006.
20. Cook, R. J., Management of the associated microbiota, in: *Plant Disease: An Advanced Treatise*, Horsfall, J. G. and Cowling, E. B, eds, p. 145, New York: Academic Press, 1977.
21. Lipps, P. E., Influence of inoculum from buried and surface corn residues on the incidence and severity of corn anthracnose, *Phytopathology*, 75, 1212, 1985.
22. Peterson, P. D., Leonard, K. J., Roelfs, A. P., and Sutton, T. B., Effect of barberry eradication on changes in populations of *Puccinia graminis* in Minnesota, *Plant Disease*, 89, 935, 2005.
23. Heichel, G. H., Sans, D. C., and Kring, J. B., Seasonal patterns and reduction by carbofuran of Stewart's bacterial wilt of sweet corn, *Plant Disease Reporter*, 61, 149, 1977.
24. Munkvold, G. P., McGee, D. C., and Iles, A., Effects of imidacloprid seed treatment of corn on foliar feeding and *Erwinia stewartii* transmission by the corn flea beetle, *Plant Disease*, 80, 747, 1996.
25. Nutter, F. W. Jr., Quantifying the temporal dynamics of plant viruses: A review, *Crop Protection*, 16, 603, 1997.
26. Nutter, F. W. Jr., Schultz, P.M., and Hill, J.H., Quantification of within-field spread of *Soybean mosaic virus* in soybean using strain-specific monoclonal antibodies, *Phytopathology*, 88, 895, 1998.
27. Soybean Rust IPM, PIPE—Pest Information Platform for Extension and Education. Available at http://sbr.ipmpipe.org (accessed January 3, 2010).
28. Nutter, F. W. Jr., Understanding the interrelationships between botanical, human, and veterinary epidemiology: The Y's and R's of it all, *Ecosystem Health*, 5, 131, 1999.
29. O'Hara, K. L., Grand, L. A., and Whitcomb, A. A., Pruning reduces blister rust in sugar pine with minimal effects on tree growth, *California Agriculture*, 64, 31, 2010.
30. Nutter, F. W. Jr. and Madden, L. V., Plant pathogens as biological weapons against agriculture, in: *Beyond Anthrax: The Weaponization of Infectious Disease*, Lutwick, L. I. and Lutwick, S. M., eds, New York: Springer Science+Business Media, LLC, 2009.
31. Nutter, F. W. Jr., Disease assessment, in: *Encyclopedia of Plant Pathology*, Maloy, O. C. and Murray, T. D., eds, p. 321, New York: John Wiley & Sons, Inc., 2001.
32. Ahmad, K., van Rij, N., Basart, J., and Nutter, F. W. Jr., Detecting and quantifying the temporal and spatial dynamics of plant pathogens using GPS, GIS, and remote-sensing technologies, *Phytopathology*, 97, S137, 2007.
33. Sinclair, W. A., Gleason, M. L., Griffiths, H. M., Iles, J. K., Zriba, N., Charlson, D. V., Batzer, J. C., and Whitlow, T. H., Responses of 11 *Fraxinus* cultivars to ash yellows phytoplasma strains of differing aggressiveness, *Plant Disease*, 84, 725, 2000.
34. Sinclair, W. A. and Griffiths, H. M., Epidemiology of a slow-decline phytoplasmal disease: Ash yellows on old-field sites in New York State, *Phytopathology*, 85, 123, 1995.

35. Walla, J. A., Jacobi, W. R., Tisserat, N. A., Harrell, M. O., Ball, J. J., Neill, G. B., Reynard, D. A., Guo, Y. H., and Spiegel, L., Condition of green ash, incidence of ash yellows phytoplasmas, and their association in the Great Plains and Rocky Mountain regions of North America, *Plant Disease*, 84, 268, 2000.
36. Gleason, M. L., Parker, S. K., Engle, T. E., Flynn, P. H., Griffiths, H. M., Vitosh, M.A., and Iles, J. K., Ash yellows occurrence and association with slow growth of green ash in Iowa and Wisconsin cities, *Journal of Arboriculture*, 23, 77, 1997.
37. Dixon, P., Testing spatial segregation using a nearest-neighbor contingency table, *Ecology*, 75, 1940, 1994.
38. Moeur, M., Characterizing spatial patterns of trees using stem-mapped data, *Forest Science*, 39, 756, 1993.
39. Penttinen, A., Stoyan, D., and Henttonen, H. M., Marked point processes in forest statistics, *Forest Science*, 38, 806, 1992.
40. Sequeira, L., Bacterial wilt of bananas: Dissemination of the pathogen and control of the disease, *Phytopathology*, 48, 64, 1958.
41. Nutter, F. W. Jr., Rubsam, R. R., Taylor, S. E., Harri, J. A., and Esker, P. D., Use of geospatially-referenced disease and weather data to improve site-specific forecasts for Stewart's disease of corn in the U. S. corn belt, *Computers and Electronics in Agriculture*, 37, 7, 2002.
42. Kelman, A. and Sequeira, L., Root-to-root spread of *Pseudomonas solanacearum*, *Phytopathology*, 55, 304, 1965.
43. Poos, F. W. and Elliott, C., Certain insect vectors of *Aplanobacter stewartii*, *Journal of Agricultural Research*, 52, 585, 1936.
44. Wegulo, S. N., Martinson, C. A., Rivera-C, J. M., and Nutter, F. W. Jr., Model for economic analysis of fungicide usage in hybrid corn seed production, *Plant Disease*, 81, 415, 1997.
45. Carlton, W. M. and Munkvold, G. P., Corn Stewart's disease, Iowa State University Extension. Available at http://www.extension.iastate.edu/Publications/PM1627.pdf, 1995 (accessed January 27, 2010).
46. Giesler, L. J., Ghabrial, S. A., Hunt, T. E., and Hill, J. H., *Bean pod mottle virus*: A threat to U.S. soybean production, *Plant Disease*, 86, 1280, 2002.
47. Moran, P. A. P., Notes on continuous stochastic phenomena, *Biometrika*, 37, 17, 1950.

5 Mapping Actual and Predicted Distribution of Pest Animals and Weeds in Australia

Peter West, Leanne Brown, Christopher Auricht, and Quentin Hart

CONTENTS

5.1 Executive Summary 92
5.2 Introduction 93
5.3 Information Needs 95
5.4 Previous Mapping Initiatives 98
5.5 Current Initiatives 99
5.6 Predicting Invasive Species Distributions 100
5.7 Methods 103
5.7.1 Agreed Data Attributes and Standards 104
5.7.2 Field Manuals for Monitoring 104
5.7.3 Consistent Data Collection Methods/Protocol 104
5.7.4 Collection, Collation, and Reporting of Information 105
5.7.4.1 Geographic Information Systems Tool 105
5.7.4.2 Stepwise Data Collection and Collation 107
5.7.4.3 Data Consolidation 110
5.7.4.4 Data Aggregation and Scaling-Up 110
5.7.4.5 Climate/Habitat Matching Methods 111
5.7.4.6 CLIMATE Software 112
5.7.4.7 Land-Use Classifications 112
5.8 Results 113
5.8.1 Challenges for Large-Scale Mapping and Monitoring Efforts 113
5.8.2 Outcomes of Australian Invasive Species Monitoring Efforts 114
5.8.2.1 Reporting Single Attribute Data 114
5.8.2.2 Multiple Attribute Maps 114
5.8.2.3 Reporting Multiple Species Data 114
5.8.2.4 Data Aggregation and Scaling-Up: Implications 114
5.8.2.5 Reporting Predictive Model Outputs Using Habitat and Climate Suitability 114

5.8.3 Limitations of Methods ... 116
5.8.3.1 Data Collation and Reporting ... 116
5.8.3.2 Climate/Habitat Matching: Potential Distribution Prediction Models ... 118
5.8.3.3 Habitat Matching Using Land Use Data ... 119
5.9 Conclusion ... 120
5.9.1 Reporting at the National Level ... 120
5.9.2 Way Forward for Invasive Species Monitoring and Reporting in Australia ... 121
Acknowledgments ... 121
Appendix 5.A ... 122
5.A.1 Monitoring Protocol for Extent, Distribution, and Abundance of Invasive Species ... 122
5.A.1.1 Step 1 Species Occurrence ... 122
5.A.1.2 Step 2 Distribution: Spatial Pattern ... 122
5.A.1.3 Step 3 Abundance: Relative Numbers ... 122
5.A.1.4 Step 4 Trend ... 122
5.A.1.5 Step 5 Data Quality ... 123
5.A.2 Classes for the Occurrence, Distribution, and Density Attributes for Pest Animals and Weeds (Modified from Queensland Government's Pest Survey Group) ... 123
References ... 125

5.1 EXECUTIVE SUMMARY

Each year in Australia, introduced pest animals cause direct impacts that are estimated to cost A$1 billion while weed impacts are estimated to cost A$4 billion. These estimates do not account for environmental costs, such as long-term degradation of vegetation, soil, biodiversity, and water.

Land managers incur considerable time and financial costs in the control of invasive species (Box 5.1). Australian governments also invest significant resources into research and management in order to reduce the impacts of invasive species. The outcomes of such investments need to be reported, and this requires ongoing, systematic collection of data on the occurrence, abundance, and distribution of invasive

BOX 5.1 DEFINITION OF INVASIVE SPECIES

An *invasive species* is a species occurring beyond its natural distribution and which has significant adverse environmental, agricultural, and/or social impacts. This chapter uses the term to cover both pest animals (including domestic species that have gone wild such as feral pigs and introduced wild species such as foxes) and weeds. Native species may also be considered pests within their natural range where they conflict with human activities or where they become established outside of their natural range.

species and their impacts. Such information can identify where management should be focused, measure the effectiveness of programs and investment, and notify managers of emerging issues.

Australia has been working toward a nationally consistent system for collecting and collating detailed information on invasive species. This is in response to a need for accurate and timely information to guide management decisions. Historically, a nationally agreed list of invasive species for which monitoring and reporting activities could be prioritized and regulated has been lacking, resulting in considerable variation in the way information on invasive species has been recorded and managed. State and national activities for monitoring and managing invasive species has been guided by jurisdictional processes and lacked national cohesion. This situation needed to be addressed because infestations of invasive species frequently occur across jurisdictional boundaries and must be managed across those boundaries.

During 2007, the Australian Government, including all state and territory governments, collaborated to develop and implement nationally consistent protocols for monitoring pest animals and weeds. Adoption of the protocols at state and national levels provided—for the first time—detailed and consistent information on the occurrence, distribution, and abundance of nationally significant pest animals and weeds, for the entire continent, using a consistent methodology and data format.

This chapter presents the approach being used throughout Australia to monitor and report the national distributions of pest animals and weeds. It describes the use of geographic information systems to collate independent state and territory datasets and to develop national datasets on invasive species. This chapter also discusses methods to predict the potential distribution of invasive species based on climate and habitat modeling.

5.2 INTRODUCTION

As an island continent, Australia's native flora and fauna have largely evolved in isolation, leading to high levels of endemicity and diversity in ecological communities. The introduction of foreign species to Australia since European settlement in 1788 has substantially altered many naturally occurring flora and fauna communities. Many exotic species were deliberately introduced for agricultural production. Some species introduced as livestock, such as pigs, have established extensive feral populations. Some crop and pasture species have become weeds; however, garden plants represent the main source of weeds in Australia (Figure 5.1). Other invasive plants and animals were deliberately introduced by acclimatization societies that aimed to make Australia more like Europe by establishing wild populations of European plants and animals. In addition, there are many invasive species that were accidentally introduced with shipping cargo or escaped from captivity or gardens—a process that continues despite today's more stringent biosecurity measures.

Invasive species such as pest animals and weeds now inhabit all regions of continental Australia and many offshore Islands. Invasive species cause considerable damage to agricultural industries and environmental values and control of

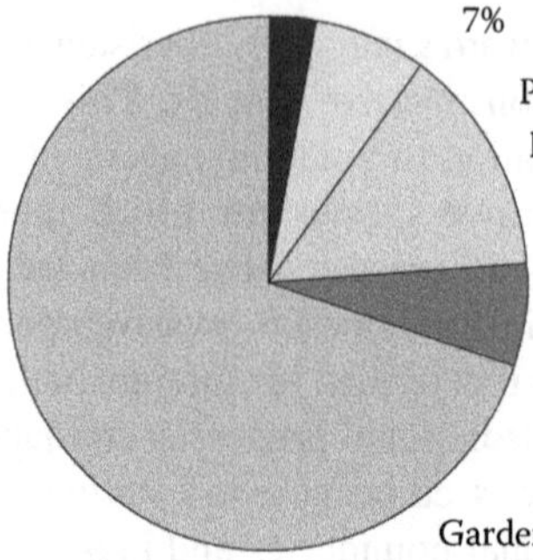

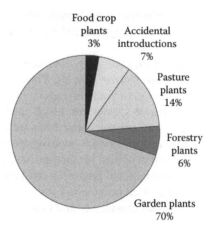

FIGURE 5.1 Sources of Australian weeds. (Adapted from Virtue, J.G. et al., Plant introductions in Australia: How can we resolve "weedy" conflicts of interest? in B.M. Sindel and S.B. Johnson (eds), *Proceedings of the 14th Australian Weeds Conference*. Weed Science Society of New South Wales, Sydney, Australia, pp. 42–48, 2004. With permission.)

invasive species imposes significant costs to farmers and conservation land managers. Some invasive species also present risks to human health, such as the spread and transmission of zoonotic diseases. Combined, weeds and pest animals cost Australian agriculture around A$5 billion annually[2–4] with the environmental costs practically unquantifiable.

Over 80 species of nonnative vertebrates—comprising mammals, birds, reptiles, amphibians, and fish—have established wild populations in Australia[5] and over 30 of these species are agricultural or environmental pests. Major agricultural impacts of pest animals include grazing and land degradation by rabbits and feral goats; livestock predation by wild dogs, foxes, and feral pigs; and damage to grain and fruit crops by mice and pest birds. Direct losses are associated with damage by pest animals and there are also considerable costs associated with their control and associated research.

Weeds are also a serious threat to Australia's agricultural industries and the natural environment. More than 2770 species of exotic plants have established wild populations and around 16% of these currently pose serious problems for agricultural production. Weeds contaminate crops, compete with crop and pasture plants for water and nutrients, displace desirable pasture plants, contaminate wool, can be toxic to livestock and reduce their condition, reduce the capacity of land to carry livestock, and harbor disease or insect pests.

The management and prevention of weeds and pest animals in Australia are largely guided by the "Australian Weeds Strategy" and the "Australian Pest Animal Strategy", respectively. Weeds are also categorized and prioritized into groups for management action. One such prioritization system identifies *Weeds of National Significance* (WONS); there are 20 listed species. A second system categorizes 17 *agricultural sleeper weeds*, some of which have been recommended for eradication. A third list, the *National Environmental Alert List*, includes 28 weeds that are potentially eradicable or containable and pose a serious environmental threat. Equivalent tools for prioritizing pest animals are under development and promote consistent management of pest animals across all jurisdictions in Australia.

Governments and communities invest considerable resources to prevent and control invasive species and their impacts. To maximize the effectiveness of this investment, a targeted and strategic approach to management is encouraged. As part of a strategic approach, managers define the problem(s), evaluate the scale of an infestation by an invasive species, select suitable management and control options, implement management, monitor the outcomes, and refine the program as needed.[6] This approach is being broadly applied throughout Australia to manage invasive species and their impacts.

Before controlling an invasive species population, it is essential for governments and managers to gather information. Historically, measurements of relative abundance have been used as an indicator of the scale of a problem. However, abundance does not always directly translate to damage or adverse impacts[7] because in some circumstances, a small number of individuals can be responsible for a large amount of damage (e.g., a single wild dog may kill or injure numerous livestock in a single night). Therefore, managers are encouraged to focus on reducing the impacts of a species rather than just the numbers or density of a species. Appropriate monitoring is needed to evaluate the consequences of management decisions on the impacts of invasive species.

An important step forward in invasive species management in Australia has been the recognition of the need for a strategic and broad-scale approach, whereby species are managed across larger rather than smaller areas. This can involve multiple land management agencies and land tenure types. Where multiple agencies or jurisdictions are involved, information needs to be collated and reported consistently to assess the effectiveness of management actions. Of similar importance is the need to monitor and evaluate temporal trends in invasive species' population size and impacts. This allows decisions to be made about the resources required to control invasive species and assessment of natural versus management-induced changes in pest populations and their impacts.

The use of geographic information systems (GIS) has become mainstream in the management of natural resources in Australia, and a range of GIS tools have been developed by governments and private industry groups to manage spatial data on invasive species. This has supported strategic approaches to managing invasive species, whereby populations and their impacts are addressed at a range of spatial scales. GIS enables the examination and integration of a wide range of data on natural resources (e.g., land use, conservation values) to address invasive species issues in a broader context.

5.3 INFORMATION NEEDS

Managers need to know where an invasive species is, how large the population is, what its adverse impacts are, and how best to control the species to mitigate adverse impacts. Importantly, managers need to know how their efforts translate into on-ground economic, environmental, and social benefits.

TABLE 5.1
Definitions of Mapping Classes Commonly Used in Australian Mapping Systems

Parameter	Descriptor and Classes
Occurrence	Presence of a species within an area (e.g., present, absent, or unknown)
Distribution	Spatial pattern or dispersion throughout an area (e.g., localized or widespread)
Abundance	Numbers of individuals, frequency or density in a defined area (e.g., occasional, common, or abundant)
Trend in abundance	Trend in the abundance of individuals of a species (e.g., increasing, decreasing, stable, or unknown)
Data quality	Quality of the underpinning datasets (e.g., high, medium, low, or no data)
Impact	Consequences of invasive species for environmental, economic and social values, assets, and services in a defined area

Population monitoring is essential to manage invasive species. Monitoring of spatial and temporal patterns in invasive species occurrence, distribution, abundance, and trend (Table 5.1) is needed to

- Identify priorities for management (e.g., planning and resource allocation)
- Evaluate previous management activities (e.g., the response of populations to control)
- Improve understanding and knowledge (e.g., the relationship between invasive species populations and their impacts)
- Educate and raise the awareness of the public on current and potential problems, as well as opportunities for prevention and control

Detailed information on the occurrence, abundance, and impacts of invasive species is needed at the local scale to guide short-term local management decisions. However, regionally and nationally consistent information on invasive species is also required to guide strategic coordinated management and the longer term allocation of resources. To obtain meaningful information at the national scale, there needs to be

- Comparability of the methods used to develop information and data layers
- Consistency across all state jurisdictions in the data reported
- Uniformity in the scale of information being measured and reported within and across jurisdictions

The most important information for managers is whether a species is present or absent in an area. Occurrence information (Table 5.1) is needed to inform policy and management at various scales and may be used to track the spread of new pest incursions at fine resolution or to track general trends over longer timescales for established species. However, establishing the *occurrence* of a species can be more difficult than it sounds. Many species are difficult to identify or locate, particularly at low densities, and some species are only visible at certain times of the year. For instance, some weed species can be difficult to identify when they are not flowering or seeding, while others resemble their native counterparts. Pest animals can move or migrate, and may exhibit cryptic or nocturnal behavior—making detection difficult. Detection of some species is also significantly influenced by seasonality and the habitat they occupy.

Information on the *abundance* of invasive species also provides meaningful information to guide management actions, including the allocation of resources and the level of control required to address invasive species problems. Information on abundance can indicate the likely extent of impacts of invasive species in a defined area, but may not always directly translate to damage levels.

A wide range of techniques are available to objectively measure and report invasive species' abundance. Some techniques are more suited to local and regional scale application than others. The Australian continent (including Tasmania) covers an area of over 7.6 million square kilometers, approximately 83% of the area of the United States (including Alaska) (Figure 5.2). Most techniques for monitoring invasive species cannot be practically or cost-effectively applied across large areas, for example, state, territory, or nation. Such broad-scale assessments need alternative solutions.

Together, information on *occurrence*, *abundance*, and the *impacts* of species can be used to assess the effectiveness of management actions. Information on the *distribution* of invasive species in a given area helps to determine the spatial pattern of a population and to coordinate on-ground control activities. Species' behavior, particularly during breeding periods, and localized resources can influence the distribution of populations. Distribution information can also be used to plan control to maximize its cost-effectiveness. Similarly, information on the *trend in abundance* of invasive species is important in establishing how populations vary over space and time, particularly in response to management.

Information on the *quality of data* of invasive species allows judgments about its usefulness to guide decision making. When the quality of data is high, managers can make decisions with a high level of confidence. When information is of low quality, or lacking, managers may be more uncertain and adopt more robust approaches to decision making, or seek to verify information before embarking on a control program.

Information about invasive species is also needed to manage a wide range of natural resources and assets—including rivers, wetlands, native vegetation, conservation, and agricultural land-use areas. The management of invasive species needs to be

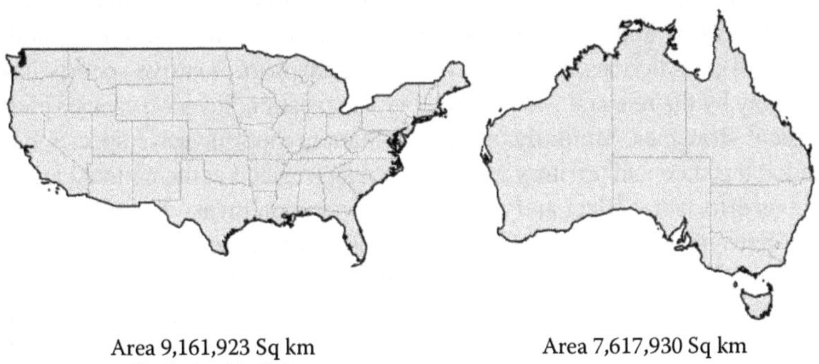

FIGURE 5.2 Administrative boundaries and proportional size of the Australian continent in contrast to the United States (including Alaska). Note: gray outlines denote state/territory boundaries.

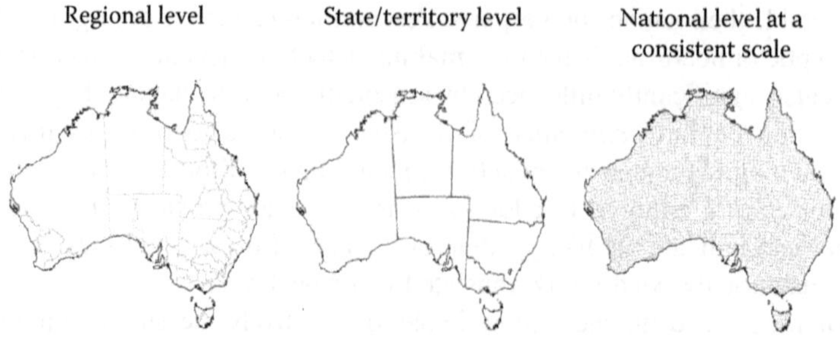

FIGURE 5.3 Scales of information compiled to develop consistent national datasets for invasive species.

considered in the context of these other resources. Australia's current emphases on developing consistent datasets for invasive species and aligning reporting requirements across jurisdictions have arisen within the context of natural resource management.

In this chapter, we present a cost-effective method to consolidate existing local and regional scale information on the *occurrence*, *abundance*, *distribution*, *trend*, and *quality of data* for invasive species into consistent and useful datasets at state and national levels (Figure 5.3).

The usefulness of invasive species information for decision making is determined by how well data are managed. Managers need local and regional data to coordinate control activities and state and national data to guide policy, investment, and planning. A consistent national database or information system is needed to provide a central source for aggregated data, and a distributed system is required along with appropriate arrangements for data sharing and governance to meet the different needs of the various stakeholders at state and national levels.[8–10]

5.4 PREVIOUS MAPPING INITIATIVES

The Australian federation is defined by eight administrative divisions consisting of six states and two territories (Figure 5.2). This poses challenges for the management of natural resources and invasive species. Many invasive species occupy more than one state or territory and the problems these species present may need to be addressed across several jurisdictions. However, the management of invasive species has been driven largely by the relevant state or territory's legislation, invasive species plans, and management strategies. Similarly, most information about invasive species has been compiled at the state and territory levels but not aggregated at the national level.

Previous efforts to collect and collate information on invasive species throughout Australia have delivered datasets and information to guide management and decision making. However, most mapping is at the local, state, and territory levels, and mapping methods differ. Therefore, most datasets cannot be compared, making it difficult to make inferences across jurisdictions or identify changes in populations over time in response to management actions.

Part of the reason for inconsistency in previous assessments is that states and territories differ in their management practices and legislation. Local factors also influence the application of on-ground monitoring techniques, such as spotlight surveys. An area's accessibility, extent of urbanization, and land tenure can influence the choice of monitoring techniques. For example, aerial surveys may be used to map large inaccessible areas (e.g., desert national parks), whereas intensive on-ground monitoring techniques may be used in densely populated areas of small private land holdings. As a result, jurisdictions' systems and processes for collecting and managing data on invasive species differ.

National initiatives to map and report invasive species information have attempted to derive national-scale information from discrete state and territory datasets. Pest animals have been mapped in "Pest Animals in Australia: A Survey of Introduced Wild Mammals"[11] and "Landscape Health in Australia."[12] Information has also been compiled through the "Natural Resource Management on Australian Farms, 2006–07."[13] Initiatives reporting on the national extent of invasive weeds have included mapping the WONS (mapped in 1998, published in Thorp and Lynch[14]); mapping of sleeper weeds;[15] the "Australia—Landscape Health Database 2001—Weeds"[16]; "Indicators of Catchment Condition in the Intensive Land Use Zone of Australia—Weed Density"[17]; "Farms with Significant Degradation Problems (Weeds), 1998–99"[18]; and "Natural Resource Management on Australian Farms, 2006–07."[13]

In the past, lack of agreement on a national list for monitoring and reporting on invasive species has led to a lack of coordination across state and territory jurisdictions in the development of datasets and information.

Jurisdictions need to agree on a consistent scale of monitoring and reporting, using comparable mapping techniques, and standardized products and outputs. Several authors have identified the need for consistency in the collection and reporting of data on invasive species across landscapes, bioregions, and management jurisdictions.[19–22] Consistent and repeatable methods would allow comparison of information across regions and across changing environments and assessment of changes in populations over time in response to government policy, investments, and on-ground management activities. In both Wilson et al.[11] pest animal survey and previous weed mapping processes, a grid-based surface of Australia was created to reflect mapsheets or smaller reporting units. The surfaces were used to consult with jurisdictions on the presence/absence and density of species in each grid area. However, in both these cases, jurisdictions vary in the method for estimating species abundance (particularly in the field), making inferences at state and national levels difficult.

Where agencies from several jurisdictions are responsible for managing invasive species, as is the case in both Australia and the United States, consistency in monitoring and reporting methods across jurisdictions is needed to compare information at regional and national levels.

5.5 CURRENT INITIATIVES

Since 2002, the Australian Government has invested in programs to support sustainable development by providing nation-wide data and information on Australia's land, water, and biological resources. Recently, the Australian Government and state and territory governments agreed to develop a joint program to provide improved national data for nationally significant pest animals and weeds.

During 2006 and 2007, the joint program was coordinated by the Australian Government in collaboration with all states and territories. The Invasive Animals Cooperative Research Centre (see www.invasiveanimals.com) was also instrumental for the parts of the program relating to pest animals. The program coordinated the collection of information on pest animals and weeds under a nationally coordinated monitoring and evaluation framework.

The program aimed to develop national methods for collection, collation, and reporting of fundamental information on invasive species and report on the status of invasive species by

- Implementing a consistent method for monitoring of invasive species populations throughout Australia
- Developing an enduring system for ongoing and consistent monitoring and reporting of invasive species
- Reporting information in a comparable and consistent format and scale across administrative boundaries

The joint program operated under the guidance of the Australian Weeds Committee and the Vertebrate Pests Committee and comprised representatives from all levels of government. The program produced a series of detailed national datasets and maps for reporting the status of nationally significant invasive species throughout Australia. In addition, wherever possible, invasive species data were made available for concurrent research and management activities. Two examples are (1) the development of a report on national management of feral camels and a tool for multi-criteria analysis to support decisions about their management[23,24] and (2) the determination of National Management Actions—2009 for Weeds of National Significance (WoNS) (refer www.weeds.org.au/WoNS/bitoubush/).

5.6 PREDICTING INVASIVE SPECIES DISTRIBUTIONS

Maps and models of the potential distribution of invasive species provide valuable information for land managers and decision makers. For example, areas with a high probability of invasion can be targeted as part of surveillance to detect incursions, or high value land at risk can be protected. The likely impacts of invasive species can be assessed by overlaying the potential distribution with maps indicating land use or protected areas, such as world heritage sites. Analysis of these overlays provides quantitative spatial information for planning programs to control invasive species.

Climate is an important factor influencing where species can exist and potential distribution maps are often generated based on the climate tolerances of species. Species are more likely to establish and spread when introduced to locations that have similar climates to their existing range.[5,25,26] The climate of areas where a species evolved or has successfully naturalized is indicative of the climatic range where that species can potentially survive in other countries if introduced.

Some computer programs for "climate matching" provide a simple method of using presence-only data to compare the climate in locations where an invasive species is present with the climate in locations where it is not yet present and identify

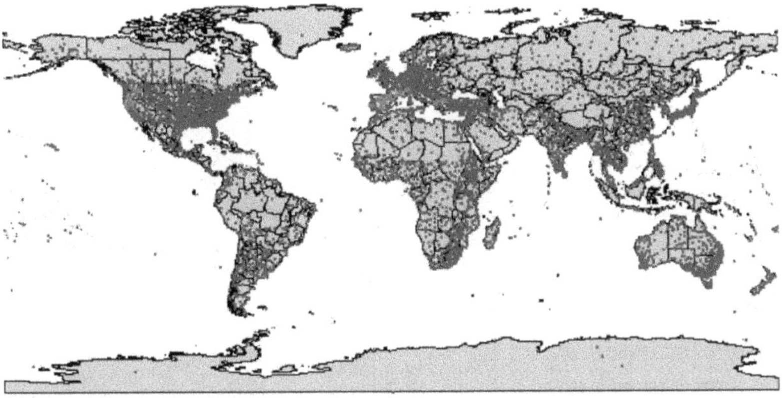

Climate v1.0
Invasive Animals CRC
Bureau of Rural Sciences 2008
(a)

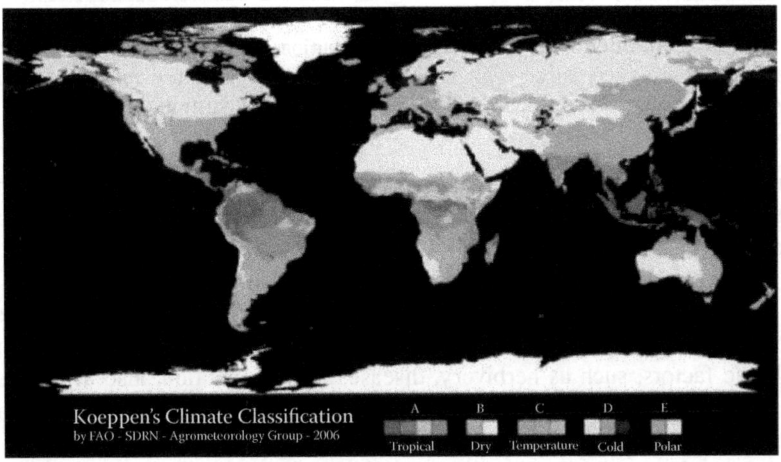

(b)

FIGURE 5.4 Diagram representing CLIMATE modeling to predict a species' potential distribution in Australia based on the species' current global range. (Modified from Cunningham, D.C. et al., *Science for Decision Makers: Managing a Menace of Agricultural Sleeper Weeds*, Bureau of Rural Sciences, Canberra, Australia, 2006. With permission.) The computer program CLIMATE uses the overseas distribution of a species (a), cross-referenced with global climate temperatures (b), to determine the potential distribution of a species in Australia (c), which are then in turn reviewed by experts. Land use types (d) at risk are then identified to guide management. An example of this process is shown in the figure where (a) the overseas distribution of a species (e.g., *Crupina vulgaris*) was determined from scientific literature; only information with a high level of certainty was considered for use (Bureau of Rural Sciences); (b) global temperatures (FAO[28]) were assessed;

(*continued*)

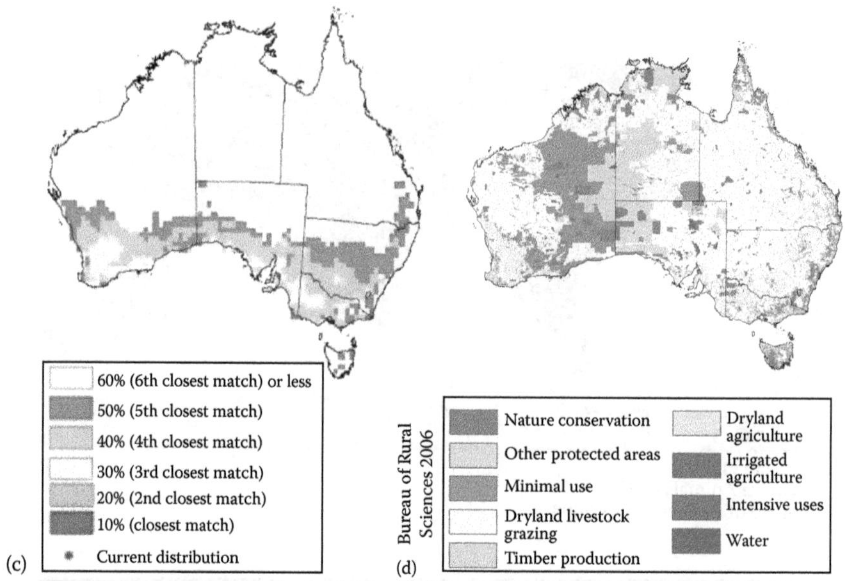

FIGURE 5.4 (continued) (c) the potential distribution of *C. vulgaris* (Bureau of Rural Sciences) was cross-referenced with temperatures and (d) land use maps of Australia 2001/2002 (Bureau of Rural Sciences, 2001/2002 Land Use of Australia, version 3).

locations with climates suited to the invader. A diagram representing "climate" modeling, used to predict a species potential distribution based on data for the species' global range, is displayed in Figure 5.4.

Climate-matching models generally assess only abiotic (nonliving chemical and physical factors in the environment) rather than biotic influences on species distribution.[29] Other abiotic factors such as soil nutrients, aspect and disturbance regimes, and biotic factors, such as herbivory, disease, and competition, also affect a species' current and potential range. Hence, predictions from climate-matching models must be interpreted within a broader ecological context.[30] When an on-ground assessment of potential distribution is required—for example, to develop detailed on-ground management plans—maps of climate matches need to be overlaid with maps of other factors such as soil or vegetation type. For planning at local scale, maps of climate matches should also reflect microclimates, wherever possible.

The intended application of the maps will determine the spatial resolution of the climate data, which may take the form of simple, point data from meteorological stations, or climate surfaces interpolated from meteorological station data using complex algorithms. Planning at national scale tends to focus on more general prioritization of the allocation of resources. Microclimatic information is usually not as relevant and a broad indication of the likely extent of a species is usually sufficient.

Maps of potential distribution that have been based on climate can be refined using overlay maps of factors that relate to the ecology of the species of interest, for example, soil, vegetation type, or land use. Overlays such as water bodies (e.g., for aquatic weeds) or soil type (e.g., for foraging pest animals such as pigs) are useful at

TABLE 5.2
General Typology of Data Used for Mapping Invasive Species at Different Spatial Scales

Scale	Relevant Data Types	Application
Local	Microclimatic habitat, soil type, waterways and catchments, local vegetation data or threatened species	Resource planning for control approaches
State	Macroclimatic, catchment scale land use, soil, state-based vegetation type, natural assets	Pest prioritization, planning for strategic broad-scale management, asset protection
National	Macroclimatic, national land use, national assets, national vegetation	Pest prioritization, planning for strategic pest management at the national level, asset protection

all scales, particularly when the target pest species are habitat specialists. Attributes affecting agricultural or environmental values also need to be considered to evaluate invasion risk and support management decisions.

At the national scale, climate-matching models can form part of broader systems to assess the risk posed by invasive species and prioritizing them. Areas of high risk—for example, where pest density is high, climatic zones are favorable, or high value assets occur—can be targeted as part of management programs.

One such risk assessment is the prioritization of species that are "sleeper weeds" in Australia. Sleeper weeds currently have small wild populations but have the potential to spread widely and have detrimental agricultural or environmental impacts.[27] Sleeper weeds were prioritized by overlaying climatic matching models and land-use data to estimate the proportion of the value of industries at risk should these species spread into their predicted potential range.

At state level, climatic matching is used in risk assessments to prioritize invasive species and to plan at the local scale. For example, climatic matching might be used to assess the suitability of a particular biological control agent or to identify the risk invasive species pose to natural or land use assets. Maps of potential distribution based on climate are refined with state-based data such as maps of vegetation or land use at catchment scale (Table 5.2).

5.7 METHODS

During 2006 and 2007, a national initiative was developed to assess fundamental information on invasive species throughout Australia. The program aimed to provide consistent information on invasive species and a repeatable process for ongoing monitoring and reporting activities. The assessment used five steps:

1. Develop and adopt nationally agreed data attributes and data standards.
2. Recommend suitable field manuals and monitoring techniques for data collection.
3. Develop consistent data collection methods for regional and state levels.

4. Collect, collate, and report information using appropriate reporting tools and techniques.
5. Report information at the national level.

Further information on the national initiative for pest animals can be found in NLWRA and IA CRC[9,10] (www.invasiveanimals.com > publications), while additional information for weeds is available in NLWRA.[8]

5.7.1 Agreed Data Attributes and Standards

National agreement was reached on a procedure to collect and report invasive species data and a minimum scale for reporting information on invasive species was nominated—that is, a 0.5° grid layer for the Australian continent and its islands (equivalent to 1:100,000 or approximately 50 km × 50 km).

Agreement was also reached on fundamental data attributes and standards for information on invasive species for reporting across the grid layer. The data attributes nominated were

- *Occurrence* of a species
- *Abundance* (relative abundance) of a species
- *Distribution* (spatial pattern) of a species
- *Trend* in the abundance of a species
- *Quality* of information and data collected on each species

5.7.2 Field Manuals for Monitoring

There are a wide range of field-based techniques available for monitoring invasive species at the local and regional levels. Therefore, to complement the above agreed procedures and ensure the accuracy and reliability of underpinning datasets, two manuals for field-based monitoring of invasive species of weeds and pest animals were identified and recommended:

1. *A Field Manual for Surveying and Mapping Nationally Significant Weeds* (Bureau of Rural Sciences 2006)[31]
2. *Monitoring Techniques for Vertebrate Pests*[32]

These documents provide guidance on techniques for field-based monitoring, detection, and reporting of weeds and pest animals.

5.7.3 Consistent Data Collection Methods/Protocol

To collect and collate data for invasive species in a consistent way, jurisdictions also agreed to adopt a standard data collection, collation, and reporting method, modified from a method previously implemented in the state of Queensland. The approach allows the collection and reporting of information on the *occurrence*, *distribution*, and *abundance* of invasive species against defined classes. Additional categories for *trend in abundance* and *data quality* provided further information for interpreting datasets. The definitions for these categories are summarized in Table 5.1 and described in more detail in Appendix A.5.1.

Specific protocols for monitoring pest animals and weeds were developed to collect, collate, and report information under these data attributes:

1. *Monitoring Protocol for the Distribution and Abundance of Significant Invasive Vertebrate Pests* (see www.nlwra.gov.au/national-land-and-water-resources-audit/vertebrate-pests)[33]
2. *Monitoring Protocol for the Extent, Density, and Distribution of Weeds* (see www.nlwra.gov.au/national-land-and-water-resources-audit/weeds)[34]

These protocols are aimed to promote consistency in the reporting of information on invasive species at the state/territory and national levels and allow the integration of a large variety of datasets from relevant sources.

5.7.4 Collection, Collation, and Reporting of Information

The data collection, collation, and reporting method and protocol (see Appendix A.5.1 and Table 5.1) enabled integration of many field-based datasets into a single data repository and classification of parcels of land according to five data attributes for invasive species:

- Occurrence—Present, absent, or unknown
- Distribution—Localized or widespread
- Abundance—Occasional, common, or abundant
- Trend—Increasing, decreasing, stable, or unknown
- Data quality—High, medium, low, or no data

A species status can therefore be evaluated using multiple attributes simultaneously.

All organizations and agencies involved in collating data were asked to provide available datasets relevant to the classification of invasive species in their respective jurisdictions. Data and information were gathered from various point, polygon, and map sheet sources and were aggregated to provide a seamless regional, state, and national data layer (Figure 5.5). The period of data collection, collation, and reporting from all state and territory jurisdictions for this project was between June 2006 and April 2007.

5.7.4.1 Geographic Information Systems Tool

To collect and collate data against the five attributes identified above, a customized GIS data-capture routine was developed within ArcView (version 3.3) ESRI.[35] The program provided a stepwise approach for reviewing and collating all available

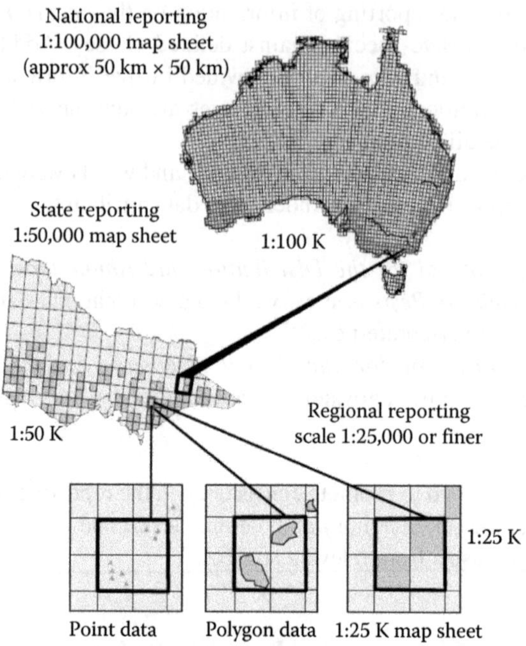

FIGURE 5.5 Process of aggregating local and regional data to produce state and national invasive species datasets. (From National Land and Water Resources Audit and Invasive Animals Cooperative Research Centre, *Assessing Invasive Animals in Australia 2008*, NLWRA, Canberra, Australia, 2008. With permission.)

spatial data and assigning a unique classification to parcels of land in a consistent format and scale. The GIS data capture routine provided a means of

- Selecting and allocating attribute data to designated parcels of land
- Ensuring all available spatial data were entered and reviewed simultaneously for reporting on the five data attributes
- Managing all available invasive species data (including point, line, and polygon), to classify mapping units according to species extent and abundance
- Reviewing data across jurisdictions to support consistent classification of land parcels and comparison of data across broad regions

The GIS tool for capturing data provided a mechanism to classify land parcels according to *a species abundance*, *distribution*, *trend*, and *data quality* (using drop-down menus) based on various point, line, and polygon data (see Figures 5.6 and 5.7). The GIS tool also supported various satellite imagery and background thematic data to support accurate classification of parcels of land (Figure 5.8).

The GIS data capture tool was used in a five-step process to classify land parcels according to the invasive species' *occurrence*, *abundance*, *distribution*, *trend in abundance*, and *data quality*.

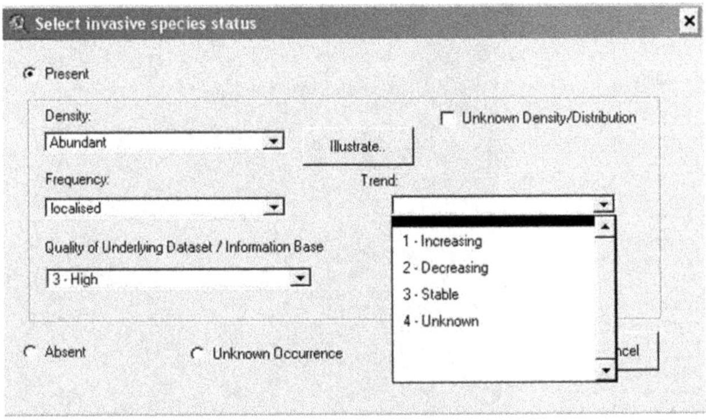

FIGURE 5.6 Image of ArcView Screen capture to select abundance (density), distribution (frequency), trend, and data quality for invasive species.

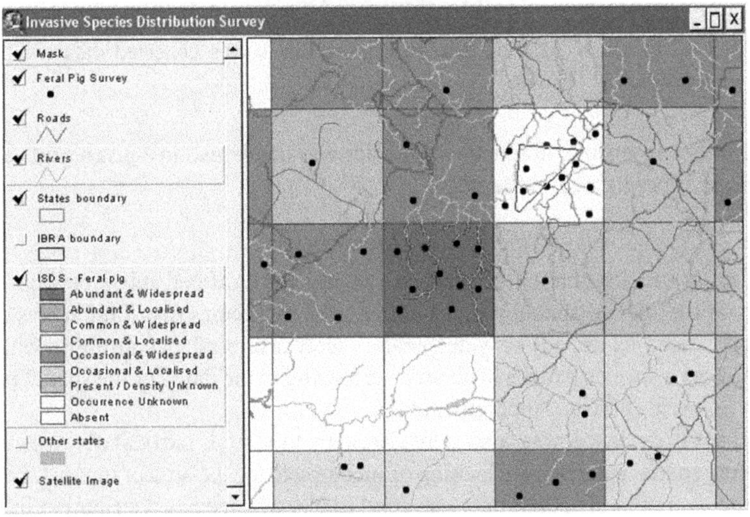

FIGURE 5.7 Image of ArcView Screen capture that shows collected data on overlay map.

5.7.4.2 Stepwise Data Collection and Collation

5.7.4.2.1 Step 1: Assessing Species Occurrence

To establish whether the species is present or absent in a given grid cell and the extent of a species, all available data layers from agencies were considered simultaneously.

Wherever a species was identified as occurring anywhere within a grid cell, the entire grid cell was assigned to the "present" class. Anecdotal evidence (expert judgment) was also considered in this step, particularly where data were lacking.

All grid cells were classed as having "unknown" status, until such time as presence or absence could be confirmed. However, given that it is often difficult to confirm if a species is absent from an area, one can assume more confidence in the presence classification for a grid cell than an absence classification.

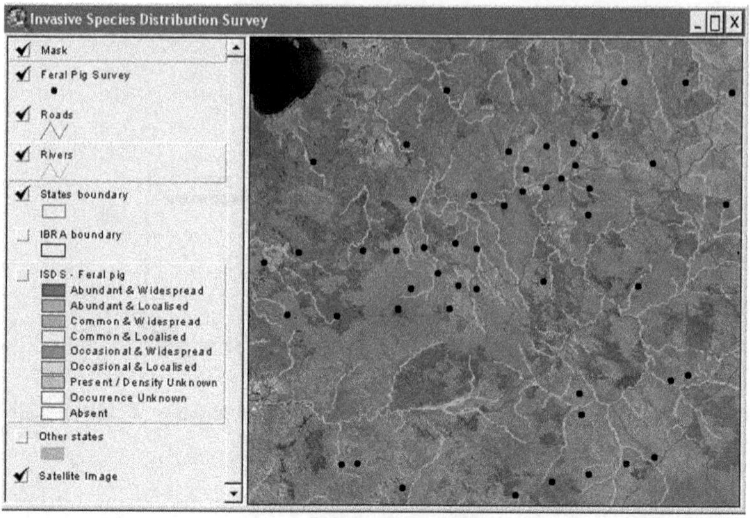

FIGURE 5.8 Image of ArcView Screen capture that shows collected data overlain on remote sensed image of the area.

When the presence or absence of a species was confirmed in a given grid cell, the GIS program progressed to Step 2.

5.7.4.2.2 Step 2: Classifying Species Distribution

Establishing the distribution of a species in a given grid cell relies on knowledge or data on the species spatial pattern in that area. A species distribution was classified as localized (clumped) or widespread (widely scattered) according to what was known about a species pattern in an area (according to the data), or the spatial pattern of data in a grid cell (see Appendix A.5.1).

Anecdotal evidence was also considered in this step, particularly where data describing spatial pattern were lacking or incomplete for the area of interest. Overall, a moderate amount of spatial data was used to classify species distribution in a grid cell. Where a species distribution could not be classified with sufficient confidence, a cell was assigned the category "Present, but distribution and abundance unknown."

When the distribution of a species was confirmed in a given grid cell, the GIS program progressed to Step 3.

5.7.4.2.3 Step 3: Classifying Species Abundance

Establishing the abundance of a species in a given grid cell requires more spatial data. A species abundance was classified as high (abundant), medium (common), or low (occasional) for each grid cell according to the known frequency of individuals in the grid cell.

The availability of data varied substantially between grid cells and anecdotal evidence was therefore also considered in this step. Some cells contained several sets of data, and other cells contained very little data. The availability of data also varied within grid cells, with some areas having detailed records and other areas

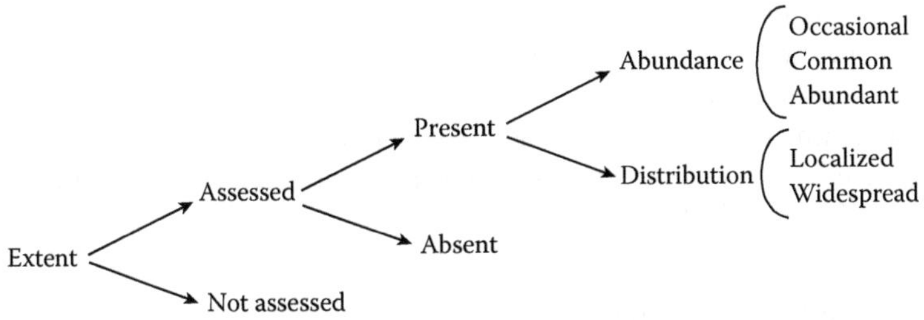

FIGURE 5.9 Stepwise procedure used to classify land parcels according to data attributes. Note: Does not show Step 4 or 5. (From National Land and Water Resources Audit and Invasive Animals Cooperative Research Centre, *Assessing Invasive Animals in Australia 2008*, NLWRA, Canberra, Australia, 2008. With permission.)

lacking data entirely. Anecdotal evidence was used to address these data gaps and was checked against reports from relevant datasets.

Overall, a large variety of spatial data were used to classify species abundance in a grid cell. Where abundance could not be classified with sufficient confidence, a cell was assigned the category "present, but distribution and abundance unknown."

At this stage, the extent of a species has been assessed or not assessed, values for abundance and distribution have been assigned in a given grid cell (Figure 5.9), and the GIS program progressed to Step 4.

5.7.4.2.4 Step 4: Assessing Trend in Species Abundance

Assessing the 5 year trend in abundance of a species relied on datasets that had been collected over time. Local knowledge from experts in the form of anecdotal evidence was also considered. Trend was classified as increasing, stable, decreasing, or unknown within each grid cell, based on data for the previous 5 years.

Where the trend in the abundance of a species was considered to vary within a grid cell, an average value was calculated for the entire cell, just as a satellite would average data across scanned areas on the ground. Future assessments of species extent and abundance will provide meaningful information on the trend of a species over time.

When the trend in abundance of a species had been classified for a given grid cell, the GIS program then progressed to Step 5.

5.7.4.2.5 Step 5: Classifying Data Quality

Grid cells were classified according to data quality based on the participants' evaluation of the quality, quantity, reliability, currency, and meaningfulness of the data for each respective grid cell. The quality of data was also reviewed by considering the purpose for which the original datasets were collected (as this can greatly influence data integrity). Cells were classified as having low, medium, or high data quality based on of the availability of anecdotal information, incidental reports, reliable expert knowledge, survey data, and scientific data (from field surveys, systematic sampling, or formal assessments). Cells recorded as containing data of high quality usually involved detailed scientific surveys for a species, where a range of data were collected using agreed monitoring protocols/techniques.

When the quality of data was assigned for a grid cell, the stepwise component of the GIS application was completed and the abundance of the species was displayed for further examination.

On completion of these five steps for entering data, data accuracy and reliability were checked by a series of expert teams who examined and reviewed each state-wide and national-scale dataset.

5.7.4.3 Data Consolidation

This procedure consolidates a large range of independent datasets into a single dataset. For pest animals, occurrence, distribution, and abundance information was compiled from the following sources:

- Field-based ground and aerial surveys using rigorous scientific sampling techniques
- Records of control and management activities in various registers and wildlife observation databases
- Historical records, verified as current
- The judgments of experts in wildlife and pest management
- Questionnaire surveys and reporting
- Anecdotal reports and subjective accounts

For weeds, information on the occurrence, distribution, and abundance of a species was compiled from the following sources:

- Data from field-based surveys
- Online data-reporting tools
- Herbarium records
- Observational databases
- Historical records of control
- Information on management activities
- Anecdotal evidence

5.7.4.4 Data Aggregation and Scaling-Up

Data collected and collated from the various regions and jurisdictions can vary considerably in spatial resolution. The generation of nationally consistent data required data to be scaled up or aggregated to a consistent scale of 0.5° (i.e., 50 km × 50 km), wherever state or territory jurisdictional datasets had been prepared at a lesser scale. The finest resolution of data made available for national reporting was property scale, followed by data at a 5 km grid, 1:25,000 and 1:50,000. The procedure described here permits the review of all datasets at a wide range of scales for consistent reporting.

A series of national data-aggregation rules were developed to scale up or aggregate (to the agreed national scale) spatial data collated from all jurisdictions. Operational rules were developed to aggregate data and a series of GIS routines customized to facilitate consistent and repeatable aggregation. All data were aggregated to the agreed national reporting scale of 0.5°. For a summary of the data aggregation rules, see National Land and Water Resources Audit (NLWRA) and Invasive Animals Cooperative Research Centre (IA CRC).[9]

5.7.4.5 Climate/Habitat Matching Methods

Climate matching is a method for determining the similarity of the climate between two locations. For invasive species, climate matching gives broad estimates of the climatic similarity between a species' current range (e.g., overseas) and potential range in a target area (e.g., in Australia).

The components of many climate matching models include species locations, meteorological stations and associated climatic variables, and an interpolated climate surface. Data describing known locations of species are analyzed with respect to climatic characteristics and locations with similar climatic characteristics plotted across the target area. Location data can be taken from a range of sources including observational data, specimen collections, or published information. Long-term global meteorological data from past climate records can be taken from various datasets.[36] Climate data are usually a set of 16 temperature and rainfall indices. The interpolated surface is generally a 0.5° resolution grid, but may range down to 0.04°. Some climate match approaches are statistically based, using classification matrices or algorithms calculating the difference in climatic variables between sites or species. A similarity index or percentile match is the output. Some systems require species' climatic tolerances to be defined by the user. Kriticos and Randal1[29] reviewed the more common systems in use and of relevance to weed risk assessment.

A climatic surface was generated from Australian meteorological station data using a procedure first tested and applied by Hutchinson and Kesteven[37] and discussed further in other publications, including Nix,[38] who applied the procedure to predict distributions of elapid snakes. A range of other climate surfaces have since been developed for Australia[37,39] with varying degrees of data availability and uncertainty.[40] Global climate surfaces have also been developed.[41,42]

The use of climate-matching models to predict the potential distribution of an invasive species based on its current worldwide distribution is well established; however, it has limitations. As with any model, generalizations and assumptions need to be defined at the outset, especially given the complexities of natural systems and the uncertainties usually associated with analyses of potential distribution and risk. Some of the limitations of climate matching are discussed in Section 5.8.3. Despite these limitations, climate matching is a useful predictor of establishment success, with recent studies having confirmed this for vertebrates[5] and birds[43] in Australia and introduced reptiles and amphibians in Britain, California, and Florida.[5,44]

The major steps to develop a climate-matching model are

- Data collection
- Generation of input data
- Calculation of matches
- Sensitivity analysis

Habitat suitability matching involves finding the environmental attributes that drive species' distribution such as watercourses, soil, or vegetation type. As a technique, habitat suitability matching is a relatively new area and requires further research. To date, its use in Australia has required complex analyses that have presented some difficulties, particularly given the generalist nature of most invasive species. There are also problems with multi-collinearity between habitat indices (e.g., there are interactions between vegetation and soil type) that can make it difficult to draw conclusions about distribution in response to discrete environmental attributes. However, when combined with climate matching, habitat matching may improve predictions of species potential distributions.

The inherent limitations of climate and habitat matching mean that output maps should be used as a broad indicator of potential distribution and not be assumed to accurately predict species' ultimate range.

5.7.4.6 CLIMATE Software

The CLIMATE software package (recently updated and released publicly as the "CLIMATCH" model—www.brs.gov.au/Climatch) matches the climates (temperature and rainfall data) of selected regions around the globe where a species is known to be present with the climate of other selected regions (the target site). The potential range of a species within the target site is represented in a series of maps. The desktop version of CLIMATE software uses the statistical program "R" and a Euclidean calculation to determine the closest matches between the overseas input data and the target site. The cane toad map presented in this chapter (Figure 5.13) uses match classes that summarize the output as high, medium, and low climate suitability.

5.7.4.7 Land-Use Classifications

Habitat data used in this method was derived from available land-use data, captured at the catchment scale, and developed through an Australian Government initiative—the Australian Land Use and Management (ALUM) classification. The ALUM classification provides nationally consistent land use information (see http://adl.brs.gov.au/mapserv/landuse/alum_classification.html).

The ALUM classification has six primary classes of land use that are distinguished in broadly decreasing order of potential impact on the natural landscape. They are

- Class 1—Conservation and Natural Environments
- Class 2—Production from Relatively Natural Environments
- Class 3—Production from Dryland Agriculture and Plantations
- Class 4—Production from Irrigated Agriculture and Plantations
- Class 5—Intensive Uses
- Class 6—Water

5.8 RESULTS

This chapter reports one approach used to map and monitor invasive species throughout Australia. It demonstrates that consistent and meaningful information can be collected and reported across administrative boundaries to inform decision makers and provide information to evaluate the effectiveness of management decisions.

5.8.1 Challenges for Large-Scale Mapping and Monitoring Efforts

Mapping and monitoring of invasive species at large landscape scales to produce state- and national-level datasets for invasive species presents challenges. These challenges include

1. Seeking agreement on species to monitor and report—Variation in state and territory legislation can make it difficult to agree on a single list of species requiring consistent monitoring and reporting. A "declared pest" in one state may not be a declared pest elsewhere.
2. Agreeing on method(s)—To promote comparability of data, it is important to adopt a consistent method for collecting and collating invasive species data. However, different circumstances in each state and territory jurisdiction may mean a single method is more suitable in one jurisdiction than another. Adopting a one-size-fits-all approach across all weed and pest animal species assumes the method is sufficiently suitable for all species.
3. Selecting an appropriate reporting scale—Consistency in reporting scale allows comparability of data across administrative boundaries. However, the suitability of the reporting scale may vary for different species or regions. Care is required to select the most appropriate scale for the species under consideration.
4. Uniform and consistent implementation of agreed methods across all jurisdictions—The effectiveness of the collection and collation of information from various regions may vary depending on variation in management budgets or the size of administrative regions. For instance, regions with more operational staff may be better able to more rigorously implement protocols for monitoring.
5. Collating discrete datasets that vary in currency, scale, and data accuracy—Where discrete datasets vary in the way they were collected, their currency, the scale at which they were collected, and the format in which they are presented, a series of operating rules or guidelines can be used to collate datasets into a uniform format. The development and agreement of these must proceed cautiously to maintain accuracy in resultant data layers.
6. Maintaining the quality and reliability of data at all scales—When large landscape scale data are collected or collated from a series of smaller datasets, maintenance of data quality and integrity can be difficult. Loss of information occurs where data (in the form of points, lines, or polygons) are aggregated into a grid-cell format. As a result, it is important to carefully select the grid-cell size most suited to representing a species and variation in their populations across space and time.

5.8.2 Outcomes of Australian Invasive Species Monitoring Efforts

Here we present some of the outcomes arising from the implementation of a consistent procedure to map and report invasive species throughout Australia. We provide examples of products developed from national datasets for invasive species.

5.8.2.1 Reporting Single Attribute Data

Data on invasive species can be reported at various scales and using combinations of attributes. Single attribute data layers have been created at a consistent reporting scale throughout Australia (Figure 5.10).

5.8.2.2 Multiple Attribute Maps

Numerous data attributes can also be presented for a single species. Figure 5.11 presents a greater range of information to managers and decision makers than maps of single attributes. Maps containing multiple attributes can combine a wide range of information types.

5.8.2.3 Reporting Multiple Species Data

A range of products can be prepared for multiple species against a range of reporting scales (Figure 5.12). These products can be used to develop and assess the effectiveness of policy, programs, or investment. Again, caution is required to ensure outputs are meaningful and easily interpreted.

5.8.2.4 Data Aggregation and Scaling-Up: Implications

Data available for invasive species varied in scale and format across the different state and territory jurisdictions. To produce national datasets in a consistent format, data from small units must be scaled up and aggregated to larger units (or from point and line to polygon). However, scaling-up of data has implications for interpretation and data accuracy.

The process of scaling-up or aggregation, results in a loss of information, because as the output dataset becomes larger in scale and incorporates more small-scale data points, the information that a single data point contributes to the classification of a large land parcel diminishes. For this reason, the appropriate output data scale should be selected with care to reflect local scale variation in data, and minimize the loss of fine-scale information.

The agreed national reporting scale of 0.5° was selected to permit national reporting on species for which detailed records were available, as well as those for which detailed data records were less consistently available.

5.8.2.5 Reporting Predictive Model Outputs Using Habitat and Climate Suitability

Datasets collated for each species also can be used to predict the species' potential range, based on data on climate and habitat suitability. For example, predictions for introduced cane toads (*Bufo marinus*) were derived from implementing national

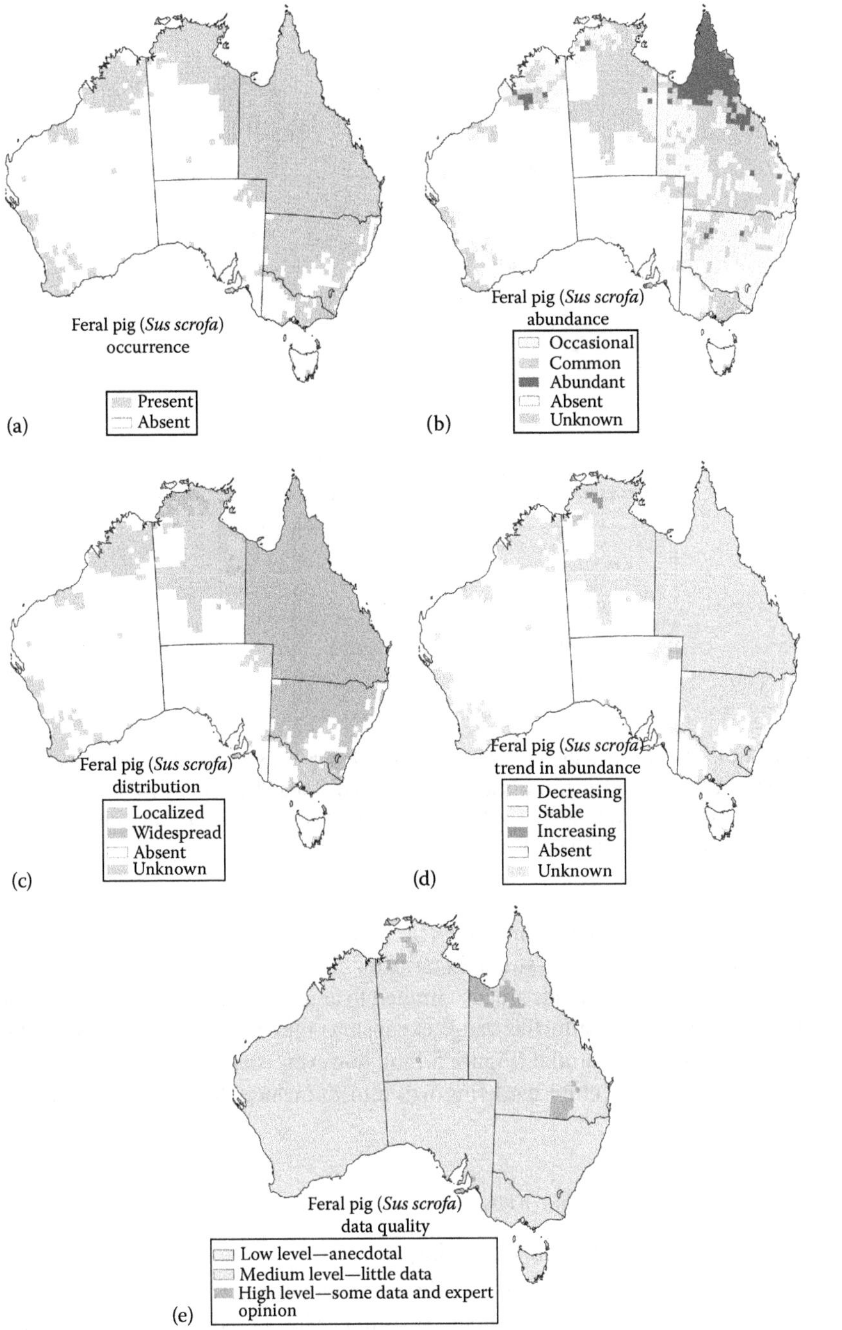

FIGURE 5.10 Example of national mapping outputs for feral pigs (*Sus scrofa*) (a) occurrence, (b) abundance, (c) distribution, (d) trend in abundance, and (e) data quality. (Modified from National Land and Water Resources Audit and Invasive Animals Cooperative Research Centre, *Assessing Invasive Animals in Australia 2008*, NLWRA, Canberra, Australia, 2008.)

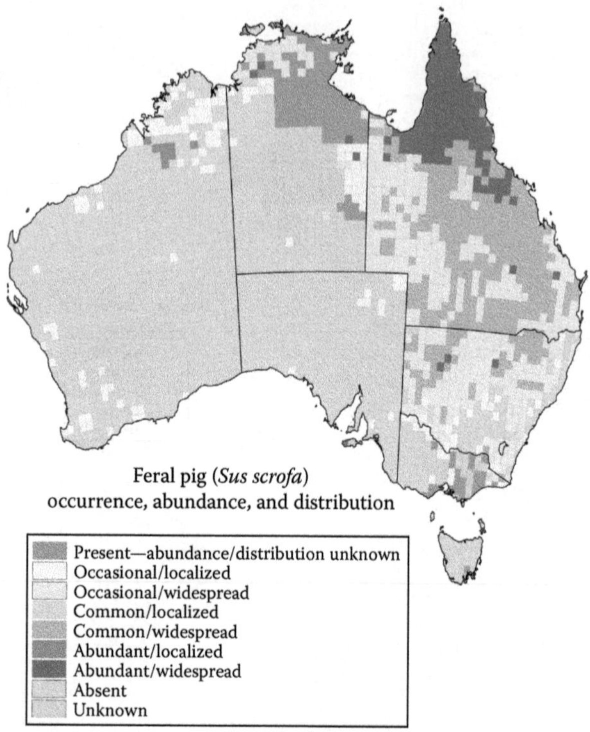

FIGURE 5.11 Example mapping output—National map of feral pig occurrence, abundance, and distribution. (Modified from National Land and Water Resources Audit and Invasive Animals Cooperative Research Centre, *Assessing Invasive Animals in Australia 2008*, NLWRA, Canberra, Australia, 2008. With permission.)

mapping and reporting procedures described herein (Figure 5.13a). Cane toads are mainly limited to northern Australia in terms of climate and to coastal areas in terms of habitat. However, cane toads are also limited to areas with access to perennial water. Model predictions suggest further range expansions may be possible in many areas of northern and eastern Australia (Figure 5.13b); however, water availability is a critical factor in this process. Future modeling will aim to include data of higher quality on water availability.

5.8.3 Limitations of Methods

5.8.3.1 Data Collation and Reporting

This chapter reports a method to rapidly assess the extent, distribution, and abundance of invasive species throughout Australia. The method efficiently and cost-effectively consolidates data on invasive species into a single format for developing national scale datasets and maps of species extent and abundance. The method uses multiple lines of evidence (consolidated data) to develop results. The validation of the results using robust, scientific sampling is recommended. To do this at a broad

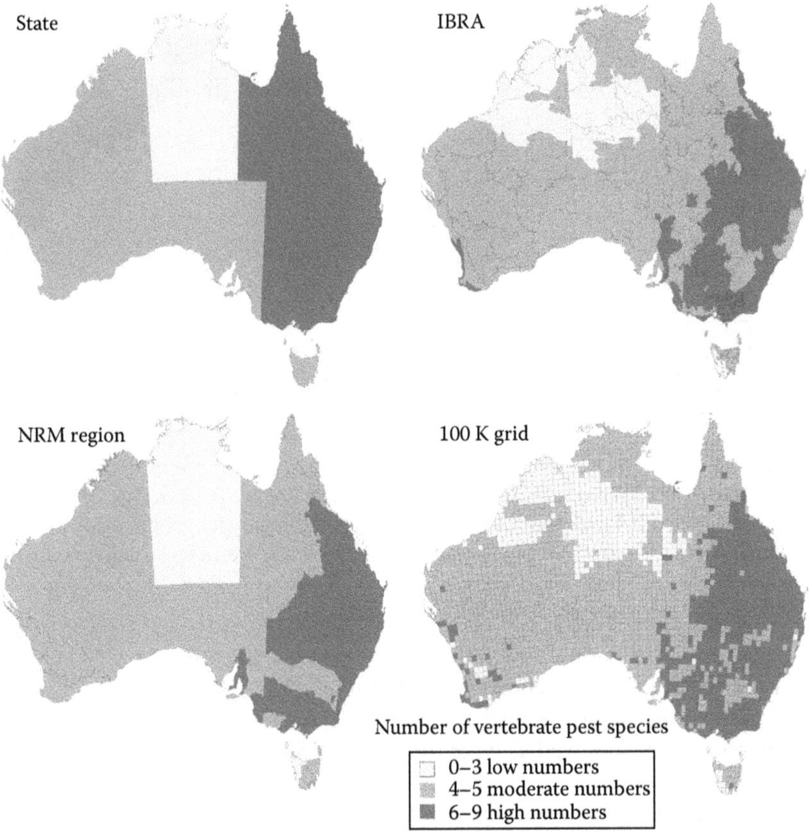

FIGURE 5.12 Example product for the number of pest animal species present in an area (of 10 assessed) mapped at state jurisdictional, national grid, and two regional scales (Modified from National Land and Water Resources Audit and Invasive Animals Cooperative Research Centre, *Significant Invasive Species (Vertebrate Pests)—Status of Natural Resources Information for Reporting against Indicators under the National Natural Resource Management Monitoring and Evaluation Framework*, NLWRA and IACRC, Canberra, Australia, 2008. With permission.)

scale would be costly. Instead validation may be strategic across broad regions with verification of data at selected areas throughout Australia, followed by an assessment of the accuracy of the datasets in reporting species occurrence, abundance, and distribution.

The information content of the data reported from this national assessment approach is influenced by five main characteristics, including the

- Suitability of the reporting format and resolution
- Currency and quality of underpinning datasets
- Appropriateness of the classes for differentiating invasive species abundance and distribution

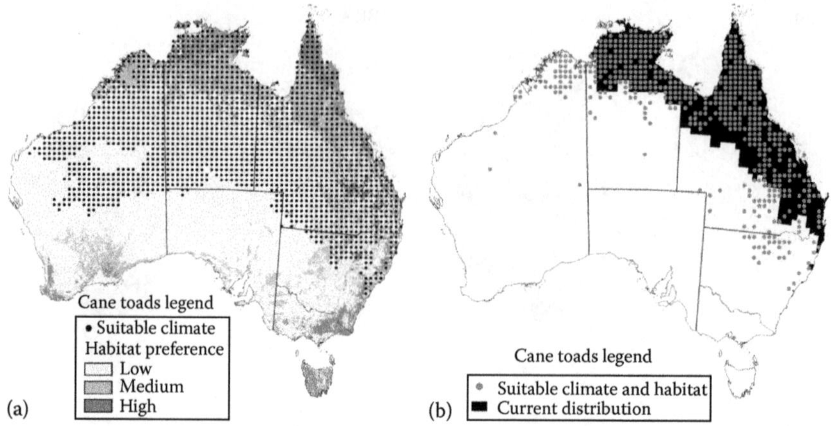

FIGURE 5.13 Example of climate prediction output derived from implementing national mapping and reporting procedures for introduced cane toads (*B. marinus*): (a) predicted suitable climate and habitat, and (b) predicted suitable climate and habitat against the current extent data (occurrence). (From National Land and Water Resources Audit and Invasive Animals Cooperative Research Centre, *Assessing Invasive Animals in Australia 2008*, NLWRA, Canberra, Australia, 2008. With permission.)

- Contribution of observer bias to final results
- Appropriateness of the method for all species simultaneously (including invasive grasses as well as large free-ranging feral herbivores)

The scale of reporting strongly influences the accuracy and precision of mapping and reporting. Here, we describe the use of a grid-based system using 1:100,000 map-sheet tiles to report on invasive species. However, this scale represents reporting units of approximately 2500 km^2 in size, in which there is usually significant variation in the population density and distribution of a species. Because finer resolution baseline data will more accurately reflect variation in a population at the local level, selecting the most suitable scale for reporting is critical.

Limiting the amount of anecdotal information is also important in maintaining data quality and accuracy. Where anecdotal information has guided the classifications of land units, perceptual bias can affect results. Observer bias results when one observer's understanding of an identical phenomenon differs from another observer's understanding. Observer bias has been reported to reduce the comparability of data based on surveys of staff knowledge.[21] Anecdotal information alone should therefore not be relied on when developing state and national level datasets.

5.8.3.2 Climate/Habitat Matching: Potential Distribution Prediction Models

Predictions of the potential distribution of invasive species using climate-matching programs (e.g., "CLIMATE" software) rely on the climatic conditions of current worldwide distribution of a species to predict its potential Australian distribution. This approach is well established—but has limitations. For example, many pest species are generalists and highly adaptable. Habitat-suitability matching has not been as fully explored, but may provide better predictions of a species' potential range, when combined with climate matching. Predictive maps based on climate and habitat suitability matching approaches should be used only as a broad indication of potential distribution.

Several factors may limit the accuracy of predictions of a species' potential distribution in Australia when based on climate matching. First, data on the species' broad geographic range are usually used as input and does not account for local presence or absence, which may be influenced by small-scale microclimates or other factors. Some species may only be capable of living within a more restricted microclimate not represented by the broad-scale inputs.

Second, the density of meteorological stations in the CLIMATE database determines how accurately the climate in a species' overseas or potential Australian range is measured. Where species have small overseas ranges, or occur in places where there are few meteorological stations, the CLIMATE software may not give an accurate measure of the climate in the species' current range.

Third, some of the meteorological data used in the current CLIMATE model are not current (up to 20 years old) and may not reflect significant shifts in climate over the past 20 years in many areas.

Fourth, a species may be capable of living within a broader range of climatic conditions than is indicated by its current geographic range. In other words, it may be restricted by other factors such as predation, competition, or geographic barriers. In such pests, climate matching may underestimate the areas with suitable climates in Australia.

Fifth, climate suitability is only one of a suite of factors that ultimately determine a species' ability to spread following their introduction to Australia. Other factors that can influence distribution include the availability of preferred habitat, food types, breeding sites, freedom from predators, competitors, and disease, and biological attributes of the species that allow rapid rates of population growth.[5,25,26,43] For example, land uses such as irrigation and the cultivation of introduced plants are likely to have a major influence on the suitability of a location for an introduced species.

In summary, climate matching alone cannot provide definitive predictions of a species potential Australian range. Predictions based on climate matching provide a broad indication of areas with suitable climate for a species and should always be considered in the context of other factors that influence distribution, including habitat suitability.

5.8.3.3 Habitat Matching Using Land Use Data

Habitat matching approaches also have limitations in predicting suitable areas for invasive species. These limitations include

- The use of broad definitions for land use classes—broad classes and variation within classes (e.g., "grasslands" may comprise a wide range of habitats ranging from alpine grasslands to tall tropical grasses) may lead to anomalous results

- Inconsistencies in land-use data—Current catchment-scale land-use data have been captured by different state jurisdictional agencies over different periods, and as a result are a mosaic of data across different years
- The large scale of data for analysis—The large reporting units used in the analysis result in large within-cell variation in outputs and results
- Lack of inclusion of species-specific requirements and temporal and spatial issues—For example, feral pigs require access to water every day
- Lack of inclusion of species requirements for multiple habitats or continuity of habitats to survive—Most species require a matrix of suitable habitats in close proximity to survive and breed
- Lack of inclusion of other factors that limit a species' ability to survive and/or spread—In addition to habitat and climate conditions, a wide range of factors may be limiting, including human activity, previous site history, and stochastic events, such as flood and fire

5.9 CONCLUSION

Consistent collection, collation, and reporting of invasive species information across all administrative jurisdictions is required to manage invasive species and their impacts at the national level. It is important to recognize that invasive species are part of our landscape and need to be managed in conjunction with other natural resources including land use, native vegetation, biodiversity, and water.

5.9.1 Reporting at the National Level

This chapter presents one method for monitoring and reporting of invasive species throughout Australia. This method captures data on the extent, relative abundance, and distribution of invasive species in a standard way. The method enables national overviews of the status of invasive species and can be used to guide government programs, policy, and investment. Importantly, developing and maintaining datasets across state jurisdictions increases the capacity of decision makers and managers to implement a coordinated cross-jurisdictional approach to invasive species management.

Decision makers and funding bodies require detailed and accurate information on invasive species. The usefulness and accuracy of national-scale products for invasive species management relies on the collection of data at local and regional levels. Promoting the use of consistent approaches to field monitoring for invasive species at local and regional levels will provide comparable data across regions and improve the quality and usefulness of data available at the state and national levels. Refining monitoring and reporting frameworks and managing invasive species data within a national information system will also lead to better quality information and more informed decision making on invasive species matters. In addition, trends in the populations of invasive species can be monitored and reported over time, allowing managers to identify where populations are spreading or contracting in relation to management effort.

5.9.2 Way Forward for Invasive Species Monitoring and Reporting in Australia

Current national-level information provides the first seamless and consistent dataset for nationally significant invasive species in Australia. However, further improvements can be made to national procedures and information to streamline the delivery of accurate and meaningful information to decision makers and land managers. These improvements include

- Reporting scale—Providing data and information at a higher resolution with more detail and accuracy.
- Monitoring protocols—Refining current monitoring and reporting methods for different species rather than adopting a "one-size-fits-all" approach. While maintaining consistency, this could include development of species-specific criteria and suitable reporting scales for different species/groups. For example, fine-scale reporting for species that usually have a small home range, and broad-scale reporting for species that occupy a wider home range.
- Data standards—Further developing data standards to ensure information and mapping products maintain accuracy and reliability at all scales.
- Species—Broadening the scope of monitoring and reporting activities to accommodate a greater range of invasive species and taxonomic groups.
- Reporting frequency—Monitoring and reporting at a suitable frequency to detect changes in populations in response to management actions.
- Verification—Recommended scientific field-based surveys and sampling should be used to verify existing species datasets. Verifying data may provide more detailed information for management and decision making, particularly where pre- and post-control information is needed at the local level.
- Improved predictive models—Improving current predictive models using up-to-date climate input data and information on species' preferred habitat throughout Australia.
- Information management—Developing a national information system or distributive system so that jurisdictional information can be periodically uploaded to deliver nationally consistent reporting with the prospect of adopting an Internet-based data collection and collation tool for future monitoring and reporting activities.

ACKNOWLEDGMENTS

This chapter was prepared in consultation with representatives from several organizations following a national initiative of the Australian Government with support from the Invasive Animals Cooperative Research Centre. The authors wish to show their appreciation to all Australian State and Territory Government representatives who have contributed to the national assessment process.

APPENDIX 5.A

5.A.1 Monitoring Protocol for Extent, Distribution, and Abundance of Invasive Species

5.A.1.1 Step 1 Species Occurrence

This criterion has the highest level of accuracy. The occurrence of a pest should be recorded as

- *Present*—Species exists in the defined area
- *Absent*—Species does not exist in the defined area
- *Unknown*—It is not known (or participants are unsure) whether the species exists

5.A.1.2 Step 2 Distribution: Spatial Pattern

When presence is confirmed, the distribution of the species (incursion or spread of a species) within the defined area can be recorded as

- *Localized*—Species occurs in a clumped pattern and occupies less than 50% of a cell
- *Widespread*—Species occurs in most areas and occupies greater than 50% of a cell

5.A.1.3 Step 3 Abundance: Relative Numbers

Abundance refers to the relative density of a species within an area and can be described as

- *Occasional or low*—Individuals spaced at wide intervals, or few or no sightings and/or little active evidence (e.g., very infrequent observations or evidence of individuals)
- *Common or medium*—A middle measure between occasional and abundant, or some individuals seen at almost any time and/or much sign of activity (e.g., frequent observations or evidence of individuals)
- *Abundant or high*—Infestations that have reached their full potential and provide little opportunity for additional individuals to survive in that area, or many individuals seen at any time and much sign of activity (e.g., very frequent observations or evidence of individuals)

5.A.1.4 Step 4 Trend

Change in abundance over time—Using anecdotal information where data trends cannot be obtained—recorded as

- *Increasing*—Populations have increased in abundance over previous 5 years
- *Stable*—Populations have remained stable in abundance over previous 5 years
- *Decreasing*—Populations have decreased in abundance over previous 5 years
- *Unknown*—No information available

Trend is particularly useful to measure change in populations over time, but this criterion has only a moderate level of accuracy.

5.A.1.5 Step 5 Data Quality

Data quality and reliability should be reported using the following classification:

- *No data*—No information about data quality
- *Low*—Anecdotal information from ad hoc sources and incidental reports; no reliable expert knowledge or survey data. Equates to "Low level—Anecdotal"
- *Medium*—Expert opinion from local specialists providing general knowledge based on observations and other sources, such as control activities. Equates to "Little data"
- *High*—Scientific data from recognized field sampling protocols, field surveys, herbarium records, systematic sampling or formal assessment. Equates to "Some data and expert opinion"

5.A.2 Classes for the Occurrence, Distribution, and Density Attributes for Pest Animals and Weeds (Modified from Queensland Government's Pest Survey Group)

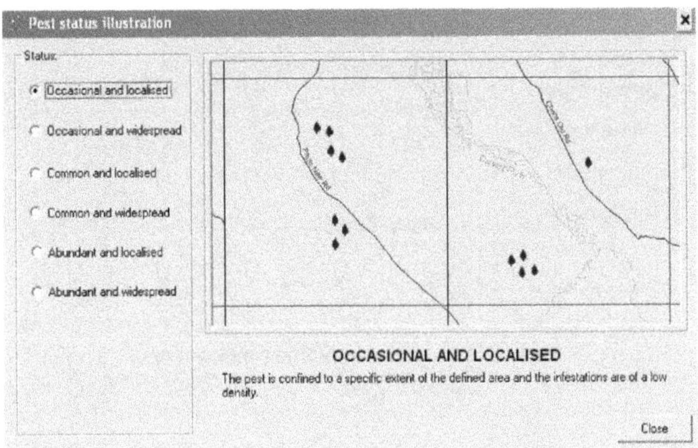

(a) Occasional and localised

(continued)

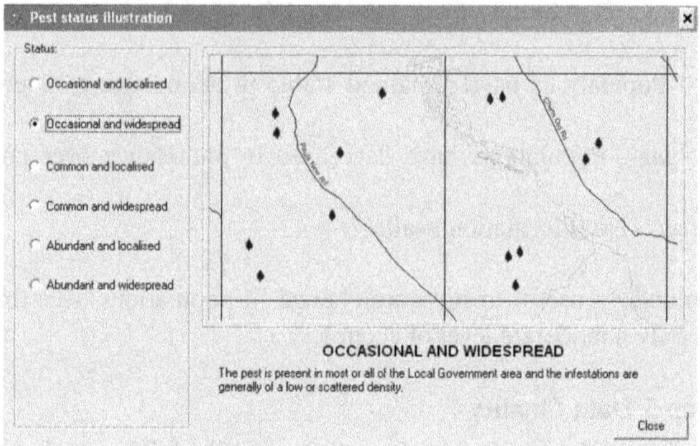

(b) Occasional and widespread

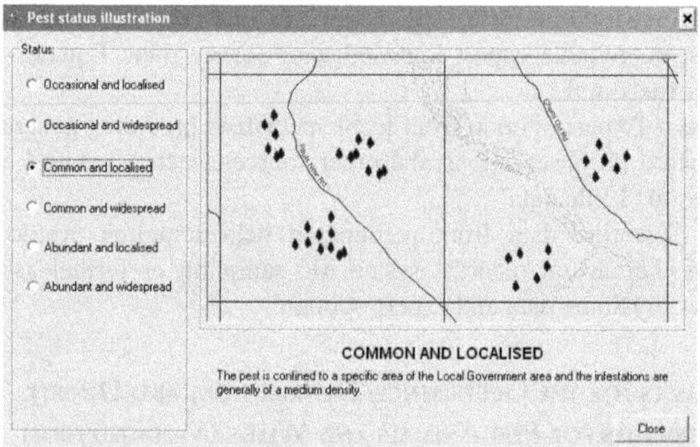

(c) Common and localised

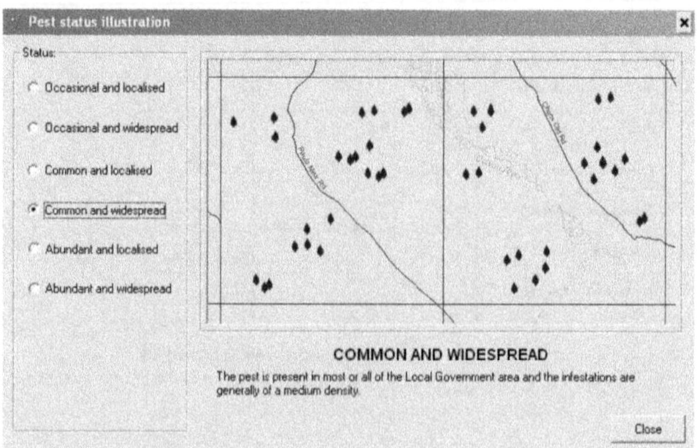

(d) Common and widespread

(continued)

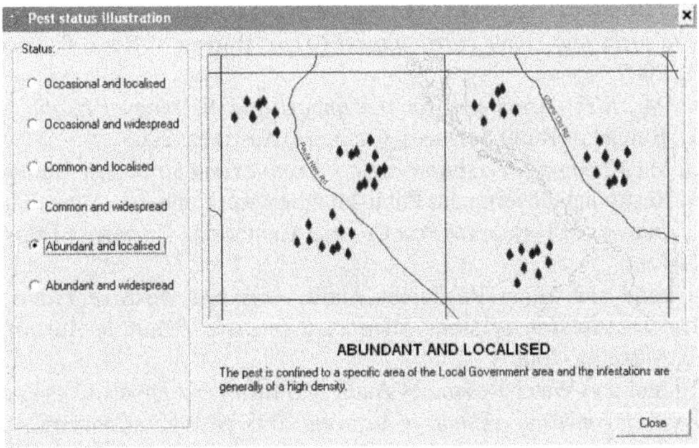

(e) Abundant and localised

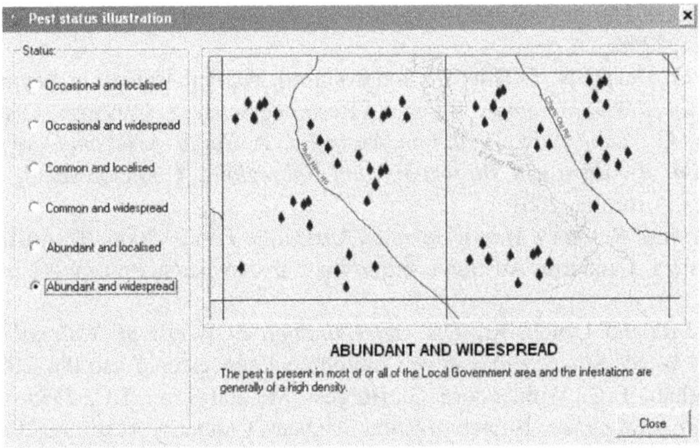

(f) Abundant and widespread

(continued)

(Screen dump images from Chen, T. and Calvert, M., Annual pest distribution survey application, accessed 2006. With permission.)

REFERENCES

1. Virtue, J.G., Bennett, S.J., and Randall, R.P., Plant introductions in Australia: How can we resolve "weedy" conflicts of interest? In: B.M. Sindel and S.B. Johnson (eds). *Proceedings of the 14th Australian Weeds Conference*. Weed Science Society of New South Wales, Sydney, Australia, pp. 42–48, 2004.
2. McLeod, R., Counting the cost: Impact of invasive animals in Australia, 2004. Cooperative Research Centre for Pest Animal Control, Canberra, Australia, 2004. http://www.feral.org.au/counting_the_cost_impact_of_invasive_animals_in_australia_2004/
3. Sinden, J., Jones, R., Hester, S., Odom, D., Kalisch, C., James, R., Cacho, O., and Griffith, G., The economic impact of weeds in Australia, *Plant Protection Quarterly*, 20, 25, 2005.

4. Tracey, J., Bomford, M., Hart, Q., Saunders, G., and Sinclair, R., *Managing Bird Damage to Fruit and Other Horticultural Crops*. Bureau of Rural Sciences, Canberra, Australia, 2007.
5. Bomford, M., *Risk Assessment for the Import and Keeping of Exotic Vertebrates in Australia*. Bureau of Rural Sciences, Canberra, Australia, 2003.
6. Braysher, M., *Managing Vertebrate Pests: Principles and Strategies*. Bureau of Resource Sciences. Australian Government Publishing Service, Canberra, Australia, 1993.
7. Hone, J., *Analysis of Vertebrate Pest Control*. Cambridge University Press, Cambridge, U.K., 1994.
8. National Land and Water Resources Audit, *Assessing Invasive Plants in Australia 2007—The Distribution of Some Significant Invasive Plants in Australia*. NLWRA, Canberra, Australia, 2008.
9. National Land and Water Resources Audit and Invasive Animals Cooperative Research Centre, *Assessing Invasive Animals in Australia 2008*. NLWRA, Canberra, Australia, 2008.
10. National Land & Water Resources Audit and Invasive Animals Cooperative Research Centre, *Significant Invasive Species (Vertebrate Pests)—Status of Natural Resources Information for Reporting against Indicators under the National Natural Resource Management Monitoring and Evaluation Framework*. NLWRA and IACRC, Canberra, Australia, 2008.
11. Wilson, G., Dexter, N., O'Brien, P., and Bomford, M., *Pest Animals in Australia: A Survey of Introduced Wild Mammals*. Bureau of Resource Sciences, Canberra, Australia, 1992.
12. Morgan, G., *Landscape Health in Australia: A Rapid Assessment of the Relative Condition of Australia's Bioregions and Subregions*. Commonwealth of Australia, Canberra, Australia, 2001.
13. ABS, *Natural Resource Management on Australian Farms, 2006–07*. Australian Bureau of Statistics, Canberra, Australia. http://www.abs.gov.au/ausstats/abs@.nsf/mf/4620.0, 2008.
14. Thorp, J.R. and Lynch, R., *The Determination of Weeds of National Significance*. National Weeds Strategy Executive Committee, Launceston, Australia, 2000.
15. Cunningham, D.C., Woldendorp, G., Burgess, M., and Barry, S.C., *Prioritising Sleeper Weeds for Eradication*. Bureau of Rural Sciences, Canberra, Australia, 2003.
16. DEHWA, Australia—Landscape Health Database 2001—Weeds. Website: http://www.environment.gov.au/metadataexplorer/full_metadata.jsp?docId=%7B1006420B-6937-4023-AC40-2B9FE68C19D9%7D&loggedIn=false, 2001.
17. National Land and Water Resources Audit, CSIRO, BRS Land & Water Australia. Assessment of catchment condition in Australia's intensive land use zone: a biophysical assessment at the national scale. Website: http://www.daffa.gov.au/_data/assets/pdf_file/0013/103801/catcon_report.pdf
18. ABARE, Farms with Significant Degradation Problems (Weeds) 1998–1999. Website: http://adl.brs.gov.au/anrdl/metadata_files/pa_fsdpwr9aa__02011a00.xml, Last updated 2009.
19. Thackway, R., Yapp, G., Cunningham, D., and McNaught, I., *Towards a National Set of Core Attributes for Mapping Weeds of National Significance*; draft discussion paper. Bureau of Rural Sciences, Canberra, Australia, 2003.
20. Thackway, R., McNaught, I., and Cunningham, D., *A National Set of Core Attributes for Surveying, Mapping and Monitoring Weeds of National Significance*. Bureau of Rural Sciences, Canberra, Australia, 2004.
21. Woolnough, A.P., West, P.B., and Saunders, G.R., Institutional knowledge as a tool for pest animal management, *Ecological Management and Restoration*, 5, 226, 2004.
22. West, P. and Saunders, G., *Pest Animal Survey. A Review of the Distribution, Impacts and Control of Invasive Animals throughout NSW and the ACT*. New South Wales Department of Primary Industries, Orange, Australia, 2007.

23. Edwards, G.P., Zeng, B., Saalfeld, W.K., Vaarzon-Morel, P., and McGregor, M. (eds). Managing the impacts of feral camels in Australia: A new way of doing business. DKCRC Report 47. Desert Knowledge Cooperative Research Centre, Alice Springs, Australia, 2008.
24. Lamb, D. and Saalfeld, W.K., A Multiple Criteria Decision Support Framework for the Management of Feral Camels, DKCRC Research Report 53. Desert Knowledge CRC, Alice Springs, Australia, 2008.
25. Duncan, R.P., Bomford, M., Forsyth, D.M., and Conibear, L., High predictability in introduction outcomes and the geographical range size of introduced Australian birds: A role for climate, *Journal of Animal Ecology*, 70, 621, 2001.
26. Forsyth, D.M., Duncan, R.P., Bomford, M., and Moore, G., Climatic suitability, life-history traits, introduction effort, and the establishment and spread of introduced mammals in Australia, *Conservation Biology*, 18, 557, 2004.
27. Cunningham, D.C., Brown, L., Woldendorp, G., and Bomford, M., *Science for Decision Makers: Managing a Menace of Agricultural Sleeper Weeds*. Bureau of Rural Sciences, Canberra, Australia, 2006.
28. Jürgen, G., Gommes, R., Cofield, S., and Bernardi, M., *New Gridded Maps of Koeppen's Climate Classification*. The Agromet Group, Environment and Natural Resources Service (SDRN) FAO of the UN, Viale delle Terme di Caracalla, Rome, Italy. http://www.fao.org/nr/climpag/globgrids/kc_classification_en.asp (modified July 2006).
29. Kriticos, D.J. and Randall, R., A comparison of systems to analyze potential weed distributions. In: R.H. Groves, J.G. Panetta, and J.G. Virtue (eds). *Weed Risk Assessment*. CSIRO, Collingwood, Australia, pp. 61–79, 2001.
30. Brown, L., Barry, S., Cunningham, D., and Bomford, M., Current practice in applying CLIMATE for weed risk assessment in Australia. In: *Proceedings of the 15th Australian Weeds Conference*, Adelaide, South Australia, p. 703, 2006.
31. McNaught, I., Thackway, R., Borwn, L., and Parsons, M., *A Field Manual for Surveying and Mapping Nationally Significant Weeds*. Bureau of Rural Sciences, Canberra, Australia, 2006.
32. Mitchell, B. and Balogh, S., *Monitoring Techniques for Vertebrate Pests: Series*. NSW Department of Primary Industries, Orange, NSW, Australia, 2007.
33. NLWRA, Distribution and abundance of significant invasive vertebrate pests. http://products.lwa.gov.au/files/PN21245.pdf, 2007.
34. NLWRA, Extent, density and distribution of weeds. http://nlwra.gov.au/products/pn21239, 2007.
35. ESRI, ArcView version 3.3. Environmental Systems Research Institute, 2002.
36. Global Ecosystems Database (GED), National Geophysical Data Center NOAA Satellite and Information Service. http://www.ngdc.noaa.gov/seg/fliers/se-2006.shtml, 2006.
37. Hutchinson, M. and Kesteven, *Monthly Mean Climate Surfaces for Australia*. Centre for Resource and Environmental Studies, Australian National University, Canberra, 1998.
38. Nix, H.A., A biogeographic analysis of Australian Elapid Snakes. In: R. Longmore (ed.). *Atlas of Elapid Snakes of Australia. Australian Flora and Fauna Series Number 7*. Australian Government Publishing Service, Canberra, Australia, pp. 4–15, 1986.
39. Rayner, D.P., Moodie, K.B., Beswick, A.R., Clarkson, N.M., and Hutchinson, R.L., *New Australian Daily Historical Climate Surfaces Using CLIMARC*. Queensland Department of Natural Resources, Mines and Energy, Queensland, Australia, 2004.
40. McBeth, B. and Roxburgh, S., Gridded Historical Monthly Climate Data for the Australian Continent: January 1990–December 2004. Technical Report. CRC for Greenhouse Accounting, Canberra, Australia, 2005.
41. New, M., Hulme, M., and Jones, P., Representing twentieth-century space-time climate variability. Part I: Development of a 1961–90 mean monthly terrestrial climatology, *Journal of Climate*, 12, 829, 1999.

42. New, M., Lister, D., Hulme, M., and Makin, I., A high-resolution data set of surface climate over global land areas, *Climate Research*, 21,1, 2002.
43. Duncan, R.P., The role of competition and introduction effort in the success of passeriform birds introduced to New Zealand, *The American Naturalist*, 149, 903, 1997.
44. Bomford, M., *Risk Assessment Models for Establishment of Exotic Vertebrates in Australia and New Zealand*. Invasive Animals Cooperative Research Centre, Canberra, Australia, 2008.
45. Chen, T. and Calvert, M., Annual pest distribution survey application (accessed 2006).

6 Use of GIS Applications to Combat the Threat of Emerging Virulent Wheat Stem Rust Races

David Hodson and Eddy DePauw

CONTENTS

6.1 Executive Summary 130
6.2 Introduction 130
6.3 Significance of the Ug99 Lineage (What Is Special about Ug99?) 131
6.3.1 Basic Biology of Ug99 131
6.3.2 Dispersal 132
6.3.3 Resistance Mechanisms and Virulence of Ug99 133
6.4 GIS Applications and Ug99 135
6.4.1 GIS-Based Surveillance and Monitoring Systems 135
6.4.2 Where Is Ug99?—Known Distribution and Range Expansion of Stem Rust (Ug99 Lineage) 136
6.4.3 Movements of Ug99 137
6.5 Deposition/Colonization Factors 143
6.5.1 Wheat Areas 144
6.5.2 Susceptibility of Wheat Cultivars 145
6.5.3 Crop Calendars/Crop Growth Stage 145
6.5.4 Climate/Environment 146
6.6 Information Tools 148
6.6.1 RustMapper 148
6.6.2 RustMapper Web 148
6.7 Challenges/Future Activities 149
6.8 Conclusion 153
References 154

6.1 EXECUTIVE SUMMARY

Historically, wheat stem rust (*Puccinia graminis* f.sp. *tritici*) was the most feared pathogen affecting wheat cultivation. For over 30 years, effective genetic resistance has kept the disease under control. The identification of new virulent races, typified by the Ug99 lineage, in East Africa during the late 1990s has made wheat stem rust once again a cause for global concern. The nature of the threat posed by new virulent races of wheat stem rust is outlined in this chapter. Using the Ug99 lineage of stem rust races as an example, the ways in which GIS/geospatial technologies are being applied to support the global efforts to combat this reemerging threat are described. GIS is playing a vital role because most of the critical factors involved with occurrence, movements, and establishment of the disease are inherently spatial in nature. International collaborative efforts are now underway to develop a global cereal rust monitoring and surveillance system, which is underpinned by GIS. Progress and challenges surrounding these international efforts are also described.

6.2 INTRODUCTION

Wheat is one of the world's principal food crops, ranking second only to rice in terms of global consumption. Globally, wheat is grown on over 200 million hectares from the equator to latitudes of 60°N–44°S and elevations ranging from sea level to over 3000m.[1] Total global production amounts to approximately 600 million tons, with developing countries accounting for nearly half of this total. Wheat accounts for a significant proportion of total calorie intake in several countries, notably in North Africa/Mediterranean, Middle East, and parts of Central Asia where annual per capita consumption rates can reach over 200kg.[2] Given the importance of wheat, control of economically damaging wheat diseases, such as rusts, has long been the focus of intense study.

Stem (or black) rust (*P. graminis* f.sp. *tritici*) is one of three fungal rust diseases that can inflict serious economic damage on wheat production. In recent years, the other rust pathogens of wheat, namely, leaf (or brown) rust (*P. triticina*) and stripe (or yellow) rust (*P. striiformis*), have caused more damage and, as a result, most of the research and breeding efforts worldwide have focused on these diseases. However, historically, stem rust has been the most feared disease of wheat, capable of causing periodic severe devastation across all continents and in most areas where wheat is grown. There is a solid foundation behind this fear as an apparently healthy crop only 3 weeks away from harvest could be reduced to nothing more than a tangle of black stems and shriveled grain by harvest. Under suitable conditions, yield losses of 70% or more are possible. Starting in the early 1900s, Midwestern States in the United States participated in the Barberry (*Berberis vulgaris*) Eradication Program to control common barberry, an important alternate host of *P. graminis*,[3] which helped mitigate rust spread. However, in the mid-1950s, over 40% of the North American spring wheat crop was lost to devastating stem rust epidemics.[4] These devastating losses were the result of the emergence of a new stem rust race named 15b, which overcame the genetic resistance in widely grown wheat cultivars at the time. The lack of epidemics in recent years, largely due to effective and durable genetic resistance and continued control of alternate hosts, has seen a shift away from stem rust research in terms of priority setting and resource allocation. This shift has been to such an extent that stem rust resistance does not feature in many breeding programs, and many wheat scientists have not even seen stem rust in the field.

As a consequence of the devastating stem rust epidemics in North America during the 1940s and 1950s, Nobel laureate Dr. N.E. Borlaug initiated his wheat improvement program in Mexico, with the development of stem rust resistant varieties being a primary goal. The resulting semidwarf wheat varieties that he developed, which underpinned the "Green Revolution" in the 1960s/1970s and subsequently became adopted on millions of hectares worldwide also had very effective resistance to stem rust. The widespread use of these resistant varieties was one of the principal factors that contributed to reduced stem rust inoculum levels worldwide.

Since the epidemics of the 1950s, the widespread use of resistant wheat cultivars worldwide has reduced the threat of stem rust to the extent that it is not a significant factor in wheat production losses. By the mid-1990s, stem rust was largely considered to be a disease under control.[5] However, with the emergence of a new virulent stem rust race lineage, popularly named Ug99, in the wheat fields of Uganda during 1998,[6] that perspective has now changed. As a result, stem rust is now very firmly back on the agenda of wheat scientists worldwide.

6.3 SIGNIFICANCE OF THE Ug99 LINEAGE (WHAT IS SPECIAL ABOUT Ug99?)

In order to understand the significance and threat posed by the new stem rust lineage, commonly termed Ug99, it is useful to consider some of the basic biology, epidemiology, and resistance mechanisms associated with wheat stem rust.

6.3.1 Basic Biology of Ug99

Stem rust, like all the wheat rusts, is a biotroph and needs a live primary host—principally wheat, barley, triticale, and related species—in the absence of alternative hosts for survival. The life cycle of the pathogen is complex as it is heteroecious, incorporating an asexual cycle entirely on primary hosts plus a sexual cycle that requires an alternate host (see http://www.ars.usda.gov/SP2UserFiles/ad_hoc/36400500Cerealrusts/PgLifecycle.jpg for a detailed life cycle). The main alternate host for stem rust is common barberry (Berberis vulgaris). The importance of the alternate host relates to its potential to act as an early-season inoculum source and/or a source for new combinations of genes and virulence. As a result, major eradication programs of barberry have been undertaken, for example, in North America.[3,7] These eradications have largely eliminated or reduced the importance of the alternate host as a source of inoculum although it should be noted that barberry remains common in parts of the Middle East and Central Asia. In most parts of the world, the life cycle of stem rust consists entirely of the continual asexual production of uredinal generations and reinfection solely on primary hosts.

Stem rust favors humid conditions and thrives in warmer temperatures (optimal range 15°C–35°C) compared to other wheat rusts.[5] Stem rust is an airborne pathogen, so disease spread occurs when spores are carried on the wind to new wheat plants in close proximity or in distant fields. An enormous number of urediniospores are produced by the pathogen as documented by one of the pioneers of rust research, E.C. Stakman, in 1957—"on an acre of moderately rusted wheat there are about 50 thousand billion urediniospores, each one capable of surviving a long air journey and starting an infection many miles from the place where it was produced."[8]

An enabling factor in this dispersion is the robust nature of rust spores ensuring protection against environmental damage and permitting them to remain viable for journeys of hundreds or even thousands of kilometers. Singh et al.[9] outlined in some detail the main dispersal mechanisms by which stem rust can spread, so only a brief summary is given in the following section.

6.3.2 Dispersal

Under normal conditions, the vast majority of rust spores will be deposited close to the source[10]; however, medium- to long-distance dispersal is well documented, and this generally occurs in one of three ways:

1. *Single event, extremely long-distance (up to several 1000 km) movements.* Movements of this type are rare, but several documented examples exist (see review by Brown and Hovmøller[11]) including at least one, and possibly up to three, wheat stem rust windborne introductions into Australia from southern/eastern Africa in the last 50 years. It is also important to note that rust spores can also travel long distances by assisted means—either on travelers clothing or on infected plant material. Despite strict phytosanitary regulations, increasing globalization and international air travel both increase the risk of pathogen spread. As an example, a race of wheat stripe rust was accidently transferred from Europe to Australia in 1979, almost certainly on travelers' clothing.[12]
2. *Stepwise range expansion.* This is more common than the previous dispersal mechanism and typically occurs over shorter distances, within country or region. Apart from very local movements, this probably represents the normal mode of dispersal for rust pathogens. A good example of this type of dispersal mechanism might include the stepwise spread of a *Yr9*-virulent race of stripe rust *P. striiformis* from Eastern Africa to South Asia over about 10 years in the 1990s and caused severe epidemics in its path.[13]
3. *Extinction and recolonization.* This occurs in areas that have unsuitable conditions for year-round survival of rusts, for example, in temperate areas with inhospitable winter conditions or seasonal absence of host plants. In North America, rust pathogens overwinter in the southern United States or Mexico and recolonize wheat areas further north following the prevailing south–north winds as the wheat crop matures. This seasonal flux has been termed the "*Puccinia* pathways"—a concept that arose from the pioneering work of Stakman and Harrar,[14] and a similar mechanism is now

being observed following the arrival of Asian soybean rust into the United States.[15] In China, a similar pattern has been observed for stripe rust. In the northern winter-wheat-growing areas of Shaanxi, Shanxi, Henan, Hebei, and Shangdong provinces, severe climate and absence of host plants preclude any pathogen survival during the winter months. However, further south in southern Gansu and northern Sichuan provinces, suitable conditions exist for year-round pathogen survival. These southern provinces then act as sources for recolonization movements into the main winter-wheat growing areas as the wheat crop develops in the summer/autumn.[11]

6.3.3 Resistance Mechanisms and Virulence of Ug99

Although fungicides may provide effective short-term control of stem rust, genetic resistance offers by far the most effective, affordable, and sustainable long-term control of the disease. Due to the devastating potential for crop damage, resistance to stem rust has been the subject of intense study for nearly a century. Currently, almost 50 stem rust resistance (*Sr*) genes have been cataloged[16] with several of these genes originating from close relatives of wheat. Alien introgressions from close relatives have made significant contributions to the control of stem rust worldwide over the last 50 years.

Of particular significance in combating stem rust has been the contribution of the *Sr2* gene and this gene in combination with other unknown minor genes, collectively termed the "*Sr2* Complex." The *Sr2* gene alone confers slow rusting resistance that is inadequate under heavy disease pressure, but effective when combined with the other minor genes in the complex. This *Sr2* complex was the primary basis behind resistance seen in the semidwarf Green Revolution wheat varieties developed by Dr. Borlaug that were subsequently adopted on millions of hectares in the 1960s/1970s.[17] This widespread adoption of semidwarf stem rust resistant varieties was one key factor in reducing the incidence of the disease worldwide. The early maturity of these varieties, hence avoiding the buildup of stem rust inoculum levels, was another important contributing factor.

As previously mentioned, another highly significant factor in the reduction of stem rust was the successful transfer of a series of alien genes into wheat from related species. Notably, these genes included *Sr24*, *Sr36* (from *Thinopyrum ponticum*), *Sr31* (from rye, *Serole cereale*), *Sr36* (from *T. timopheevi*), and *Sr38* (from *T. ventricosum*). Incorporation of these resistance genes by many breeding programs into many popular wheat cultivars worldwide during the 1970s and 1980s further reduced stem rust survival and populations. So successful were these global wheat improvement efforts that by the 1990s, stem rust had ceased to be a disease having any real significant negative impact on wheat production.[18] Despite the large-scale global deployment of the *Sr31* gene, used in most wheat improvement efforts with the exception of Australian wheat breeding, resistance remained unbroken until Ug99 was detected in Uganda in 1998.

Ug99 is the only known race of stem rust that has virulence for *Sr31*, a unique characteristic that facilitated its original identification. However, in addition, it also shows virulence to most of the stem rust resistance genes originating from wheat, plus virulence to gene *Sr38*, also of alien origin. This unique combination of virulence to both known and unknown resistance genes in wheat is what makes Ug99 special and why it is considered a potential major threat to global wheat production. Results from Kenya, now, show that the pathogen is continuing to change, resulting in variants that exhibit differing virulence, and render further *Sr* genes ineffective. Two additional new variants of Ug99 are now recognized from Kenya, all very closely related and thought to have arisen through single-step mutations.[19] These new variants have rendered additional important stem rust resistance genes ineffective, namely *Sr24* and *Sr36*. These new variants of Ug99 are not adequately differentiated by the existing standard North American nomenclature system for stem rust and hence have prompted a revision in scientific nomenclature. This has been achieved through the introduction of an additional set of wheat differentials specifically screening for the *Sr* genes of interest. The original "race Ug99," found in Uganda, was initially designated as TTKS using the standard North American system[20]; however, under the new expanded nomenclature system, the three recognized individual races within the Ug99 lineage are now termed: TTKSK (original "Ug99" formerly known as TTKS, i.e., virulent to *Sr31*), TTKST ("*Sr24* variant," i.e., virulent to *Sr31* and *Sr24*), and TTTSK ("*Sr36* variant," i.e., virulent to *Sr31* and *Sr36*).[19]

These new variants in the Ug99 lineage represent an increased level of threat to global wheat varieties. The *Sr24* gene was a valuable source of resistance worldwide, effective against most races of stem rust. This gene is present in many cultivars in South America, Australia, the United States, and CIMMYT germplasm. The *Sr24* variant of Ug99 (TTKST) was only detected in Kenya in 2006 and at low frequency,[21] but, by 2007, it had increased to such levels that major stem rust epidemics were observed on the popular Kenyan variety "Mwamba" covering approximately 30% of the Kenyan wheat area. At the time of writing, both the *Sr24* and *Sr36* variants of Ug99 have not been recorded outside of Kenya, but migration to other areas, as has been observed for the original TTKSK race, is considered to be virtually inevitable. The appearance of the *Sr24* variant of Ug99 is considered particularly significant as the additional breakdown of this *Sr* gene now halves the estimated number of current varieties previously considered resistant to Ug99.

The information above provides a clear indication of the dangers and consequences of reliance on single race-specific genes to control stem rust, especially in areas where the disease is endemic. Consequently, major breeding efforts (e.g., at CIMMYT and ICARDA) are focusing on durable resistance resulting from the combination of diverse sources of resistance and the accumulation of complex resistance from the combination of four to five minor resistance genes. Such approaches are already bearing fruit, and several high-yielding wheat lines having potential durable resistance to Ug99 have now been identified.[18] However, there is an inevitable time lag before any such promising new elite material emerging from breeding programs gets released and then adopted by farmers on significant scales. That leaves a major immediate unanswered question—just how much of the current global wheat acreage in farmer's fields is susceptible to Ug99?

Extensive screening of tens of thousands of wheat varieties from all over the world has been undertaken at key sites in Kenya and Ethiopia since 2005.

Singh et al.[9] and Jin and Singh[22] have provided summaries of this information, all of which highlight a very low frequency of resistant materials in wheat varieties originating from 22 countries. Initial estimates from Reynolds and Borlaug[23] considered that approximately 50 million ha (about 25% of the world wheat area) was susceptible and at risk from Ug99. Changes in the pathogen virulence, notably the breakdown of the *Sr24* gene, and increased information from screening trials have caused an upward revision of these estimates. It is now considered likely that approximately 80% or more of the current global wheat varieties are susceptible to stem rust.

Concerns surrounding the emergence of the Ug99 lineage of stem rust should now be apparent. This is a highly mobile pathogen, capable of devastating losses if conditions are suitable, and one that has overcome the genetic resistance possessed by a large proportion of the world's wheat cultivars.

6.4 GIS APPLICATIONS AND Ug99

The emergence of the Ug99 lineage of stem rust in East Africa has prompted a global and concerted effort by wheat scientists to try and mitigate the threat posed. Nobel laureate Dr. N.E. Borlaug was at the forefront of efforts to raise the alarm surrounding the potential threat of Ug99, convening an expert panel that published an assessment report in 2005.[24] Following on from the 2005 expert panel assessment, an international global consortium termed the Borlaug Global Rust Initiative (BGRI) (http://www.globalrust.org/) has been formed bringing together institutions interested in the mitigation of wheat rust diseases.

In the 2005 expert assessment, one of the key recommendations of the panel was that GIS could play an important role in global efforts to combat the reemerging threat of wheat stem rust. A direct quotation from the 2005 expert panel report states:

> **Recommendation #1.** Because the stem rust pathogen is airborne and genetically variable, the Panel **recommends** (1) population monitoring by means of trap nurseries and limited sampling for race analysis for the Kenya—Ethiopia region, adjacent areas, and beyond; (2) the establishment of a warning system based on the above data and modeling, using GIS and other appropriate tools.

This recommendation recognized the fact that many of the factors surrounding the threat posed by Ug99 were inherently spatial—distributions of various stem rust races, movements, risk zones, etc.—hence, GIS could be a valuable tool. The clear risk posed by the Ug99 lineage makes it a priority focus, but the overall goal is to develop a global cereal rust monitoring and surveillance system that has GIS technology as a fundamental component. This section now describes some of the advances that have been made in the use of GIS relating to Ug99.

6.4.1 GIS-Based Surveillance and Monitoring Systems

In order to provide effective information to decision makers on the current status and potential future threat of Ug99, several key components need to be integrated in order to form the basis of any effective information or monitoring system. Important elements are considered to include the following:

- Actual locations at which stem rust is present or absent
- Information on which race of stem rust is present at any location
- Location of important wheat areas and other key crops
- Stage of development of the wheat crop
- Susceptibility of wheat varieties being grown
- Likely or potential movement—direction and distance—of rust spores from known sources
- Climatic conditions to favor spore deposition and development or survival of the pathogen

In addition, an effective means to integrate and analyze the information in a timely manner and then communicate the results in an effective way are an absolute requirement.

GIS technology is now forming the backbone of an emerging rust monitoring and surveillance information system being developed collaboratively by two international agricultural research centers (CIMMYT and ICARDA), the UN Food and Agriculture Organization (FAO), several advanced research institutes, and a network of national partners. Although several challenges remain before a fully operational system is created, considerable advances have already been made that address many of the important components needed. The current status of these advances and remaining challenges are now described.

6.4.2 Where Is Ug99?—Known Distribution and Range Expansion of Stem Rust (Ug99 Lineage)

Since the initial discovery and identification of Ug99 (race TTKSK) in Uganda during 1998/1999, the known distribution has expanded considerably. By 2002, reports had been received covering all of the main wheat-growing areas in Kenya. Following the spread across Kenya, reports were received from Ethiopia, and by 2003 the presence of Ug99 (race TTKSK) had been confirmed from several locations dispersed across the main Ethiopian wheat belt. All of the initial reports originated from experimental research stations, where wheat scientists detected the presence of the new race due to infections on varieties known to carry the *Sr31* resistance gene. Despite the fact that no GPS data were recorded with these early reports, presence at established research sites with known coordinates permitted the subsequent retrospective incorporation into a spatial database and corresponding mapping. All of the available data compiled indicated that Ug99 (race TTKSK) had exhibited a gradual stepwise expansion across the highlands of East Africa following the predominant west–east airflows. At the end of 2005, the first preliminary assessment of the Ug99 situation was undertaken using GIS.[25] In this assessment, generalized regional monthly wind vectors[26] were used along with documented movements of another wheat rust pathogen that originated in East Africa in the late 1980s[13] and general climatic information. Results from this initial rapid assessment indicated the potential for rust spores to cross to the Arabian Peninsula and move across the Middle East and onward toward South Asia.

In 2006, reports of stem rust were received from a site close to New Halfa in eastern Sudan and subsequently from at least two sites in Western Yemen. Analysis of rust samples collected from these locations subsequently confirmed the presence of Ug99 (race TTKSK). The observed continued expansion of Ug99 was in line with the prior predictions arising from the initial GIS study and once again indicated stepwise movements following regional winds. Crossing of the Red Sea into Yemen was considered particularly significant as several lines of evidence indicated that this might prove to be a gateway for onward movement into important wheat areas of the Middle East and Asia. In 2008, confirmatory race analysis data were obtained from stem rust samples that had been collected at two sites in Iran—Borujerd and Hamadan—at the end (i.e., July) of the 2007 wheat season.[27] Stepwise expansion of the Ug99 (race TTKSK) range had continued and the pathogen had now reached major wheat-producing areas. In the monitoring of Ug99 to date, a fundamental and basic element of the GIS work has been the creation of a centralized spatial database containing confirmed locations of Ug99 and survey data, plus the subsequent production of regularly updated maps showing current distribution. Figure 6.1 shows current known locations.

Growing awareness of the potential threat posed by the Ug99 lineage of stem rust has attracted the attention of several major donors, resulting in successful projects being developed under the BGRI umbrella to counter the threat of Ug99. These projects are now permitting a more coordinated approach to tracking and monitoring the spread of Ug99. An important part of this emerging approach is the development of standardized field survey protocols, provision of GPS units, and capacity building for survey teams in priority countries. These ongoing activities are already resulting in a very significant increase in the amount of geo-referenced field survey data being collected for stem rust, with subsequent incorporation of the data into the existing centralized spatial database.

6.4.3 Movements of Ug99

Recorded known locations of Ug99 over time illustrate the mobility of the pathogen and highlight the possibilities for long-distance airborne transmission. Obviously, understanding and, if possible, predicting likely movements is a critical component of any monitoring system for stem rust. For Ug99, this is vitally important as there is a dual role. Firstly, there is an immediate need to understand potential onward movements of the original TTKSK race, for example, where next after Iran? Secondly, can we learn and apply useful knowledge gained from known TTKSK movements in respect to the new variants that might follow? Both the *Sr24* and *Sr36* variants are at present only known from Kenya, but will they stay there? Obviously, nothing about predicting stem rust, or other airborne pathogen movements, is easy, and any assumptions of "fixed repeatable pathways" must be approached with utmost caution as significant deviations may well occur. Airborne particle movement is a challenging area due to the inherent complexity and variability of the underlying system, so it

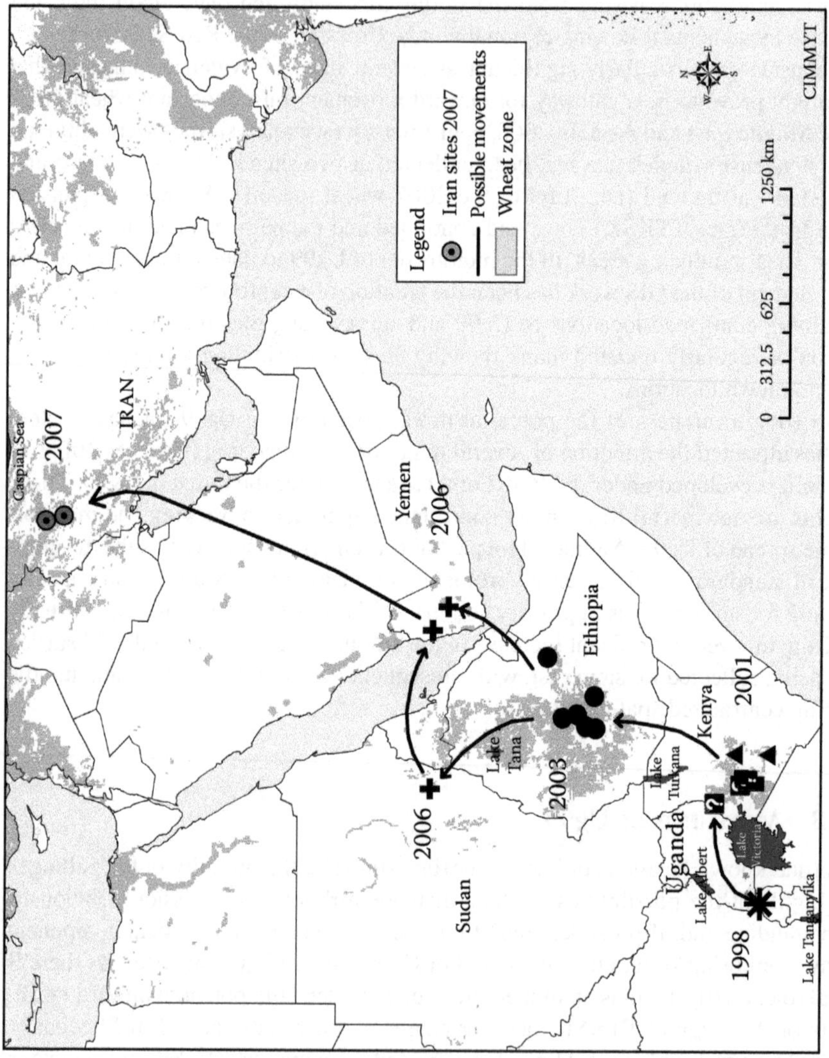

FIGURE 6.1 Current known distribution map of stem rust lineage Ug99.

must always be borne in mind that there are considerable uncertainties associated with any such pathogen prediction studies.

Initial attempts to understand the potential migration routes of Ug99 relied solely on generalized monthly "normal" wind trajectories.[26] Despite obvious limitations, these data, once integrated into a GIS and coupled with other data, did however provide some useful initial insights.[9] When Ug99 first appeared in East Africa, there was very little information available on where or how quickly it might spread but huge demand and interest for such information. Analysis of the general wind vectors indicated the possibility for movement from the Horn of Africa across to the Arabian Peninsula during May–September (the main wheat-growing season). If the crossing of the Red Sea did occur, then the combination of year-round wheat production and apparently suitable climatic conditions could favor pathogen survival in the coastal areas of the Arabian Peninsula, so that spores could be transported up the Peninsula on the prevailing north/north–west winds during November–February (main wheat season in this region). If any spores did move in this direction, then they would likely encounter the predominant west–east winds around the Mediterranean basin, which would facilitate movement into the Middle East and onward toward South Asia. There was also a remote possibility that the same wind systems that could move spores from the Horn of Africa across to the Arabian Peninsula might be capable of transporting spores all the way to southern Pakistan and western India. Based on available evidence at the end of 2005, two potential generalized migration routes for Ug99 were postulated. These are illustrated in Figure 6.2.

Actual recorded observations of Ug99, outlined in the previous section, obtained only after the first postulated migration routes were produced have been supportive, not contradictory, of the initial GIS-based predictions.

Based on these encouraging results, and given the obvious importance of airflows for the movement of stem rust, it was seen as high priority to improve the spatial and, more critically, the temporal resolution of the airflow predictions. This resulted in a search for suitable, accessible models and final implementation of the HYSPLIT (Hybrid Single-Particle Lagrangian Integrated Trajectory) model developed by the Air Resources Laboratory at NOAA and the Australian Meteorology Bureau.[28] This impressive public-domain model permits the generation of air parcel trajectories from user-defined source locations for specified start times, release heights, and durations. The HYSPLIT model also permits the generation of "back trajectories," which allows the exploration of potential source locations for, and routes to, known final destination sites. Critically important is that the model uses near-real-time data inputs and also permits access to a historical data archive. Another big advantage of the HYSPLIT model is that outputs are produced in GIS format; hence, integration with other spatial data is easily achieved.

Implementation of the HYSPLIT model using Ug99 data has resulted in an improved understanding of observed movements recorded to date and offers the potential to gain insights into potential future spread. Following the confirmed spread of Ug99 (race TTKSK) into both Sudan and Yemen during 2006, an analysis of trajectories from these sites during the main wheat-growing season was undertaken using data from December 2005 to April 2006 and December 2006 to April 2007. The aim was to determine if any consistent seasonal trends in trajectory patterns

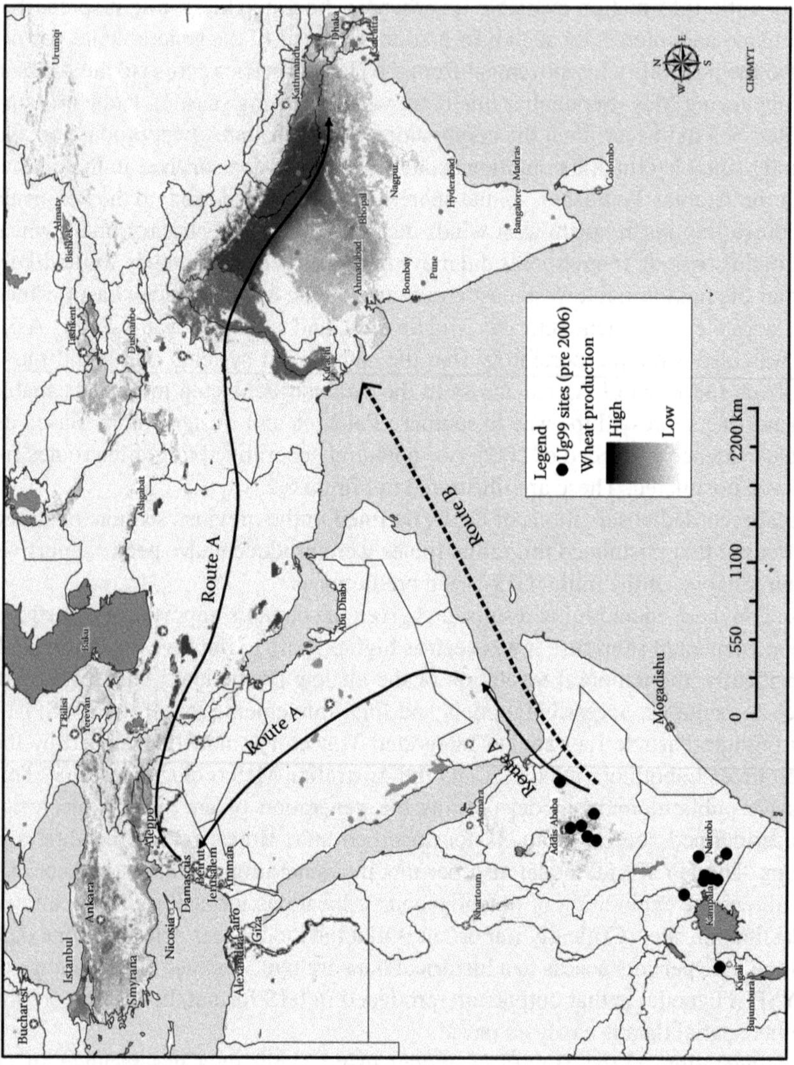

FIGURE 6.2 Initial migration routes prediction map. (After Singh, R.P. et al., *CAB Rev. Persp. Agric. Vet. Sci. Nutr. Nat. Resourc.*, 1, 1, 2006.)

were apparent that might be indicative of onward movements. Both seasons provided near-identical outcomes. The results obtained supported the previous hypothesis that crossing of the Red Sea was a significant event and that Yemen could act as a gateway to the important wheat areas of the Middle East and Asia. Trajectories from Yemen indicated the potential for spore movements in a predominantly northeasterly direction, across the Arabian Peninsula and toward Iran and Iraq. Trajectory model results implied the possibility for movements into both of those countries in 72 h or less. Conversely, trajectory data from the New Halfa site in eastern Sudan indicated a predominantly southwesterly direction for potential spore movements. Hence, two locations (New Halfa, Sudan and Al Kedan, Yemen) on equivalent latitudes and less than 850 km apart produced almost diametrically opposing airflow patterns during the main wheat-growing season. Summary trajectory patterns are illustrated in Figure 6.3.

The observed south-westerly trajectory pattern originating from New Halfa in Sudan was not totally unexpected, as generalized regional wind vector data had already indicated that the prevailing winds during this period were likely to be in a southerly direction moving up the Nile valley from the Mediterranean. This was one factor in the decision to exclude Egypt from the initial postulated potential migration routes (in Figure 6.2). The trajectory model results obtained from New Halfa added support to this original hypothesis. However, this does not in any way imply that the important wheat areas of Egypt are free of risk from infection by Ug99. Wind patterns around the Red Sea are complex; dust storms moving from Saudi Arabia across into Egypt have been captured by MODIS satellite imagery on a regular basis (e.g., http://www.redorbit.com/images/gallery/modis_moderate_resolution_imaging_spectroradiometer/dust_storm_across_the_red_sea/156/307/index.html). If dust can move in this direction, then rust spores could do the same. In addition, entry of Ug99 spores at the Nile delta following movement up the Arabian Red Sea coast and subsequent movement up the Nile valley at some point in the future cannot be ruled out.

Interestingly, the trajectory model data also indicated a direct connection between New Halfa in Sudan and Al Kedan in Yemen (results not shown). These trajectory paths occurred outside of the "main" wheat-growing season, but given the presence of wheat virtually year-round in Yemen, the possibility of a route across the Red Sea, originating in eastern Sudan, cannot be excluded.

The implications of the trajectory model using the Yemeni sites as sources are interesting. They clearly indicated the potential for rust movements across the Arabian Peninsula in the direction of Iran, rather than directly up the Peninsula in the direction of Jordan and Syria as had been implied by the original generalized NOAA wind vector data. After the trajectory model results had been obtained, reports from Iran indicated that stem rust—potentially Ug99—had been observed on the 2007 wheat crop. Final published race analysis in 2008[27] confirmed that Ug99 (race TTKSK) was indeed present at two sites, Borujerd and Hamadan, in western Iran in 2007. The two confirmed Iranian sites were not exactly in the trajectory paths, previously produced by the model, but in reasonable proximity—being no more than 500–600 km distant from the closest trajectory paths. Further analysis is currently ongoing to determine a more precise reconstruction of the possible route that Ug99 (race TTKSK) may have taken to arrive in Iran. However, initial indications are that

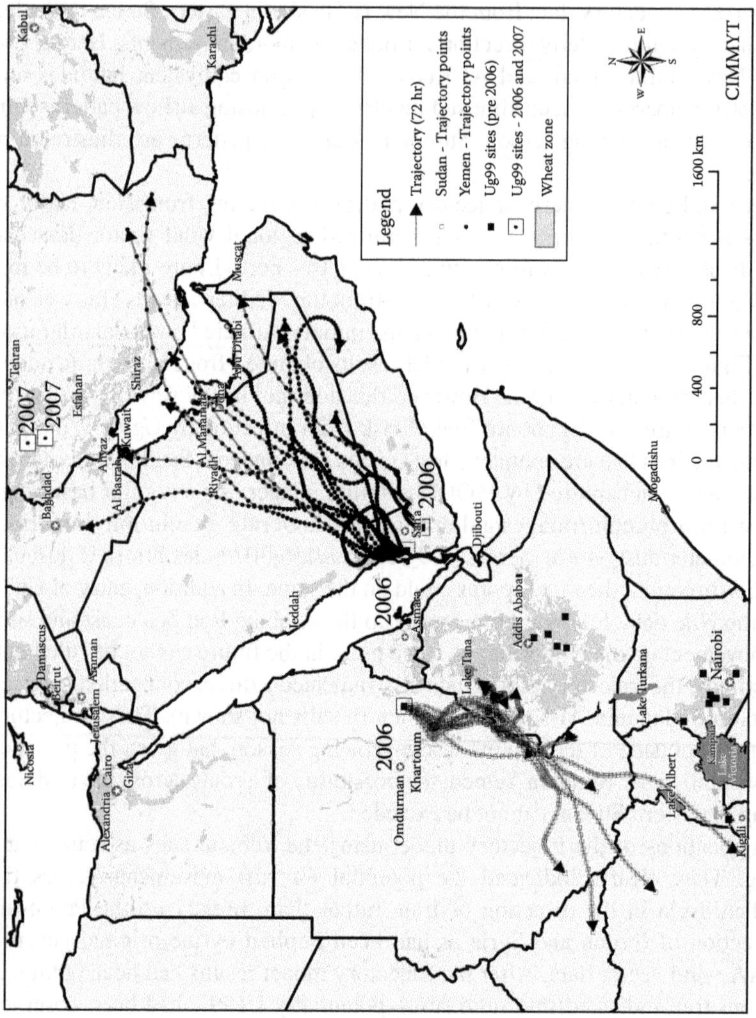

FIGURE 6.3 Selected wind trajectories (72 h) from the HYSPLIT model from December 2006 to March 2007 originating at confirmed Ug99 sites in Yemen and Sudan.

the Al Kedan area in Yemen was a possible source and that staging at a previously undetected and unreported intermediate location in southern Iran or Iraq occurred earlier in the 2007 crop season, prior to onward movement to the confirmed sites at Borujerd and Hamadan at the end of the crop season. In this reconstruction, the forward and back trajectory models of HYSPLIT are being combined with daily rainfall and climatic data within a GIS. Rainfall data are especially important, since rainfall events are one of the principal means by which rust spores contained within an air parcel are deposited into new areas, a process termed "rain scrubbing"[29] (see Section 6.5.4 for more details).

Experience to date with the HYSPLIT model and Ug99 data indicates that the model can produce valuable insights, and these increasingly appear to coincide with observations in the field. This implies that some advances might be possible in terms of forward predictions for Ug99. However, given the number of coincident factors needed in order for a rust infection to occur, many uncertainties will almost certainly remain. It also must be borne in mind that accidental, assisted movement, for example, on contaminated clothing or via infected plant material is always possible. At present, based on known information, this does not appear to have occurred with Ug99; however, detection or prediction of any such occurrence is beyond the scope of any model.

Another area of potential concern is that currently nothing/very little is known about the potential status of Ug99 in areas to the south of Kenya or Uganda in Africa. Although wheat is a minor crop in much of this region, with the notable exception of South Africa, the basis of concern is the region's potential as a source for onward movement (albeit at low probability) to either Australia or the Americas. Documented historical evidence indicates the potential threat is real. Nagarajan and Singh,[30] Brown and Hovmøller,[11] Isard et al.,[31] and Prospero et al.[32] all cite convincing examples supported by high-altitude balloon data, wind trajectory model data, and sample analysis of cross-continental rust or bacteria movements originating from southern or western Africa. The possibility of a similar rare event type movement involving Ug99 cannot be totally excluded; hence, vigilance and surveillance activities in southern Africa appear warranted.

Movement of rust, despite the obvious complexity, is only one factor important for effective monitoring, and many other elements also need consideration. In order for colonization of new areas to occur, several other factors apart from the potential for movement have to occur simultaneously in time and space.

6.5 DEPOSITION/COLONIZATION FACTORS

A number of factors influence the ability of stem rust to colonize new areas. These are in line with the classical "disease triangle" concept, with the three elements being susceptible plants, a suitable environment, and the presence of the pathogen—all of which have to occur simultaneously for the disease to occur. Without exception, all of them have a spatial dimension and are hence amenable for inclusion into a GIS-based analysis. Major factors currently being considered in the assessment of Ug99 include the following:

6.5.1 Wheat Areas

Stem rust is a biotroph, so it needs living green material in order to survive. Hence, a primary factor for the establishment of stem rust is the presence of suitable hosts. Crop species of primary importance include bread wheat, durum wheat, barley, and triticale, although it must be noted that many additional wild grass species can also serve as hosts.[5] In the existing spatial database that has been created for Ug99, efforts so far have focused on the development and incorporation of a data layer for major wheat production zones.

Despite the global importance of wheat as a primary cereal crop, surprisingly, few datasets exist in the public domain that provides a detailed representation of global or regional wheat distribution in a format suitable for GIS. Many OECD countries make available regular census data on production, area planted, and yield at a disaggregated level, for example, county level in the United States.[33] However, in many parts of the world, including those that are priority regions for Ug99, no equivalent data are readily available. This lack of knowledge relating to accurate distributions of major crops is seen as a major shortcoming, although a few initiatives are now trying to address this. The study undertaken by the SAGE group at the University of Wisconsin-Madison, reported by Leff et al.,[34] was one of the first efforts to describe global distributions of major crops at subnational scales. This work has continued, and a new updated version has just been released.[35] In both cases, it is commendable that actual GIS raster data have been made publicly available on the Internet (see http://www.sage.wisc.edu/pages/datamodels.html). The basic approach used by the SAGE group has been to use satellite imagery to determine crop zones and then combine that with available subnational agricultural census information to produce global crop distribution maps at a 5 arcmin resolution (approximately 9 km grid at the Equator). A very similar approach has been pursued by the spatial analysis group at the International Food Policy Research Institute (IFPRI) using a cross-entropy approach that triangulates and optimizes crop allocations across multiple sources of relevant information, including satellite imagery (to identify croplands), subnational agricultural census data, biophysical crop suitability assessments, and population density.[36,37] This methodology simultaneously allocates distributions of 20 major crops, including wheat, again producing global distributions at a 5 arcmin grid resolution. These raster-based crop distribution mapping efforts have been a huge advance, but problems remain, and the quality of the final outputs can be variable, that is, in some areas, the crop distributions appear accurate, but this is not the case in all areas. Major problems are seen in areas where satellite imagery is unable to accurately detect croplands, for example, in complex small-holder farming systems in Africa and other parts of the developing world. The recent release of the new high-resolution (300 m grid based on MODIS imagery) global landcover map "GlobCover"[38] might help resolve some of the problematic results observed in the current distributions, and at least the IFPRI group plan to evaluate this information in order to guide their future crop allocations.

At present, the wheat distribution information being used in the Ug99 analysis work is a composite—based primarily on the IFPRI dataset, but incorporating some of the SAGE data and also some expert assessments by CIMMYT wheat scientists.

Work is ongoing to try and improve the quality of this data layer, but despite noted limitations, the available data are considered to provide a reasonable initial indication of the major wheat-growing areas.

6.5.2 Susceptibility of Wheat Cultivars

Another key set of information, beyond knowing where wheat is being grown, relates to the susceptibility of wheat cultivars present in farmers' fields. Given the constantly changing nature of varieties, obtaining this information in a timely fashion and at a spatially disaggregated level is extremely challenging. At present, for Ug99, only country-level data, relating to varietal area estimates reported in 2002, have been obtained to date. Improving the timeliness and resolution of this information is a high priority.

Information obtained to date results from two sources: First, the extensive Ug99/stem rust screening nurseries that are being undertaken at key sites in Kenya (Njoro in the rift valley) and in Ethiopia (Kulumsa and Debre Zeit, again in the rift valley) by national agricultural programs, that is, KARI in Kenya and EIAR in Ethiopia. Intensive screening efforts started in 2005, and by 2007 tens of thousands of different wheat varieties, originating from 22 different countries, had been evaluated. Second, these extensive screening data have been linked via known pedigrees to estimates of areas planted to known varieties held in a CIMMYT database—these latter estimates being obtained from surveys of in-country wheat experts, based on their own personal experience and knowledge of the wheat varietal releases in country. All of the available information indicates that resistance to the Ug99 lineage of races occurs at a very low frequency, with up to 80% of the wheat varieties currently grown in farmers' fields being susceptible. In 2006, only an estimated 5% of a total estimated area of 75 m ha was thought to be planted with resistant varieties.[9] In some countries, for example, South Asia, extremely large areas of wheat are planted to popular, but highly susceptible cultivars, and there is a clear and urgent need to replace these with durable resistant varieties. Repeat surveys of wheat experts are planned as a priority, and it is hoped to obtain updated estimates on areas planted to specific varieties at the subnational level. This information will be vitally important in order to improve impact assessments for Ug99 and for entry into early warning or monitoring systems.

6.5.3 Crop Calendars/Crop Growth Stage

The growth stage of the crop is a critical factor in the establishment of the pathogen and subsequent infection. Living green tissue is a primary requirement for the pathogen, but local temperature and moisture conditions must be conducive for pathogen survival and infection. Obviously, if no susceptible crop is in the ground, no infection is possible, although "green bridges," that is, a year-round wheat crop (as occurs in east Africa and Yemen), off-season volunteer plants, and presence of alternative hosts all play a vital role in the provision of inoculum sources for infection once the main wheat crop reaches a susceptible stage. Stem rust is more important late in the growing period, primarily as a functional requirement for warmer temperatures compared to other rusts, that is, in many areas, optimal temperatures only occur late in the growing season. A classic example of this is the "Puccinia pathway" of North America in which stem rust finds suitable year-round survival conditions only in the extreme south of the United States or Mexico. Then, a northward expansion is observed following prevailing winds as the wheat crop matures and temperatures become optimal as the season progresses. Hence, in northern U.S. states like North Dakota, optimal stem rust temperatures occur only during later maturity stages late in the season, and in Manitoba, Canada, stem rust would only start to appear late June to mid-July.[39] As a result, any factors that prolong the growing season, for example, late-sown crops or slow-maturing varieties, increase the risk of infection by stem rust. The optimal timing of field surveys for stem rust is to coincide with approximate heading of the wheat crop.

General crop calendars based on recorded/known or estimated planting, heading, and harvesting dates provide important information for planning national field surveys and also serve as indicators when major production areas are likely to be at risk. At present, expert local knowledge combined with information held in international wheat trial databases are being compiled to provide broad indications of when specific wheat-producing areas are likely to be at increased risk of infection by stem rust. This information will be incorporated as spatial layers in the geo-database.

In the future, it may be possible to refine the growth stage estimates using a model-based approach. The Foreign Agricultural Service (FAS) division of USDA has already implemented a winter wheat growth model based on growing degree days.[40] A similar approach, but using crop simulation models, has been implemented as part of the Pest Information Platform for Extension and Education (PIPE) developed for Asian Soybean Rust monitoring in the United States by USDA APHIS, Penn State University and ZedX Inc. (see http://sbr.ipmpipe.org/cgi-bin/sbr/public.cgi).

6.5.4 Climate/Environment

Climate plays a major role in the deposition and subsequent increase of pathogen populations. The role of winds in the movement of the pathogen has already been outlined in some detail in previous sections, but several additional factors are also important.

In order for spore deposition to occur, there are two principal mechanisms: dry deposition and wet deposition. Dry deposition, where the spores simply land following sedimentation and gravitational pull after being blown by the wind, is an important mechanism for local short-distance movements, but potentially less important for long-distance movements. Wet deposition, or rain scrubbing, is the principal mechanism for deposition over longer distances. Rainfall is a key factor in bringing rust spores down to earth and can provide the required moisture needed for infection to occur. However, intense rainfall events may wash spores off the plant material.[41] As little as 2 mm of rain can be effective in removing spores from the air.[42]

Moisture and temperature both play a vital role in spore germination, infection, and disease development. Spores that land on a suitable host germinate 1–3 h after contact with free moisture and require 6–8 h of dew or free moisture (e.g., from rain or irrigation) to complete infection.[5] Optimal infection conditions occur when nighttime dew lasts for 8–12 h, nighttime temperatures are around 18°C, the plant and pathogen are exposed to daylight, dew dries slowly, and daytime temperatures rise to 30°C.[43] In general, stem rust is favored by hot days (25°C–30°C) and mild nights (15°C–20°C) with adequate moisture for nighttime dews. Despite the fact that stem rust favors warmer conditions compared to other rusts, it should be noted that a wide range of temperatures are tolerated. Roelfs et al.[5] report minimum, optimum, and maximum temperatures for urediniospore germination as 2°C, 15°C–24°C, and 30°C; and for sporulation 5°C, 30°C, and 40°C. Within 10–15 days, a typical asexual cycle is completed, and fungal pustules burst to release millions more urediniospores into the atmosphere. Spore release is favored by warm dry days with low humidity.

At present, only generalized long-term normal climatic data have been incorporated into assessments for Ug99. Hodson et al.[25] used monthly long-term normal temperature and relative humidity data to outline areas with potential for year-round survival of stem rust. Incorporation of daily weather data, such as the precipitation estimates originating from the Climate Prediction Center Morphing Technique (CMORPH)[44] and delivered through the IRI/LDEO climate data library at Columbia University as a component of the Desert Locust monitoring program at FAO (see http://ingrid.ldeo.columbia.edu/maproom/.Food_Security/.Locusts/.Regional/.Rainfall/), is already being initiated for Ug99 work.

Despite documented environmental conditions that favor stem rust development and increasing access to relevant datasets, it is extremely difficult to predict exactly when or where an epidemic may occur. No reliable climatic predictor of stem rust epidemics has been documented to date. Presence of the pathogen and seemingly favorable conditions on the ground may still not result in a disease epidemic. Singh et al.[18] succinctly outlined some of difficulties and the complex interactions between time, pathogen, host, and environment. Observed outbreaks of stem rust associated with the Ug99 lineage in East Africa are illustrative of the complexity. Ug99, race TTKSK, has been present in both Kenya and Ethiopia for over 5 years, yet no major losses to the wheat crop have been reported from this race despite the presence of susceptible hosts. In contrast, in 2007, Kenya experienced major stem rust epidemics resulting from the new *Sr24* variant of Ug99, race TTKST—only 1 year after the race was first detected. Although the exact triggers of the 2007 Kenyan epidemics remain unknown, the specific combination in time and space of several factors is likely to have been important. Changing virulence patterns of the pathogen and large areas planted to a susceptible cultivar (Mwamba) were important; in addition, higher-than-normal rainfall proceeding and during the main wheat season was probably favorable for stem rust. Additional unrecorded microclimatic factors will almost certainly have also played a role. The overall favorable environmental conditions are assumed to have resulted in an inoculum buildup in the off-season and subsequent early infection of the main wheat crop. Despite similar general climatic conditions in Ethiopia at the same time, no epidemics were observed.

Although major climatic factors, for example, above-average rainfall at critical times, have no absolute predictive power for epidemics, the generally favorable conditions they may create are undoubtedly a potential influence. It is not unrealistic to speculate, therefore, that the extensive and detrimental regional drought in the Middle East during 2008[45] may have some influence on reducing inoculum levels of stem rust in drought-affected rainfed wheat areas.

6.6 INFORMATION TOOLS

All of the preceding sections have outlined some of the key spatial datasets and models that are being used in the work on the Ug99 lineage of stem rust. However, to be useful, all this information needs to be integrated and made available in a timely and targeted manner. GIS/geographic-based tools are proving to be extremely valuable for this task. Several tools have already been developed that attempt to present synthesized information, with other products planned in the near future. Publicly available tools are briefly described in the following sections.

6.6.1 RustMapper

Google Earth is one of the most widely known and used virtual globes, with over 350 million downloads claimed by Google (http://google-latlong.blogspot.com/2008/02/truly-global.html). Google Earth provides excellent neo-geographic visualization capacity and offers many opportunities for customization using the KML scripting language. For these reasons, Google Earth was chosen as the platform for an initial information tool named RustMapper developed by the GIS unit at CIMMYT. RustMapper is publicly available at http://www.cimmyt.org/gis/RustMapper/index.htm as a KMZ download file for Google Earth; it is a networked link, so automatically updates after download. The idea behind RustMapper is to provide synthesized information regarding the current status of Ug99 with clear visualization. Key information incorporated into RustMapper includes all known sites for Ug99 or variants and recent survey sites—indicating the presence or absence of stem rust, near-real-time wind trajectories from the HYSPLIT model originating from known Ug99 sites or sites recording stem rust (these trajectories are run for 24, 48, 72 h durations and updated every 5 days), and major wheat-growing areas in Africa and Asia (see Figure 6.4). In addition, country-level summary information is provided on susceptibility estimates to Ug99 and basic wheat production statistics. A complete archive of wind trajectories back to April 2007 is also included. Any new information that is obtained and cleared for public release, for example, sites, wind trajectories, etc., is automatically incorporated into RustMapper.

6.6.2 RustMapper Web

Following the release in 2008 by Google of the free "Google Earth Plug In," options to embed Google Earth within a web browser were created. Based on positive reactions to RustMapper, it was decided to implement the Google Earth Plug In to broaden the range of access options. Key components of RustMapper were migrated into a browser-based tool. As a result, RustMapper Web is a derived "lite" version of the original RustMapper running within a browser environment. Like RustMapper, it is publicly available and updated every 5 days (see http://www.cimmyt.org/gis/

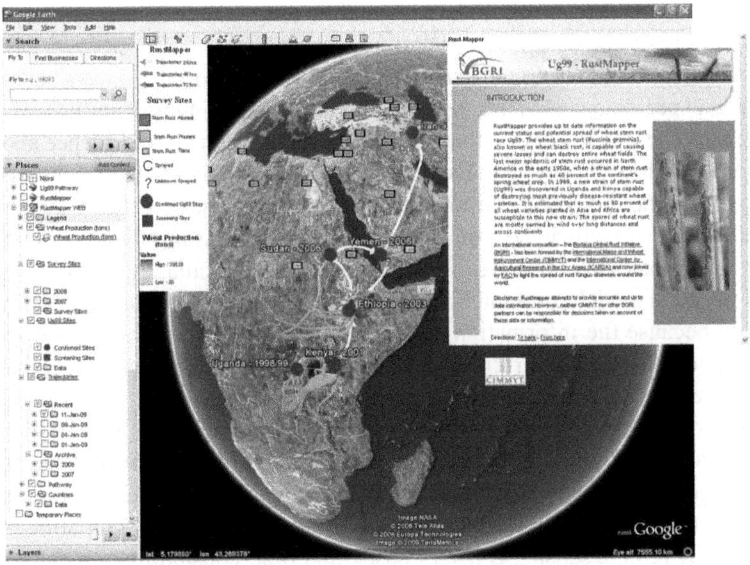

FIGURE 6.4 RustMapper—a Google Earth–based information tool.

rustmapper/RustMapper_Web.html). All of the primary components of RustMapper are included in the web version, although only the most recent wind trajectories are presented. RustMapper Web now functions on all major browsers on both Windows and Mac.

The above examples illustrate how geography-based visualization platforms are being used to present integrated information relating to the Ug99 lineage of stem rust in a timely manner. Despite the progress that has been made, further improvements and advances are still needed if the goal of a fully operational monitoring and surveillance system is to be achieved.

6.7 CHALLENGES/FUTURE ACTIVITIES

One major challenge is the determination of host crop zones, and this is a priority area for future work. As explained earlier, GIS-based surveillance and monitoring systems for stem rust need to address not only the challenge of tracking the pathogens but also the uncertainty about the presence of suitable host crops in the target region. Whereas the broad wheat regions are well known, the diversity in ecological conditions and farming systems makes it difficult to assess at a fine resolution where the wheat is actually grown and at what growth stage it is. Such uncertainty about the areas at risk makes it difficult to target rust surveys or cost-effective spraying campaigns.

Without expensive crop distribution surveys, involving farming systems research, agricultural surveys, remote sensing, and field validation, it is indeed difficult to identify crop zones in more than a sketchy pattern. Yet there is an affordable approach that allows compiling a base of "circumstantial evidence" that may point to the likelihood that a given crop occurs within a certain area. In GIS, it is possible to adopt such a method, similar to a criminal investigation, which could lead to the most likely crop to be designated as "prime suspect."

The application of Bayesian statistics in GIS, using "evidence layers" combined with local expert knowledge, has much potential for mapping probabilities of wheat occurrence. The way it works is that each GIS layer brings a bit of evidence about the likelihood of wheat occurrence. Each layer in isolation may not be conclusive in its own right, but the integration through expert knowledge and statistics may present a compelling case for either accepting or rejecting the hypothesis of wheat occurrence.

Obvious choices for evidence layers would be, for example, climate surfaces, digital elevation models, and satellite imagery. These are prime candidates, first and foremost, because the information they contain is of direct relevance to decide on the likelihood of wheat occurrence. Gridded data on the distribution of precipitation, temperature, and humidity indicate whether an area is not too wet, too dry, too cold, too warm, or too humid to grow wheat. A digital elevation model indicates whether an area is too steep, and satellite imagery whether an area is cultivated or not, and, if the images are taken at the right time of the year, whether the crops are grown under either rainfed or irrigated conditions. However, direct differentiation of specific crops by remote sensing is a very costly and time-consuming exercise and can realistically not be undertaken for the huge target region.

Another good reason for taking this approach seriously is that, owing to a huge recent advance in accessibility of geospatial technology, this information is now available free of charge for most parts of the globe to anyone with an Internet connection. For example, the WorldClim dataset[46] (http://www.worldclim.org/) is a set of global climate layers (climate grids) with a spatial resolution of about a square kilometer. The high-resolution global digital elevation model SRTM (Shuttle Radar Topographic Mission; http://srtm.csi.cgiar.org/), which was released in 2000, has a resolution of 3 arcsec (90 m). The GeoCover dataset, a global coverage of Orthorectified Landsat Enhanced Thematic Mapper Compressed Mosaics, is available (https://zulu.ssc.nasa.gov/mrsid/), showing the earth around the year 2000 at 14 m resolution. Electronic access to the entire USGS Landsat 5 and 7 archives enables users since late 2008 to download at no charge for selected areas' more recent scenes. Google Earth's growing database of very high-resolution images even constitutes a mechanism for ground truthing in sample areas, as individual objects such as trees can be distinguished, offering additional clues on what kind of crops are grown.

Whereas these datasets are useful to identify major biophysical constraints that may exist in an area and thus allow some assessment of how *unlikely* it is that wheat is grown in an area, they do not confirm its presence. For that goal, *local* knowledge is needed, and the most useful sources of knowledge about crop distribution exist in the form of crop statistics and farming system studies. Every country in the target region issues annual reports with statistics on crop area, production, and yield at subnational level, mostly at the provincial but sometimes at the district level as well. The quality of this information varies considerably between countries, but as a rule of thumb, the reliability decreases with the areal unit of data aggregation (the provincial data are more reliable than the district-level data, etc.). Although crop statistics are the only objective information source about the presence of wheat or any other crop, the main disadvantage is that these data are usually aggregated at a high administrative level, usually the province, often including areas with different ecological conditions, some of which may be physically unsuitable for wheat. As an exception to this general rule, in Syria, the province (Muhafaza)-level crop statistics are further disaggregated according to precipitation zones, a.k.a. Agricultural Stability Zones.[47]

Studies of farming systems or agricultural systems, if reasonably up-to-date, are potentially a goldmine of information about the crops being grown in an area, the people who grow them, and the economics of growing them. The characteristics of the described systems may encompass a wide range of attributes, related to population, integration within markets, resource access, culture, agricultural practices, input–output relationships, public investment, poverty, and tenure systems. However, for the purpose of mapping crop zones, common problems to be expected are that, like gold, farming system studies are relatively rare and, usually, cover small areas that are not necessarily representative. Moreover, these studies are, in general, not spatially explicit and therefore difficult to integrate in a GIS.

Remote sensing may help overcome these difficulties in spatializing farming system studies. Wattenbach[48] describes the development of a farming systems map for Syria in which a tentative initial sketch map is integrated with a spatially explicit map of agricultural regions in Syria, obtained by the interpretation of Landsat imagery, resulting in a geographically correct map of the farming systems.

Soil or land capability maps for the target region are another useful addition to the suite of evidence layers that can be integrated into the probabilistic approach for mapping the wheat-growing areas, especially for defining areas suitable for specific crops. Soil maps need, however, to be interpreted with caution. Many maps are either at too general scale to be useful, or are out of date, with most surveys dating from the 1960s to 1980s. Often, they fail to incorporate land improvements, which may raise the land quality for crop production, such as terracing, new irrigation development, saline land reclamation, stone and rock removal, or alternatively land degradation trends, particularly due to salinization of irrigated areas. However, also in this case, remote sensing can help with updating the soil information. A more serious problem is that many soil maps use a taxonomic classification, such as Soil Taxonomy[49] or the FAO classification system,[50] which do not necessarily have a direct linkage to the physical and chemical soil properties that determine the suitability for wheat. For this reason, in most cases, it will be necessary to undertake a land suitability evaluation first by matching the biophysical factors affecting wheat growth (climate, terrain, and soils) to the requirements of the crop. Adapting text book methods for land evaluation[51,52] to the available databases and local conditions may be the best way forward.

How can these information sources be combined and yield country-level maps of the probability of wheat at a fine (e.g., 1 km^2) resolution? A good way to start would be by the elimination of areas where wheat is highly unlikely to be present. The first step would be to remove from the analysis those statistical mapping units in which no wheat has been reported. The second step, to be performed through image analysis of satellite imagery, is to extract the areas with cropland. These areas, which contain the full mix of crops reported for the statistical unit, can then be differentiated into rainfed and irrigated croplands, again through remote sensing. Using local knowledge and farming systems studies, one could then obtain information which crops are only grown under either irrigated or rainfed conditions, or both.

The next step is to determine the comparative advantage of each crop in the statistical reporting unit from two perspectives: biophysical potential and social preference. Biophysical potential can be assessed for each crop through an adapted land evaluation method, leading toward a crop-specific classification of the croplands into one of four classes, "highly suitable," "moderately suitable," "marginally suitable," and "unsuitable,"[53] which in turn can be linked to average yield levels, again obtained from local knowledge and farming systems studies. Social preference for a crop will depend on its ability to maximize farm income itself determined by the average expected yield level and the price for the crop.

The final step for preparing a probability map for wheat (or any other crop) is to use a Bayesian approach of convergence of the available evidence from a prior to a posterior probability. Bayes Theorem states that

$$p(h/\varepsilon) = \frac{p(\varepsilon/h) \times p(h)}{\sum_i p(\varepsilon/h_i) \times p(h_i)}$$

where

$p(h/\varepsilon)$ is the probability of the hypothesis being true given the evidence (posterior probability)

$p(\varepsilon/h)$ is the probability of finding that evidence given the hypothesis being true (conditional probability)

$p(h)$ is the probability of the hypothesis being true regardless of the evidence (prior probability)

is a useful basis for combining different types of evidence, as derived from direct sampling, local knowledge, and secondary sources. Starting with a prior probability determined by the share of wheat in the crop mix of the statistical reporting unit, Bayesian statistics allows adjusting this first guess in the light of the maps used as predictors. Land suitability is a particularly useful layer, as the modeling involved allows integrating the key factors of the environment (climate, soils, and topography) into a single crop-specific score. Spatialization of crop prices, even if tentative, results in another useful evidence layer. Policy-related evidence layers could also be considered where relevant. For example, in Syria, rainfed cropping below 200 mm annual precipitation is prohibited. Relatively simple implementations of Bayesian statistics in a GIS environment are available,[54] which can be adapted for calculating the conditional probabilities arising from the possible combinations of land being in a particular suitability class with a given crop.

Another major challenge is the changing nature of the pathogen. Two major new variants of Ug99 have been detected in less than 5 years. For germplasm deployment strategies, it is essential to know which pathotypes are present in which areas and how these are likely to move. Race analysis of rust pathogens requires bioassays performed on collected rust samples under controlled conditions, using sets of differential testers, that is, specific wheat varieties that have known stem rust resistance genes. At present, not all countries have the capacity to undertake such analysis, and transfer of samples to advanced research laboratories in North America is required.

Strict quarantine regulations introduce a considerable time lag (up to 6 months or more) between sample collection in the field and final race analysis confirmation. This time lag, combined with the multi-institutional and cross-continental elements of race analysis, implies the requirement for a very stringent procedure to track samples and trace race analysis results back to source collection sites. At present, race analysis data have not been integrated into the centralized rust database, but this needs to occur. Strict sample coding, or even bar coding, will be required to effectively connect pathotype data to field collection sites.

6.8 CONCLUSION

Historically, wheat stem rust was the most feared plant disease capable of devastating epidemics and crop losses. By the mid-1990s, widespread use of resistant cultivars had reduced disease incidence to nonsignificant levels worldwide. Stem rust research and resistance breeding ceased to be a priority activity. The emergence of a new virulent stem rust race lineage in Uganda in 1998/1999, popularly named Ug99, and subsequent variants have rendered 80% or more of global wheat varieties stem rust susceptible. Emergence and spread of the Ug99 lineage have put stem rust firmly back on the agenda of wheat scientists worldwide.

The response of the global wheat community in relation to the threat raised by Ug99 has been positive and effective. Already, significant progress has been made in the identification and development of new resistant materials, but several challenges still remain before those varieties find their way into farmer's fields on a large scale. Considerable progress has also been made in raising awareness of the threat posed by Ug99 and in providing access to reliable and timely information on current status.

In line with the recommendations of an expert panel convened to assess the threat of Ug99, GIS technology forms the backbone of an emerging monitoring and surveillance system for cereal rusts. This embryonic system is initially focused on the emerging stem rust threat but over time plans to incorporate other cereal rusts and is being developed collaboratively by CIMMYT, ICARDA, and FAO. Few, if any, other technologies apart from GIS possess the capacity to seamlessly integrate the multitude of factors relevant to stem rust movements, establishment, and impact. Online, geographic-based visualization tools are already playing a critical role in the dissemination of complex datasets in a timely fashion.

A series of factors have been identified as being important if an effective surveillance and monitoring system for stem rust is to be created. These include location information regarding presence or absence of the disease and the specific race involved, location of important wheat-growing areas and the susceptibility of cultivars being grown, potential direction and distance of spore movements, and local climatic conditions favoring spore deposition pathogen survival and development. Without exception, all of these key elements have a spatial component and as such are amenable for incorporation into a geo-database and analysis using GIS. Absolutely critical is the requirement for timely, geo-referenced field survey data.

Through a successful international collaboration, several of the required elements of the monitoring and surveillance system are already starting to be addressed. Standardized field protocols have been developed; provision of, and training in the use of, GPS has been initiated via existing and expanded national partner networks—resulting in a substantial increase in the amount of survey data incorporated into a centralized geo-database. As a result, regularly updated known distribution maps for Ug99 are now being produced. Routine incorporation of wind trajectory models is providing improved information on potential movements, with results so far corresponding closely to actual confirmed observations in the field. Results obtained to date indicate movements and range expansion of Ug99 in-line with predicted regional airflows, and generally following previously reported movements of other rust races originating in east Africa.[13] However, there is neither room for complacency regarding future movements nor any substitute for regular, timely field surveys of the key wheat areas. Factors associated with spore deposition and pathogen establishment have started to be addressed, but many challenges remain. Predominant wheat-producing areas have been identified, and initial, albeit somewhat outdated, estimates have been made regarding susceptibility and areas of existing wheat varieties. However, both sets of information are seen as being high priorities for future improvement, and efforts are underway to address these aspects. Similarly, crop growth stages and key climatic factors are starting to be addressed, but, significantly, more progress is required. Obtaining reliable and timely data for all of the key climate variables at the required temporal resolution over large data sparse regions presents some challenges, although the increasing availability of remotely sensed weather data in near real time might provide some useful options.

Good progress has been made in the development and release of initial information tools that draw upon existing centralized data and provide near-real-time information on the current status of Ug99. The ready availability of powerful geographic-based visualization options, such as provided by Google Earth, has been a key factor in the successful presentation of information in a clear and flexible way. In the future, an expansion of the type and range of information products is planned. These will be targeted very closely to wheat scientists and decision and policy-makers in at-risk countries. Using the successful FAO desert locust monitoring system as a model (see http://www.fao.org/ag/locusts/en/info/info/index.html), an improved and expanded web presence will be created issuing status reports and alerts, along with lightweight rapid mapping capacity, plus regular summary bulletins.

GIS has already proved to be a useful and critical tool in providing support to the ongoing global efforts addressing the threat posed by emerging new races of wheat stem rust. As these efforts continue and expand, GIS technology will undoubtedly underpin a large proportion of the monitoring and surveillance activities required for these economically important fungal diseases of wheat.

REFERENCES

1. Hodson, D.P. and White, J.W. Use of spatial analyses for global characterization of wheat-based production systems. *J. Agric. Sci.*, 145, 115, 2007.
2. FAOSTAT FAO Statistical Databases. Available online at http://faostat.fao.org/, 2003. (verified September 1, 2009).
3. USDA-APHIS. Barberry. Available online at http://www.aphis.usda.gov/plant_health/plant_pest_info/barberry/background.shtml, 2009 (verified December 1, 2009).

4. Leonard, K.J. Stem rust—Future enemy? In: P.D. Peterson, Ed., *Stem Rust of Wheat: From Ancient Enemy to Modern Foe*. APS Press, St. Paul, MN, pp. 119–146, 2001.
5. Roelfs, A.P., Singh, R.P., and Saari, E.E. *Rust Diseases of Wheat: Concepts and Methods of Disease Management*. CIMMYT, Mexico, D.F., 1992.
6. Pretorius, Z.A., Singh, R.P., Wagoire, W.W., and Payne, T.S. Detection of virulence to wheat stem rust resistance gene Sr31 in *Puccinia graminis* f. sp. *tritici* in Uganda. *Plant Dis.*, 84, 203, 2000.
7. Leonard, K.J. Black stem rust biology and threat to wheat growers. *Presentation to Central Plant Board Meeting*, February 5–8, 2001, Lexington, KY. Available online at http://www.ars.usda.gov/Main/docs.htm?docid=10755, 2001.
8. Stakman, E.C. Problems in preventing disease epidemics. *Am. J. Bot.*, 44, 259, 1957.
9. Singh, R.P., Hodson, D.P., Jin, Y., Huerta-Espino, J., Kinyua, M., Wanyera, R., Njau, P., and Ward, R.W. Current status, likely migration and strategies to mitigate the threat to wheat production from race Ug99 (TTKS) of stem rust pathogen. *CAB Rev. Persp. Agric. Vet. Sci. Nutr. Nat. Resourc.*, 1, 54, 2006.
10. Roelfs, A.P. and Martell, L.B. Uredospore dispersal from a point source within a wheat canopy. *Phytopathology*, 74, 1262, 1984.
11. Brown, J.K.M. and Hovmøller, M.S. Aerial dispersal of pathogens on the global and continental scales and its impact on plant disease. *Science*, 297, 537, 2002.
12. Steele, K.A., Humphreys, E., Wellings, C.R., and Dickinson, M.J. Support for a stepwise mutation model for pathogen evolution in Australasian *Puccinia striiformis* f. sp. *tritici* by use of molecular markers. *Plant Pathol.*, 50, 174, 2001.
13. Singh, R.P., William, H.M., Huerta-Espino, J., and Rosewarne, G. Wheat rust in Asia: Meeting the challenges with old and new technologies. In: *New Directions for a Diverse Planet: Proceedings of the 4th International Crop Science Congress*, September 26–October 1, 2004, Brisbane, Australia, 2004. Available online at http://www.cropscience.org.au./icsc2004/symposia/3/7/141_singhrp.htm (accessed on January 9, 2009).
14. Stakman, E.C. and Harrar, J.G. *Principles of Plant Pathology*. Ronald Press, New York, 581 p., 1957.
15. Isard, S.A., Gage, S.H., Comtois, P., and Russo, J.M. Principles of the atmospheric pathway for invasive species applied to soybean rust. *BioScience*, 55, 851, 2005.
16. McIntosh, R.A., Wellings, C.R., and Park, R.F. *Wheat Rusts: An Atlas of Resistance Genes*. CSIRO Publications, Victoria, Australia, 1995.
17. Rajaram, S., Singh, R.P., and Torres, E. Current CIMMYT approaches in breeding wheat for rust resistance. In: N.W. Simmonds and S. Rajaram, Eds., *Breeding Strategies for Resistance to the Rust of Wheat*. CIMMYT, Mexico, D.F., pp. 101–118, 1988.
18. Singh, R.P., Hodson, D.P., Huerta-Espino, J., Jin, Y., Njau, P., Wanyera, R., Herrera-Foessel, S.A., and Ward, R.W. Will stem rust destroy the world's wheat crop? *Advances in Agronomy*, 98, 271, 2008.
19. Jin, Y., Szabo, L.J., and Pretorius, Z.A. Virulence variation within the Ug99 lineage. In: R. Appels, R. Eastwood, E. Lagudah, P. Langridge, and M. Mackay Lynne, Eds. *Proceedings of the 11th International Wheat Genetics Symposium*. Available online at http://ses.library.usyd.edu.au/bitstream/2123/3435/1/O02.pdf, 2008.
20. Wanyera, R., Kinyua, M.G., Jin, Y., and Singh, R.P. The spread of stem rust caused by *Puccinia graminis* f. sp. tritici, with virulence on Sr31 in wheat in Eastern Africa. *Plant Dis.*, 90, 113, 2006.
21. Jin, Y., Pretorius, Z.A., and Singh, R.P. New virulence within race TTKS (Ug99) of the stem rust pathogen and effective resistance genes. *Phytopathology*, 97, S137(Abstract), 2007.
22. Jin, Y. and Singh, R.P. Resistance in U.S. wheat to recent eastern African isolates of *Puccinia graminis* f. sp. *tritici* with virulence to resistance gene *Sr31* in US wheats. *Plant Dis.*, 90, 476, 2006.

23. Reynolds, M.P. and Borlaug, N.E. Applying innovations and new technologies from international collaborative wheat improvement. *J. Agric. Sci.*, 144, 95, 2006.
24. CIMMYT. *Sounding the Alarm on Global Stem Rust.* CIMMYT, Mexico, D.F. Available online at http://www.globalrust.org/uploads/documents/SoundingAlarmGlobalRust.pdf (accessed on January 9, 2009), 2005.
25. Hodson, D.P., Singh, R.P., and Dixon, J.M. An initial assessment of the potential impact of stem rust (race Ug99) on wheat producing regions of Africa and Asia using GIS. In: *Abstracts of the 7th International Wheat Conference*, Mar del Plata, Argentina, p. 142, 2005.
26. NOAA. NCEP/NCAR 40-year reanalysis project. Available online at http://ingrid.ldeo.columbia.edu/maproom/Global/.Climatologies/.Vector_Winds/ (verified January 9, 2009), 2005.
27. Nazari, K., Mafi, M., Yahyaoui, A., Singh, R.P., Park, R.F., and Hodson, D. Detection of wheat stem rust race "Ug99" (TTKSK) in Iran. In: R. Appels, R. Eastwood, E. Lagudah, P. Langridge, and M. Mackay Lynne, Eds., *Proceedings of the 11th International Wheat Genetics Symposium.* Available online at http://ses.library.usyd.edu.au/bitstream/2123/3434/1/O05.pdf, 2008.
28. Draxler, R.R. and Rolph, G.D. *HYSPLIT (HYbrid Single-Particle Lagrangian Integrated Trajectory).* NOAA Air Resources Laboratory, Silver Spring, MD, 2003. Available online at http://www.arl.noaa.gov/ready/hysplit4.html (accessed on January 9, 2009).
29. Rowell, J.B. and Romig, R.W. Detection of uredospores of wheat rusts in spring rains. *Phytopathology*, 56, 807, 1966.
30. Nagarajan, S. and Singh, D.V. Long-distance dispersion of rust pathogens. *Ann. Rev. Phytopathol.*, 28, 39, 1990.
31. Isard, S.A., Main, C., Keever, T., Magarey, R., Redlin, S., and Russo, J.M. Weather-based assessment of soybean rust threat to North America. Final report to APHIS, July 15, 2004. Available online at http://www.aphis.usda.gov/ppq/ep/soybean_rust/sbrfinal15july.pdf (verified 1 June 2006), 2004.
32. Prospero, J.M., Blades, E., Mathieson, G., and Naidu, R. Interhemispheric transport of viable fungi and bacteria from Africa to the Caribbean with soil dust. *Aerobiologia*, 21, 1, 2005.
33. USDA. National Agricultural Statistics Service. *All Wheat: Planted Acreage by County.* Available online at http://www.nass.usda.gov/Charts_and_Maps/Crops_County/aw-pl.asp (verified January 9, 2009), 2007.
34. Leff, B., Ramankutty, N., and Foley, J.A. Geographic distribution of major crops across the world. *Global Biogeochem. Cycles*, 18, GB1009, 10.1029/2003GB002108, 2004.
35. Monfreda, C., Ramankutty, N., and Foley, J.A. Farming the planet. Part 2: The geographic distribution of crop areas and yields in the year 2000. *Glob. Biogeochem. Cycles*, 22, GB1022, doi:10.1029/2007GB002947, 19, 2008. Available online at http://www.sage.wise.edu/pubs/articles/m-z/Monfreda,Monfreda6BC2008.pdf (verified September 22, 2010).
36. You, L. and Wood, S. An entropy approach to spatial disaggregation of agricultural production. *Agric. Syst.*, 90 (1–3), 329, 2006.
37. You, L., Wood, S., Wood-Sichra, U., and Chamberlin, J. Generating plausible crop distribution maps for Sub-Sahara Africa using a spatial allocation model. *Inf. Dev.*, 23, 151, 2007.
38. Bicheron, P., Defourny, P., Brockmann, C., Schouten, L., Vancutsem, C., Huc, M., Bontemps, S., Leroy, M., Achard, F., Herold, M., Ranera, F., and Arino, O. GLOBCOVER: Products description and validation report. Available online at ftp://uranus.esrin.esa.int/pub/globcover_v2/global/GLOBCOVER_Products_Description_Validation_Report_I2.1.pdf (verified January 9, 2009), 2008.
39. MAFRI stem rust in wheat, barley and oats. Available online at http://www.gov.mb.ca/agriculture/crops/diseases/fac15s00.html (verified February 9, 2010), 2008.

40. USDA FAS. Commodity intelligence report, China winter wheat update—April 2006. Available online at http://www.pecad.fas.usda.gov/highlights/2006/04/china_18apr2006/ (verified January 9, 2009), 2006.
41. Sache, I. Short-distance dispersal of wheat rust spores by wind and rain. *Agronomie*, 20, 757, 2000.
42. Gregory, P.H. *The Microbiology of the Atmosphere*, 2nd edn. John Wiley & Sons, New York, 377 p., 1973.
43. Rowell, J.B. Controlled infection by *Puccinia graminis* f. sp. *tritici* under artificial conditions. In: A.P. Roelfs and W.R. Bushnell, Eds., *The Cereal Rusts Vol. I.: Origins, Specificity, Structure, and Physiology*. Academic Press, Orlando, FL, pp. 291–332, 1984.
44. Joyce, R.J., Janowiak, J.E., Arkin, P.A., and Xie, P. CMORPH: A method that produces global precipitation estimates from passive microwave and infrared data at high spatial and temporal resolution. *J. Hydrometeorol.*, 5, 487, 2004.
45. USDA FAS. Commodity intelligence report. Middle East: Deficient rainfall threatens 2009/10 wheat production prospects. Available online at http://www.pecad.fas.usda.gov/highlights/2008/12/mideast/ (verified January 9, 2009), 2008.
46. Hijmans, R.J., Cameron, S.E., Parra, J.L., Jones. P.G., and Jarvis, A. Very high resolution interpolated climate surfaces for global land areas. *Int. J. Climatol.*, 25, 1965, 2005.
47. MAAR. *Agricultural Statistical Abstract* (several years). Ministry of Agriculture and Agrarian Reform Statistics Department, Damascus, Syrian Arab Republic, 2000.
48. Wattenbach, H. *Farming Systems of the Syrian Arab Republic*. National Agricultural Planning Commission, Damascus, Syria, 184 p., 2005.
49. Soil Survey Staff. *Soil Taxonomy: A Basic System of Soil Classification for Making and Interpreting Soil Surveys*, 6th edn. United States Department of Agriculture, Soil Conservation Service, Washington, DC, 1999.
50. FAO. World reference base for soil resources. World soil resources report 84. Food and Agriculture Organization of the United Nations, Rome, 88 p., 1998.
51. Dent, D. and Young, A. *Soil Survey and Land Evaluation*. George Allen and Unwin, New South Wales, Australia, 1981.
52. Sys, C., Van Ranst, E., and Debaveye, J. *Land Evaluation*. Agricultural Publications No. 7, General Administration for Development Cooperation, Brussels, Belgium, 1991.
53. FAO. A framework for land evaluation. *FAO Soils Bull.*, 32, FAO, Rome, 1976.
54. Corner, R.J., Hickey, R.J., and Cook, S.E. Knowledge based soil attribute mapping in GIS: The expector method. *Trans. GIS*, 6, 383, 2002.

7 Online Aerobiology Process Model

Joseph M. Russo and Scott A. Isard

CONTENTS

7.1 Executive Summary ... 159
7.2 Introduction ... 160
7.3 Principles of an Aerobiology Process Model ... 160
7.4 Configuration of the Aerobiology Process Model ... 162
7.5 Online Simulation of the Aerobiology Process Model ... 163
7.6 Conclusion ... 166
Acknowledgment ... 166
References ... 166

7.1 EXECUTIVE SUMMARY

Models have been used to describe the behavior of phenomena for centuries. Beginning in the 1950s, models in the form of computer programs began to proliferate in a number of sciences. As interactive programming on the Internet became popular in the early twenty-first century, there was a growing interest among professionals in government, academia, and business to access models online in order to monitor the day-to-day behavior of some phenomenon, such as pest movement in agriculture. This chapter introduces the reader to the principles of an aerobiology process model, including the components of preconditioning in source area, take-off/ascent, horizontal transport, descent/landing, and impact at destination. These model principles are put into practice for one pest—the fungus *Puccinia graminis* f.sp. *tritici* (*Pgt*), which is known by its common name "wheat stem rust." Instructions are provided in the chapter on how a reader as a user can go to a Web site and configure one of the components of the online aerobiology process model. After configuring the online model, the reader can execute the model and create a simulation for a geographic area and year. By comparing simulations between locations and years, a user can appreciate the different geographic and seasonal behaviors of the wheat stem rust disease. With the chapter as a guide and the online experience, a reader will also gain an appreciation of the role model simulations play in describing the spread of a disease in the conterminous United States.

7.2 INTRODUCTION

Models had humble beginnings centuries ago when pen and paper were the means for executing equations that described the behavior of phenomena. With the advent of computers in the mid-twentieth century, models evolved in both size and complexity and in their execution speed. In the 1950s, only large institutions, like federal governments and international corporations, could afford to run models as national services. By the early 1990s, even small companies had the resources to offer similar model-based services. As the number of computers grew through the twentieth century, the appearance of the Internet became a major incentive for encouraging the use of models. The Internet allowed individuals on desktop computers to communicate with each other in real time. Starting with attached modems, and later with wireless routers, individuals became "users" of "online" models over the Internet. The ability of people to interact in real time allowed for more personal input in both the setup and execution of models. Furthermore, the use of models as an interactive tool allows their output to be used for timely decision making.

Dr. Scott Isard, of the Pennsylvania State University, in a coauthored book entitled *Flow of Life in the Atmospheres* laid out the basic principles for an "aerobiology process" model.[1] Dr. Joseph Russo as a cofounder of ZedX, Inc. in 1987 has dedicated his career toward building models for practical applications. ZedX as a commercial entity has been offering model-based services to government, universities, and industry in the agriculture sector since the mid-1990s. This chapter represents a collaboration of principles and practices. It combines the aerobiology process model principles with state-of-the-art, online information technology (IT) tools for practical model applications. The web-enabled exercises in this chapter represent an evolutionary step in modeling by allowing users a richer experience through the configuration, execution, and analyses of model simulations.

The chapter is organized into six sections, three of which reflect the steps for setting up and executing an online aerobiology process model. After an initial reading to become familiar with the programming steps, the user is expected to log on to the specified Web site and put to practice the steps in this chapter (note this Web site will be active until March 2011). While the principles and practices outlined in this chapter would apply to most pests in a global modeling domain, the online exercises are limited to the conterminous United States and to one pest—the fungus *Pgt*, which is known by its common name "wheat stem rust." Stem rust has been one of the most important diseases of wheat since the emergence of western civilization.[2]

7.3 PRINCIPLES OF AN AEROBIOLOGY PROCESS MODEL

The aerobiology process model supports a deterministic "systems" approach to modeling. That is, the mathematical formulations used to define a model simulation attempt to mimic the actual processes and properties associated with the movement of biota in nature. One of the main advantages of a deterministic model design is that the simulated sequence of processes and properties of biota movement and behavior can be compared one-to-one with field observations. This comparison between "simulated" and "observed" values quantifies the precision and accuracy of a model and, depending on the outcome of a comparison, gives a user confidence in the performance of a model.

An aerobiology process model consists of five components or stages:[3,4]

- Preconditioning in source area
- Take-off/ascent
- Horizontal transport
- Descent/landing
- Impact at destination

"Preconditioning in source area" is the most important component of the aerobiology process model. It sets the stage for the initial launch or movement of a pest from its starting point or source area. The preconditioning involves both the state of the pest, its host, and the surrounding ambient environment. In the aerobiology process model, preconditioning is accounted for by simulating the state of host, the pest, and the ambient environment. A pest may have multiple hosts with different canopy architectures and a range of favorable environments. While acknowledging this diversity, the modeling exercise in this chapter is limited to wheat as the host, the fungus *Pgt* as the pest, and an ambient environment defined only by canopy temperature and leaf wetness.

The "take-off/ascent" component refers to the movement of a pest out of its source habitat, which in the case of the modeling exercise is a canopy. This movement can range from passive lifting by the wind of fungal spores off a leaf, or it can be the powered flight of an insect. The take-off/ascent is probably the most complex and least understood component of the aerobiology process model. The passive lift of fungal spores from a wheat canopy depends on the characteristics of the biota (especially those that determine fall velocity of particles), characteristics of the canopy (density, height and arrangement of foliage/stems), characteristics of the air flow (unsteady coherent motions), and take-off/release height position in the canopy. As a canopy matures and leaves begin to die in the understory, more ventilation takes place favoring the movement of disease vertically through the vegetation, with the eventual escape of spores through the topmost layer. The ambient conditions and vegetative state for the lifting of spores out of a canopy are accounted for in the aerobiology process model.

The "horizontal transport" component refers to the vertical and horizontal movements of a pest through the atmosphere. After leaving the environs of a canopy, a pest is either passively transported by vertical and horizontal currents of wind or by powered flight. In the case of the wheat stem rust pathogen, plumes of fungal spores are carried three dimensionally by the wind through the atmosphere. Dispersed across an arc centered on a wind vector, spores can reach high altitudes where they may encounter lethal environmental conditions. The aerobiology process model accounts for the exposure of spores to high-altitude environments by simulating the ultraviolet (UV) radiation conditions at different heights.

The "descent/landing" component refers to the deposition of pest to the surface either passively by prevailing atmospheric conditions or by a powered landing. In the case of a fungal pathogen, transported spores are either dry deposited to the surface due to gravity and/or descending wind currents or wet deposited by precipitation. The simulation of downstream dry periods and rain events in the aerobiology process model trigger the spore descent phase of the transport process.

The "impact at destination" is the last component of the aerobiology process model. It represents the final landing point on the surface after transport. A pest may find favorable or unfavorable host and environmental conditions after landing. The touching down of spores in another canopy distant from the source area marks the end of one cycle and the beginning of the next. Depending on the micrometeorological conditions within the new canopy, the preconditioning in source area component is started anew with the latest deposited spores.

With multiple cycles of preconditioning in source area, takeoff/ascent, horizontal transport, descent/landing, and impact at destination, the aerobiology process model simulates the movement of a pest across the U.S. landscape. In the case of wheat stem rust, the chosen pest to illustrate the principles, the online model simulates the movement and survival of spores from overwintering source areas in the southern United States to more northerly wheat canopies during the growing season.

7.4 CONFIGURATION OF THE AEROBIOLOGY PROCESS MODEL

The five components of the aerobiology process model consist of a set of parameters, whose values are either preset or can be selected online from drop-down menus by a user. The particular configuration of values is described for each component below.

For the "preconditioning in source area" component, a user can select values from drop-down menus for each of the following parameters:

- State and county having wheat acreage (i.e., potential source area)
- Crop stage of wheat acreage
- Percent of wheat acreage that is infected in the selected county
- Severity level of the infection

For the "takeoff/ascent" component, values are preset for the

- Daytime maximum hourly spore release rate from a wheat canopy
- Nighttime maximum hourly spore release rate from a wheat canopy
- Minimum wind speed threshold to initiate spore escape from a canopy
- Maximum spore escape rate as a percentage of the release rate and wind speed

For the "horizontal transport" component, values are preset for the

- Spore transport originating from wheat acreage designated as the source area
- Spore transport arc centered on the compass bearing of wind direction
- Maximum transport height of spores in the atmosphere
- Spore mortality as a function of lethal UV radiation exposure

For the "descent/landing" component, values are preset for the

- Percent of spores landing as a function of downward vertical motion during a dry deposition event during a specified period
- Percent of spores landing as a function of an hourly rainfall total during a wet deposition event
- Minimum hourly rainfall total for a wet deposition event

For the "impact at destination" component, values are preset for the

- Percent of coverage of spore deposition on land or on wheat acreage. In reality, transported spores could land on alternative hosts other than wheat, such as barley, and surfaces not subject to infection. Also, after impact on a destination canopy, local environmental conditions affect the level of infection and subsequent disease progress.

7.5 ONLINE SIMULATION OF THE AEROBIOLOGY PROCESS MODEL

Online simulations of the aerobiology process model require that the reader as a user configure values for the source area component of the aerobiology process model and choose weather data by selecting a year from a drop-down menu.

This section provides a step-by-step description of one possible simulation of the wheat stem rust disease using weather data from a single year. The goal of this section is to provide a "walk through" of both configuring and executing the model for one year and one county in the conterminous United States. The user is encouraged to follow along online by making the same selections as shown in the example simulation.

To begin an online simulation of the wheat disease, the user should type in the following Web site address in their favorite browser and click the keyboard **Enter** key.

http://crcmodel.zedxinc.com

Upon seeing the opening screen to the Web site, the user is required to type in the password provided by CRC Press into the **Registration Code** text box and click the online **Enter** key. Please note that the case-sensitive password requires the reader to type capital letters and characters correctly.

After successfully entering a password, the user will see a main screen similar to the one depicted in Figure 7.1. The sections of the main screen will be described from top to bottom beginning with the title **Online Aerobiology Process Model**. At any time, a user can minimize or exit the Web site by clicking the appropriate icons of the web browser.

A line of drop-down menus are shown just below the title section of the screen (Figure 7.1). A user can select a particular model from the **Model** drop-down menu. In CRC version of the Web site, there is only one model selection for **Wheat Stem Rust**. Moving from left to right across the line, a user selects a state (**TX**), county (**Floyd**), and year (**2008**) from drop-down menus. Model simulations are executed

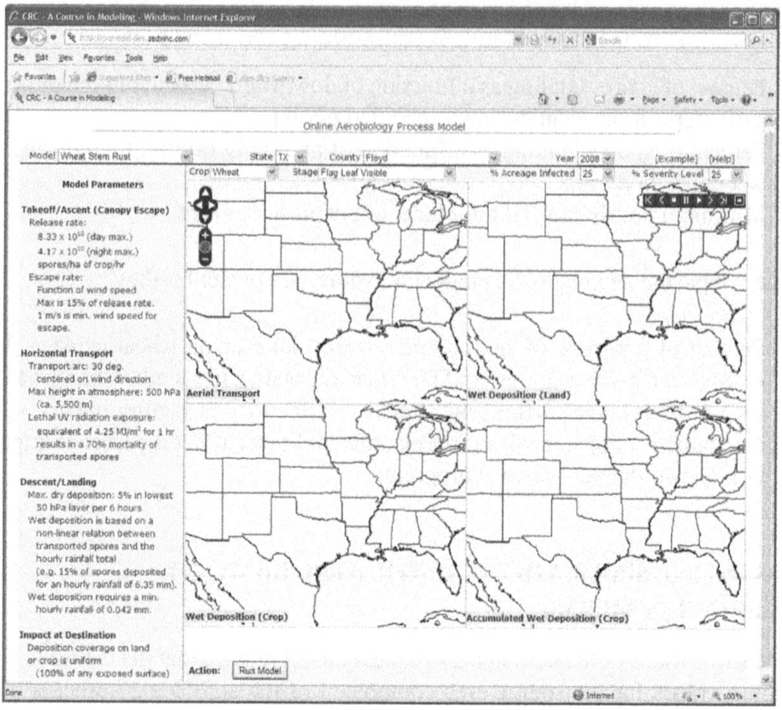

FIGURE 7.1 Main screen for online aerobiology process model.

one year at a time. There are two commands on the first menu line. A user can click **Example** to see a demonstration of a model run or click **Help** to view step-by-step instructions on how to set up and execute the model. The **Help** command also provides definitions and explanations of model components and entries.

A second drop-down menu line is displayed just right of center and below the first menu line and just above the map section of the main screen (Figure 7.1). A user can select a crop (**Wheat**) and its stage (**Flag Leaf Visible**) from the first two drop-down menus. The choice of a stage sets the start date of a simulation, which uses the seasonal weather conditions of the selected year in the county. A user can next select the percent of crop acreage infected (**25**) and the percent disease severity level (**25**) in the selected county from the remaining two drop-down menus.

The map section of the main screen provides for user interaction with a navigation map and displays results associated with a model execution or **run**. In its default setting, the map section depicts a map of the United States with some elementary Geographic Information Systems (GIS) tools, in the upper left corner, for zooming in and out. After each model run, the map section displays the model output as multiple panel maps.

A listing of model parameters is provided in a section just to the left of the map section. The model parameters are organized around four of the five components of the aerobiology process model, which were described in previous sections of this chapter. The four components include "takeoff/ascent" (canopy escape), "horizontal transport," "descent/landing," and "impact at destination." The listing of parameters is for informational purposes only and cannot be changed by the user.

The last section of the main screen is the **Action** bar below the map section. By clicking on the appropriate button, a user can execute a model (**Run Model**) and print model results (**Print**) displayed in the map section. After clicking the **Run Model** button, a **Player Control** bar appears in upper right corner of the map panel. The **Player Control** bar allows a user to initiate and control the animation of the model output. A user starts the animation by clicking the start icon (▶) on the bar. The same user can also stop, pause, forward, fast forward, backward, and fast backward an animation by clicking the appropriate icon (Figure 7.1).

Model output is in the form of four panels displaying maps of daily aerial transport, daily wet deposition on land, daily wet deposition on crop, and seasonal accumulation of wet deposition on crop (Figure 7.2). It is synchronized by date, which is displayed in the top center of the main section of the screen. When a user clicks the start icon on the **Player Control** bar, daily model output will be animated in the four panels. As evidenced by the changing date at the top center of the section, a user will see the daily patterns of transport and wet deposition on land and crop, and the progressively increasing area of accumulated wet deposition on a crop. A user can "freeze" a frame of the animation by clicking the stop icon on the **Player Control** bar and then click the **Print** button on the **Action** bar to get a hardcopy of the model output being displayed on the screen.

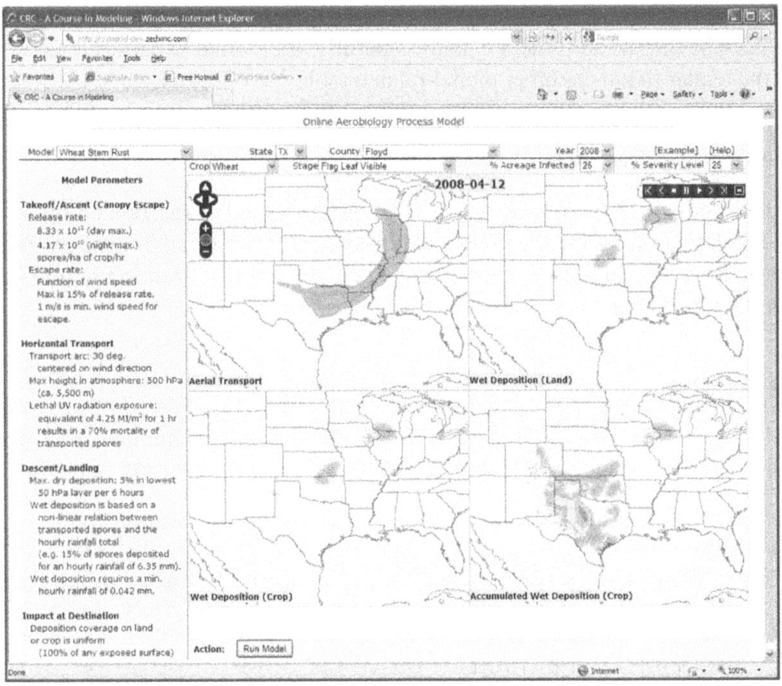

FIGURE 7.2 Example of wheat stem rust model output for April 12, 2008.

An example of a frame of wheat stem rust model output is depicted in Figure 7.2 for Floyd, Texas, on April 12, 2008. The aerial transport of spores on that single day extends from Texas up through the Midwest United States. While the number of transported spores is large, only a few are wet deposited on land, and fewer still are deposited on a crop. The daily deposited spores in the left lower plane contribute to the seasonal accumulation in the right lower panel (Figure 7.2).

In summary, the main screen of the **Online Aerobiology Process Model** allows a user to interactively set up the preconditioning in source area at the scale of a county. The same user can execute the wheat stem rust model for different years. The daily output of each model run can be animated and printed on demand. The selections made in the drop-down menus during one session must be retyped each time a user logs out and returns to the Web site.

7.6 CONCLUSION

This chapter has introduced a reader to the principles of an aerobiology process model. It has also introduced a reader to the practice of modeling through an interactive Web site called the **Online Aerobiology Process Model**. Through the Web site, a reader as a user can minimally configure one component (**Preconditioning in source area**) of the aerobiology process model and run simulations for individual years and selected counties in the conterminous United States. By comparing simulations between locations and years, a user can appreciate the different geographic and seasonal behaviors of the wheat stem rust disease. While the simulations are limited to one model of one disease on one crop, the principles of aerobiology modeling have been put into practice, and, hopefully, the online experience has stimulated the reader to pursue other model-related topics.

ACKNOWLEDGMENT

The development of the online aerobiology process model was funded by a grant from the USDA Cooperative States Research, Education and Extension Service National Research Initiative Animal and Plant Biosecurity grant program (Award No. 2009-55605-05049).

REFERENCES

1. Isard, S.A. and Gage, S.H. *Flow of Life in the Atmosphere: An Airscape Approach to Understanding Invasive Organisms*. Michigan State University Press, East Lansing, MI, p. 240, 2001.
2. Peterson, P.D. Stem rust of wheat: Exploring the concepts. In: Peterson, P.D. (ed.), *Stem Rust of Wheat: From Ancient Enemy to Modern Foe*. APS Press, St. Paul, MN, p. 1, 2001.
3. Benninghoff, W.S. and Edmonds, R.L. (eds.). Ecological systems approaches to aerobiology I. Identification of component elements and their functional relationships. *US/IBP Aerobiology Program Handbook Number 2*. University of Michigan, Ann Arbor, MI, p. 158, 1978.
4. Aylor, D.E. A framework for examining inter-regional aerial transport of fungal spores. *Agricultural and Forest Meteorology*, 38, 263, 1986.

8 Site-Specific Management of Green Peach Aphid, *Myzus persicae* (Sulzer)

Ian MacRae, Matthew Carroll, and Min Zhu

CONTENTS

8.1 Executive Summary 167
8.2 Introduction 168
8.3 Methods 174
8.3.1 Exercise 1: Describe the Spatiotemporal Colonization Patterns of *M. persicae* in Seed Potato 174
8.3.1.1 Description of Dataset 175
8.3.1.2 Assessing Spatial Autocorrelation Using Semivariograms in GS+ 176
8.3.1.3 Plot the Data from the Dataset in ArcMap 179
8.3.1.4 Discussion of Observed Colonization Patterns 182
8.3.2 Exercise 2: Using the HYSPLIT Model to Determine If the Wind Vectors at Specific Dates Provided a Significant Risk of Aphid Immigration into the Red River Valley 183
8.3.2.1 Using HYSPLIT to Examine LLJ to Facilitate Movement of Aphids into the Red River Valley 183
8.3.2.2 Discussion of HYSPLIT Results 186
8.4 Conclusions 186
References 186

8.1 EXECUTIVE SUMMARY

Integrated pest management (IPM) is an environmentally and economically sustainable method of managing pests based on applying pesticides only when pest populations surpass a point where expected resource loss exceeds control costs. Site-specific pest management extends IPM principles by determining the spatial and temporal distribution of pest species and optimizes agricultural inputs by spatially adjusting location and/or application rate based on varying conditions within a field. The green peach aphid, *Myzus persicae* (Sulzer), is a candidate for site-specific management due to colonization patterns that begin at field edges and progress inward. While aphid populations may not cause extensive damage, viruses that the aphids vector may reduce yield, cause the loss of virus-free certification in seed potato (*Solanum tuberosum*) production, and result in significant monetary loss for producers. This case study shows how to summarize the spatial and temporal distribution of *M. persicae* colonizing seed potato fields based on 3 years of data. In addition, the Hybrid Single-Particle Lagrangian Integrated Trajectory (HYSPLIT) model will be demonstrated to predict the seasonal abundance of *M. persicae* based on wind-mediated dispersion as low-level jet (LLJ) events can identify when scouting efforts of seed potato field margins should begin.

8.2 INTRODUCTION

IPM is an environmentally and economically sustainable method of managing pests. One of the fundamental principles of IPM is to apply pesticides only when pest populations surpass a point where expected resource loss exceeds control costs. This implies that (1) the level of the pest population and the level where damage occurs are known and (2) the area of infestation can be delimited. A logical extension of IPM, site-specific pest management examines the spatial and temporal distribution of pest species and optimizes agricultural inputs by spatially adjusting location and/or rate of applications based on varying conditions within a field.[1,2] By focusing application of agrochemicals, especially fertilizers and pesticides, on only those areas where needed rather than the entire management unit, often significantly less material is used to attain results similar to a full application. Not surprisingly, decreasing the amounts of applied pesticides has obvious environmental and economic benefits. However, precision management technologies also provide opportunities to realize economic benefits from a wider range of agricultural inputs than just agrochemicals.

Integrated and site-specific management techniques require scouting to monitor pest population levels. Any method that refines and focuses scouting efforts to improve scouting efficiency would provide economic savings in time and labor. Refined scouting methods in conjunction with a targeted pesticide application program allow for the creation of more comprehensive and flexible IPM programs.

Both IPM and site-specific management techniques have been used against several pests of potato. The Colorado potato beetle, *Leptinotarsa decemlineata* (Say), and green peach aphid have been identified as candidates for site-specific management due to colonization patterns that begin at field edges and progress inward.[3–6] Blom et al.[5] had some experimental success in using site-specific management to manage *L. decemlineata* in Pennsylvania by treating the field perimeter with a targeted band of systemic insecticide and periodic, supplemental applications of a foliar insecticide to within field populations. A comprehensive site-specific management program involving targeted application and refined scouting efforts has been developed for *M. persicae*.[6–10]

Myzus persicae originates from Europe but can now be found throughout the world, including all regions of North America.[11] Its successful establishment in crop-growing regions has made it a highly successful invasive species, resulting, not in a small part, from its ability to feed on a wide variety of plant hosts. In various parts of the United States, green peach aphid may undergo either a holocyclic or nonholocyclic life cycle (and in some locations, both). Holocyclic life cycles in aphids involve overwintering as eggs on primary hosts, in the case of green peach aphid, plants in the genus *Prunus* (e.g., peach, *Prunus persicae* Miller and prune, *Prunus domestica* L.). Aphids will then move to a secondary host in the summer and reproduce parthenogenetically, giving birth to live daughters. In the fall, aphids will move back to the primary host. At this time, a sex shift occurs with some individuals becoming males. Sexual reproduction occurs, and the offspring lay overwintering eggs. In areas where winter temperatures do not preclude the ongoing production of one of the 875 secondary hosts utilized by green peach aphid (including potato, *S. tuberosum* L.[12,13]), this insect will persist over the entire year parthenogenetically, with no sexual reproduction occurring (i.e., a nonholocyclic life cycle).[11,14] Consequently, through the growing season, all *M. persicae* in a population are females and parthenogenetically produce live offspring, the immature forms of the next generation being within every newborn aphid (the concept of telescoping generations).

In addition to differences in life cycle, there are two forms of the aphid, alate (winged) forms, which disperse to new hosts, and apterous (nonwinged) forms, which feed on a plant and have higher rates of reproduction. Because the next generation of nymphs is present and their wing phase is determined at the time of their mother's birth, population changes resulting from individual changes (like the development of dispersal forms) happen two generations *after* the stimulus that triggered the change (e.g., decrease in food quality triggering the development of dispersive forms). Nymphs born from alate mothers do not themselves become alate adults, thus ensuring a period during which colony expansion occurs through apterous forms.[11,14] These unusual reproductive characteristics explain the rapid increase in *M. persicae* populations. Although high populations of *M. persicae* can damage a number of crops, it is more serious as a vector of a number of virus species that infect plants.

The Red River Valley (RRV) of northwest Minnesota, eastern North Dakota, and southern Manitoba produce over 3 million MT of potatoes on 92,000 ha.[15,16] This requires the support of continuous production of disease-free seed potatoes for planting. While commercial potato production for chipping and table stock represents greater hectarage, seed potatoes are by far a more valuable production system. In the past 2 decades, the acreage of seed potatoes in the RRV has decreased precipitously, mostly due to aphid-vectored virus diseases, the most important of which are the non-persistently transmitted Potato Virus Y (PVY) (*Potyviridae: Potyvirus*) and persistently transmitted Potato Leafroll Virus (PLRV) (*Luteoviridae: Polerovirus*), both of which are most effectively vectored by *M. persicae*.[17–20] These aphid-transmitted viruses reduce yield and, in certain cultivars, reduce tuber quality.[21] Non-persistently transmitted viruses require only that the mouthparts of the vector penetrate the plant to achieve infection and are often flushed from the vector when it feeds. Persistently transmitted viruses require longer, uninterrupted feeding by the vector for successful transmission. For a field to be certified as virus free and receive the highest market price, it must have a viral infection rate among its plants of less than 0.05%. Failure to attain certification results in a significant monetary loss for a producer. Consequently, aphid control, especially for *M. persicae*, is an important consideration in commercial seed potato production.

Unless present at high densities, aphids are not generally yield limiting in potatoes. However, their role in vectoring disease makes their control vital. The epidemiology of the disease depends heavily on the seasonal population dynamics of the vectors and local inoculum levels. *M. persicae* does not overwinter in the RRV; seasonal populations are reestablished annually by spring immigrants from the south. Once in the region, the aphids colonize weeds and other crops hosts as potatoes have not yet emerged at this time and are unavailable for colonization. Aphids populations increase on these alternate hosts, and movement of *M. persicae* into seed potato fields occurs in late July–early August.[22–24] Because seed potato fields are established with disease-free seed stock, infection of seed potato fields most frequently results from colonization by viruliferous aphids emigrating from an inoculum reservoir. In the RRV, the most frequent source of viruliferous aphids is commercial potato fields produced for chipping, flakes, fries, and table stock.[25]

It has been noted that aphids colonizing potato fields tend to first establish at field borders prior to dispersing across the field.[23,24] Potato production requires significant tillage throughout the season, not only for weed control but to create and maintain the hills in which potato plants grow. Turning tillage equipment at end rows precludes the establishment of plants in the fields' headlands and results in a large band of bare soil bordering the potato field. The resulting reflectance contrast between bare soil and green fields is suspected to be attractive to colonizing aphids.[23] The use of border crops was effective in reducing PVY spread to small plots of seed potato; winged aphids were captured in greater abundance in crop borders than in the protected interiors.[23] It was postulated that this alighting preference at the crop border was a response to the contrast provided by the interface of fallow ground and crop canopy and that reduction of PVY spread occurred because aphids lost virus from their mouthparts when they probed plants within the border crop. Similarly, in large seed potato fields, distributions of *M. persicae* were found to be concentrated within a few meters of field margins in days immediately following inflights.[7,24] The initial distribution of *M. persicae* along field borders provides an opportunity to reduce insecticide usage by directing initial control efforts at field borders.

The temporal distributions and densities of *M. persicae* populations colonizing seed potato were assessed over a 3 year period. Different fields were selected each year and a standard pattern of sample points established in each. Aphid populations were sampled from each sample point weekly throughout the summer. A number of digital mapping techniques were used to create maps estimating the weekly within-field distribution and density of *M. persicae*. The result of these efforts assessed and described the spatial and temporal distribution of *M. persicae* colonizing seed potato fields and is the basis for Exercise #1 (see below).

Insecticides are widely used to control the current season spread of potato viruses in seed potatoes[26] but tend to be more effective in controlling the spread of persistently transmitted viruses than non-persistently transmitted viruses.[27] By the time insecticides are effective in killing the vector of a non-persistently transmitted virus, the virus has already been transmitted to the host plant. While generally ineffective in controlling the spread of non-persistent virus diseases like PVY,[28] insecticides can prevent transmission of PLRV by maintaining aphid densities below 3–10 per 100 leaves.[22,29] A working threshold for insecticide application has been established at 1 aphid per 10 plants sampled. This very low treatment threshold underscores the impact that even a few virus vectors can have on a seed potato field. To be most effective, aphids must be controlled early upon their arrival, requiring insecticides to be applied during aphid colonization of the field; timing this application, in turn, requires knowledge of when initial colonization occurs. Typically in seed production, insecticides will be applied when aphids are first colonizing fields and then again within 2 weeks. To accurately judge colonization, fields must be scouted. Because aphids enter fields as a result of environmental cues rather than calendar date, scouting must be spread over several weeks; refining this effort can result in a significant economic saving. Regional scouting has been attempted, but the time required to scout multiple fields restricts the area that can be sampled. Suction trap networks have been used to monitor the immigration of regional populations but are expensive to maintain, and their contents still must be sorted and identified. An alternate method of predicting the possibility of potential colonization was needed.

As noted, *M. persicae* does not overwinter in the RRV; its populations are annually reestablished by immigrating aphids. Colonization of seed potato fields results after other secondary hosts have begun to senesce and a decrease in food quality has triggered the development of winged forms.[22–24] It was this colonization behavior that facilitated the establishment of a network of suction traps providing an early warning system for immigrating aphids. This network, the *Aphid Alert Network*, ran over 9 years providing detailed data on the regional spatial and temporal occurrence of aphids in the RRV. When combined with the results of winter virus tests, it was possible to correlate years with a high occurrence of viral disease (i.e., high aphid populations) with environmental variables, providing the groundwork for identifying the overwintering locations (and source populations) of *M. persicae* immigrating into the RRV.

Aphids are weak fliers, and their long-range dispersal or migration is highly dependent on wind events, such as LLJ streams (Figure 8.1). LLJs are narrow, horizontal bands of air in the lower atmosphere moving faster than the surrounding air. These jet streams travel a considerable distance and most commonly originate east of the Rocky Mountains near Oklahoma.[31,32] They quickly transport warm Gulf air and moisture to the northern Great Plains at rates of 45–140 km/h.[32–38] There are two different kinds of LLJs: nocturnal LLJs and cyclone-induced LLJs. Nocturnal LLJs are caused by cooling of air at high elevations relative to air at the same height relative to the earth to the east. The resulting pressure gradient causes the eastern warmer air to flow toward the cooler western air and is turned north by the Coriolis effect. Cyclone-induced LLJs are caused by the significant temperature gradient following the cold front to the warm front of a cyclone. Cyclone-induced LLJs are typically stronger than nocturnal LLJs and tend to occur at a higher altitude, last longer, and have higher wind speeds.[39] Consequently, the movement of aphids into the northern Great Plains is more of a result of cyclone-induced LLJs than nocturnal LLJs.

Being a weak flier, once caught in an LLJ, *M. persicae* would be unable to fly out of the wind event and self-direct an landing. Rather, the aphids remain in the LLJ until some physical factor terminates their flight. The migration of aphids or other biota typically is terminated by a number of factors including gravity, changing temperature gradients, precipitation, depletion of energy resources, impaction, or some

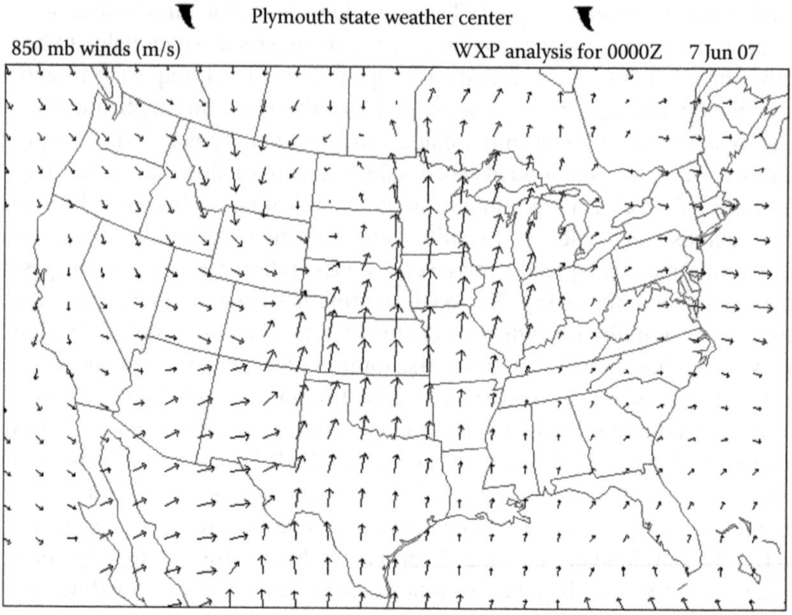

FIGURE 8.1 Wind vector plot illustrating an LLJ from June 18, 2007. Plot prepared by Plymouth State Weather Center (From Plymouth State Weather Center, Make your own product generator for archived data, available online at: http://vortex.plymouth.edu/u-make.html, 2009.)

combination of these factors.[40,41] Convergent downward air currents caused by the rapid decreasing wind speed at the edge of LLJs and their associated rainfall are both ideal deposition mechanisms for migratory aphids.[37,42]

The transport of various aphid species into and occurrence of aphid-transmitted viruses in Iowa and South Dakota in spring have been associated with southerly LLJs.[43–45] Spring migration of insects into the Great Plains also seems to depend more on the persistence and predominance of southerly LLJs than on wind maxima.[46,47] In addition, the LLJs most commonly associated with insect migration are associated with cold fronts arising from the south.[48,49] Given their characteristics for moving and depositing aphids into the Great Plains, it seems LLJs are the ideal transport mechanism to reestablish aphid populations in the RRV. By assessing annual activity of LLJs from areas where *M. persicae* overwinters, it is possible to identify years wherein there is a higher probability of aphid presence and, therefore, the potential for viral disease.

Determining overwintering locations was instrumental in developing a model that would predict the potential for *M. persicae* infestation. Overwintering locations were identified by backtracking to their origin LLJs that occurred in years with high virus occurrence (i.e., years which must have had high vector populations). These LLJ origin points were then correlated with the environmental conditions necessary for successful overwintering: presence of appropriate overwintering hosts for holocyclic or non-holocyclic populations, winter temperatures, frost occurrence, etc.[6]

It was found that populations of *M. persicae* infesting seed potato in Minnesota and North Dakota originated in southerly states west of the Mississippi River, most frequently from an area bordered 25°–35°N, 95°–105°W, an area including most or all of Arkansas, eastern Colorado, Kansas, Louisiana, Missouri, Oklahoma, and northern Texas.[8] A complete explanation of the process of identifying overwintering locations can be found in Zhu[6] and Zhu et al.[8]

Once overwintering potential locations had been determined, it was then possible to describe the timing, duration, and geographic origin of LLJ events most suitable for transporting green peach aphid migrants from overwintering areas to the northern Great Plains. The occurrence of an LLJ fitting this description can be used to predict the potential for *M. persicae* populations to immigrate into the region in a particular year. While it cannot predict the actual presence of aphids, such a model can provide a warning of the potential of aphid infestations.

Biological data from the literature and historical aphid population data were used to develop the criteria with which to identify LLJs that could possibly transport aphids from overwintering areas to the northern Great Plains.[6] Maximum wind velocities lower than 12 m/s would not keep aphids airborne over the entire dispersal length (i.e., these slower winds would eventually result in impaction prior to arrival in the RRV). In addition, insects are ectothermic and rely on environmental conditions to regulate body temperature, and because their small size and high surface to volume ratio result in rapid cooling, aphids caught in LLJs that exceeded 1500 m above ground level (AGL) have a greater probability of being exposed to low and perhaps lethal temperatures. In addition, LLJ events greater than 12 h were more commonly associated with aphid establishment.[6] Wind events obviously had to originate from areas wherein *M. persicae* can overwinter (generally southerly direction) and because aphids are weak fliers, wind events had to decrease in velocity or end in a precipitation event to release aphids from the LLJ. Successful criteria, therefore, for LLJ events believed to have the potential for bringing aphids into the RRV were (1) wind maxima >12 m/s at 1500 m AGL, (2) event durations >12 h, (3) wind direction from SE (1358) to SW (2258), and (4) decreased wind speed (>4 m/s) area over the northern Great Plains. The first and the last date that LLJ criteria were satisfied and the duration (h) that these criteria were continuously satisfied were recorded.[8]

A number of models are available to predict the movement of a wind event, but one was determined to be especially useful. The HYSPLIT is a trajectory analysis model developed by the National Oceanic and Atmospheric Administration and Australia Bureau of Meteorology, to track airborne particulate pollutants to and from their source. The wind-mediated dispersal of insects closely mirrors that of airborne dust;[50] consequently, HYSPLIT is an excellent tool to model this aspect of insect movement. It uses a three-dimensional and geometric approach to predict simple trajectories of wind events and the dispersion and deposition of airborne particles.[51–53] HYSPLIT uses Lagrangian trajectory methods, which are popular approaches in the atmospheric chemistry community, to pinpoint potential source areas of atmospheric pollutants.[51] The National Oceanic and Atmospheric Association (NOAA) collects and archives meteorological data and makes them available as gridded sets generated by the National Climatic Data Center. The Nested Grid Model (archived climatic data covering the continental United States at 2 h intervals for dates prior to 1998) and Eta Data Assimilation System (climatic data covering the continental United States at 3 h intervals for dates from 1997 on) are available from HYSPLIT's Web site and can be used to calculate trajectories and air concentrations.[52,54] The HYSPLIT model also uses meteorological forecasting models to give real-time prediction of dispersal.[52] The accuracy of HYSPLIT has been tested in several different experiments.[53,55]

HYSPLIT has been successfully used in measuring and forecasting movement of dust storms, wildfires, volcano ash, and other aerosols, as well as the causative agents of several plant diseases and pollen transport in the atmosphere.[56] Possibly the best-known agricultural application of the HYSPLIT model is forecasting transport of spores and dispersal of tobacco blue mold, *Peronospora tabacina* Adam, in North America.[57]

The HYSPLIT model was used to predict the seasonal abundance of *M. persicae*, and extensive scouting of both field edges and centers validated the utility of HYSPLIT to accurately predict movement of green peach aphid. LLJ events were used to identify when scouting efforts of seed potato field margins should begin. It was demonstrated that the LLJs that are known to move green peach aphids coincide with when green peach aphids were first found at field margins. Growers were advised to apply insecticide at the field's edge following sampling that verified presence of green peach aphids. In addition, at three sites, fields of comparable size and variety and containing similar aphid population densities and distributions were treated by whole field application. All fields treated in the 2006 trial were >20 ha in size; most were ~40 ha. In these trials, targeting insecticide treatments to field edges was found to be as effective as whole field applications at suppressing aphid populations[10] (Exercise #2).

8.3 METHODS

8.3.1 Exercise 1: Describe the Spatiotemporal Colonization Patterns of *M. persicae* in Seed Potato

The potential loss of value associated with aphid infestation of seed potato fields is considerable. Because quality and monetary losses would be unacceptable, commercial fields (i.e., fields whose yield goes to potato chips, French fry, or table vegetable industries) were used to initially determine the pattern of infesting *M. persicae*. The same agronomic practices are used in both seed and commercial fields; however, commercial fields generally are not treated for aphids as the presence of virus is less of an issue.

Preliminary data indicated aphids preferentially colonize fields generally within 25 m of their edge[24]; our initial interest was to define the area (i.e., width) of this preferred colonization zone. Sample points were arranged in each field in grids. To obtain sufficient sample resolution of the edge, the sample point grids were stratified so that grid lines were denser at the field's edge than within the field. From a practical viewpoint, targeted application in seed potato fields is likely to be applied aerially (to avoid having to drive over the rows). Because the swath width of aerial applications is approximately 19 m (which is less than that indicated by preliminary data) and aerial applications cannot be conducted in less than complete swaths, the outer 38 m (i.e., two complete application swaths) of each field was stratified with denser sample grids. To simplify spatial analyses, sample locations within the fields were arranged in a staggered, equilateral pattern to homogenize the lag distances between points. They were marked using a flag and later georeferenced using a differentially corrected global positioning system, and the geographic coordinates of the sample locations were recorded.

Fields were sampled along the margins daily to detect colonizing *M. persicae* alatae; their presence would initiate full field sampling. Once *M. persicae* was present in a field, it was sampled twice weekly, weather permitting. All life stages of *M. persicae* were sampled by taking one leaf from the lower third of two randomly selected plants in a 1 cm radius around each sample point and recording the total number of alate and apterous aphids detected in the two-leaf sample.

The supplied dataset (see Chapter 8 data file, **SPUD1.txt**, on accompanying CD) represents the *M. persicae* data collected from one sampled field. The following description will become clear after you prepare a point map with the dataset. Rows within the field ran east to west, resulting in a contrast between the potato crop canopy and the open bare soil at the ends of the rows. In addition, the presence of recently harvested crops in the adjacent field to the north and road surfaces to the south meant there also was significant contrast between canopy and bare ground. Consequently, stratified dense sample grids ringed this particular field. There was a total of 365 sample points in this field.

Point maps were constructed from these data and interpolated surfaces created to estimate the distribution of *M. persicae* populations throughout the fields. Interpolated surfaces predict values at unsampled locations based on the values of surrounding sampled locations. It is based on the assumption of spatial autocorrelation that values at locations closer together are more similar than those from points further apart.[58] A common method of assessing spatial autocorrelation is the semivariogram. Semivariograms (sometimes referred to simply as variograms) plot half the squared variance of the difference between values at data points against the distance between data points (essentially plot a measure of data variance by distance). If spatial autocorrelation is present, then interpolative techniques can be used to create predictive estimates of distributions. Different interpolative techniques use different methods of weighting to determine influence of close and distant values on individual point estimates. When all unsampled points have been estimated, the interpolation is presented as a surface, the elevation of which predicts the value at the underlying geographical point.

Myzus persicae population data were assessed for spatial autocorrelation using GS+ V.5.0 (Gamma Design Software, Plainwell, Michigan), and interpolated maps were prepared with ArcMap V.9.2 (Environmental Sciences Research Institute, Redlands, California) to visually assess the within-field distribution and density of *M. persicae* using appropriate techniques.

8.3.1.1 Description of Dataset

The dataset, **SPUD1.txt** found on the accompanying CD in Chapter 8 file, is a tab-delimited text file arranged with features in columns and the data for each sample point in rows. The data for each sample point include its location and *M. persicae* population counts from that point at each of seven dates (August 05, 08, 13, 16, 20, 23, and 26). The coordinates are in the columns titled XFIELD and YFIELD (X and Y coordinates, respectively). In addition, grid sample point names and numbers are supplied. The population data are split between alate and apterous *M. persicae*, resulting in two columns of population data for each sample point at each date (those columns titled "APT_" followed by a date refer to counts of apterous aphids, while those titled "ALAT_" followed by a date refer to counts of alate aphids). There are 18 columns titled as follows: Pointname, PointNum, XFIELD, YFIELD, APT_AUG5, ALAT_AUG5, APT_AUG8, ALATE_AUG8, APT_AUG13, ALATE_AUG13, APT_AUG16, ALAT_AUG16, APT_AUG20, ALAT_AUG20, APT_AUG22, ALAT_AUG22, APT_AUG26, and ALAT_AUG26. The population counts are #aphids/2plants. Keep this in mind when making comparisons to treatment thresholds (~1/10 plants).

8.3.1.2 Assessing Spatial Autocorrelation Using Semivariograms in GS+

Variograms of aggregated (i.e., spatially autocorrelated) data often take the form of a line function that reaches an asymptote (Figure 8.2). The Y intercept (referred to as the nugget) represents the inherent variance in the system (not related to distance); the maximum of the asymptote (the Sill) reflects the maximum amount of variance in the dataset, while the distance at which the asymptote occurs (the Range) is the distance beyond which variance is not related to distance between sample points. The semivariogram is a graphic representation of the structure of the autocorrelation of a dataset.

Variograms are not always straightforward and may sometimes be more challenging to interpret. Some of the semivariograms of the dataset provided provide examples of more complex structures than the idealized result shown in Figure 8.2.

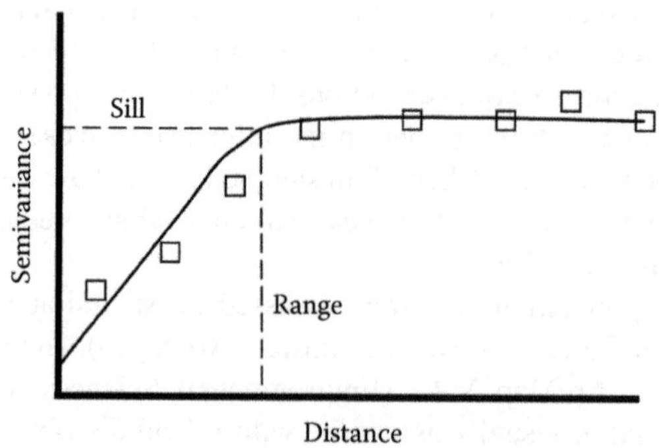

FIGURE 8.2 A standardized semivariogram measuring the structure of spatial autocorrelation of a dataset. The Y axis reflects the semivariance and X axis is the distance between sample points. The Y intercept is referred to as the nugget, the height of the asymptote is the sill, and the distance between sample points at which the asymptote occurs is the range.

1. Select **Data** from the **Menu Bar** and then click **Import Data File (text or binary file)**.
2. Highlight "**Spud1.txt**" in the **Select File** dialogue box. Click **OK** on the **Import file** dialogue box.
3. Double click on the columns for **XFIELD**.
4. In the **Field (Column) Assignment** dialogue box, click the box next to **X Coordinate** to set **XFIELD** as the X coordinate. At this time, you can ensure the decimal place and column width settings are appropriate for the dataset.
5. Click **OK**. You may receive a notice that before the program can build arrays, you need to set the other coordinates, simply click **OK** to move on.
6. Repeat the above steps for the **YFIELD** column, setting it as the Y coordinate.
7. You can set any of the population data columns as the Z coordinate to assess the spatial autocorrelation for that date's dataset. For the purposes of this exercise, assign column 7 (**APT_AUG16**) as the Z coordinate.
8. Select **Autocorrelation** from the **Menu Bar**, click on **Variogram Analysis**, and then select **Primary Variate (Z)** from the choices.
9. When the resulting dialogue box opens, click on **Calculate**.
10. In the dialogue box windows, semivariograms will appear, click on the **Expand** button underneath the **Isotropic Variogram**. Isotropic variograms assume there is no directional trend in the data. Because we know the structure of the dataset, we can opt for this selection. But a variogram that is calculated in all directions may hide or minimize some of the autocorrelation. Consequently, GS+ provides you with Anisotropic Variograms, calculated in specific directions. The strongest pattern for variance, being similar at close distances, dissimilar at medium distances, and similar again at long distances, is seen at 0°, straight across the field. Both 90° and 135° show interesting patterns as well. Can you explain these patterns? (think about the other field edges …)
11. Examine the semivariogram graphs (Figure 8.3).

Note that the variance increases with distance between points (i.e., data at closer locations tend to be more similar). But in this dataset, variance at first increases with distance, then decreases as data points get even further apart, and then increases again. This reflects the doughnut shape of the population distribution; points that are on the same field edge have similar values and are dissimilar to the data points in the center of the field (the population is clustered at the edges of the field). Unfortunately, the values at data points at all areas of the field (including the far edge) are similar because of this edge effect. Consequently, the data values are similar on the same edge, get less similar at distances from an edge to the middle, but get similar again at the distance to the far edge, resulting in the wave-like pattern of the semivariogram. There are a number of solutions to this problem (e.g., conducting semivariance analysis on a subset of the data—perhaps half the field).

Further exercises—GS+ allows for a number of other spatial analyses under **Autocorrelation** on the **Menu Bar**, such as Moran's I (a standard measure of aggregation) and Madogram analyses (similar to semivariograms analysis but less sensitive to

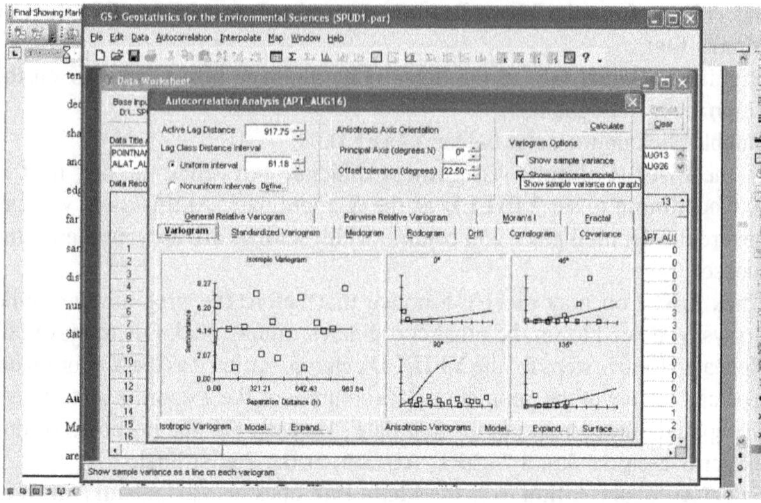

FIGURE 8.3 Semivariogram output from GS+ measuring spatial autocorrelation of the distribution of apterous *M. persicae* on August 16 from the Spud1.txt dataset.

extreme values than are semivariograms). In addition, this program will produce interpolated maps, much like ArcMap, under **Interpolate** on the **Menu Bar**. Before interpolating, you will first have to conduct the autocorrelation analysis as above. **Interpolate** menu has several choices including **krig**, **inverse distance weighting (IDW)**, and **simulation**. Each menu has specific inputs that can be modified at the user's discretion. When the inputs have been modified as desired, click **Calculate**. The results will open in a new screen (e.g., Figure 8.4). The **Map** function on the main menu bar allows for **3d Map**, **2d Map**, or **1d Transect** output display of the results. Other data in the dataset can be analyzed in similar manners. The Z field can be specified from any of

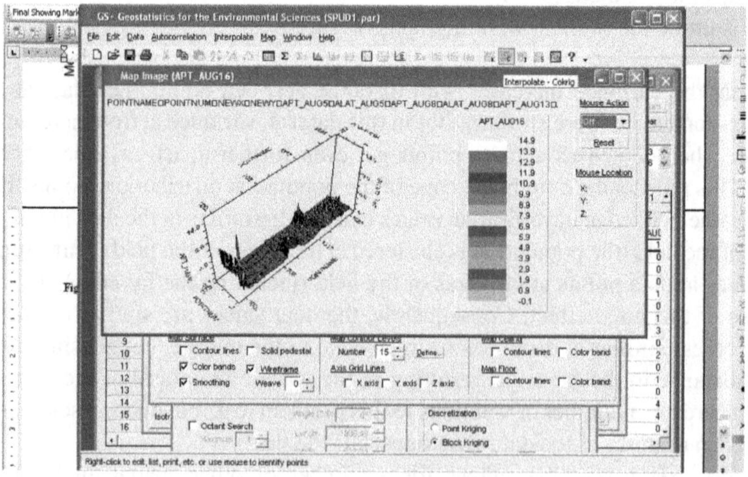

FIGURE 8.4 Interpolated surface output from GS+ showing the modeled distribution of apterous *M. persicae* on August 16 from Spudtxt1 dataset.

the columns containing insect data and new analysis conducted. The infestation levels at different dates and across areas of the field can then be compared.

8.3.1.3 Plot the Data from the Dataset in ArcMap

Using the **NEWX** and **NEWY** variables in the dataset **SPUD1.txt** as **X** and **Y** coordinates, create a point map of the sample points of the experimental field. Individual population levels at each date then can be assessed by labeling the sample points using the count data for an individual date. Open **ArcMap**.

1. Creating a point map of the data.
   a. Starting with a new map, add **SPUD1.txt** as a data file by
      i. Click the **ADD DATA** button (a plus sign on a yellow button) in the **Menu Bar** above.
      ii. Select **SPUD1.txt** and click **OK**.
   b. After **SPUD1.txt** has been added, no map will yet be displayed. Right click on the **SPUD1.txt** in the left column of **ArcMap** and select **Display XY Data**.
   c. Use the dropdown menus in the dialogue box to assign the column *NEWX* in **SPUD1.txt** as the XFIELD and *NEWY* as the YFIELD for the map coordinates. Click **OK**. Note—there will be an error message regarding the lack of an Object-ID file, ignore this message and click **OK**.
   d. A map of all the sample sites (dots) will appear (Figure 8.5).

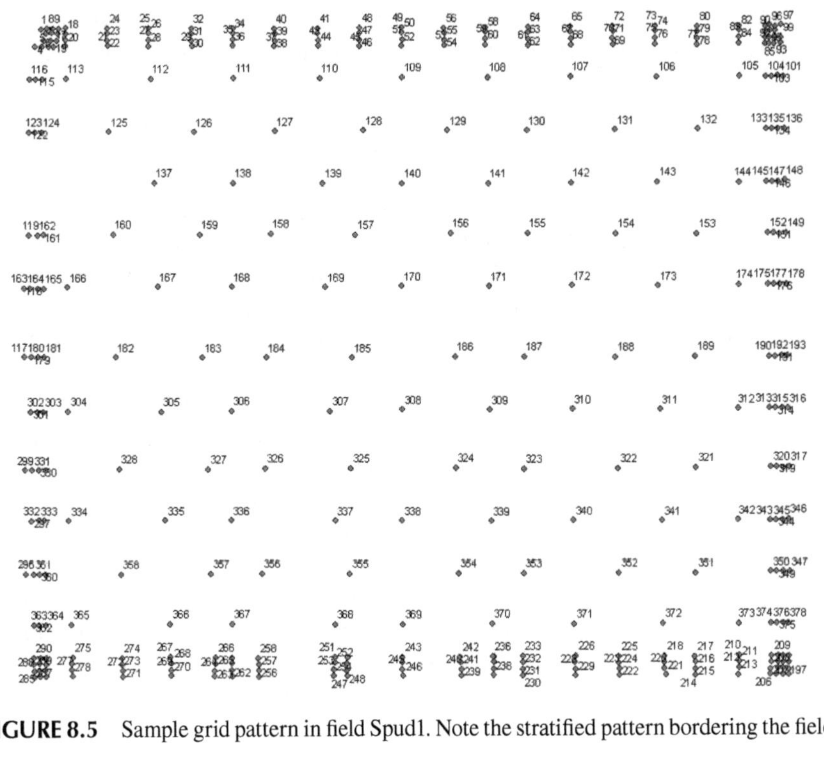

FIGURE 8.5 Sample grid pattern in field Spud1. Note the stratified pattern bordering the field.

   e. Right click on ***SPUD1.txt Events*** in the left hand column and select **Properties**.
   f. Select **Labels** in the dialogue box. Locate the dropdown menu box under **Text String**. Selecting ***POINTNUM*** in this dropdown menu will label the points on the map with their assigned sample number. Selecting the data column for any date will label the points of the map with the values at that sample date.
      i. Note the stratified pattern in the outer sample grids. These were designed to gain greater resolution of population count in the preferred colonization area. What impact does this design have on spatial analyses?
      ii. Note the more equilateral spacing of the interior sample points, what spatial analytical process does this pattern facilitate?
2. Creating an interpolated map of the data using IDW. (You will need to add the **3D Analyst** button [if not already present] to the Tool Box for this exercise.)
   a. Click on **3D Analyst** button. Select **Interpolate to Raster** in the **3D Analyst** dialogue box.
   b. Select **Inverse Distance Weighted**. In the dialogue box, select ***APT_AUG8*** in the **Z value field**: drop-down selection.
   c. Set the **Output Cell Size** to *1*. This reflects the 1 m^2 area sampled at each point.
   d. An IDW-interpolated map will appear (Figure 8.6A).
3. Creating an interpolated map of the data using the triangular irregular network (TIN). (You will need to add the **3D Analyst** button [if not already present] to the Tool Box for this exercise.)
   a. Click on the check mark beside ***IDW of SPUD1.txt Events*** in the left hand columns to turn off the IDW surface.
   b. Select **Create/Modify TIN** under the **3D Analyst** button.
   c. Select **Create TIN from Features** in the dialogue box.
   d. In the **Create TIN from Features** dialogue box, click on ***SPUD1.txt Events***, set the **Height Source** as ***APT_AUG8***, and click **OK**.
   e. A TIN-interpolated map will appear (Figure 8.6B).
   f. Right click on **TIN** in the left hand column and select **Properties**. In the **Properties** dialogue box, select **Symbology**.
   g. In the **Classification** dialog box, set the number in **Classes** to *8* and click the **Classify** button. In the next **Classification** dialog box, change the classification to **Manual**. The **Break Values** can now be modified. For this exercise, use the values **0, 0.5, 1.0, 1.5, 2.0, 2.5, 3.0**, and **6.0**. (Note: for other dates, other classification breaks may be more appropriate and can be modified as needed using this method.) Click **OK**.
   h. In the **Symbology** dialog box, **Color Ramps** can be modified, and by right clicking in the symbol/range/label area, symbols may be flipped (**Flip symbols**), classes removed (**Remove classes**), or other properties modified.
   i. To examine aphid density at other sampling dates, change the selection in the **Z value field**: drop-down selection box.

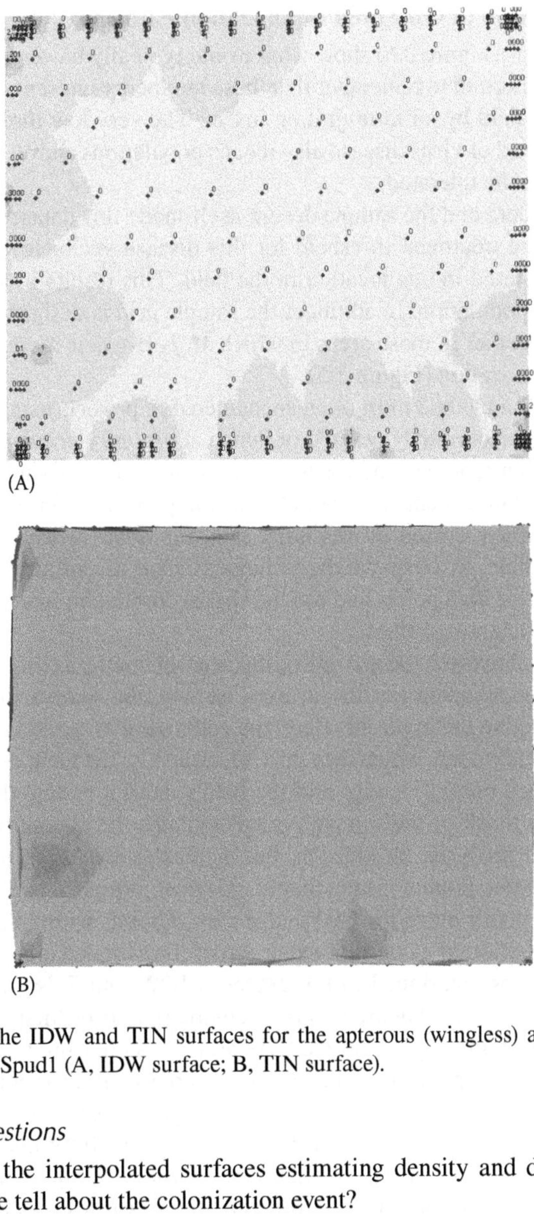

FIGURE 8.6 The IDW and TIN surfaces for the apterous (wingless) aphids sampled on August 8 in field Spud1 (A, IDW surface; B, TIN surface).

Suggested Questions

a. What do the interpolated surfaces estimating density and distribution of each stage tell about the colonization event?
b. At what date should targeted application be recommended? Keep in mind the sample data show number of aphids per two/plants.
c. At what date is targeted application likely to be ineffective?
d. Considering that alate aphids fly and have fewer offspring but on more plants than do apterous aphids, discuss which are more important as vectors of disease.
e. Research *Indicator Kriging* and determine if it is an effective tool for this exercise.

8.3.1.4 Discussion of Observed Colonization Patterns

The resulting maps (Figure 8.6) show what would typically be very low populations of aphids. It is important to remember that these numbers represent the first stages of colonization of a field by an immigrating insect. The very low threshold necessary to control the spread of virus disease also means populations above 1/10 plants at the field's edge cannot be tolerated.

The resulting data and the sample design itself made this dataset difficult to spatially analyze. The treatment threshold for this disease vector is very low, and its population is clustered in one location in the field. This results in most of the field area having zero population. In addition, the sample grid is designed so that sample sites are closer together in those areas in which *M. persicae* is more likely to first be found during colonization (Figure 8.5).

The interpolation method most often encountered in projects using geospatial data is *kriging*; the most commonly utilized of which is *ordinary kriging*, which assumes an unknown but constant mean across the sampled area, regularly utilized. Kriging is a group of interpolative techniques but all, including ordinary kriging assume some constant mean, which is used in weighting sampled values when estimating values for unsampled points. In other words, techniques such as ordinary kriging assume normality among the data points and use the spatial correlation structure of the entire dataset to assign weighting values.

With the current dataset, the use of kriging is problematic; a constant mean across the field cannot be assumed (recall our exercise with the semivariograms). In fact, the data indicate quite the opposite effect, the colonizing *M. persicae* were assumed to cluster at the field's edge, which they do. Consequently, the great disparity of aphid population between the field's edge and the field's interior (where the population is basically zero for much of the sample period) violates the assumptions for kriging as an appropriate analysis. In addition, the increased resolution of sample effort (i.e., clustered sample points) means that *M. persicae* presence is more likely to be found at sampled points along the field's edge than at points within the field. This too impacts the assumption of a constant mean across the sampled area.

The two methods mentioned in the exercise, IDW and TIN, also are not perfectly suited for these data but are less problematic than is ordinary kriging. When estimating values at unsampled points, IDW assigns decreasing weight to sampled values from known points as the known point increases in distance from the estimated point but was specifically designed for irregularly spaced data.[59] While the sample grid for the fields is stratified, it is not irregular. IDW also is sensitive to clustering (of both sample points and data) and outliers. While in some analytical situations this characteristic might be problematic, it was seen as beneficial when estimating the establishment of the aphid population. TINs are response surfaces prepared by triangulating and connecting closest points in such a manner that there is no overlap in point connections. This results in a network of connected, nonoverlapping triangles, the nodes of which are known sampled points, forming a surface representing estimated values at unsampled points. The interpolation weighting is dependent on the values of direct neighbors and can provide great resolution in areas with highly variable data.[60] Consequently, while neither method perfectly fits the current dataset, the resulting interpolations from both are more valid than those derived from ordinary kriging.

Note the difference between the two methods. Surfaces interpolated by IDW (e.g., Figure 8.6A) show higher population predictions in areas where no aphids were actually sampled than do surfaces interpolated with TIN (e.g., Figure 8.6B). This is partially due to the influence clustered data on the opposite side of a field have on IDW-interpolated surfaces, a problem with all interpolation techniques when attempting to model a "doughnut"-shaped population distribution[61,62] and the fact that TIN surfaces have less weighting associated with the estimated values between sampled points.[63]

While this exercise deals only with using GIS to visually assess the patterns of colonizing *M. persicae* populations, these data were also spatially analyzed numerically. A nonlinear, decaying exponential equation was used to model the response of distance from the field's edge on *M. persicae* population distribution, and additional analyses were used to determine the independent predictor and coefficient of determination. A more complete discussion of the mapping and statistical methodologies and results, including interpretation and management benefits, are available in Carroll.[7]

8.3.2 Exercise 2: Using the HYSPLIT Model to Determine If the Wind Vectors at Specific Dates Provided a Significant Risk of Aphid Immigration into the Red River Valley

Exercise 1 and other research, for example,[7,10,23] have demonstrated that *M. persicae* has a preference for colonizing field edges at the beginning of immigration. This preference facilitates site-specific application of insecticide and also provides economic savings by focusing field scouting and sampling efforts. Because *M. persicae* will first be found at the field's edge, it is now necessary to only scout the field periphery instead of the entire field, saving time, effort, and scouting costs. And when the first aphids are found, a targeted application of insecticide will control the newly immigrating population. This technique has been demonstrated as being effective in controlling the spread of aphid-vectored viral diseases when compared to full-field application of insecticide.[10] There are, however, opportunities to further refine sampling effort. Because *M. persicae* immigration is facilitated by predictable wind patterns, it is possible to use meteorological models to predict if there is the possibility of high or low *M. persicae* seasonal populations.

The model HYSPLIT will be used to identify potential LLJs that might facilitate *M. persicae* immigration. Using the criteria outlined in the chapter, you will examine a number of dates and assess if the wind vectors could potentially move *M. persicae* into the region.

8.3.2.1 Using HYSPLIT to Examine LLJ to Facilitate Movement of Aphids into the Red River Valley

1. Plot the LLJs.
    a. Using your browser, go to the HYSPLIT Web site (http://ready.arl. noaa. gov/HYSPLIT.php). The page should be titled **HYSPLIT** – Hybrid Single Particle Lagrangian Integrated Trajectory Model.

b. Select **Run HYSPLIT Trajectory Model**, your browser will be directed to a page titled **HYSPLIT Trajectory Model.**
c. Select **Compute *Archive* Trajectories**, and the next page will be **HYSPLIT Trajectory Model—Choose the Number and Type of Trajectory(ies)**. Select **Number of Trajectory *1*** and **Type of Trajectory** *Normal.* Click **Next**.
d. The next page is as follows: **Select the Meteorological Data and Starting Location(s)**. A selection box below will allow you to select the archived meteorological dataset with which to run to the HYSPLIT model. Select **EDAS 40km (U.S. 2004–present)**. EDAS40 is the Eta Data Assimilation System, a 3h, 40km resolution meteorological dataset generated by high frequency observations (wind profiler, NEXRAD, and aircraft data).[55] In the starting location box on the same page, enter ***48.01*** as the latitude and ***–97.25*** as longitude, and click **Continue.**
e. You will be directed to a page titled **Select the EDAS40 File**. You will need to select data archived over a specific time period. Select **edas.may07.001** (this is the first half of the dataset for May 2007) and click **Next.**
f. Your browser will be directed to a page titled **Model Run Details** with a box to **Change Default Model Parameters and Display Options**. This is the page on which you will enter the parameters for your LLJ plot. Select the following variables (italics indicate variable values to be entered):
    i. Trajectory direction: ***Backward*** (this plots where the LLJ originated)
    ii. Vertical Motion: ***Model vertical velocity***
    iii. Start time (UTC): year: *7*, month: ***05***, day: ***06***, hour: ***00***
    iv. Total run time (hours): ***48***
    v. Start a new trajectory: every ***0*** h, maximum number of trajectories: ***24***
    vi. Start latitude and longitude should already be set to ***48.01*** and ***–97.25***
    vii. Leave other latitudes and longitudes blank
    viii. Start height 1: ***500*** (plotting LLJs at 500m AGL)
    ix. Start height 2: ***1000*** (plotting LLJs at 1000m AGL)
    x. Start height 3: ***1500*** (plotting LLJs at 1500m AGL)
    xi. Display options can remain as defaults, and click the **Request Trajectory** button
g. Your browser will be directed to a page titled **HYSPLIT Model Run Submitted.** On this page will be a job number. If you want to download the results of the model run at a later date, you will need this number. Be aware the graphic plot will be deleted from the system after 6h.
h. Your browser will be directed to a page titled **HYSPLIT MODEL RESULTS FOR JOB NUMBER** with the job number. This page shows the progression of the model calculations, and after ~10–30s, a list of options will appear. Select **Trajectories *GIF***, a graphic of the resulting backtrack trajectory will open in a separate browser window. Note that the map results show three lines, the first for the 500m height (origin Louisiana/Mississippi border), the second for the 1000m height (mid-Texas), and the third for the 1500m height (Ohio/Kentucky border).

i. Repeat the exercise with the following 2007 dates: May 23, May 29, June 16, and June 18 (you will need to go back and select the appropriate dataset for the new dates). The results from June 18 are presented in Figure 8.7.

2. Compare the plots of LLJ movement on the five dates. Previous research indicated aphids immigrating into the RRV originate from an area bordered

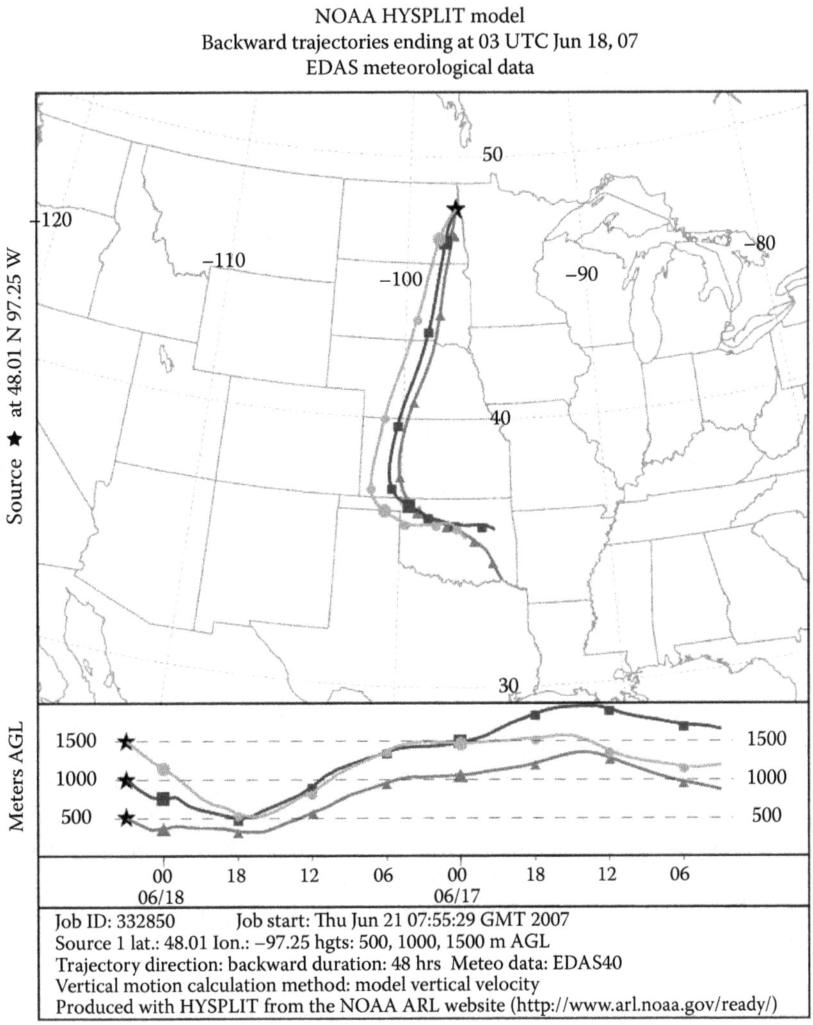

FIGURE 8.7 A backtrack plot from the HYSPLIT model showing LLJ events occurring on June 18, 2007. The star represents the endpoint of the 48 h LLJ event and the other end of the lines represents where the LLJs began. Plot lines denoted triangles are LLJs occurring at 500 m AGL, squares at 1000 m AGL, and circles at 1500 m AGL. Note that LLJ events initiated at a specific altitude do not necessarily remain at that altitude throughout the 48 h duration. An LLJ that terminates at altitude requires either a precipitation event or downdraft to deposit aphids at the endpoint.

25°–35°N, 95°–105°W, an area including most or all of Arkansas, eastern Colorado, Kansas, Louisiana, Missouri, Oklahoma, and northern Texas.

a. Which dates had LLJs that would facilitate immigration of *M. persicae* into the RRV?
b. Which date's LLJs would have no effect on *M. persicae* populations in the RRV?

8.3.2.2 Discussion of HYSPLIT Results

In a given year, using HYSPLIT modeling technique involves adding the cumulative hours of LLJs that could facilitate *M. persicae* immigration. In 2007, there were two additional LLJs that contributed to aphid immigration as well as the five positive events you just plotted. All seven LLJs ended either in precipitation events or in conditions that would have resulted in impaction. In 2007, *M. persicae* populations were high, and there were significant PVY and PLRV present in the RRV. Seed potato producers, crop consultants, and other agricultural professionals involved in the seed potato industry were advised to scout the margins of their fields carefully to detect colonization by aphids and target initial insecticide applications at the fields' edges. Several major seed potato producers in the region have adopted this technique, and their losses to virus disease in 2007 were no lower than those using full-field insecticide applications.[10]

A complete description of the HYSPLIT model, its documentation, and uses can be found on its Web site. This useful tool has been expanded, and an updated version and a desktop resident version are now available. To access the full options of HYSPLIT, a user must register with the site, but there is no fee, and model use is free to the public.

8.4 CONCLUSIONS

This case study indicated that using site-specific scouting for aphids at field edges was a successful method to save time, labor, and chemical input in seed potato production. However, along with the ground-truthed data, additional information about when immigration of the aphids could occur is needed. The aphids do not overwinter in the RRV but are moved by LLJs from southern areas of the United States. The HYSPLIT model can be used to examine wind currents in near real time and is a useful tool in predicting when aphid immigration(s) may occur during the current season. This information can then, in turn, be used as a trigger to start an edge-of-field monitoring program.

REFERENCES

1. Larson, W.E. and Robert, P.C. Farming by soil. In: Lal, R. and Pierce, F.J. (eds). *Soil Management for Sustainability*. Soil and Water Conservation Society, Ankeny, IA, p. 103, 1991.
2. Midgarden, D., Fleischer, S., and Smilowitz, Z. Site-specific integrated pest management impact on development of esfenvalerate resistance in Colorado potato beetle (Coleoptera: Chrysomelidae) and on densities of natural enemies. *J. Econ. Entomol.*, 90, 855, 1997.

3. French, N.M., Follet, H.P., Nault, B.A., and Kennedy, G.G. Colonization of potato fields in eastern North Carolina by Colorado potato beetle. *Entomol. Exp. Appl.*, 68, 247, 1993.
4. Weisz, R., Smilowitz, Z., and Fleischer, S. Evaluating risk of Colorado potato beetle (Coleoptera: Chrysomelidae) infestation as a function of migratory distance. *J. Econ. Entomol.*, 89, 435, 1996.
5. Blom, P.E., Fleischer, S.J., and Smilowitz, Z. Spatial and temporal dynamics of Colorado potato beetle (Coleoptera: Chrysomelidae) in fields with perimeter and spatially targeted insecticides. *Environ. Entomol.*, 31,149, 2002.
6. Zhu, M. Weather patterns influence seasonal abundance of green peach aphid, *Myzus persicae* (Sulzer), and spread of Potato leafroll virus in the northern Great Plains. PhD dissertation, University of Minnesota, St. Paul, MN, 2004.
7. Carroll, M.W. Spatial distribution of the green peach aphid, *Myzus persicae* (Sulzer), and management applications in seed potato. PhD dissertation, University of Minnesota, St. Paul, MN, 2005.
8. Zhu, M., Radcliffe, E.B., Ragsdale, D.W., MacRae, I.V., and Seeley, M.W. Low-level jet streams associated with spring aphid migration and current season spread of potato viruses in the U.S. northern Great Plains. *Agric. For. Meteorol.*, 138, 192, 2006.
9. Radcliffe, E.B., Ragsdale, D.W., Suranyi, R.A., DiFonzo, C.D., and Hladilek, E.E. Aphid Alert: How it came to be, what it achieved, and why it proved unsustainable. In: Koul, O. et al. (eds). *Areawide IPM: Theory to Implementation*. CAB International, Wallingford, U.K., p. 227, 2008.
10. Carroll, M.W., Radcliffe, E.B., MacRae, I.V., Ragsdale, D.W., Olson, K.D., and Badibanga, T. Border treatment to reduce insecticide use in seed potato production: Biological, economic, and managerial analysis. *Am. J. Potato Res.*, 86, 31, 2009.
11. van Emden, H.F., Eastop, V.F., Hughes, R.D., and Way, M.J. The ecology of *Myzus persicae. Annu. Rev. Entomol.*, 14, 197, 1969.
12. Leonard, M.D., Walker, H.G., and Enari, L. Host plants of *Myzus persicae* at the Los Angeles State and County Arbouretum, Arcadia, California. *Proc. Entomol. Soc. Wash.*, 72, 294, 1970.
13. Tamaki, G. Exploiting the ecological interaction of the green peach aphid on peach trees. *U.S. Dep. Agric. Sci. Tech. Bull.*, 1640, 1981.
14. Blackman, R.L. Life-cycle variation of Myzus persicae (Sulz.) (Hom., Aphididae) in different parts of the world, in relation to genotype and environment. *Bull. Entomol. Res.*, 63, 595, 1974.
15. NASS (U.S. Department of Agriculture National Agricultural Statistics Service). *Crop Production 2007 Summary*, 2007. Available online at: http://www.nass.usda.gov/index.asp
16. STATCAN (Statistics Canada). Potato area and yield survey, 2007. Available online at: http://www.statcan.ca/
17. Carden, P.W. and Golightly, W.H. Potato aphids, MAFF. Advisory Leaflet 575 (Rev), 1977.
18. Folwell, R.J., Fagerlie, D.L., Tamaki, G., Ogg, A.G., Comes, R., and Baritelle, J.L. Economic evaluation of selected cultural methods for suppressing the green peach aphid as a vector of virus diseases of potatoes and sugarbeets, Bulletin 900, College of Agriculture Research Center, Washington University. 1981.
19. Ragsdale, D.W., Radcliffe, E.B., and Dizonzo, C.D. Epidemiology and field control of PVY and PLRV. In: Loebenstein, G., Berger, P.H., Brunt, A.A., and Lawson, R.H. (eds). *Virus and Virus-Like Diseases of Potatoes*. Kluwer, Amsterdam, the Netherlands, pp. 237–270, 2001.
20. Radcliffe, E.B., Ragsdale, D.W., and Suranyi, R.A. IPM case studies: Seed potato. In: van Emden, H.F. and Harrington, R. (eds). *Aphids as Crop Pests*. CABI, Wallingford, U.K., 2007.

21. Harrison, B.D. Potato leafroll virus, CMI/AAB Descriptions of Plant Viruses, July 1984, No. 291 (No. 36 Rev). Commonwealth Agricultural Bureaux/Association of Applied Biologists, 1984.
22. Hanafi, A., Radcliffe, E.B., and Ragsdale, D.W. Spread and control of potato leafroll virus in Minnesota. *J. Econ. Entomol.*, 82, 1201, 1989.
23. DiFonzo, C.D., Ragsdale, D.W., Radcliffe, E.B., Gudmestad, N.C., and Secor, G.A. Crop borders reduce potato virus Y incidence in seed potato. *Ann. Appl. Biol.*, 129, 289, 1996.
24. Suranyi, R.A., MacRae, I.V., Ragsdale, D.W., and Radcliffe, E.B. Site-specific management of green peach aphids in seed potatoes. *Am. J. Potato Res.*, 77, 420 (abstr), 2000.
25. Radcliffe, E.B., Ragsdale, D.W., and Flanders, K.L. Management of aphids and leafhoppers. In: R.C. Rowe (ed.), *APS Plant Health Management in Potato Production.* American Phytopathological Society, St. Paul, MN, pp. 117–126, 1993.
26. Radcliffe, E.B. and Ragsdale, D.W. Invited review. Aphid transmitted potato viruses: The importance of understanding vector biology. *Am. J. Potato Res.*, 79, 353, 2002.
27. Perring, T.M., Gruenhagen, N.M., and Farrar, C.A. Management of plant viral diseases through chemical control of insect vectors. *Ann. Rev. Entomol.*, 44, 457, 1999.
28. Ragsdale, D.W., Radcliffe, E.B., DiFonzo, C.D., and Connelly, M.S. Action thresholds for an aphid vector of potato leafroll virus. In: Zehnder, G.W. et al. (eds). *Potato Pest Biology and Management*, American Phytopathological Society, St. Paul, MN, p. 99, 1994.
29. Flanders, K.L., Radcliffe, E.B., and Ragsdale, D.W. Potato leafroll virus spread in relation to densities of green peach aphid (Homoptera: Aphididae): Implications for management thresholds for Minnesota seed potatoes. *J. Econ. Entomol.*, 84, 1028, 1991.
30. Plymouth State Weather Center. Make your own product generator for archived data, 2009. Available online at: http://vortex.plymouth.edu/u-make.html (last accessed August 31, 2009).
31. Bonner, W.D. Climatology of the low-level jet. *Mon. Wea. Rev.*, 96, 735, 1968.
32. Stensrud, D.J. Importance of low-level jets to climate: A review. *J. Clim.*, 9, 1698, 1996.
33. Chen, T.C. and Kpaeyeh, J.A. The synoptic-scale environment associated with the low-level jet of the Great Plains. *Mon. Wea. Rev.*, 121, 416, 1993.
34. Cotton, W.R., Lin, M.S., McAnelly, R.L., and Tremback, C.J. A composite model of mesoscale convective complexes. *Mon. Wea. Rev.*, 117, 765, 1989.
35. Helfand, H.M. and Schubert, S.D. Climatology of the simulated Great Plains low-level jet and its contribution to the continental moisture budget of the United States. *J. Clim.*, 8, 784, 1995.
36. Higgins, R.W., Yao, Y., Yarosh, E.S., Janowiak, J.E., and Mo, K.C. Influence of the Great Plains low-level jet on summertime precipitation and moisture transport over the central United States. *J. Clim.*, 10, 481, 1997.
37. Schubert, S.D., Helfand, H.M., Wu, C., and Min, W. Subseasonal variations in warm-season moisture transport and precipitation over the central and eastern United States. *J. Clim.*, 11, 2530, 1998.
38. Uccellini, L.W. On the role of upper tropospheric jet streams and leeside cyclogenesis in the development of low-level jet in the Great Plains. *Mon. Wea. Rev.*, 108, 1689, 1980.
39. Mitchell, M.J., Arritt, R.W., and Labas, K. A climatology of the warm season Great Plains low-level jet using wind profiler observations. *Wea. Forecast.*, 10, 576, 1995.
40. Pedgley, D.E. *Windborne Pests and Diseases: Meteorology of Airborne Organisms.* Ellis Horwood, Chichester, U.K., 1982.
41. Westbrook, J.K. and Isard, S.A. Atmospheric scales of motion for dispersal of biota. *Agric. For. Meteorol.*, 97, 263, 1999.
42. Wu, Y. and Raman, S. The summertime Great Plains low level jet and the effect of its origin on moisture transport. *Bound. Layer Meteorol.*, 88, 445, 1998.

43. Kieckhefer, R.W., Lytle, W.F., and Spuhler, W. Spring movement of cereal aphids into South Dakota. *Environ. Entomol.*, 3, 347, 1974.
44. Wallin, J.R., Peters, D., and Johnson, L.C. Low-level jet winds, early cereal aphid and barley yellow dwarf detection in Iowa. *Plant Dis. Rep.*, 51, 527, 1967.
45. Wallin, J.R. and Loonan, D.V. Low-level jet winds, aphid vectors, local weather and barley yellow dwarf virus outbreaks. *Phytopathology*, 61, 1068, 1971.
46. Berry, R.E. and Taylor, L.R. High-altitude migration of aphids in maritime and continental climates. *J. Anim. Ecol.*, 37, 713, 1968.
47. Drake, V.A. Insects in the sea-breeze front at Canberra: A radar study. *Weather*, 37, 134, 1982.
48. Drake, V. A. and Farrow, R.A. The influence of atmospheric structure and motions on insect migration. *Annu. Rev. Entomol.*, 33, 183, 1988.
49. Medler, J.T. and Smith, P.W. Greenbug dispersal and distribution of barley yellow dwarf virus in Wisconsin. *J. Econ. Entomol.*, 53, 473, 1960.
50. Johnson, C.G. *Migration and Dispersal of Insects by Flight*. Methuen, London, U.K. 1969.
51. Draxler, R.R. Boundary layer isentropic and kinematic trajectories during the August 1993 North Atlantic Regional Experiment Intensive. *J. Geophys. Res.*, 101, 29255, 1996.
52. Draxler, R.R. *Hybrid Single-Particle Lagrangian Integrated Trajectories (HYSPLIT): User's Guide*, Version 4. NOAA Technical Memorandum ERL ARL-230, 1, Silver Spring, 1999.
53. Draxler, R.R. and Hess, G.D. An overview of the Hysplit 4 modeling system for trajectories, dispersion and deposition. *Aust. Meteorol. Mag.*, 44, 295, 1998.
54. Rolph, G.D. Real-time environmental applications and display system (READY), 2003. NOAA Air Resources Laboratory, Silver Spring, MD. Available online at: http://www.arl.noaa.gov/ready/hysplit4.html
55. Draxler, R.R. and Hess, G.D. *Description of the HYSPLIT_4* modeling system. NOAA Technical Memorandum ERL ARL-224. Silver Spring, MD. 1996.
56. Air Resource Laboratory. NOAA. *Hybrid* single-particle Lagrangian integrated trajectory (HYSPLIT) model, 2003. Available online at: http://www.arl.noaa.gov/ready/hysplit4.html
57. North America Plant Disease Forecasting Center (NAPDFC). Blue Mold Forecasting Center, 2003. Available online at: http://www.ces.ncsu.edu/depts/pp/bluemold (last accessed August 20, 2009).
58. Griffith, D. *Spatial Autocorrelation: A Primer*. Association of American Geographers, Washington, DC, 1987.
59. Shepard, D. A two-dimensional interpolation function for irregularly-spaced data. *Proceedings of the 1968 Association of Computing Machinery National Conference*, New York, p. 517, 1968.
60. Environmental Sciences Research Institute (ESRI). ArcGIS Desktop 9.3 Help. ESRI, Redlands, CA, 2009. Available online at: http://webhelp.esri.com/arcgisdesktop/9.3/index.cfm?TopicName=About_TIN_surfaces (last accessed October 31, 2009).
61. Isaaks, E.H. and Srivastava, R.M. *An Introduction to Applied Geostatistics*. Oxford University Press, Inc., New York, 1989.
62. Cressie, N.A.C. *Statistics for Spatial Data*. John Wiley & Sons, Inc., New York, 1993.
63. Krajewski, S.A. and Gibbs, B.L. *Understanding Contouring: A Practical Guide to Spatial Estimation Using a Computer and Basics of Using Variograms*. Gibbs Associates, Boulder, CO, 2001.

9 Analysis of the 2002 Equine West Nile Virus Outbreak in South Dakota Using GIS and Spatial Statistics

Michael C. Wimberly, Erik Lindquist, and Christine L. Wey

CONTENTS

9.1 Executive Summary 192
9.2 Introduction 192
9.3 Methods 193
9.3.1 Mapping WNv Cases in a GIS 193
9.4 Results 194
9.4.1 Smoothed Maps of Disease Risk 194
9.4.2 Spatial Autocorrelation Analysis 197
9.4.3 Spatiotemporal Clustering 198
9.5 Summary and Conclusions 201
9.6 Step-by-Step Exercise Using GeoDa 0.9.5 Software 202
9.6.1 Opening a Shapefile in GeoDa 202
9.6.2 Computing a Spatial Weights File 203
9.6.3 Creating a Map of Raw Disease Rates 203
9.6.4 Creating a Map of Disease Rates Using Empirical Bayes Smoothing 203
9.6.5 Adding Calculated Rates to the Attribute Table 203
9.6.6 Computing the Global Moran's I Index of Spatial Autocorrelation 204
9.6.7 Computing the Local Moran's I Index of Spatial Autocorrelation 204
References 205

9.1 EXECUTIVE SUMMARY

Emerging infectious diseases can impact the health of humans, domestic animals, and wildlife. Understanding the risk factors for disease emergence in agriculture is important because it impacts both crop and livestock productivity. West Nile virus (WNv) is indigenous to Africa, Asia, Europe, and Australia. It was first identified in New York during the summer of 1999. The virus rapidly spread through the Midwestern States with the first WNv cases in horses reported in South Dakota in 2002. This case study examines the pattern of WNv disease development in SD horse populations in one of the first years of its reported incidence using geographic information and spatial statistics.

9.2 INTRODUCTION

In the past several decades, the threat of emerging infectious disease has come to the forefront as an important concern in the fields of human and veterinary medicine as well as wildlife conservation.[1,2] Emerging infectious diseases include those diseases that have recently appeared for the first time, are rapidly increasing in prevalence, or have expanded into new geographic areas. Infectious disease emergence often occurs as the result of a species invasion, in which a pathogen, possibly in combination with one or more arthropod vectors or mammalian hosts, is introduced into a novel environment by humans. Understanding the risk factors for infectious disease emergence is important from an agricultural perspective because of the potential impacts on both crop and livestock productivity.[2,3]

As an example, H5N1, a novel subtype of highly pathogenic avian influenza, was first detected in poultry in Southeast China in 1996. In 2004, H5N1 began a rapid geographic expansion throughout Asia. The pathogen spread eastward and ultimately reached Western Europe and Africa in 2006. Although migratory birds were initially implicated as the primary drivers of this disease expansion,[4] a more recent assessment provides strong evidence that the emergence of H5N1 was in fact more likely driven by human activity via the poultry trade.[5] Most of the public's attention has been focused on the spillover risk from poultry to humans and the potential for resulting emergence of a global influenza pandemic in the human population.[6] However, the spread of H5N1 has already had a significant impact on agriculture within the newly invaded areas as a result of direct mortality caused by the disease, the culling of millions of birds to prevent disease spread, and indirect effects on poultry consumption and the poultry trade.[7,8]

One of the best-known examples of infectious disease emergence in the United States is the introduction of WNv and its subsequent spread across the entire North American continent. WNv is indigenous to Africa, Asia, Europe, and Australia and was first identified in North America in the New York City metropolitan area during the summer of 1999.[9] In 2002, widespread WNv cases occurred in the Midwest and south-central states, and the first WNv cases were reported in South Dakota. WNv is an arthropod-transmitted virus, or arbovirus, that is maintained in an enzootic cycle with birds as the primary reservoir hosts and mosquitoes as the primary vectors. Humans and horses are incidental hosts; both species can acquire the virus from infected mosquitoes but do not transmit the virus to uninfected mosquitoes and are not necessary for the maintenance of the pathogen.

In horses, symptoms of WNv include stumbling, weakness, muscle spasms, fever, and in some cases partial paralysis. Fatality rates are high, particularly for non-vaccinated horses. In the northern Great Plains, fatality rates were 22% during a 2002 outbreak in North Dakota,[10] and 44% during a 2003 outbreak in Saskatchewan.[11] Although horses in the United States are primarily leisure animals rather than agricultural products, the economic impacts of equine WNv outbreaks can still have a large impact in rural areas.[12] Our main goal in this chapter is to use the 2002 outbreak of equine WNv in South Dakota as a case study to demonstrate the ways in which geographic information systems (GIS) and spatial statistics can be used in the epidemiological study of an emerging infectious disease. The tools and techniques described herein can be extended to study other emerging infectious diseases of agricultural importance.

9.3 METHODS

9.3.1 Mapping WNv Cases in a GIS

Data on equine WNv cases occurring in 2002 were obtained from South Dakota State University Animal Disease Research and Diagnostic Laboratory (William Epperson, personal communication). Each case had associated information on the county, city, and zip code where the case occurred as well as the date of diagnosis. Postal addresses were not available, so it was not possible to geocode the cases and analyze the distribution of their individual locations. Instead, the total number of cases was summarized for two areal units: zip codes and counties. The resulting tables were then joined to ESRI shapefiles containing the boundaries of counties and zip code tabulation areas (ZCTAs) in South Dakota (Figure 9.1).

Linking disease cases to geographic areas based on the county in which the case occurred is relatively straightforward. However, mapping disease cases based on ZIP codes is more complicated. ZIP codes do not actually represent stable, precisely bounded geographic areas but are instead collections of street addresses that can change over time.[13] Data on ZIP code boundaries are generally not publically available. Instead, ZCTAs are commonly used areal representations of ZIP codes that were developed by the U.S. Census Bureau. ZCTAs are actually aggregations of census blocks that approximate ZIP code boundaries. The spatial and temporal mismatches between the ZIP codes that are reported with disease cases, and the ZCTAs that are used for mapping may lead to errors in associating cases with ZCTAs, which in turn may affect the results of spatial analyses.[13]

A comparison of maps of the total numbers of equine WNv cases at the county and zip code levels (Figure 9.1) highlights the modifiable areal unit problem (MAUP), a fundamental challenge in the spatial analysis of areal data.[14] The essence of the MAUP is that any study can be subdivided in an infinite number of ways to create areal units of varying sizes (the scale component of the MAUP) and shapes (the zonation component of the MAUP). The results of a spatial analysis, such as cluster detection or spatial regression, typically will vary when carried out on the same dataset

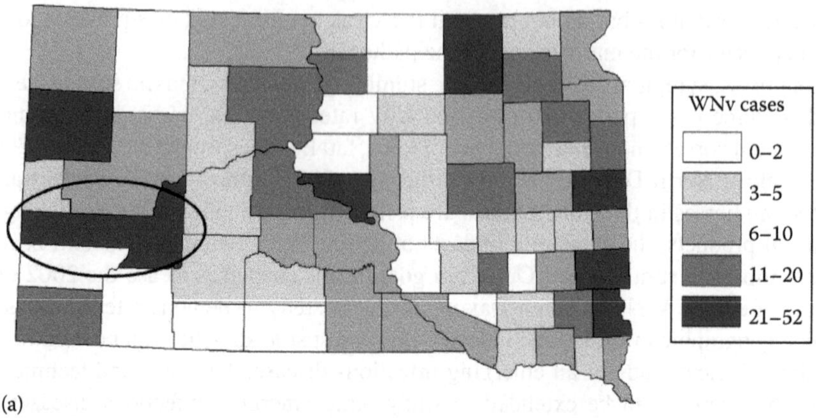

(a)

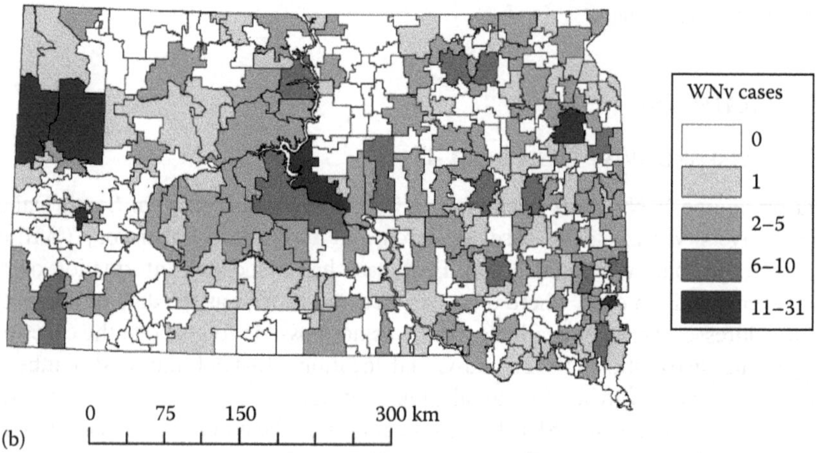

(b)

FIGURE 9.1 2002 equine WNv cases in South Dakota: case counts summarized (a) at the county level (Pennington county in circled) and (b) ZCTA level.

that is aggregated in different ways. In the present example, the smaller ZCTAs clearly reveal more fine-scale spatial variability than the county-level map of equine WNv cases (Figure 9.1). For example, in the county map, Pennington County is shown to have had one of the highest counts of equine WNv cases in 2002. However, the ZCTA map provides additional spatial detail that is masked at the county level. Most of the cases in Pennington County were concentrated in a few small ZCTAs in Rapid City, whereas cases were uncommon in the rest of Pennington County.

9.4 RESULTS

9.4.1 Smoothed Maps of Disease Risk

Maps of raw case counts are difficult to interpret because it is not possible to distinguish variability related to environmental risk factors from variability related to the size of the population at risk. Data on the numbers of horses per county were obtained from the 2002 U.S. Census of Agriculture and were joined to the South

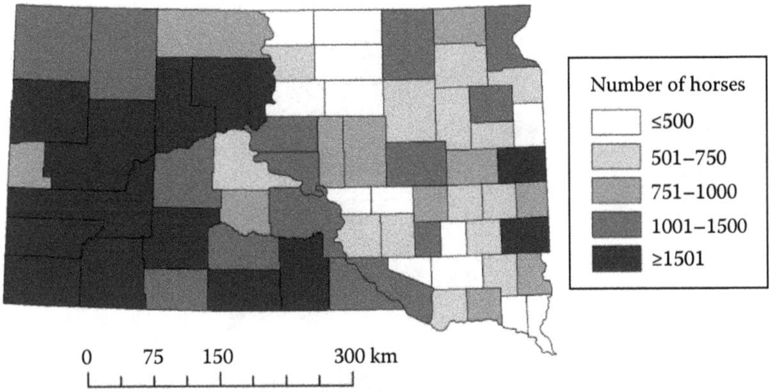

FIGURE 9.2 Numbers of horses per county from the 2002 U.S. Census of Agriculture.

Dakota county shapefile to map horse populations across the state. Counties in western South Dakota tended to have more horses than counties in eastern South Dakota. Data on horse populations were not available at the zip code level (Figure 9.2). Therefore, we proceeded with mapping of disease rates only at the county level. Raw rates were computed as

$$R_i = \left(\frac{C_i}{P_i}\right) \times 1000 \tag{9.1}$$

where

R_i was the raw rate for county i expressed as the number cases per 1000 horses
C_i was the number of equine cases
P_i was the number of horses

In contrast to the map of case counts, the map of disease rates showed a clear pattern in which 2002 equine WNv rates were higher in eastern South Dakota than in western South Dakota (Figure 9.3a). Although the map of raw rates provides a clearer picture of equine WNv risk patterns during the 2002 outbreak than the map of case counts, mapping of disease rates presents additional challenges. When the population at risk is highly variable, as is the case with the population of horses across South Dakota, variability in the rate estimates will be highest in areas with low populations.[15] Thus, we will have the highest confidence in rate estimates from counties with high populations. In contrast, rates are often inflated for areas with very low populations.

There are a number of methods for smoothing disease rates to produce more reliable maps of disease risk. One commonly used approach is empirical Bayes smoothing, in which local rate estimates are stabilized by "borrowing" information from a broader geographic area. The empirical Bayes estimate can be expressed mathematically as the weighted average of a local estimator, computed for the county of interest, and a global estimator computed using data pooled across the entire study area. Counties with low populations are thus "shrunk" toward the global estimator,

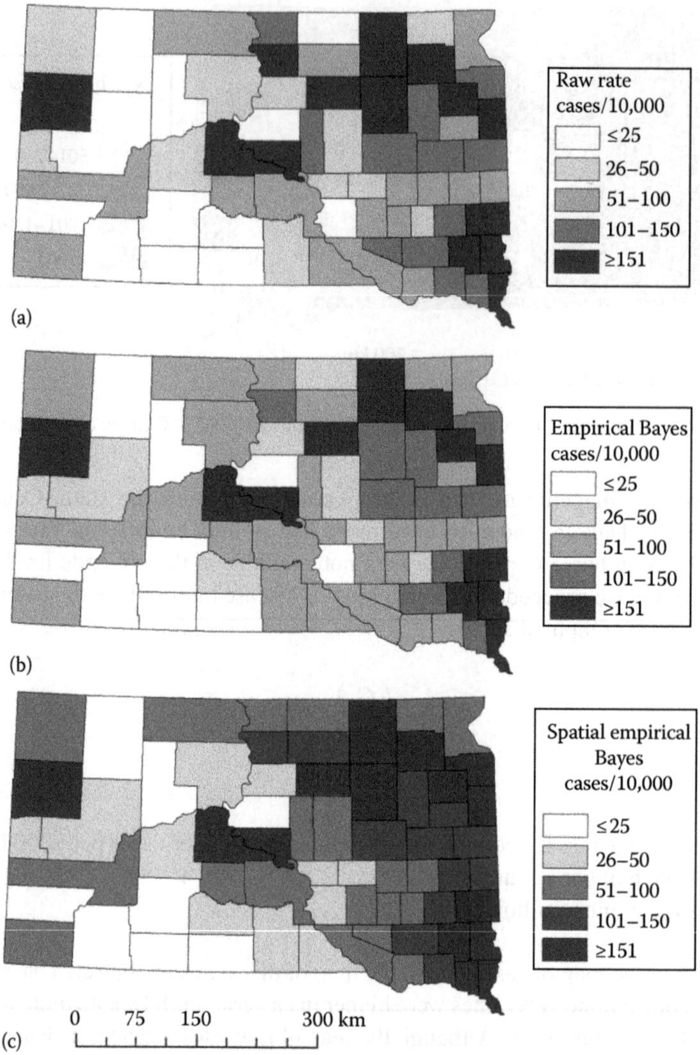

FIGURE 9.3 2002 equine WNv incidence rates in South Dakota summarized at the county level: (a) Raw rates, (b) empirical Bayes smoothing, and (c) spatial empirical Bayes smoothing.

whereas counties with high populations retain values close to the local estimator. The spatial empirical Bayes estimator operates in a similar fashion, except that local estimators are adjusted using the values in neighboring counties rather than the entire study area. A mathematical derivation of these estimators is provided in Marshall.[16] Further discussion of the broader topic of mapping smoothed disease rates is provided in Waller and Gotway.[15]

We used the GeoDa software package[17] to compute both empirical Bayes and spatial empirical Bayes rates for the 2002 equine WNv outbreak in South Dakota. The map of empirical Bayes rates generally was similar to the map of raw rates, but several of the counties in the highest and lowest rate categories shifted into the intermediate rates categories (Figure 9.3b). The counties that shifted were mostly in eastern South Dakota, where horse populations were relatively low. The map of spatial empirical Bayes rates resulted in even greater smoothing, emphasizing the statewide pattern of relatively high rates in the eastern portion of the state and low rates in the western portion of the state (Figure 9.3c).

9.4.2 Spatial Autocorrelation Analysis

To quantify the spatial pattern of 2002 equine WNv cases, we used both the global and local variants of the Moran's I statistic. The global Moran's I statistic is computed as

$$I = \frac{n}{\sum_{i=1}^{n}\sum_{j=1}^{n} w_{ij}} \times \frac{\sum_{i=1}^{n}\sum_{j=1}^{n} w_{ij}(x_i - \bar{x})(x_j - \bar{x})}{\sum_{i=1}^{n}(x_i - \bar{x})^2} \tag{9.2}$$

where

n is the total number of counties

x is the disease rate for each county

w_{ij} is a matrix of spatial weights containing values >0 when counties i and j are neighbors, and 0 otherwise

We defined the neighborhood of each county based on the queen's rule, in which all counties that shared either a line segment or a vertex were considered to be neighbors. Because the variable of interest was a disease rate, GeoDa was used to compute a variant of Moran's I that adjusts for the variability in population density across counties.[18]

Statistical significance was assessed using a permutation test, in which the x values were randomly assigned to counties using sampling without replacement, and a Moran's I value was computed for the permuted data. This permutation algorithm was repeated 999 times, and the observed I statistic was compared to the resulting distribution to test the null hypothesis that the disease rates are randomly distributed across the study area. The resulting value of Moran's I was 0.33, and the p-value was <0.001, indicating positive spatial autocorrelation—spatial clustering of counties with high and low disease rates.

Another way to visualize spatial autocorrelation is through a Moran scatterplot.[19] Each point in the Moran scatterplot represents a county (Figure 9.4). The horizontal axis represents the empirical Bayes rate for each county. The vertical axis represents the spatially lagged rate, which is equivalent to the weighted average of empirical Bayes rates for the neighbors of each county. When a line fitted to this scatter of points has a positive slope, it indicates that counties with high disease rates tend to be surrounded by neighbors with high disease rates, and counties with low disease rates tend to be surrounded by neighbors with low disease rates. When the weights for the neighbors of each county are row standardized to sum to unity, the slope of this line is equivalent to the Moran's I statistic.[19]

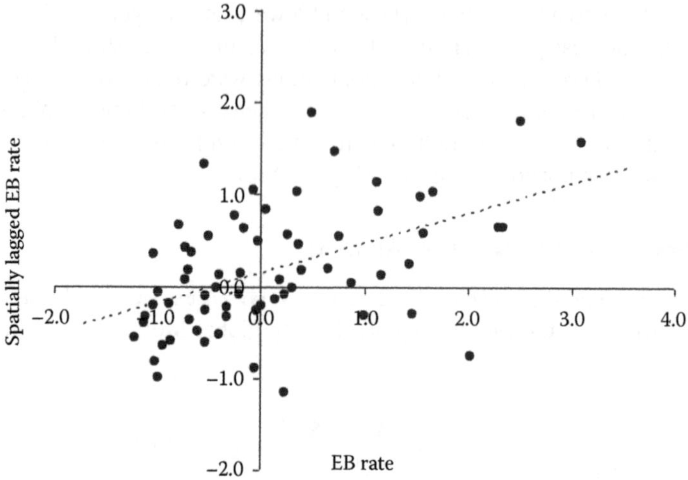

FIGURE 9.4 Moran scatterplot illustrating spatial autocorrelation of 2002 WNv incidence rates in South Dakota summarized at the county level.

A separate, local variant of the Moran's *I* statistic can also be computed for each county as

$$I_i = (x_i - \bar{x})\sum_{j=1}^{n} w_{ij}(x_j - \bar{x}) \tag{9.3}$$

Counties with positive local Moran's *I* statistics are those that have above-average rates and are surrounded by neighbors that have above-average rates, and those that have below-average rates and are surrounded by neighbors with below-average rates. These are the counties that are located in the upper right and lower left quadrants of the Moran scatterplot (Figure 9.4). Counties with negative local Moran's *I* statistics are those that have below-average rates and are surrounded by neighbors that have above-average rates, and those that have above-average rates and are surrounded by neighbors with below-average rates. These are the counties that are located in the upper left and lower right quadrants of the Moran scatterplot (Figure 9.4).

As with the global Moran's *I*, we used GeoDa to compute a modified variant of local Moran's *I* that adjusts for the variability in population density across counties.[18] High positive values of the local Moran statistic identified two zones of high disease rates in northeastern and southeastern South Dakota, and a zone of low disease rates in southwestern South Dakota (Figure 9.5a). A permutation test was used to identify counties where local Moran statistics were significantly higher or lower than zero at the $p = 0.05$ level (Figure 9.5b).

9.4.3 Spatiotemporal Clustering

Purely spatial analyses such as global and local Moran's *I* do not consider the temporal aspect of disease occurrence. In 2002, the first equine WNv cases occurred in

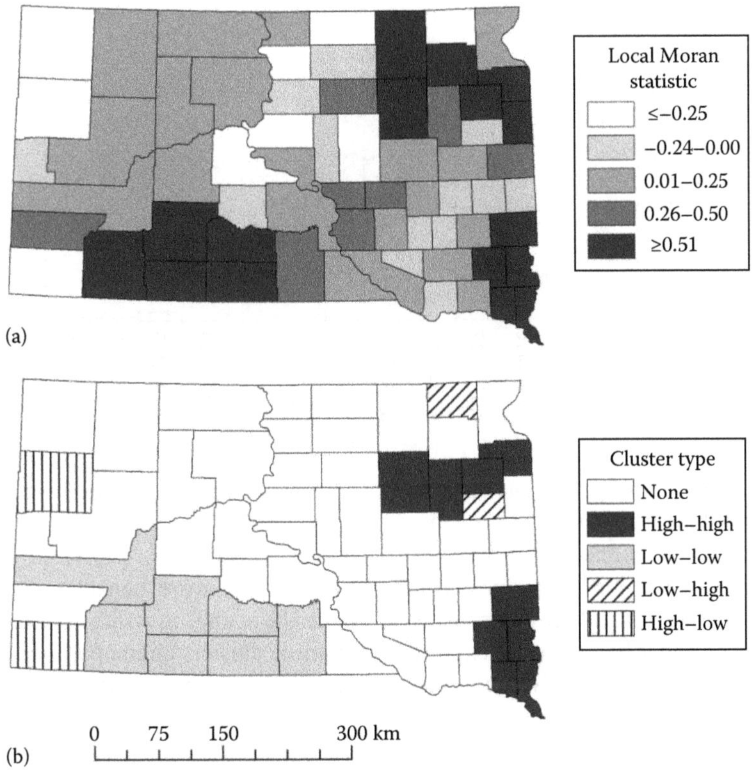

FIGURE 9.5 Local Moran's *I* analysis of 2002 WNv incidence rates in South Dakota summarized at the county level. (a) Local Moran statistics. (b) Counties identified as having statistically significant local Moran statistics ($p < 0.05$) along with their cluster types: high-rate counties with high-rate neighbors, low-rate counties with low-rate neighbors, low-rate counties with high-rate neighbors, and high-rate counties with low-rate neighbors.

late July, cases peaked during the last week of August, and the last cases occurred in mid-October (Figure 9.6). Thus, across the entire state, there was clear evidence of temporal clustering, with the majority of cases occurring between mid-August and mid-September. A further question is whether there is evidence of spatiotemporal clustering. Did the progression of equine WNv cases follow a similar temporal trajectory at all locations in South Dakota, or are there distinctive spatial clusters of cases that occurred at different times?

To carry out the spatiotemporal clustering analysis, we used the space–time permutation statistic, a variant of the spatial scan statistic.[20] This statistic was developed to test for groups of cases that are close to one another in both space and time. One advantage to this statistic is that it is based only on the distributions of case counts and does not require data on underlying population. Therefore, we were able to carry out a spatiotemporal analysis using the ZCTA level data. The space–time permutation statistic uses a variable-sized cylindrical "window" to scan for possible clusters. The base of the cylinder was centered on each ZCTA, and a range of different-sized circles was examined to consider potential clusters

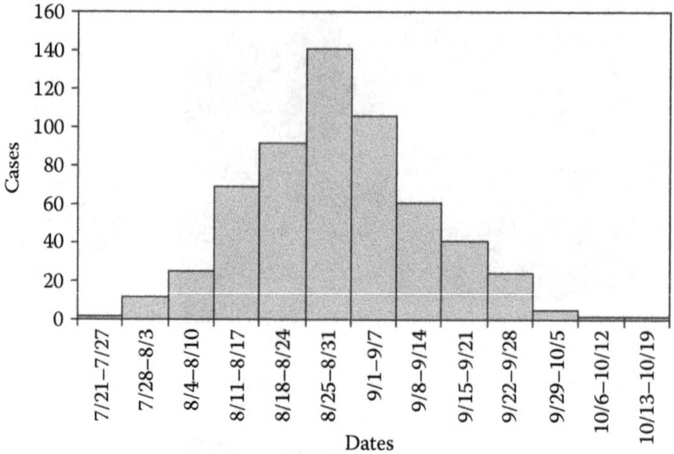

FIGURE 9.6 Temporal distribution of 2002 WNv cases in South Dakota expressed as the number of cases per week.

of varying locations and sizes. Neighboring ZCTAs were considered to fall within the cluster if their centroids were inside the scanning window. Similarly, varying cylinder heights were explored to examine clusters encompassing different time periods. For a given cylindrical window, a likelihood ratio statistic was defined as

$$L = \left(\frac{c_A}{\mu_A}\right)^{c_A} \left(\frac{C - c_A}{C - \mu_A}\right)^{C - c_A} \tag{9.4}$$

where

- C was the total number of cases in the study area
- c_A was the number of cases in the window
- μ_A was the expected number of cases in the window under the null hypothesis that there is no space–time interaction[20]

This likelihood was maximized across all possible spatial cluster locations and time periods to identify the most likely disease cluster. A test of statistical significance was carried out using a Monte Carlo simulation of 999 random datasets generated under the null hypothesis. The space–time permutation analysis was carried out using the SaTScan™ software package.[21]

We identified two statistically significant clusters of equine WNv cases in eastern South Dakota. The locations of these two clusters were generally similar to the areas of high disease rates identified in the county-level analyses (Figure 9.5). However, the spatiotemporal analysis also showed that the cases in these two areas occurred at different times. Cases within the larger spatial cluster located in northeastern South Dakota mostly occurred at the peak of the WNv season in early August (Figure 9.7). In contrast, cases within the smaller spatial cluster located

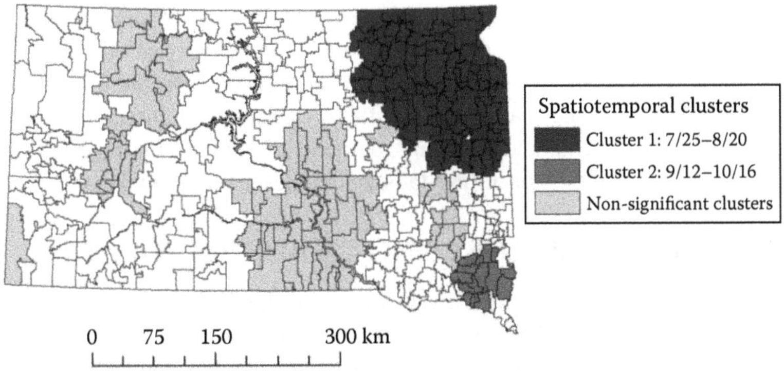

FIGURE 9.7 Spatiotemporal clusters of 2002 WNv cases in South Dakota identified using the space–time permutation statistic. Statistically significant clusters ($p < 0.05$) identify areas where groups of cases are close to one another in both space and time.

in southeastern South Dakota occurred near the end of the WNv season in late September and early October, after the overall rate of WNv cases in the state had decreased dramatically.

9.5 SUMMARY AND CONCLUSIONS

The results presented in this study provide a descriptive overview of the spatial and temporal patterns of the 2002 equine WNv outbreak in South Dakota. We have demonstrated some of the strengths and limitations of spatial analysis of disease patterns in a GIS, emphasizing how applying appropriate spatial analytical techniques can help clarify important patterns. One of the challenges in analyzing veterinary cases is that reliable data on populations at risk are seldom available for spatial units smaller than counties. Although we were able to proceed with analysis of equine WNv rates at the county level, a comparison of the ZCTA and county-level maps of case counts showed that there were potentially important, finer-scale patterns masked at the county level. Smoothing of the rates using empirical Bayes methods and spatial autocorrelation analysis using Moran's I indicated that there was spatial clustering of the county-level disease rates, with two distinct areas of high rates in eastern South Dakota. A further spatiotemporal clustering analysis identified WNv clusters in the same two areas, but also showed that they occurred during different time periods.

These exploratory analyses can serve as a basis for hypothesis development and further explanatory analyses. For example, we are typically interested in determining the environmental risk factors that lead to the emergence of disease clusters in certain areas. One hypothesis might be that particular environmental conditions, such as higher rainfall or a greater prevalence of breeding sites, are responsible for higher WNv rates in eastern South Dakota. Alternately, dispersal limitation might explain the higher WNv rates in eastern South Dakota, since the disease spread across the country from east to west and may not yet have been widely established in wild bird communities in 2002. Similar hypotheses might be developed to explain the spatiotemporal clustering pattern. The emergence of the late-season cluster in southeastern South Dakota may have been the result of different weather patterns from other parts of the state, or might have occurred because the virus did not disperse into this geographic area until much later in the season. Testing these types of hypotheses would require additional GIS datasets characterizing environmental variability in space and time, as well as appropriate methods for fitting statistical models using spatial data.

9.6 STEP-BY-STEP EXERCISE USING GEODA 0.9.5 SOFTWARE

The following describes how to perform several of the analyses from this chapter. All of the analyses described below were carried out using the GeoDa 0.9.5 software. This freeware package can be downloaded from the GeoDa Center for Geospatial Analysis and Computation at Arizona State University.

http://geodacenter.asu.edu/

GeoDa can read and modify data in the ESRI shapefile format. For the analyses in this chapter, we used GeoDa to carry out the computations, output the results to the shapefile attribute table, and then used ArcGIS 9.3 to make the final maps.

WNV case counts were provided by the South Dakota Animal Disease Research and Diagnostic Laboratory. The shapefile *sd_equine_wnv.shp* contains county-level data for South Dakota in the following fields:

FIPS: State/county FIPS code

cnty_name: County name

equine_case: Total number of equine WNV cases occurring in 2002

pophorses: Total number of horses on farms from the 2002 U.S. Census of Agriculture.

9.6.1 Opening a Shapefile in GeoDa

1. Choose **File-> Open Project**
2. Select ***sd_equine_wnv.shp*** as the **Input Map,** select ***FIPS*** as the **Key Variable**, and click **OK**. The results of opening the shapefile in GeoDa are presented in Figure 9.8.

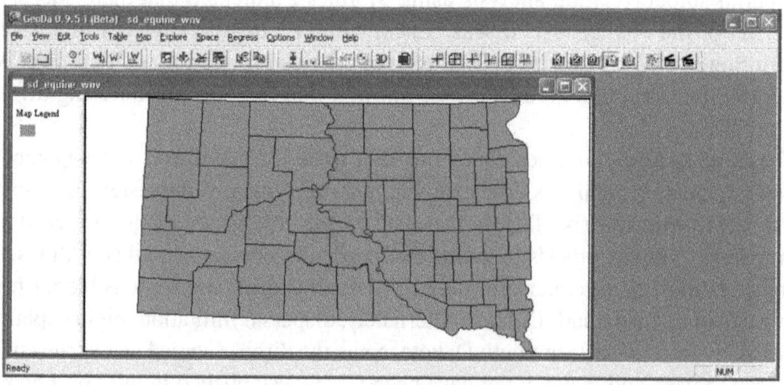

FIGURE 9.8 Results of opening the shapefile in GeoDa.

9.6.2 Computing a Spatial Weights File

1. Select **Tools-> Weights-> Create**
2. Select ***sd_equine_wnv.shp*** as the **Input File**, select an output directory and ***sd_cnty_queen.gal*** as the file name under **Save output as**, click the **Queen Contiguity** button under **Contiguity Weight**, and click the **Create** button at the bottom of the window.

9.6.3 Creating a Map of Raw Disease Rates

1. Select **Map-> Smooth-> Raw Rate**
2. Select ***EQUINE_CAS*** as the **Event Variable,** ***POPHORSES*** as the **Base Variable**, **Quantile Map** as the **Map Theme**, and click **OK**
3. Select ***4*** as the **# of Classes/Groups**, click **OK**. The results of creating a raw disease map are shown in Figure 9.9.

9.6.4 Creating a Map of Disease Rates Using Empirical Bayes Smoothing

1. Select **Map-> Smooth-> Empirical Bayes**
2. Select ***EQUINE_CAS*** as the **Event Variable**, ***POPHORSES*** as the **Base Variable**, **Quantile Map** as the **Map Theme**, and click **OK**.
3. Select ***4*** as the **# of Classes/Groups**, click **OK**. The map created using empirical Bayes smoothing is shown in Figure 9.10.

9.6.5 Adding Calculated Rates to the Attribute Table

To save the calculated rates to the attribute table,

1. Click the **Table** button on the button bar.
2. Select **Table-> Add Column** to add a new field.
3. Select **Table-> Field Calculation** and choose the **Rate Operations** tab.

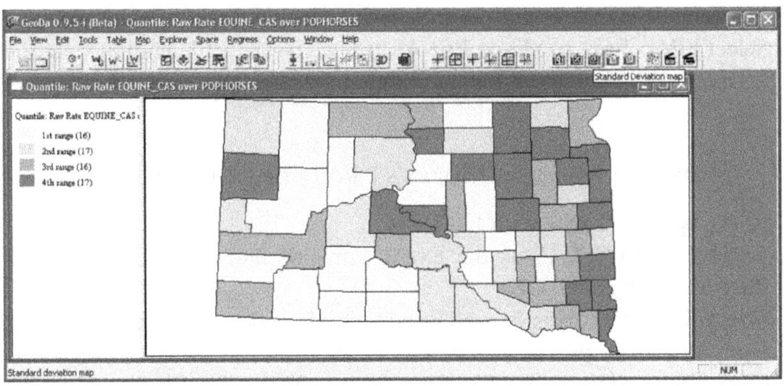

FIGURE 9.9 Results of creating a raw disease map using GeoDa.

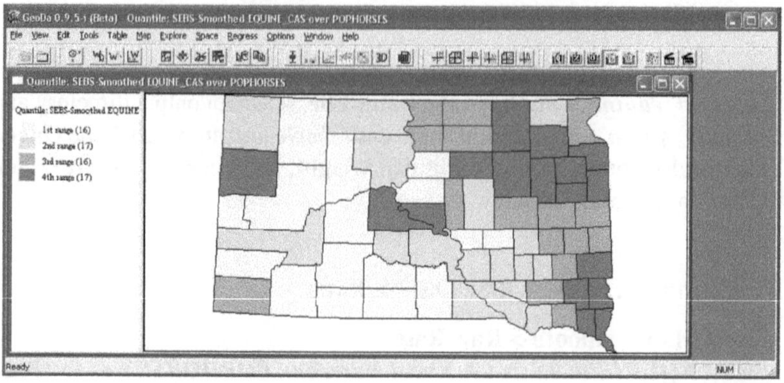

FIGURE 9.10 Results of creating a map of disease rates using empirical Bayes smoothing.

4. Select the field that will contain the calculated values under **Result**.
5. Select the type of rate that you want to calculate under **Methods** and select the **Event Variable** (***EQUINE_CAS***) and **Base Variable** (***POPHORSES***). Click **OK** to update the calculated values in the Result field.
6. Select **File-> Save to Shape File As** to save the shapefile with the modified attribute table.

9.6.6 Computing the Global Moran's I Index of Spatial Autocorrelation

1. Select **Space-> Moran's I with EB Rate**
2. Select ***EQUINE_CAS*** as the **Event Variable** and ***POPHORSES*** as the **Base Variable**, click OK.
3. Select either the currently used weights or ***sd_cnty_queen.gal*** under **Select from file,** click **OK**.
4. To compute a *p*-value, right click in the **Moran's I Scatterplot Window** and select **Randomization->** ***999*** **Permutations**.
5. To output the results to the attribute table, right click in the **Moran's I Scatterplot Window** and select **Save Results**.
6. Select **File-> Save to Shape File As** to save the shapefile with the modified attribute table. Results of this calculation are shown in Figure 9.11.

9.6.7 Computing the Local Moran's I Index of Spatial Autocorrelation

1. Select **Space-> LISA with EB Rate**
2. Select ***EQUINE_CAS*** as the **Event Variable** and ***POPHORSES*** as the **Base Variable**, click **OK**.
3. Select either the currently used weights or ***sd_cnty_queen.gal*** under **Select from file**, click **OK**.
4. Click all check boxes for **The Significance Map**, **The Cluster Map**, **The Box Plot**, and **The Moran Scatter Plot**.

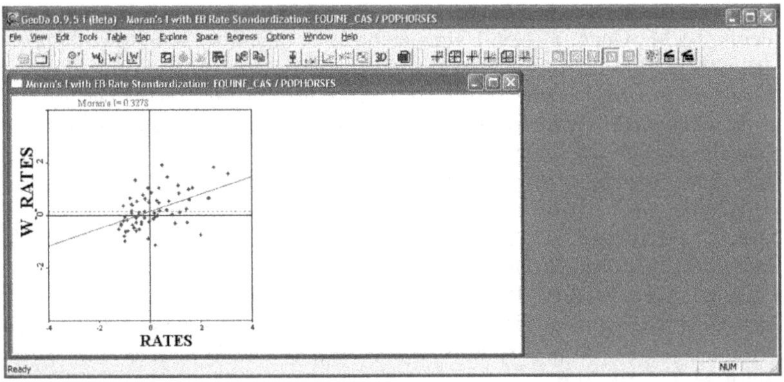

FIGURE 9.11 Results of computing the global Moran's I Index of spatial autocorrelation.

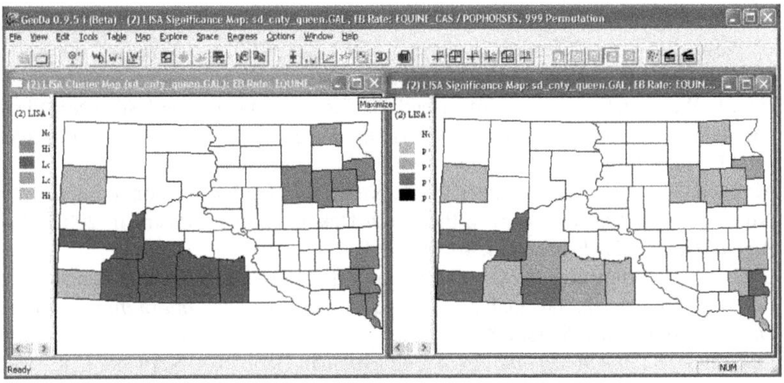

FIGURE 9.12 Results of computing the local Moran's I Index of spatial autocorrelation.

5. To output the **LISA** results to the attribute table, right click on one of the **output windows** and select **Save Results**.
6. Select **File-> Save to Shape File As** to save the shapefile with the modified attribute table. The results from computing the local Moran's I Index of spatial autocorrelation are presented in Figure 9.12.

REFERENCES

1. Daszak, P., Cunningham, A.A., Hyatt, A.D., Wildlife ecology—Emerging infectious diseases of wildlife—Threats to biodiversity and human health, *Science*, 287, 443, 2000.
2. Cleaveland, S., Laurenson, M.K., Taylor, L.H., Diseases of humans and their domestic mammals: Pathogen characteristics, host range and the risk of emergence, *Philosophical Transactions of the Royal Society of London Series B—Biological Sciences*, 356, 991, 2001.
3. Anderson, P.K., Cunningham, A.A., Patel, N.G., Morales, F.J., Epstein, P.R., Daszak, P., Emerging infectious diseases of plants: Pathogen pollution, climate change and agro-technology drivers, *Trends in Ecology & Evolution*, 19, 535, 2004.

4. Chen, H., Smith, G.J.D., Zhang, S.Y., Qin, K., Wang, J., Li, K.S., Webster, R.G., Peiris, J.S.M., Guan, Y., H5N1 virus outbreak in migratory waterfowl, *Nature*, 436, 191, 2005.
5. Gauthier-Clerc, M., Lebarbenchon, C., Thomas, F., Recent expansion of highly pathogenic avian influenza H5N1: A critical review, *Canadian Veterinary Journal—Revue Veterinaire Canadienne*, 149, 202, 2007.
6. Katz, J.M., The impact of avian influenza viruses on public health, *Avian Diseases*, 47, 914, 2003.
7. Capua, I., Alexander, D.J., Animal and human health implications of avian influenza infections, *Bioscience Reports*, 27, 359, 2007.
8. Sharkey, K.J., Bowers, R.G., Morgan, K.L., Robinson, S.E., Christley, R.M., Epidemiological consequences of an incursion of highly pathogenic H5N1 avian influenza into the British poultry flock, *Proceedings of the Royal Society B—Biological Sciences*, 275, 19, 2008.
9. Hayes, E.B., Komar, N., Nasci, R.S., Montgomery, S.P., O'Leary, D.R., Campbell, G.L., Epidemiology and transmission dynamics of West Nile Virus disease, *Emerging Infectious Diseases*, 11, 1167, 2005.
10. Schuler, L.A., Khaitsa, M.L., Dyer, N.W., Stoltenow, C.L., Evaluation of an outbreak of West Nile virus infection in horses: 569 cases (2002), *Journal of the American Veterinary Medical Association*, 225, 1084, 2004.
11. Epp, T., Waldner, C., West, K., Townsend, H., Factors associated with West Nile virus disease fatalities in horses, *Canadian Veterinary Journal—Revue Veterinaire Canadienne*, 48, 1137, 2007.
12. Mongoh, M.N., Hearne, R., Dyer, N.W., Khaitsa, M.L., The economic impact of West Nile virus infection in horses in the North Dakota equine industry in 2002, *Tropical Animal Health and Production*, 40, 69, 2008.
13. Grubesic, T., Matisziw, T., On the use of ZIP codes and ZIP code tabulation areas (ZCTAs) for the spatial analysis of epidemiological data, *International Journal of Health Geographics*, 5, 58, 2006.
14. Openshaw, S., *The Modifiable Areal Unit Problem.* Geobooks, Norwich, U.K., 1984.
15. Waller, L.A., Gotway, C.A., *Applied Spatial Statistics for Public Health Data.* John Wiley & Sons, Hoboken, NJ, 2004.
16. Marshall, R.J., Mapping disease and mortality-rates using empirical Bayes estimators, *Journal of the Royal Statistical Society. Series C, Applied Statistics*, 40, 283, 1991.
17. Anselin, L., Syabri, I., Kho, Y., GeoDa: An introduction to spatial data analysis. *Geographical Analysis*, 38, 5–22, 2006.
18. Assuncao, R.M., Reis, E.A., A new proposal to adjust Moran's I for population density, *Statistics in Medicine*, 18, 2147, 1999.
19. Anselin, L., Local indicators of spatial association—LISA, *Geographical Analysis*, 27, 93, 1995.
20. Kulldorff, M., Heffernan, R., Hartman, J., Assuncao, R., Mostashari, F., A space-time permutation scan statistic for disease outbreak detection, *PLoS Medicine*, 2, e59, 2005.
21. Kulldorff, M., SaTScan™ v7.0.3: Software for the spatial and space-time scan statistics. Information Management Services, Inc, 2006.

10 Designing a Local-Scale Microsimulation of Lesser Grain Borer Population Dynamics and Movements

J. M. Shawn Hutchinson, James F. Campbell, Michael D. Toews, Thomas J. Vought, Jr., and Sonny B. Ramaswamy

CONTENTS

10.1 Executive Summary 208
10.2 Introduction 208
  10.2.1 Lesser Grain Borer Economic Impact and Management 209
  10.2.2 Behavior and Ecology outside Grain Storage 211
10.3 Geocomputation 212
  10.3.1 Agent-Based Simulation and Modeling 212
  10.3.2 About NetLogo 213
10.4 Methods 214
  10.4.1 Overview: Creating a NetLogo Model 214
  10.4.2 Model Setup 215
    10.4.2.1 Setup and Go Buttons and Energy Switch 215
    10.4.2.2 Turtle Variables and Setup Procedure 216
    10.4.2.3 Defining Initial Variables with Sliders 218
    10.4.2.4 Go Procedure 219
    10.4.2.5 Bug Movement, Eating, Reproduction, and Death Procedures 220
    10.4.2.6 Setting Up Gain-from-Grain, Bug-Birth-Energy Procedures, and Control Sliders 222
    10.4.2.7 Forest Regrowth Procedure and Control Slider 222
    10.4.2.8 Show Energy and Display Labels 223
  10.4.3 Plot Procedure 225
    10.4.3.1 Create Plot Window and Monitors 226
10.5 Conclusions 228

Acknowledgments..229
References...230

10.1 EXECUTIVE SUMMARY

Many systems, including those studied in the agricultural sciences and biogeography, are inherently complex and require the use of many different spatial analysis and modeling techniques to understand underlying spatial patterns and processes. In this context, the methods encompassed by the term "geocomputation" represent a rapidly evolving, and increasingly effective, suite of tools to both analyze and visualize problems that cannot be addressed adequately with the use of standard aspatial or spatial statistics. Geocomputation methods include agent-based models, genetic algorithms, cellular automata, and neural networks. Agent-based models, in particular, are increasingly being used to examine a variety of questions involving spatiotemporal dynamics (e.g., changes in landuse/landcover, urban population growth) and the temporal paths of mobile entities (e.g., vehicular traffic analysis, pedestrian flows, disease dispersal, animal movements). Presented here is the rationale behind, and instructions for, constructing a simple agent-based model that can be used to form or revise theory regarding the early season infestation of grain storage bins by a small (3–5mm) beetle called the lesser grain borer (LGB) (*Rhyzopertha dominica*). LGB is a devastating, long-lived pest of stored cereal grains worldwide and the most damaging pest of stored wheat in the United States. Extending our understanding of LGB populations to include an explicit spatial component via agent-based simulation provides an opportunity to better evaluate the connectivity of grain storage bins on neighboring farms and more fully appreciate the role of landuse, landcover, and landscape configuration in shaping insect source-sink dynamics.

10.2 INTRODUCTION

LGB, *Rhyzopertha dominica* (F.) (Coleoptera: Bostrichidae), is a small beetle, but its mobility and appetite for stored cereal grains make it one of the most damaging pests of stored wheat (*Triticum aestivum*) worldwide, including the United States (Figure 10.1).

FIGURE 10.1 LGB-infested rice. (Photo courtesy of J.F. Campbell.)

Immature LGBs develop inside kernels, obscured from the naked eye, until they emerge as adults and chew out of the kernel causing insect damaged kernels (IDK), a widely used measure of grain quality.

LGB is the major cause of wheat IDK in the United States and is a frequent source of insect fragments in milled products. Given its importance as a grain pest, it is perhaps surprising that it has a potentially broad host range and may have evolved from a wood-boring life cycle.[1] There is evidence that LGB, and a closely related stored-product pest, the larger grain borer (*Prostephanus truncates*, Coleoptera: Bostrichidae), continues to be associated with forested areas, can feed on wood and tree seeds such as acorns, is commonly captured far from grain storage locations, is a strong flier, and actively immigrates into grain storage structures. These findings suggest that processes occurring over broader spatial scales than just the inside of a grain bin are important in LGB population dynamics. However, ecological processes that affect establishment and spread of LGB in landscapes are poorly understood.

It is unclear, for example, in what regions LGB successfully overwinters, and whether overwintering sites are inside grain bins, in nearby forested areas, or a combination of both sites. Addressing such questions will help determine what parts of the country serve as sources and sinks for dispersing insects. This is especially important in the context of identifying the origin of LGB adults responsible for establishing early season infestations (local versus long-range dispersal). Also unknown is how LGBs exploit broader landscapes around grain storage bins and, in turn, how these patterns of usage influence population dynamics. Spatiotemporal distributions of LGB occur at a local scale in real-world landscape matrices comprising agricultural land and natural landcover types. These landscapes contain a range of food resources that could be exploited by LGB as either overwintering sites or alternative food sources.

Extending our understanding of LGB populations to include an explicit spatial component will help better evaluate the connectivity of grain storage bins on neighboring farms and more fully appreciate the role of landuse/landcover type and configuration in shaping source-sink dynamics within agricultural landscapes.

10.2.1 Lesser Grain Borer Economic Impact and Management

Grain production, storage, and processing are a significant part of the U.S. economy, and LGB is an important pest of stored cereals such as wheat and rice (*Oryza sativa*). In 2008, for example, 2.5 billion bushels of wheat (valued at $17 billion) and 0.2 billion bushels of rice (valued at $3 billion) were produced in the U.S.[2] After harvest, grain is moved through a storage and processing supply chain. For wheat, grain moves from relatively small-scale farm storage bins (Figure 10.2) and rural county elevators to large grain storage elevators located at distribution centers, and finally to mills and food processing facilities where it is converted into food and/or feed products for people or animals. Stored grain is susceptible to degradation and quality losses due to stored-product insects at all points in this distribution channel. Often, insects that enter the grain early in the supply chain (local farm or county elevator) are transported through the system where they feed, reproduce, and spread to previously clean grain stocks.

FIGURE 10.2 Small-scale grain storage bins typical of many farms. (Photo courtesy of J.F. Campbell.)

Insect infestations have an economic impact through both their influence on the value of the grain and acceptability by buyers and the costs of pest management tactics (time, labor, and materials). There are market penalties set by grain buyers in the United States for the presence of insects that approach $1.51/metric ton.[3] If a load of grain is rejected at the processor, the cost to the supplier can be as much as 10%–20% of the total value of the commodity.[4] The economic consequences of insect infestations after the grain is processed are even greater.[5] Pest management in stored grain typically involves the application of grain protectants, fumigation, and temperature management. A USDA National Agricultural Statistics Service (NASS) survey of grain elevator managers indicated that 11.6% of wheat stored in elevators was fumigated with phosphine and 2.9% treated with grain protectants (chlorpyrifos-methyl and malathion), with a total annual estimated cost of $5.7 million.[4] Similar management tactics are used in farm storage situations.

Generally, grain is infested from one of three sources: (1) residual populations inside storage structures,[6] (2) commingling of infested and uninfested grain, or (3) invasion by individual insects originating from outside sources. Management of stored-product insects can therefore focus on elimination of insects from storage structures prior to adding the grain, preventing insects from entering the storage structure by eliminating live insects from grain being added to bins, eliminating routes of entry for immigrating insects, treating grain with a protectant prior to infestation to prevent or decrease population growth, or killing established populations using a fumigant. Targeting intervention strategies that are focused on sources of invasion and preventing grain from becoming infested in the first place may be a superior strategy than simply responding to established infestations.

10.2.2 Behavior and Ecology outside Grain Storage

It is unlikely that LGB infestation of grain occurs in the field prior to harvest in the United States.[6,7] LGB is capable of persisting in grain residues left in bins and elevators and appears capable of moving from these sources into clean grain later added to the bins.[6] A number of studies have also reported that LGB, and other stored-product pest species, may immigrate into structures from the outside.[8] Entry into grain bins through the eaves, vents, and poorly sealed bottoms can occur at a rate of six beetles per day from sources outside bins.[7] Toews et al.[9] demonstrated that LGB could enter grain storage warehouses through the gaps around overhead doors. Once established within a bulk storage bin, populations can increase to large numbers, and beetles have been reported to emigrate from bins as well.[7] The source(s) of beetles immigrating into structures are not well known, although one source is likely individuals dispersing from other infested bins.

LGBs have been captured in flight traps outside of grain storage structures in many parts of the United States and Canada.[10–13] In an Australian study, trap captures tended to decline with distance from storage structures (e.g., Sinclair and Hadrell[14]), although beetles were captured several kilometers from grain silos. However, LGBs have also been captured in traps in diverse types of habitats often far from grain storage sources of beetles.[9,12,14–17] These findings suggest that the beetles disperse far from grain storage sites or that beetles are exploiting food sources in the landscape in addition to grain storage facilities.

Adult LGB may survive for less than 2 weeks without food but can live for months when food is available.[18,19] Some effort has been made to determine alternate hosts of the LGB, but proof for sustained use of non-grain hosts is lacking. Wright et al.[20] assessed the ability of LGB to live and reproduce in fruits and seeds from rodent burrows in Kansas. Although they did not observe feral LGB in samples, LGB progeny developed on acorns (*Quercus* sp.), hackberry (*Celtis laevigata*), and buckbush (*Ceanothus cuneatus*) fruits, and adults survived on fruits of sandhill plum (*Prunus angustifolia*), chinquapin oak (*Quercus muehlenbergii*), hackberry, buckbrush, and black walnut (*Juglans nigra*). Recent studies demonstrated that LGB can develop on acorns as well as on wheat.[21,22] LGB feeds and reproduces on roots and branches of some tree species in West Africa.[23] In the United States, older USDA publications list LGBs as attacking oak, hickory, pine (*Pinus* sp.), black walnut, and ash (*Fraxinus* sp.),[24,25] but these reports were not supported by data. A recent evaluation of LGB survival and reproduction on live and dead twigs from tree species in Kansas revealed tunneling behavior by beetles but generally poor survival and no reproduction.[22]

Historically, population ecology and pest management of LGB have emphasized processes within the grain bin, but based on the information presented above there is considerable evidence that processes working at spatial scales much larger than that occurring within an individual bin have even more important influences on this pest's populations. We hypothesize that LGB has spatiallystructured populations and a high dispersal ability that contributes to its ability to successfully exploit spatially patchy and ephemeral resources such as stored grain. Forested areas may function as a reservoir or overwintering location that helps LGB persist in these landscapes and, thus, as a source of individuals for seasonal invasion of grain bins. LGB is the primary pest of stored grain in the United States, and tangible and practical approaches could move the focus of management from treating bulk storage *after* infestation using pesticides such as the fumigant phosphine to *preventing* problems from developing. This may be through more effective exclusion, better timing of management, or targeted pest management at sources of infestation. Application of geocomputation to the biology and management of LGB may offer innovative approaches to explaining the origins and dispersal of organisms such as LGB at a range of spatial scales and evaluate the factors that influence spatial pattern. Additionally, its application may offer new approaches to develop integrated pest management methods to mitigate the economic impacts of this insect.

10.3 GEOCOMPUTATION

Because many systems, including those studied in biogeography, are so complex, the use of many different spatial analysis and modeling techniques is often required in order to improve our understanding of particular patterns and processes. Dragicevic[26] describes spatiotemporal modeling methods as borrowing heavily from the knowledge domains of geographic information science (GIScience),[27] geocomputation,[28] and geosimulation.[29]

The University Consortium for Geographic Information Science (UCGIS) describes the knowledge area of geocomputation as "a series of methods designed and used to simulate, model, analyze, and visualize a range of highly complex, often nondeterministic, nonlinear problems."[30] Such methods include, but are not limited to, agent-based models, genetic algorithms, cellular automata, and neural networks. It is important to recognize that geocomputation is distinct from geographic information system (GIS) or GIScience. Rather, geocomputation focuses on spatial analysis by supplementing the number and variety of methods that can be used to both analyze and visualize spatiotemporal questions with, or without, the use of GIS software that cannot otherwise be appropriately addressed using more traditional statistical approaches.[31,32] Here, geocomputation excels through its emphasis on process, change, and interaction.[33,34]

Critics of geocomputation point out that a fascination with "computational 'black box' methods" may be preventing adequate development of the science of geocomputation.[34] However, others have cited the importance of computer-based methods of inductive generalization for improved predictive success,[35] that the ability to visualize spatial patterns improves our understanding of the processes that created those patterns,[36] and that the computers in which geocomputation methods are performed provide a kind of digital laboratory for the testing of theory.[33]

10.3.1 Agent-Based Simulation and Modeling

Over the past 20 years, significant progress has been made in the representation of spatiotemporal dynamics and design of data structures that can capture both the geographic and temporal paths of mobile objects.[37,38] Concurrently, agent-based modeling and simulation (ABMS), or microsimulations, have increasingly been used as a tool to replicate the spatiotemporal behavior of individual entities (agents) to provide insight into a number of biological, ecological, and cultural processes.[39,40]

An agent can be considered a "software robot" that lives in, and reacts to, the virtual world in which it is placed.[41] While ABMS focuses on individual (agent) behavior, it is possible to evaluate how both individual- and system-level patterns emerge from rules based on theories of the individual. An agent-based system yields results that are arguably more representative of reality than relatively simple mathematical models. Agent-based systems have been used to study the spread of disease,[42] the flow of people through urban environments,[43] changes in landuse and landcover,[44] and the movement of animals.[45,46]

The use of ABMS permits local-scale experimentation with LGB populations in realistic environments and promises to better illustrate the spatial distribution of the beetle over time. In particular, it is possible to examine how overwintering beetles might be capable of moving from forest habitats to nearby grain storage bins during early season infestations. Further, it is also possible to investigate the role of changing storage bin conditions (e.g., full, empty) and local landcover types in determining the number and spatial distribution of beetles during the course of a typical year.

In this demonstration, individual beetles will serve as agents, which are represented by a series of state and bioenergetic variables. Values used to parameterize each variable are based on literature from laboratory and field studies on LGB biology, expert knowledge, and "real-world" environmental data. Agents can be designed to sense, respond to, and interact with their environment and with other beetles, with decision-making being a reactive response to both local stimuli and agent state.[47] Beetle movements in the model that follow are random (simplest) but could be modified to be based upon bounded rationality heuristics (more difficult) or controlled by a combination of rational decision-making and environmental controls (e.g., wind speed and direction) (most difficult). Results from, and visualization of, multiple simulations can be used as the initial basis to form/revise theory regarding early season LGB infestation and to inform data collection needs for further research conducted in the field.

10.3.2 About NetLogo

The ABMS software used here is NetLogo, a cross-platform multiagent programmable modeling environment for simulating natural and social phenomena.[48,49] NetLogo was created by Uri Wilensky at the Center for Connected Learning and Computer-Based Modeling at Northwestern University. The first version of NetLogo was published in 1999 and was followed in 2002 by an updated NetLogo 1.0. Since then, three additional versions have been produced, with the latest version 4.0 introduced in 2007. With the addition of a participatory tool called HubNet, multiple users can interact with one another in the same simulation across a computer network for both research and educational purposes.

The principal actors within any ABMS are agents, and NetLogo incorporates four types of agents: Turtles, patches, links, and observer. Turtles are mobile agents that move within a gridded spatially referenced landscape and interact with other agents and the environment. Patches are stationary agents consisting of individual grid cells in the raster landscape. Links are "optional" agents that can be used to functionally connect two turtles. Finally, the observer agent controls the overall simulation and exists outside of the system of turtles, patches, and links. Turtle and patch agents can perceive information about the surrounding environment (e.g., presence and state of agents, including self-state) and operate autonomously by following rules informed by static or dynamic turtle and patch variables. Increasingly complex modeling environments can be created by the addition of different types of turtles, additional agent variables, and more sophisticated rules to govern agent behavior.

The NetLogo environment provides users with a powerful, yet simple to learn, programming language to design, run, and visualize dynamic simulations ranging from one agent acting by itself to a multiagent program with thousands of interacting agents responding to one another and the environment. Creation of user-friendly graphical user interfaces (GUI) is straightforward, and users have access to plotting capabilities, a broad and powerful set of primitives (i.e., set of values and corresponding operations), and support for floating point mathematics.[48]

A considerable community of users has developed around NetLogo. Users are invited to share their models via the Community Models link on the NetLogo Web site. The Community Models library caters to both programmers and nonprogrammers alike. Because the program is a Java-based application, nonusers may view and run the user-created content through web browser applets. On the other hand, experienced NetLogo users can go a step further and download the source code of user-created content from the Community Models library, allowing them to view the behavior rules firsthand or modify the models for their own needs.

10.4 METHODS

10.4.1 Overview: Creating a NetLogo Model

The NetLogo software can be downloaded free of charge at http://ccl.northwestern.edu/netlogo/ for both the Microsoft Windows and Mac operating systems. Download the appropriate NetLogo version and install it on your computer. The tutorial in this chapter, as well as the supplemental models included with the book, was created using NetLogo version 4.0.4 (November 24, 2008). After installing the program, it is recommended that you complete the three online tutorials to familiarize yourself with NetLogo. The tutorials can be found on the same Web site, by clicking on the "Web Version" link for the Users Manual. Once you have accessed the User Manual, you will find links to Tutorials #1–#3 listed along the left side of the page.

The tutorials included with this publication provide step-by-step guides for creating two agent-based simulation models designed to gain insight on LGB population dynamics. Of particular interest is the exploration of how LGBs survive and move within a natural landscape mosaic of agricultural and forested landuses/landcovers, possibly resulting in early season infestations of grain storage bins. The first model allows users to modify a number of environmental and biological variables over numerous simulations to evaluate the impact these changes have on LGB populations. The second model incorporates LGB predation into the procedures of the first model. The instructions that follow, assume that you are familiar with, and understand, basic NetLogo commands and procedures.

10.4.2 Model Setup

10.4.2.1 Setup and Go Buttons and Energy Switch

After starting NetLogo, you are presented with an empty model. As a first step, create **setup** and **go** buttons (called "**setup**" and "**go**") that, upon completion of the model, will be used to run the procedures.

1. From the **Interface** tab of Netlogo, create a button by either clicking on the **Button** icon in the toolbar or by right clicking anywhere in the empty white space next to the display window.
2. The button dialog window for editing the button then opens. Type ***setup*** in the **Command** box and select **OK**.
3. Following the same steps, create a go button. Type *go* in the command box, and check the forever button in the dialog box. Select **OK**.

Next, create a switch button that will be used to turn on (or off) the energy labels of our future insects. The code for making this optional switch operational is outlined below under "**Show Energy and Display Labels**."

1. Still in the **Interface** tab, create the switch by clicking on the **Button** icon and selecting **Switch** from the list that appears. A switch can also be created by right clicking anywhere in the empty white space next to the display window.
2. The switch dialog window opens. Type ***show-energy?*** in the global variable box. Select **OK**.

The newly created **setup** and **go** buttons will be red (Figure 10.3) since the actual **setup** and **go** procedures have yet to be written.

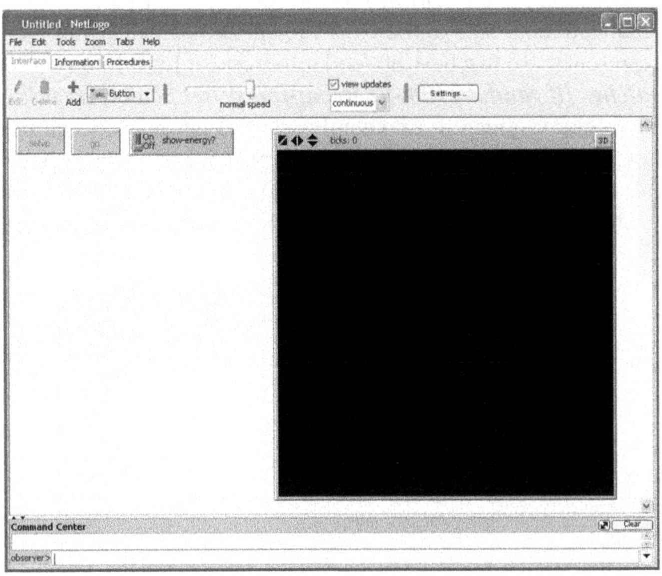

FIGURE 10.3 NetLogo interface with initial buttons and switches created.

At this point, it is a good practice to save your new NetLogo file, and to do so frequently in the coding process,

1. Select **File** from the NetLogo main menu, then **Save As** from the context menu that appears.
2. In the **Save As** dialog box, navigate to where you would like the new file to be saved, then type in a ***filename***. Be sure to include the NetLogo file suffix (***.nlogo***) with the ***filename*** or an error message will appear.

10.4.2.2 Turtle Variables and Setup Procedure

Before writing the **setup** and **go** procedures, first establish some **turtle** variables that will be important for model operation. Click on the **Procedures** tab and add the following code (Figure 10.4) by typing in italicized commands:

1. ***breed [bugs bug]*** defines bugs or bug as valid type of turtle.
2. ***turtles-own [energy age]*** associates the variables energy and age as belonging to each turtle. The **energy** variable will be used to control when bugs reproduce and die and the age variable kills any bug that reaches a specific age (and which will be defined later).

With these variables defined at the beginning of the model, write the necessary code for the **setup** procedure (Figure 10.5) by typing the italicized commands:

1. ***to setup*** signals that a procedure called **setup** is about to be defined.
2. ***clear-all*** reinitializes the model to an empty state in preparation for a new series of instructions.
3. ***ask patches [set pcolor brown]*** sets the color of every patch (pixel) within the simulated landscape to the color brown. Brown patches in the simulation represent those that have no food resources for LGB (bugs) to consume.
4. ***ask patches [if random 100 < percent-forest [set pcolor green]]*** changes a patch's color to green if a randomly selected integer between the values of 0 and 99 is less than the variable **percent forest**. The **percent forest** variable allows the simulation to be based on different areas of forest landcover. The user control for this variable will be added to the interface later.

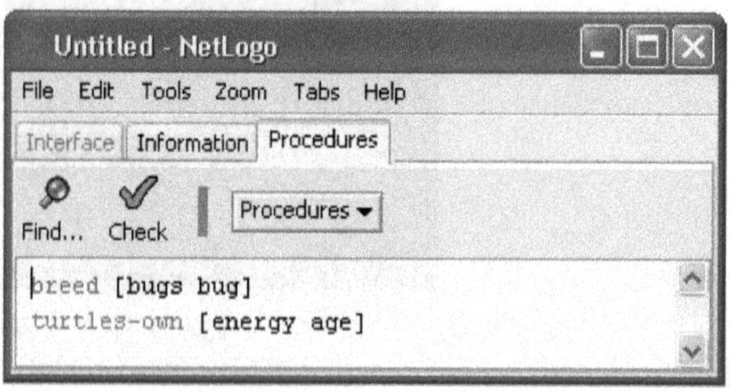

FIGURE 10.4 Initial list of turtle variables.

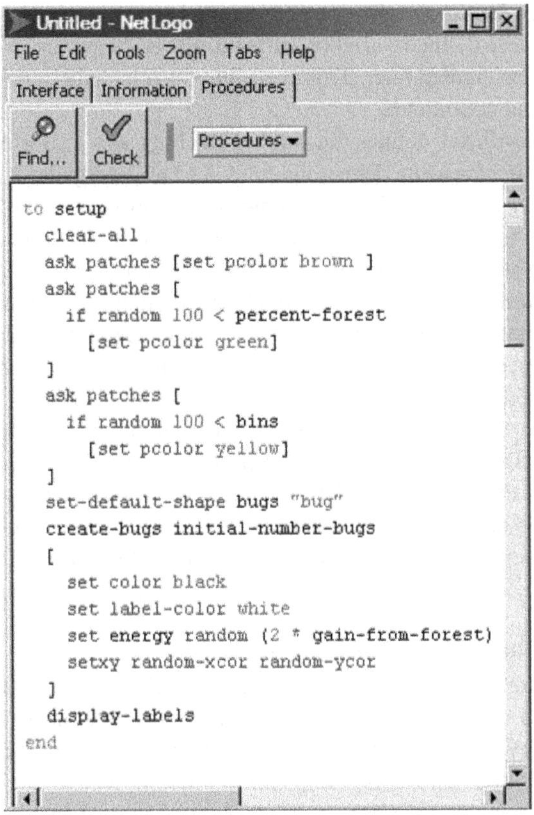
Untitled - NetLogo
File Edit Tools Zoom Tabs Help
Interface
Information
Procedures
Find...
Check
Procedures
to setup
clear-all
ask patches [set pcolor brown]
ask patches [
if random 100 < percent-forest
[set pcolor green]
]
ask patches [
if random 100 < bins
[set pcolor yellow]
]
set-default-shape bugs "bug"
create-bugs initial-number-bugs
[
set color black
set label-color white
set energy random (2 * gain-from-forest)
setxy random-xcor random-ycor
]
display-labels
end

FIGURE 10.5 Initial **setup** procedure.

5. ***ask patches [if random 100 < bins [set pcolor yellow]]*** changes a patch's color to yellow if a randomly selected integer between 0 and 99 is less than the variable **bins**. The **bins** variable allow a specified proportion of the simulated landscape to contain grain storage bins. As with the **percent forest** variable, the user control for **bins** will be added to the interface in a later step.
6. ***set-default-shape bugs "bug"*** defines the shape of the bug turtles as the NetLogo library shape of "bug" rather than the default "triangle." A bug, or bugs, was previously defined as a **breed** of turtle and is a NetLogo library shape. If you receive an error message related to this line of code, you will need to import the "**bug**" graphic from the **NetLogo Turtle Shapes Library**:
   a. In NetLogo, click on **Tools** in the main menu, then select **Turtle Shapes Editor**.
   b. Click on the button **Import from Library**.
   c. Scroll down the list of shapes until you find the correct graphic (in this case "**bug**").
   d. Select the correct shape and click the button **Import**. The imported turtle shape is now available to you within the current NetLogo session and will remain so until deleted.

7. ***create-bugs initial-number-bugs*** creates a certain number of bugs according to the user-defined variable **initial-number-bugs**, the user control for which will be created later. This procedure also includes a number of other variables and commands.
   a. ***[set color black*** defines the color of the bugs as a uniform black color.
   b. ***set label-color white*** defines the color of a bugs energy-level label as white.
   c. ***set energy random (2 * gain-from-forest)*** provides newly created bugs with an initial energy level twice that which can be gained by consuming forest resources (the **gain-from-forest** variable to be set later).
   d. ***setxy random-xcor random-ycor]*** the command setxy runs two **reporters** called random-xcor and random-ycor. **Reporters** report a result, with these reporters returning a random number from the range of coordinates along the x- and y-axes of the simulated landscape. Afterwards, each turtle executes the ***setxy*** command using the random x- and y-coordinates generated by the reporters and moves to that location.
8. ***end*** defines the end of the **setup** procedure. (*Note*: since the variables have not yet been defined, a yellow bar will appear that states that the variables have not been defined as you navigate from the **Procedures** tab. Continue to the next section, as the variables will be defined later in the exercise.)

10.4.2.3 Defining Initial Variables with Sliders

In writing the preceding **setup** procedure, several variables were declared but remain undefined. These variables include **percent-forest, bins, initial-number-bugs**, and **gain-from-forest**. Each variable is user-defined at the beginning of the simulation, rather than hard-coded into the simulation, which allows different scenarios to be constructed and LGB population dynamics and dispersal capability to be assessed. Create **slider** buttons for each variable and add them to the user interface:

1. Select the **Interface** tab and create the **slider** by clicking on the **Button** icon and selecting **Slider** from the list that appears. A slider can also be created by right clicking anywhere in the empty white space next to the display window.
2. The slider dialog window opens. Type ***percent-forest*** in the global variable box. Specify a minimum value of ***0***, increment of ***1***, and maximum value of ***100***. Also specify a (starting) value of ***20*** and add ***%*** to the optional units section. Select **OK**.
3. Similarly, create a slider for **bins**. Type ***bins*** in the global variable box, and specify ***0***, ***1***, and ***20*** as the minimum, increment, and maximum values, respectively. A good starting value is ***2*** and, again, use ***%*** as the optional units entry.
4. Create a third slider, this time for the variable **initial-number-bugs**. Type ***initial-number-bugs*** in the global variable box, and specify ***0***, ***2***, and ***750*** as the minimum, increment, and maximum values, respectively. Type ***300*** in the value box.
5. Create a fourth slider for the variable **gain-from-forest**. Type ***gain-from-forest*** in the global variable box, and specify ***0***, ***1***, and ***50*** as the minimum, increment, and maximum values, respectively. Start by specifying a value of ***10***.

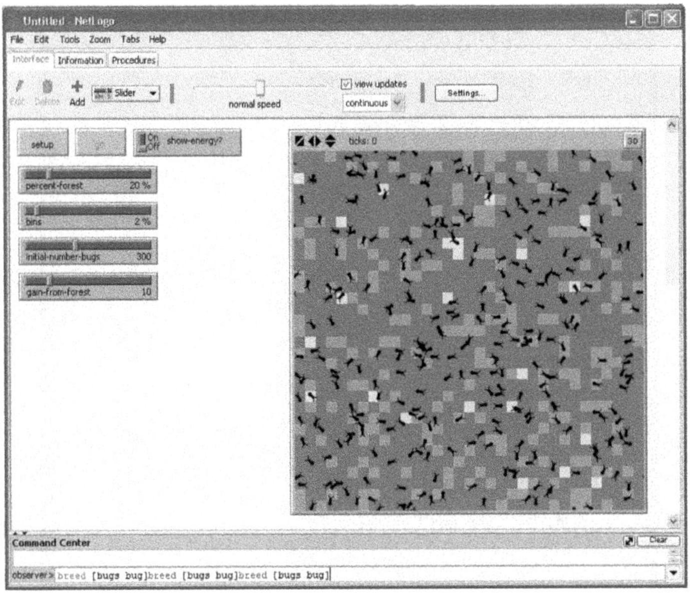

FIGURE 10.6 NetLogo interface with operational **setup** procedure.

With these modifications, click the **setup** button, and the screen of the **Interface** tab should look similar to Figure 10.6. Note that the button for the **go** procedure remains red, since it is still undefined. However, the **setup** procedure is operational and should produce varying displays of bugs superimposed on brown, green, and yellow patches. Feel free to modify the slider-based variable settings by clicking the **setup** button after changes in slider settings to verify that the sliders work as planned.

10.4.2.4 Go Procedure

Next, create the **go** procedure by typing in the following italicized code immediately after the end of the **setup** procedure (Figure 10.7) in the **Procedure** tab:

1. ***to go*** signals that a procedure called **go** is about to be defined.
2. ***if not any? turtles [stop]*** will stop the simulation if the number of turtles (only bugs in this model) reaches zero.
3. ***ask bugs*** indicates that the turtle breed bugs should run the commands in brackets that follow.
   a. ***move-bugs*** specifies a new procedure that will control bug movement.
   b. ***set energy energy - 2*** reduces the energy level of each bug by two units after the **move-bugs** procedure is complete.
   c. ***eat-forest*** is a procedure that allows bugs to gain energy from forest patches.
   d. ***eat-grain*** is a procedure that allows bugs to gain energy from grain storage bins (bins).
   e. ***reproduce-bugs*** is a procedure that controls bug reproduction.
   f. ***death*** procedure that controls bug mortality.

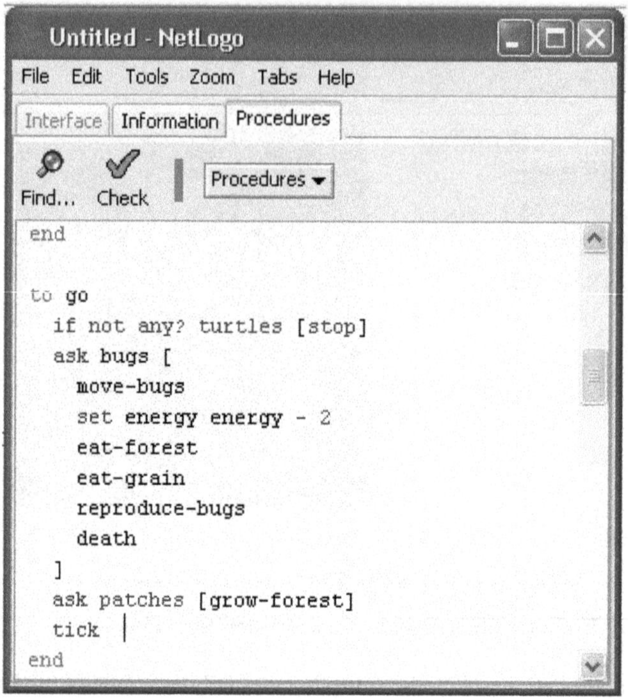

FIGURE 10.7 Initial **go** procedure.

4. ***ask patches [grow-forest]*** is a patch procedure that allows forest patches to replenish its food resources for later LGB consumption.
5. ***tick*** moves the simulation forward one time step.
6. ***update-plot***.
7. ***end*** defines the end of the **go** procedure.

10.4.2.5 Bug Movement, Eating, Reproduction, and Death Procedures

In the steps above, five bug procedures (**move-bugs, eat-forest, eat-grain, reproduce-bugs,** and **death**) and one patch procedure (**grow-forest**) were introduced, but none have yet been defined. Define the new bug procedures by adding the following code that is italicized below in the **Procedures** tab immediately after the go procedure (Figure 10.8):

1. ***to move-bugs*** signals that a procedure called **move-bugs** is about to be defined.
   a. ***rt random 360 lt random 360*** rt (right) and lt (left) are two commands that also use reporters. Here, each bug selects a random integer between the values of 0 and 359 and turns right that number of degrees. This process is repeated for turning left.

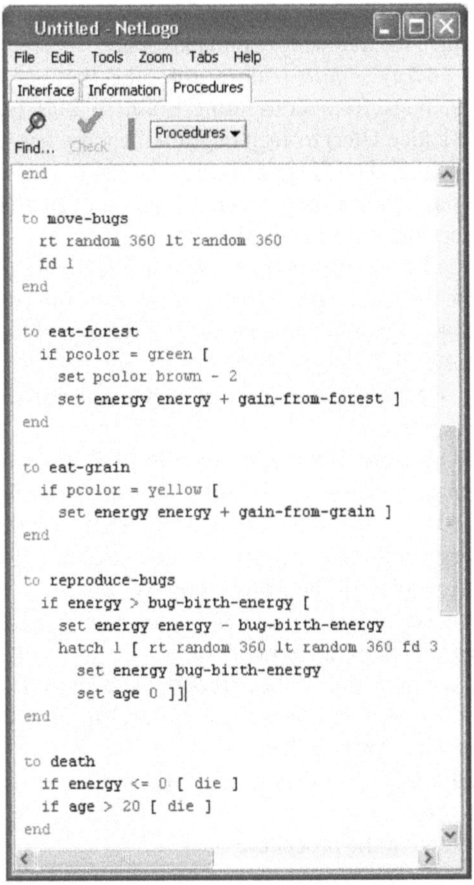

FIGURE 10.8 Additional bug procedures for movement, eating, reproduction, and death.

   b. ***fd 1*** moves forward one patch after the bug turns right and left to arrive at a final new heading.
   c. ***end*** defines the end of the procedure.
2. ***to eat-forest*** starts the procedure called **eat-forest** that allows bugs to consume resources (and gain energy) from forest patches.
   a. ***if pcolor = green*** if a bug, after moving, lands on a green (forest) patch, then the following commands apply.
   b. ***[set pcolor brown - 2*** changes the patch color from green to a dark brown (i.e., brown - 2).
   c. ***set energy + gain-from-forest]*** adds the user-specified energy available to bugs from forest resources (the **gain-from-forest** slider was added above) to each bug's total energy level.
   d. ***end*** defines the end of the procedure.
3. ***to eat-grain*** starts the procedure called **eat-grain** that allows bugs to consume resources (and gain energy) from grain found inside grain storage bins.

a. ***if pcolor = yellow [***if a bug, after moving, lands on a yellow (grain storage bin) patch, then the following commands apply:
b. ***set energy energy + gain-from-grain]*** adds the user-specified energy available to bugs from consuming grain in bins (**gain-from-grain** slider to be added later) to each bugs total energy level.
c. ***end*** defines the end of the procedure.

4. ***to reproduce-bugs*** starts the procedure called **reproduce-bugs** that controls bug reproduction in the simulation.
   a. ***if energy > bug-birth-energy*** when a bug's energy level exceeds that which the user specifies is necessary for reproduction (the **bug-birth-energy** slider to be set later), then the following commands apply:
   b. ***set energy - bug-birth-energy*** subtract **bug-birth-energy** from the bug's energy level.
   c. ***hatch 1 [rt random 360 lt random 360 fd 3]*** gives birth to one new bug, determine a movement heading by randomly selecting a number of degrees to turn right then left, then move forward three patches.
      1. ***set energy bug-birth-energy*** sets the initial energy level to be equal to that specified for **bug-birth-energy**.
      2. ***set age 0*** sets the initial age of the new bug to zero.
   d. ***end*** defines the end of the procedure.
5. ***to death*** starts the procedure called **death** that determines when bugs die.
   a. ***if energy <= 0 [die]*** if a bug's energy level reaches 0 (or less after a movement), then that bug dies.
   b. ***if age > 20 [die]*** a bug dies after 20 ticks, even if its energy level is greater than 0.
6. ***end*** defines the end of the procedure.

10.4.2.6 Setting Up Gain-from-Grain, Bug-Birth-Energy Procedures, and Control Sliders

In the process of adding these new bug procedures, three new user-specified variables were introduced (**gain-from-grain** and **bug-birth-energy**). Click on the **Interface** tab and create new slider buttons for each new variable.

1. Create a **gain-from-grain** slider. Type ***gain-from-grain*** in the global variable box. Specify a minimum value of ***0***, increment of ***1***, and maximum value of ***50***. The (starting) value should be set at ***20***.
2. Create a **bug-birth-energy** slider. Type ***bug-birth-energy*** in the global variable box. Specify a minimum value of ***1***, increment of ***1***, and maximum value of ***100***. The (starting) value should be set at ***40***.

10.4.2.7 Forest Regrowth Procedure and Control Slider

Next, add code for a new patch procedure that will allow some, or all, forest (green) patches whose resources have been consumed by a bug to "regenerate" and be available to bugs in the future (Figure 10.9).

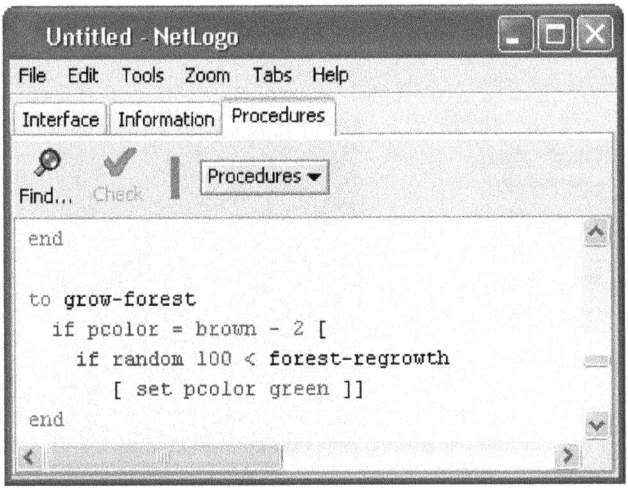

FIGURE 10.9 Patch procedure controlling forest "regeneration."

1. ***to grow-forest*** signals that a procedure called **grow-forest** is about to be defined.
   a. ***if pcolor = brown - 2 [*** if the color of a patch is dark brown (i.e., brown - 2), meaning that a bug has consumed that patch's resources during a previous time step, then apply the following commands:
   b. ***if random 100 < forest-regrowth*** for dark brown patches, if a randomly selected integer between the values of 0 and 99 is less than the variable **forest-regrowth**, then the following command applies. The **forest-regrowth** variable is user-controlled with a slider button and will be created later.
   c. ***[set pcolor green]]*** changes the dark brown patch to green, making it a "resource full" forest patch once again.
2. ***end*** defines the end of the procedure.

The previous patch procedure introduced the new variable **forest-regrowth**. Click on the **Interface** tab and create one more new slider button for **forest-regrowth**.

1. Type ***forest-regrowth*** in the global variable box, and specify ***0***, ***1***, and ***100*** as the minimum, increment, and maximum values, respectively. Enter a starting value of ***20*** and ***%*** in the optional units.

At this point, the Interface tab of your model should look similar to Figure 10.10. Note that in this example, the slider button for **forest-regrowth** was placed near the top of the array of buttons, just after **percent-forest**.

10.4.2.8 Show Energy and Display Labels

The model is nearly complete but would benefit from a few more minor modifications. The first modification addresses the, as yet, nonfunctional **show-energy?** switch created earlier in the tutorial. To accomplish this, a new procedure called **display-labels**

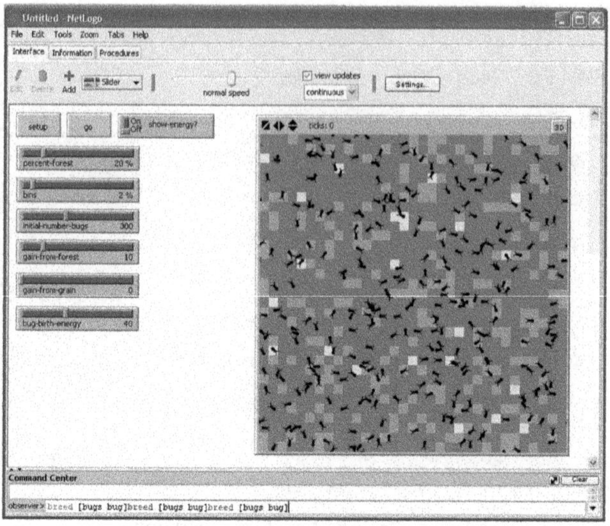

FIGURE 10.10 NetLogo interface with operational sliders.

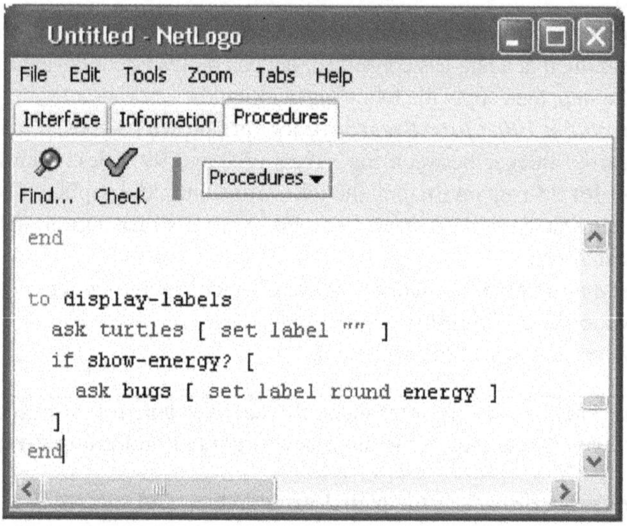

FIGURE 10.11 Procedure to create **energy** labels for each turtle.

will be created, then a new command **display-labels** will be added to the existing setup procedure. In the **Procedures** tab, enter the following code that is shown in italics immediately after the **grow-forest** procedure (Figure 10.11):

1. ***to display-labels*** signals that a procedure called **display-labels** is about to be defined.
2. ***ask turtles [set label ""]*** creates an empty label for each turtle (just bugs in this model).

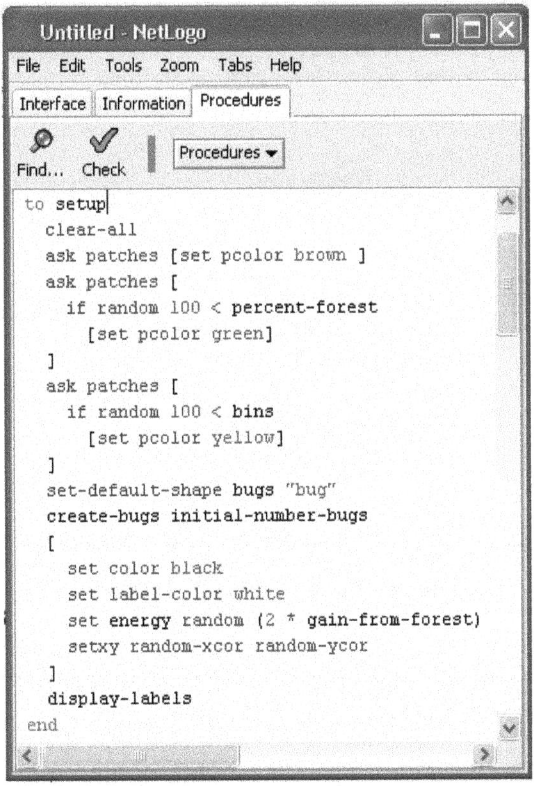
Untitled - NetLogo
File Edit Tools Zoom Tabs Help
Interface Information Procedures
Find... Check
Procedures
to setup
clear-all
ask patches [set pcolor brown]
ask patches [
if random 100 < percent-forest
[set pcolor green]
]
ask patches [
if random 100 < bins
[set pcolor yellow]
]
set-default-shape bugs "bug"
create-bugs initial-number-bugs
[
set color black
set label-color white
set energy random (2 * gain-from-forest)
setxy random-xcor random-ycor
]
display-labels
end

FIGURE 10.12 Revised **setup** procedure that references the **display-labels** procedure.

3. ***if show-energy? [***if the **show-energy?** switch is "on," then the following command applies:
   a. ***ask bugs [set label round energy]]*** has each bug turtle (again, just bugs in this model) display a label that reflects their current energy level. Recall that it was previously declared that each turtle "owns" the attribute "energy."
4. ***end*** defines the end of the procedure.

Next, return to the section in the **Procedures** tab where the setup procedure is located. Add the code ***display-labels*** just before the end command. This invokes the **display-label** procedure just coded. The **setup** procedure should now look like the example given in Figure 10.12.

10.4.3 Plot Procedure

The next modification will add a plot window to the **Interface** tab. The plot window will display continuous graphs of key variables (e.g., number of bugs, number of forest patches, etc.) throughout the duration of the simulation. As with the first modification, this requires adding an additional procedure. In the **Procedures** tab,

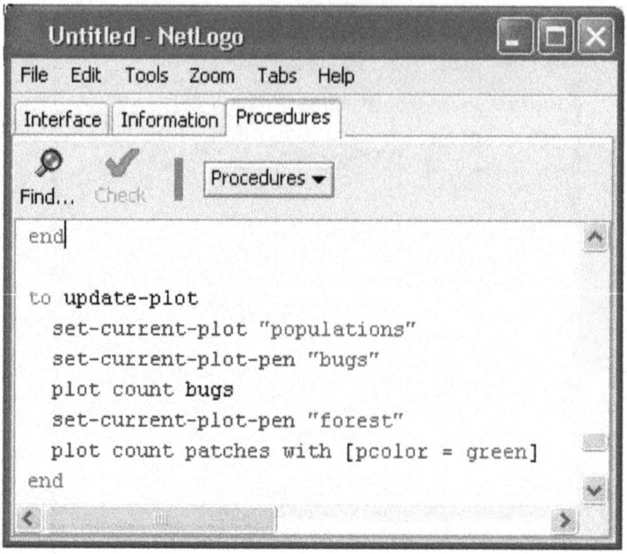

FIGURE 10.13 Procedure to create and update a plot window to track the number of bugs and forest patches.

add the code shown in italics for a new procedure called **update-plot** immediately after the **display-labels** procedure (Figure 10.13):

1. ***to update-plot*** signals that a procedure called **update-plot** is about to be defined.
2. ***set-current-plot "populations"*** defines the name of the new plot as **populations**.
3. ***set-current-plot-pen "bugs"*** defines one of the graphs to appear on the plot as the variable **bugs**.
4. ***plot count bugs*** at the end of each model iteration, or tick, count the number of bugs to be graphed using the pen **bugs**.
5. ***set-current-plot-pen "forest"*** defines one of the graphs to appear on the plot as the variable **forest**.
6. ***plot count patches with [pcolor = green]*** the total number of forest patches will be counted each tick.
7. ***end*** defines the end of the procedure.

Click once again on the **Procedures** tab and locate the **go** procedure. Add the code ***update-plot*** just before the end command. This invokes the **update-plot** procedure just coded. The **go** procedure should now look like the example given in Figure 10.14.

10.4.3.1 Create Plot Window and Monitors

With the **update-plot** procedure defined and called from the **go** procedure, the plot itself must be added to the **Interface** tab. The method for adding the plot is similar to that of adding a button, switch, or slider.

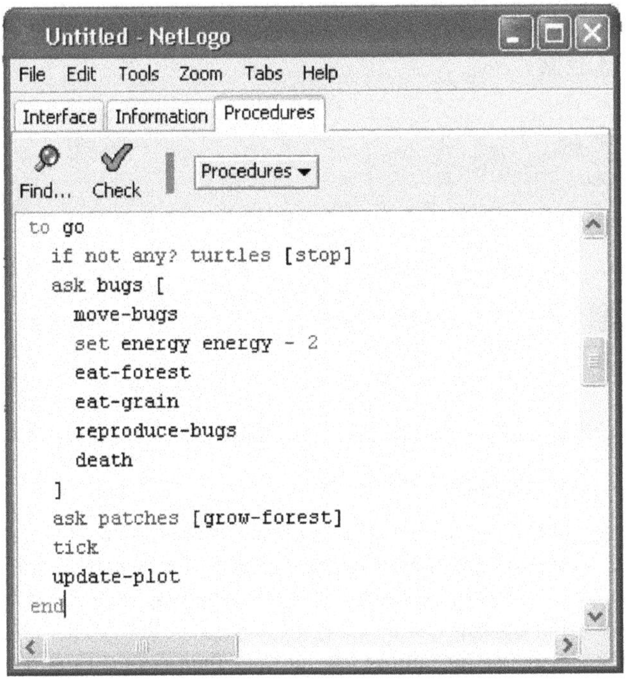

FIGURE 10.14 Revised **go** procedure that references the **update-plot** procedure.

1. From the **Interface** tab, create a plot by either clicking on the **Button** icon in the toolbar or selecting **plot** from the options that appear. Alternatively, right clicking anywhere in the empty white space next to the display window brings up the same list of options.
2. The dialog window for editing the plot then opens. First, type ***Populations*** in the name box.
3. Enter ***time*** and ***pop.*** for the names of the x- and y-axes, respectively. Also change the default x and y max values from ***10*** to ***100***.
4. Make sure that the **Autoplot** and **Show legend** boxes are checked.
5. In the **Plot Pens** section, click on the **Rename** button and enter ***bugs*** as a pen name to replace **default**. Accept the default options for color, mode, and interval, and ensure that the **Show in legend** box is checked (Figure 10.15).
6. Again in the **Plot Pens** section, click on the **Create** button and enter ***forest*** as a pen name. Change the color to green, but accept the defaults for mode and interval. Also make sure that the **Show in legend** box is checked.
7. Select **OK** and the plot appears on the **Interface**. Adjust the size and position of the plot window as desired.

As a complement to the plot, individual **monitors** may be added that will report specific turtle counts (bugs, types of patches, etc.) at the end of each tick. The method for adding the **monitor** is similar to that of adding a button, switch, or slider.

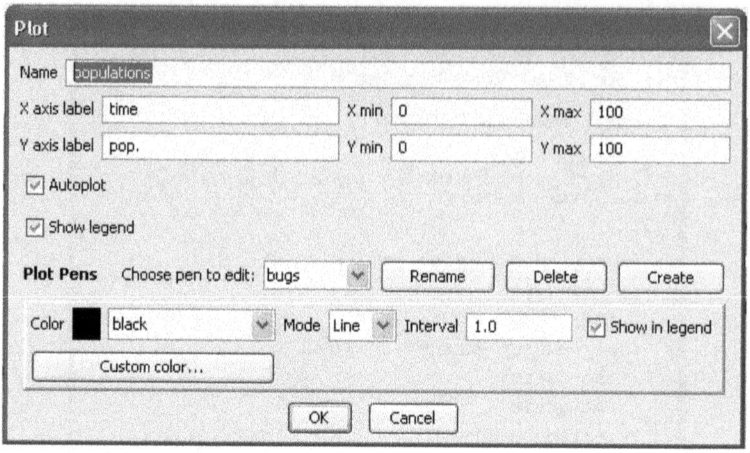

FIGURE 10.15 Example of the **Plot** dialog box.

1. Create a monitor by either clicking on the **Button** icon in the toolbar or selecting **monitor** from the options that appear. Alternatively, right clicking anywhere in the empty white space next to the display window brings up the same list of options.
2. The dialog window for editing the monitor then opens. First, type ***count bugs*** in the reporter box, then ***bugs*** in display name. Enter ***0*** for the number of decimal places, and accept the default font size. Select **OK**.
3. The **bug monitor** appears and can be moved as desired.
4. Create monitors for both **bins** and **forest**. In the reporter box, enter ***count patches with [pcolor = yellow]*** for the **bins** monitor and ***count patches with [pcolor = green]*** for the **forest** monitor.

10.5 CONCLUSIONS

The model is now complete and should look similar to Figure 10.16 (after a few iterations). The simulation may now be run (click **setup** and then **go**) and the impact of the various model settings on LGB population numbers examined. With the initial variable sliders set to the values specified in the tutorial, run the simulation (click **setup** and then **go**) and see what happens to the LGB population over a time period of approximately 2000 ticks.

Does the LGB population go extinct or does it settle into a regular "boom and bust" cycle? Re-run the simulation many times, and observe what happens. Speculate as to why the results are similar or different. Experiment with the variable settings by changing the slider values in a sequential manner. What variable(s), and value(s), most influence an LGB population increase or decline? In terms of maintaining a viable LGB population, is it more important to have many grain bins or a large extent of forest? What happens to the LGB population if the grain bins are fumigated or emptied (test by sliding the **gain from grain** slider to zero)?

Although the model presented in the tutorial represents a simplified version of the simulated system and features many assumptions about LGB biology and how these

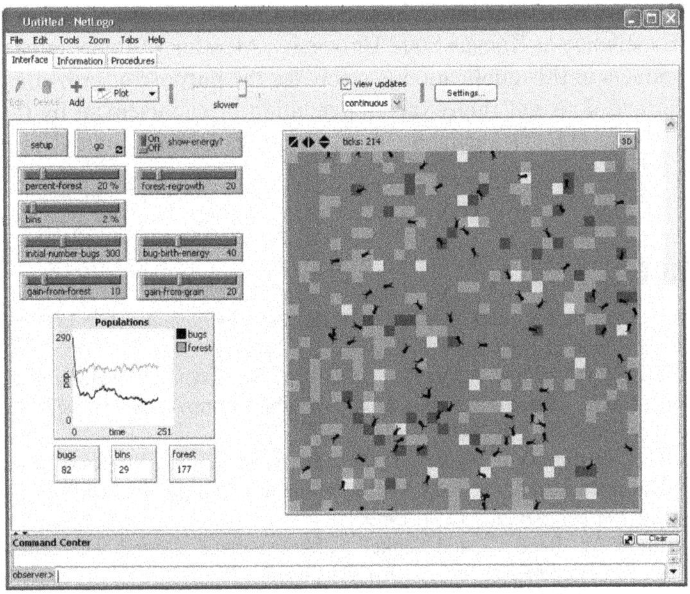

FIGURE 10.16 Final view of the NetLogo interface after all coding and interface construction is complete. **Plot** and **display** windows show results after 214 time steps, or **ticks**.

insects interact with their environment, any number of biological and environmental variables and associated procedures can be added, removed, and revised over time as one's knowledge of the system becomes more complete. This approach is also modular in the sense that it can be used to study the origins and dispersal characteristics of similar organisms and to evaluate the anthropogenic and environmental factors that influence their spatial distribution. In more comprehensive form, this and other agent-based simulation models can also provide scientists with a new kind of "virtual" laboratory for assessing the biological and geographical importance of many variables when striving to form theory, revise theory, to better inform the design of field data collection strategies, and potentially inform the development of new approaches to managing pest insects.

In addition to this tutorial, two additional NetLogo models have been included as supplemental material with this chapter. The first model (LGB ABS Demo A.nlogo) is nearly identical to the one outlined here but features a larger simulation landscape. The second model (LGB ABS Demo B.nlogo) includes all of the features of the first, plus the addition of **bug-seeking** birds that serve the role of predator. Each model includes documentation and fully commented code, so that users can read through and understand how changes to the basic tutorial code were implemented to create the more advanced simulations.

ACKNOWLEDGMENTS

Support for this effort was provided by the Kansas State University Geographic Information Systems Spatial Analysis Laboratory (GISSAL), Department of Entomology, and Department of Geography. Partial funding was provided by the "GIScience Infrastructure Enhancement—Phase 2" project, a program of Targeted Excellence at Kansas State University. Mention of trade names or commercial products in this publication is solely for the purpose of providing specific information and does not imply recommendation or endorsement by the Kansas State University, U.S. Department of Agriculture, Oregon State University, or University of Georgia

REFERENCES

1. Potter, C., The biology and distribution of *Rhizopertha dominca* (Fab.), *Transactions of the Royal Entomological Society of London*, 83, 449, 1935.
2. USDA, USDA National Agricultural Statistics Service, *2008 Agricultural Statistics Annual*, 2008. http://www.nass.usda.gov/Publications/Ag_Statistics/2008/index.asp (accessed September 22, 2009).
3. Reed, C.R., Wright, V.F., Pedersen, J.R., and Anderson, K., Effects of insect infestation of farm-stored what on its sale price at country and terminal elevators, *Journal of Economic Entomology*, 82, 1254, 1989.
4. Hagstrum, D.W., Reed, C., and Kenkel, P., Management of stored wheat insect pests in the USA, *Integrated Pest Management Reviews*, 4, 127, 1999.
5. Arthur, F.H. and Phillips, T.W., Stored-product insect pest management and control. In: *Food Plant Sanitation*, Yui, Y.H. et al., Eds., Marcel Dekker, New York, p. 341, 2003.
6. Reed, C.R., Hagstrum, D.W., Flinn, P.W., and Allen, R.F., Wheat in bins and discharge spouts, and grain residues on floors of empty bins in concrete grain elevators as habitats for stored-grain beetles and their natural enemies, *Journal of Economic Entomology*, 96, 996, 2003.
7. Hagstrum, D.W., Immigration of insects into bins storing newly harvested wheat on 12 Kansas farms, *Journal of Stored Products Research*, 37, 221, 2001.
8. Campbell, J.F. and Arthur, F.H., Ecological implications for post-harvest IPM of grain and grain based products. In: *Ecologically Based Integrated Pest Management*, Koul, O. and Cuperus, G.W., Eds., CABI Publishing, Wallingford, U.K., p. 406, 2007.
9. Toews, M.D., Campbell, J.F., Arthur, F.H., and Ramaswamy, S.B., Outdoor flight activity and immigration of *Rhyzopertha dominica* into seed wheat warehouses, *Entomologia Experimentalis et Applicata*, 121, 73, 2006.
10. Dowdy, A.K. and McGaughey, W.H., Seasonal activity of stored-product insects in and around farm-stored wheat, *Journal of Economic Entomology*, 87, 1351, 1994.
11. Dowdy, A.K. and McGaughey, W.H., Stored-product insect activity outside of grain masses in commercial grain elevators in the Midwestern United States, *Journal of Stored Products Research*, 34, 129, 1998.
12. Fields, P.G., Van Loon, J., Dolinski, M.G., Harris, J.L., and Burkholder, W.E., The distribution of *Rhyzopertha dominica* (F.) in western Canada, *Canadian Entomologist*, 125, 317, 1993.
13. Throne, J.E. and Cline, L.D., Seasonal abundance of maize and rice weevils (Coleoptera: Curculionidae) in South Carolina, *Journal of Agricultural Entomology*, 8, 93, 1991.
14. Sinclair, E.R. and Haddrell, R.L., Flight of stored products beetles over a grain farming area in southern Queensland, *Australian Journal of Entomology*, 24, 9, 1985.
15. Edde, P.A., Phillips, T.W., and Toews, M.D., Response of *Rhyzopertha dominica* (Coleoptera: Bostrichidae) to its aggregation pheromone as influenced by trap design, trap height, and habitat, *Environmental Entomology*, 34, 1549, 2005.
16. Ching'oma, P.G., Spatial distribution and movement of the lesser grain borer, *Rhyzopertha dominica* (F.). Unpublished PhD dissertation, Kansas State University, Manhattan, KS, 2006.

17. Campbell, J.F., Ching'oma, G.P., Toews, M.D., and Ramaswamy, S.B., Spatial distribution and movement patterns of stored-product insects. In: *Proceedings of the 9th International Working Conference of Stored Product Protection*, Lorini, I. et al., Eds., Campinas, Sao Paulo, Brazil, Brazilian Post-Harvest Association—ABRAPOS, p. 361, 2006.
18. Daglish, G.J., Survival and reproduction of *Tribolium castaneum* (Herbst), *Rhyzopertha dominica* (F.) and *Sitophilus oryzae* (L.) following periods of starvation, *Journal of Stored Products Research*, 42, 328, 2006.
19. Edde, P.A. and Phillips, T.W., Longevity and pheromone output in stored-product Bostrichidae, *Bulletin of Entomological Research*, 96, 547, 2006.
20. Wright, V.F., Fleming, E.E., and Post, D., Survival of *Rhyzopertha dominica* (Coleoptera, Bostrichidae) on fruits and seeds collected from woodrat nests in Kansas, *Journal of the Kansas Entomological Society*, 63, 344, 1990.
21. Edde, P.A. and Phillips, T.W., Potential host affinities for the lesser grain borer, Rhyzopertha dominica (F.) (Coleoptera: Bostrichidae): Behavioral responses to host odors and pheromones and reproductive ability on non-grain hosts, *Entomologia Experimentalis et Applicata*, 119, 255, 2006.
22. Jia, F., Toews, M.D., Campbell, J.F., and Ramaswamy, S.B., Survival and reproduction of *Rhyzopertha dominica* (F.) (Coleoptera: Bostrichidae) on flora associated with native habitats in Kansas, *Journal of Stored Products Research*, 44, 366, 2008.
23. Nansen, C., Meikle, W.G., Tigar, B., Harding, S., and Tchabi, A., Nonagricultural hosts of *Prostephanus truncates* (Coleoptera: Bostrichidae) in a West African forest, *Annals of the Entomological Society of America*, 97, 481, 2004.
24. Craighead, F.C., Insect enemies of eastern forests, USDA Miscellaneous Publication Number 657, U.S. Government Printing Office, Washington, DC, 1950.
25. Fisher, W.S., A revision of the North American species of beetles belonging to the family Bostrichidae. Miscellaneous Publication Number 698, USDA, Washington, DC, 1950.
26. Dragicevic, S., Geocomputation: Modeling with spatial agents, *Computers: Environment and Urban Systems*, 32, 415, 2008.
27. Goodchild, M.F., Geographic information science, *International Journal of Geographical Information Science*, 6, 11, 1992.
28. Openshaw, S. and Abrahart, R.J., Geocomputation. In: *Proceedings of the 1st International Conference on Geocomputation*, Abrahart, R.J., Ed., Plenum Press, New York, p. 665, 1996.
29. Benenson, I. and Torrens, P.M., *Geosimulation: Automata-Based Modeling of Urban Phenomena*, John Wiley & Sons, London, U.K., 2004.
30. UCGIS, *Geographic Information Science & Technology Body of Knowledge*, DiBiase, D. et al., Eds., Association of American Geographers, Washington, DC, 2006.
31. Burrough, P.A., Dynamic modeling and geocomputation. In: *Geocomputation: A Primer*, Longley, P.A. et al., Eds., Wiley, New York, p. 165, 1998.
32. Gahegan, M., What is geocomputation? *Transactions in GIS*, 3, 203, 1999.
33. Albrecht, J., A new age for geosimulation, *Transactions in GIS*, 9, 451, 2005.
34. Longley, P.A., Foundations. In: *Geocomputation: A Primer*, Longley, P.A. et al., Eds., Wiley, New York, p. 1, 1998.
35. Openshaw, S., Building automated geographical analysis and exploration machines. In: *Geocomputation: A Primer*, Longley, P.A. et al., Eds., Wiley, New York, p. 95, 1998.
36. Batty, M. and Longley, P., *Fractal Cities*, Academic Press, London, U.K., 1994.
37. Langran, G., *Time in Geographic Information Systems*, Taylor & Francis, Bristol, PA, 1992.
38. Yuan, M., Representing complex geographic phenomena in GIS, *Cartography and Geographic Information Science*, 28, 83, 2001.
39. Grimm, V., Ten years of individual-based modelling in ecology: What we have learned and what could we learn in the future? *Ecological Modelling*, 115, 129, 1999.

40. Railsback, S.F., Concepts from complex adaptive systems as a framework for individual-based modeling, *Ecological Modelling*, 139, 47, 2001.
41. Remondino, M. and Cappellini, A., Agent based simulation in biology: The case of periodical insects as natural prime number generators, *Vet On-Line: The International Journal of Veterinary Medicine*, 2005. http://www.priory.com/vet/cicada.pdf (accessed August 31, 2009).
42. Barrett, C.L., Eubank, S.G., and Smith, J.P., If smallpox strikes Portland, *Scientific American*, 292, 54, 2005.
43. Batty, M., Desyllas, J., and Duxbury, E., The discrete dynamics of small-scale spatial events: Agent-based models of mobility in carnivals and street parades, *International Journal of Geographical Information Science*, 17, 673, 2003.
44. Parker, D.C., Manson, S.M., Janssen, M.A., Hoffmann, M.J., and Deadman, P., Multi-agent systems for the simulation of land-use and land-cover change: A review, *Annals of the Association of American Geographers*, 93, 314, 2003.
45. Ahearn, S.C., Smith, J.L.D., Joshi, A.R., and Ding, J., TIGMOD: An individual-based spatially explicit model for simulating tiger/human interaction in multiple use forests, *Ecological Modelling*, 140, 81, 2001.
46. Bennett, D.A. and Tang, W., Modelling adaptive, spatially aware, and mobile agents: Elk migration in Yellowstone, *International Journal of Geographical Information Science*, 20, 1039, 2006.
47. Russell, S.J. and Norvig, P., *Artificial Intelligence: A Modern Approach* (2nd edn.), NJ, Prentice Hall, Englewood Cliffs, 2002.
48. Sakellariou, I., Kefalas, P., and Stamatopoulou, I., Teaching intelligent agents using NetLogo. Paper presented at the *ACM-IFIP Informatics Education Europe III Conference, IEEIII 2008*, Venice, Italy, December 4–5, 2008.
49. Wilensky, U., NetLogo, Center for Connected Learning Comp.-Based Modeling, Northwestern University, Evanston, IL, 1999. http://ccl.northwestern.edu/netlogo (accessed August 30, 2009).

11 Geographic Information Systems in Corn Rootworm Management

B. Wade French, Kurtis D. Reitsma, Amber A. Beckler, Laurence D. Chandler, and Sharon A. Clay

CONTENTS

11.1 Executive Summary ... 233
11.2 Introduction ... 234
11.3 Materials and Data Collection ... 235
 11.3.1 Field and Insect Trap Locations ... 235
 11.3.2 System Requirements ... 235
11.4 Getting Started with ArcGIS™ ... 236
 11.4.1 Importing Latitude–Longitude Trap Data ... 237
 11.4.2 Symbolizing Map Layers ... 238
 11.4.3 Coordinate Systems and ESRI® Shapefiles ... 240
11.5 Analysis of Adult CRW Population and Distribution ... 240
 11.5.1 Spatial Autocorrelation, Moran's *I* ... 241
 11.5.2 Interpolation, Inverse Distance Weighting ... 243
 11.5.3 Comparing CRW Population with Soil Texture ... 245
11.6 Conclusion ... 251
Acknowledgments ... 252
References ... 252

11.1 EXECUTIVE SUMMARY

Corn rootworms (CRW) (*Diabrotica* spp. Coleoptera: Chrysomelidae) are serious pests of corn (*Zea mays*) in the United States and Europe. Control measures for CRW were historically based upon chemical pesticides and crop rotation. Pesticide use created environmental and economic concerns. In 1997, a 5 year areawide CRW management program was established in five states to manage CRW populations on a multi-field scale to help producers better manage these pests. Spatial data were used to predict areas of infestation, leading to more site-specific management techniques. The goal was to more fully understand the spatial relationships between CRW infestations and physical features of the landscape. Geographical information systems (GIS) and spatial analytical techniques were used to examine relationships between CRW populations and soil texture. This chapter uses CRW (adult) data collected in 1998 from cornfields within a 41,400ha study area in Brookings, County in eastern South Dakota to demonstrate the procedures and techniques used to examine CRW populations and soil texture relationships. Procedures and techniques demonstrated in this chapter include interpolation, spatial autocorrelation, and comparative analysis.

11.2 INTRODUCTION

The western (*Diabrotica virgifera virgifera* LeConte) and northern (*D. barberi* Smith and Lawrence) CRW (Coleoptera: Chrysomelidae) are significant agricultural insect pests of corn in the United States Corn Belt. The western CRW recently has become a serious corn pest in Europe.[1,2] Larval feeding on roots results in severe injury to the plant by reducing physiological efficiency at the sixth leaf to the tassel development stages and by promoting stalk lodging in mature plants. Physiological mechanisms negatively impacted by feeding injury to the seminal and adventitious root system include plant dry matter accumulation, nutrient and water uptake, and photosynthesis. The reduction in the physiological efficiency as a result of larval root feeding directly causes yield loss. Feeding also weakens the adventitious brace root system, which leads to stalk lodging and indirect yield losses due to inefficient harvesting.[3] Although adults cause less damage, populations are indicative of larval populations in the current and upcoming year.

In this chapter, we describe the use of GIS and spatial analysis to interpret patterns of spatial variation in CRW abundance in the South Dakota CRW areawide management site (Figure 11.1). Examples presented in this chapter are intended to

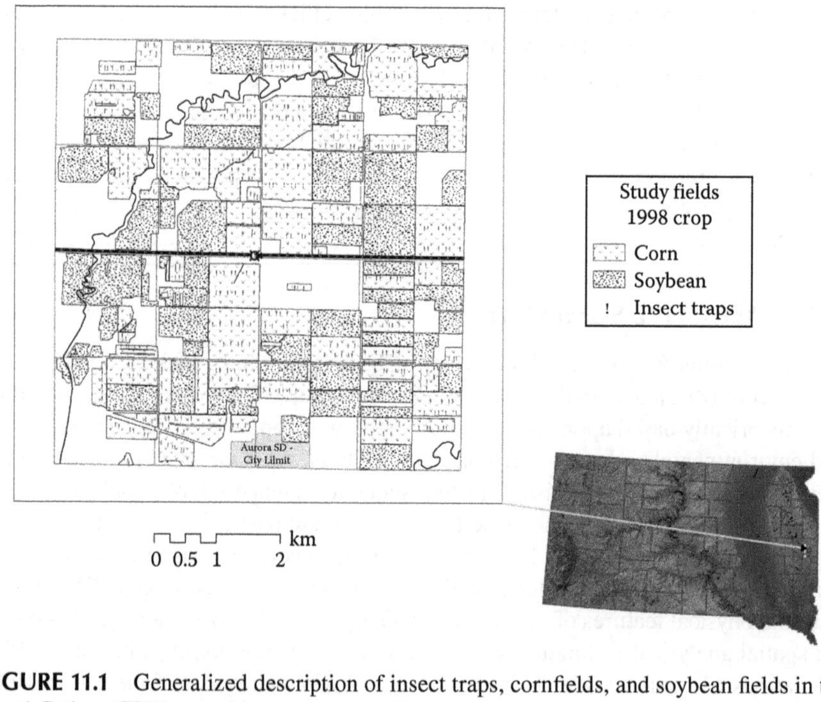

FIGURE 11.1 Generalized description of insect traps, cornfields, and soybean fields in the South Dakota CRW areawide management site.

serve as a guide for using GIS to estimate spatial distributions of CRW populations in multiple cornfields. For more information on further spatial analysis of these data used in this chapter, refer to Beckler et al.[4] and for western and northern CRW biology to Beckler et al.[5] and French et al.[6]

The data analysis presented in this chapter specifically examines the spatial relationships between adult CRW populations and soil texture for 1 year. Several spatially explicit procedures are demonstrated for analyzing these relationships including interpolation, spatial autocorrelation, and comparative analysis. A better understanding of the spatial interactions that occur with insect pests and the landscape may generate information leading to regional and site-specific models to predict future populations and provide help in creating well-informed management decisions.

11.3 MATERIALS AND DATA COLLECTION

Adult CRW beetles were captured in 417 Pherocon® AM yellow sticky traps (Trécé Inc., Adair, Oklahoma) attached to wooden laths placed approximately 1 m above the soil surface in 61 cornfields that were located within the 41,400 ha study region.[7,8] Traps were placed along two transects ≈120 m apart in each field prior to adult emergence. The number of traps placed in each field varied from 3 to 12 depending on field size.[4] Sticky traps were replaced each week for ≈10 weeks and returned to the laboratory and stored in freezers for later processing (see French et al.[8] for more information on traps and sampling). For the purposes of this chapter, the total numbers of northern and western CRW adult beetles captured over the entire growing season were combined to represent one CRW metapopulation. Because rotated corn comprised ≈70% of all corn production, and given the prominence of 2 year egg diapause, northern CRW made up ≈83% of the total beetles collected, which is typical for the area.[9,10]

11.3.1 Field and Insect Trap Locations

Corn and soybean (*Glycine max*) fields were georeferenced in 1997 using handheld Trimble GeoExplorer II (Trimble, Sunnyvale, California) global positioning system (GPS) units. Trap locations were georeferenced in 1998, 1999, and 2000 and classified by trap type. All georeferenced data were differentially corrected using Trimble Pathfinder Office 2.01 (Trimble, Sunnyvale, California). For this exercise, the coordinate system at the time of collection is assumed as GCS, North American Datum, 1983 (NAD-83). Soybean fields were classified because of the prominent 2 year rotation with corn in the study area and their ability to harbor northern CRW eggs resulting from the prominence of 2 year egg diapause in this region of the Corn Belt.[9,10]

11.3.2 System Requirements

Minimum system recommendations include IBM-PC Pentium® 4 or higher processor, with 1 Gb RAM, operating system, Microsoft Windows (Windows 7, Vista, or XP Service Pack 2), installed software, MS Excel ver. 2003 or higher, ESRI® ArcGIS™ (ver. 9.3) with Spatial Analyst extension.[11–13]

On the CD ROM, in the Chapter 11 folder, there are two data folders, **ArcGIS** and **Tab Data**. The **ArcGIS** folder contains four folders entitled **Geodatabase, Raster, Shape**, and **Temp**. Some of these file folders contain other files and folders that will be used in the exercise below. The **Temp** and **Shape** folders are empty at the beginning of the exercise but will be populated by the user as the exercise progresses. The **Tab Data** folder contains the Excel® file that has the longitude and latitude coordinates of the traps and the total CRW numbers collected by trap location.

11.4 GETTING STARTED WITH ArcGIS™

The objective of this exercise is to determine if adult CRW population and surface soil texture are related. If a strong relationship exists, then these data could be used to improve site-specific management for these insects. Data for this exercise are included on the accompanying compact disk (CD). To begin this exercise, copy the entire "*X*:\Chapter_11" (*X*:\is the CD/DVD drive) folder to the root directory of your hard drive (C:\). **Important: Failure to save example data to the proper location will increase the difficulty of recreating the resultant map and data.**

The majority of the data for this project is stored in a geodatabase (**C:\Chapter_11\ArcGIS\Geodatabase\Ch_11_GeoDB.mdb**). To begin an ArcMap™ session, add base data to the map view.

1. Begin a session of ArcMap™, select **New Empty Map**, and click **OK**.
2. Add the study boundary (**stdy_bnd**) by clicking the **+ button** (add data), and then navigate to ***C:\Chapter_11\ArcGIS\Geodatabase\Ch_11_GeoDB***; select ***stdy_bnd***, and click **Add**.
3. Change the name of this layer by right clicking on the name, select **Properties>General**, and in the **Layer name** box, type ***Study boundary.***
4. While still in **Properties**, click on the **Source** tab. This dialog box provides information about the layer including data source and coordinate system. The coordinate system for this layer is **NAD_1983_UTM_Zone_14N** indicating a **Universal Transverse Mercator, Zone 14 North, North American Datum—1983** coordinate system with a **Transverse Mercator** projection. Click **OK** to close the **Properties** box.
5. Compare the coordinate system information that you observed in the **Properties>Source** box with the ArcMap™ data frame by right clicking in the dataframe and selecting **Data Frame Properties>Coordinate System**. Coordinate systems and projections of all layers in this dataframe should be identical, as ArcMap™ defaults the dataframe to the first layer added.
6. Next, add the study field boundary (***stdy_fld_bnd_all***) and section line (***stdy_sec***) boundary layers from the **ArcGIS\Geodatabase\Ch_11_GeoDB** folder. Note that map layers in the **Table of Contents** appear in the same order that they are added. Change the names of these two layers by right clicking on the layer, selecting **Properties>General** and changing the name of the ***std_fld_bnd_all*** layer to ***study fields*** and the ***stdy_sec*** layer to ***section line*** in the **Layer name** box.

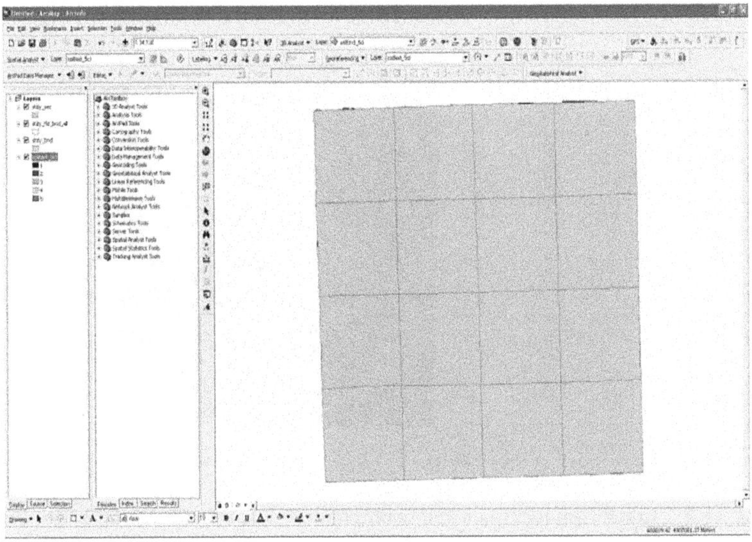

FIGURE 11.2 Initial layers shown in the ArcMap™ environment. (The ArcGIS® graphical user interface is the intellectual property of ESRI and is reproduced herein by permission. Copyright © 1999–2009 ESRI. All rights reserved.)

7. Add the soil textural class from the raster layer folder (*C:\Chapter_11\ArcGIS\Rasters\soiltext_5cl*). Change the layer name to ***Soil textural class***.
8. Arrange map layers in the order (from top to bottom) of ***section line***, ***study fields***, ***study boundary***, and ***soil textural class*** by clicking and dragging layers to the desired location in the **Table of Contents**. The map should resemble Figure 11.2. Save the file as **(C:\Chapter_11\ArcGIS\Rootworm.mxd)**, and periodically save throughout this exercise.

11.4.1 Importing Latitude–Longitude Trap Data

Longitude (X) and latitude (Y) values for locations of insect sticky traps were recorded using GPS. Coordinates of each trap and total CRW adult numbers captured in that trap are imported into ArcMap™ from an MS Excel spreadsheet in the **Tab_Data** folder as follows:

1. Select **Tools** from the main menu, and click **Add XY Data**, so the **Add XY Data** dialog box appears.
2. Click the **open file symbol** next to the combo box under **Choose a table from the map or browse for another table**.
3. Navigate to ***C:\Chapter_11\Tab_Data***, and select the MS Excel Workbook, ***GPS_CRW_TRAP.xls***, click **Add**, select the ***CRW_TRAPS$*** worksheet, and click **Add**.
4. Select ***LONGITUDE*** for the X Field and ***LATITUDE*** for the Y Field.
5. Note that the coordinate system in the **Add XY Data** dialog box is undefined. The coordinate system needs to be defined as GCS, NAD-83, so that

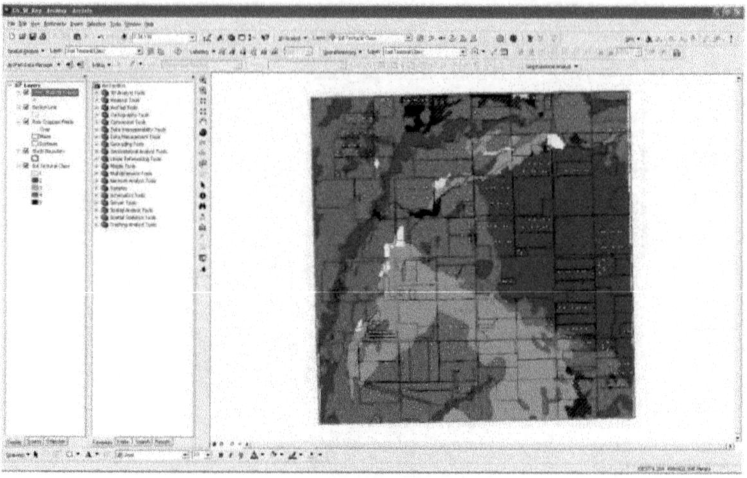

FIGURE 11.3 Insect trap locations and base data layers shown in the ArcMap™ environment. (The ArcGIS® graphical user interface is the intellectual property of ESRI and is reproduced herein by permission. Copyright © 1999–2009 ESRI. All rights reserved.)

it is properly displayed. To define the coordinate system, in the **Coordinate System of Input Coordinates** box, click **Edit>Select**. Choose the **GCS** (either double left click or select **Add)>North America>North American Datum 1983.prj**. Click **Add**. Click **OK** (this then goes to the **Add XY** data dialog box), and click **OK**. Note that a **Table Does Not Have Object ID Field** message box appears. Click **OK** in this box to close the message. The locations of the CRW traps are now added as an event theme. Points representing trap locations and a layer named ***CRW_TRAPS$ Events*** should appear in the **Table of Contents** (Figure 11.3).

11.4.2 Symbolizing Map Layers

Symbology defines how features are represented on a map and aids in map interpretation. Map features may influence the size, color, shape, and label of a symbol. ArcMap™ includes a symbology editor so that maps can be customized in the each layer's **Properties** dialog box. The base layers ***Study boundary*** and ***Section lines*** are symbolized using the following procedure:

1. Double click the ***study boundary*** layer to activate the **Properties** dialog box. Click **Symbology**.
2. Click the box in the **Symbol** section to open the **Symbol Selector** dialog box. Under **Options>Fill color**, select **No Color**.
3. While still in the **Symbol Selector** dialog box, click **Properties>Outline...** and scroll to select the symbol for **Boundary, County**. Click **OK>OK>OK>OK** to exit all dialog boxes and return to the main menu.
4. Repeat the above procedure for ***Section lines***; only select the **Boundary, Township** as the **Outline** style.

The ***study fields*** layer has data to differentiate between corn and soybean fields and change the symbology for visually identifying the locations of the crops. However, this layer depicts boundaries, and it may be necessary to view features (e.g., traps or soils information) within the boundary. Therefore, hollow polygons with distinguishing patterns will be used to show the location of the two crops. It is suggested that the ***study fields*** layer be symbolized as follows:

1. Double click ***study fields*** layer to activate **Properties** dialog box, and click **Symbology**.
2. In the **Show** dialog box, click **Catagories>Unique values.**
3. Select **CROP** in the **Value Field** combo box, click **Add All Values** to populate the legend with ***Corn*** and ***Soybean***, and uncheck the **All other values box.**
4. Double click ***Corn***, and scroll the options to select **10% Crosshatch** in the **Symbol Selector** window.
5. In the **Options** area of the **Symbol Selector** window, click the arrow next to **Outline Color**, and select a color of your choice for the outline (this exercise uses black). Click the arrow next to **Fill Color**, and select a color of your choice as the fill color (this exercise uses black). Click **OK>Apply**. The symbology for corn is now a crosshatched pattern with the color you selected.
6. Repeat this procedure for ***Soybean***, selecting a new hatching pattern, outline color, and fill color. After the selections, click **OK** to return to the main menu. This exercise used 10% simple hatch, black as the outline color, and black for the fill color. Click **OK** to close the symbology editor. The map in gray scale is shown in Figure 11.4 (see CD figure to see color map).

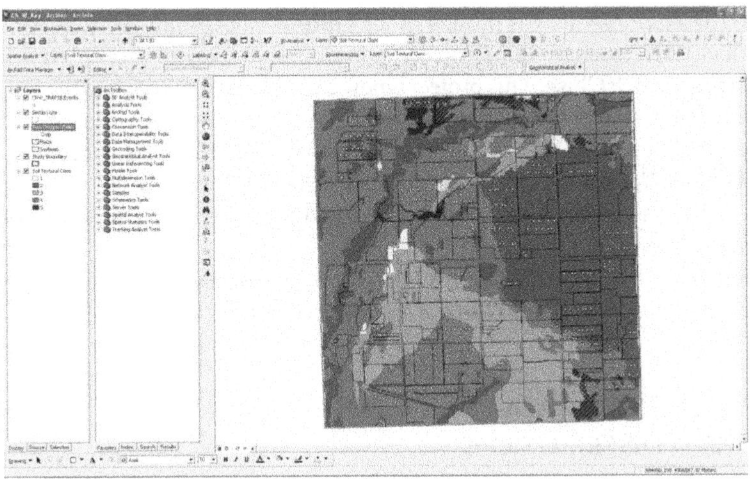

FIGURE 11.4 Symbolized map base layers shown in ArcMap™ environment. (The ArcGIS® graphical user interface is the intellectual property of ESRI and is reproduced herein by permission. Copyright © 1999–2009 ESRI. All rights reserved.)

11.4.3 Coordinate Systems and ESRI® Shapefiles

Shapefiles store non-topological geometry and attribute data for spatial features (point, line, or polygon) in a dataset. The geometry for the feature is stored as a shape comprising a set of vector coordinates. Shapefiles consist of three to seven files where the **.shp* (stores shape geometry), **.shx* (spatial index file), and **.dbf* (attribute file) are mandatory. A fourth file, **.prj* stores coordinate and projection data. In this exercise, the coordinate system of the event theme is defined as a geographic coordinate system, NAD 83 (GCS North American_1983), with no linear units. Conversion of geographic data to a shapefile and projection are needed to continue the data analysis. The **Projections and Transformations** tool of ArcMap™ simultaneously converts these data to a shapefile and projection (although each operation can be done separately) using the following procedure.

1. Open the ArcMap™ toolbox by clicking the red toolbox symbol on the menu bar.
2. Expand **Data Management Tools>Projections and Transformation>Feature**, and double click **Project** to open the **Project** dialog box.
3. In the **Input Dataset or Feature Class** combo box, select ***CRW_TRAPS$ Events***.
4. Click the **Open Folder** symbol next to **Output Dataset or Feature Class**, navigate to ***C:\Chapter_11\ArcGIS\Shapefiles***\enter ***stdy_crw_trap*** in the input box next to **Name:**, and click **Save**.
5. Click the symbol next to **Output Coordinate System**, then click **Select**, double click **Projected Coordinate Systems**, double click **UTM**, double click **NAD 1983**, select ***NAD 1983 UTM Zone 14N.prj***, and click **ADD**.
6. Click **OK** in the spatial reference properties dialog box, and ***NAD 1983 UTM Zone 14N*** now is present in the **Output Coordinate System** box.
7. Click **OK** to close the **Project** dialog box. A **Project Completed** box appears; click **Close**. This project is now complete and has created the shapefile that is stored in the **shapefile** folder on the C drive. A similar shapefile is also present on the CD in the Chapter 11 ArcGIS folder (***Chapter_11\ArcGIS\Geodatabase\Ch_11_GeoDB.mdb, stdy_crw_trap***).
8. Click the **+ button** to add the newly created shapefile ***stdy_crw_trap.shp*** as a map layer.
9. Remove ***CRW_TRAPS$ Events*** by right clicking on it and selecting **Remove**.
10. Rename ***stdy_crw_trap, Corn Rootworm Trap.***

11.5 ANALYSIS OF ADULT CRW POPULATION AND DISTRIBUTION

Total number of adult CRW beetles was recorded for each sticky trap location (***TOT_CRW*** in the MS Excel datasheet). We are interested to know if there are any spatial patterns with respect to population of CRW and will use geostatistical procedures to provide insight into the distribution. In this exercise, an analysis for spatial autocorrelation (Moran's *I*) is performed to measure and analyze the degree of spatial dependency among observations.

11.5.1 Spatial Autocorrelation, Moran's *I*

The Moran's *I* coefficient describes the degree of similarity of neighboring points. When adjacent points are very dissimilar (negative spatial autocorrelation), the coefficient approaches −1 (uniformly spaced dispersion pattern), when they are very much alike (positive spatial autocorrelation), the coefficient approaches +1 (aggregated dispersion pattern), coefficient values of 0 indicate a random dispersion pattern.[14] Mathematically, Moran's *I* is expressed as

$$I = \frac{N}{\sum_i \sum_j w_{ij}} \frac{\sum_i \sum_j w_{ij}(x_i - \bar{x})(x_j - \bar{x})}{\sum_j (x_i - \bar{x})^2} \tag{11.1}$$

where

N is the number of points
w_{ij} is the weight of relationship of neighboring points
x_i is the value of point i
x_j is the value of point j
$\bar{x}$ is the mean value of point

Values of Moran's *I* can be transformed to *Z* scores with values >1.96 or <−1.96 indicating spatial autocorrelation significance at $P \leq 0.05$.

ArcMap™ provides a tool for calculating Moran's *I* and corresponding *Z* value returning graphic and text results. Calculate Moran's *I* using the following procedure:

1. Open ArcToolbox by clicking the red toolbox symbol on the menu bar.
2. Expand **Spatial Statistics Tools>Analyzing Patterns>Spatial Autocorrelation (Morans I)**.
3. Select ***Corn Rootworm Trap*** from the **Input Feature Class** combo box.
4. Select ***TOT_CRW*** from the **Input Field** combo box.
5. Check the **Display Output Graphically (optional)** box, and allow default values for all other fields.
6. Click **OK**. The results of Moran's *I* analysis are displayed graphically and as text (Figure 11.5). Note that the Moran's *I* value for the CWR distribution is ≈0.80 and a *Z* score of ≈16.8. These results indicate that the data have a high degree of spatial autocorrelation and the correlation is highly significant, as the *Z*-score value ≫1.96. Interpolation of these data should reveal clustered patterns.
7. Click **Close** in the graphic display and in the text window when finished viewing the results.

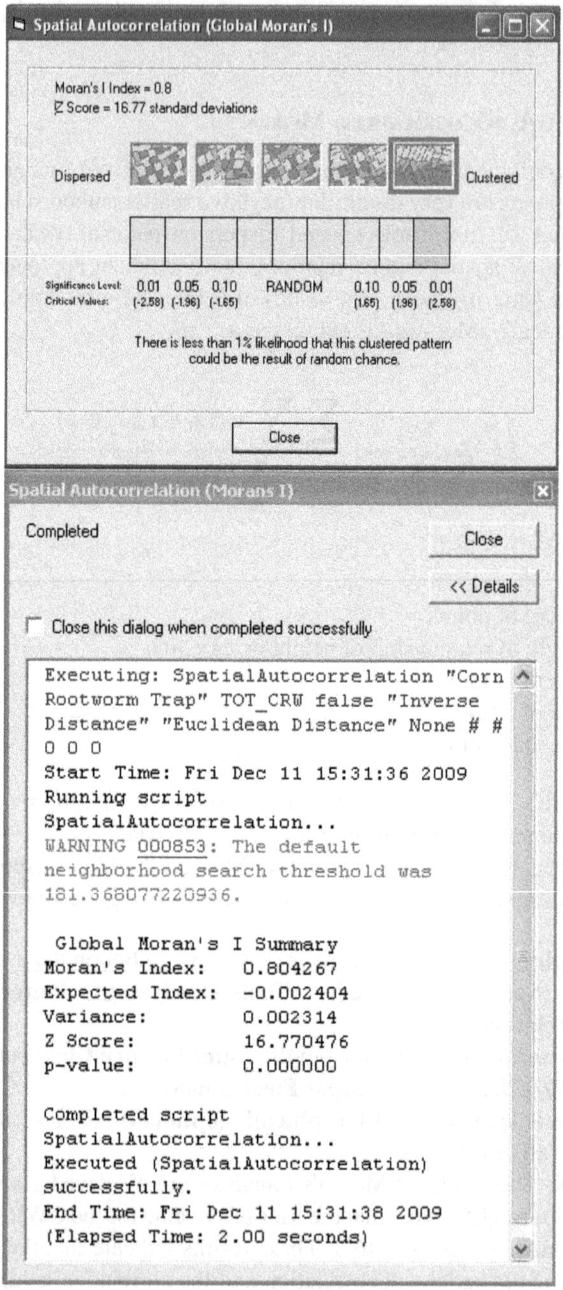

FIGURE 11.5 Graphical and text output of Moran's *I* analysis in the ArcMap™ environment. (The ArcGIS® graphical user interface is the intellectual property of ESRI and is reproduced herein by permission. Copyright © 1999–2009 ESRI. All rights reserved.)

11.5.2 Interpolation, Inverse Distance Weighting

Interpolation is a geostatistical process that assigns values to unknown points based on values from known points. In this example, CRW populations are predicted throughout the study area based on trap counts in the study area. While there are several methods of interpolation, this example uses the inverse distance weighting (IDW) method to predict CRW distribution and abundance. The IDW interpolation estimates the cell value using a linearly weighted value from a combination set of sample points. The function (or weight) is determined based on the inverse distance between known and estimated points. Therefore, values of closer known points to the cell of interest have greater influence on the value than points farther from a known cell. With IDW, the exponent or power value controls the significance of known georeferenced points upon the interpolated values, based upon the distance from the output point.[15–17]

Interpolation of CRW population is performed using the following procedure:

1. Expand **Spatial Analyst Tools>Interpolation>IDW**.
2. The **IDW** input dialog box appears in a new window. Select ***Corn Rootworm Trap*** in the **Input Point Features** combo box and ***TOT_CRW*** in the **Z value field** combo box.
3. Click **Open** next to **Output raster input** box, navigate to ***C:\Chapter_11\ArcGIS\Temp\', enter 'crw_idw_temp***, and click **Save**.
4. Right click on the ***Soil Textural Class*** layer, and select **Properties**. click the **Source** tab; note that cell size is equal to **26.398** m. Therefore, the cell size of the CRW abundance interpolation should have a similar size for later comparison purposes. Enter ***26.4*** in the **Output cell size (optional)** input box. Cell size values can be sourced directly from the ***Soil Textural Class*** raster file by using the navigation tool at the right of the cell size input box.
5. Click **Environment>General Settings>**, select ***Same as layer Soil Textural Class*** from the **Extent** combo box, and select ***Soil Textural Class*** from the **Snap Raster** combo box.
6. Expand **Raster Analysis Settings**, select ***Study Boundary*** in the **Mask** combo box, and click **OK**.
7. The ArcMap™ Spatial Analyst tools, IDW input dialog box should resemble Figure 11.6, click **OK**. When complete, an **IDW** dialog box will appear and a new layer ***crw_idw_temp*** is created. Click **Close** in the **IDW** box. The new layer should resemble Figure 11.7. CRW counts are divided into several classes, with 2 the lowest count and 690 the highest count.

Note that when right clicking on ***crw_idw_temp***, the **Open Attribute Table** option is grayed out and unavailable. This dataset must be reclassified before the attribute table is available. Data can be classified using a number of methods. In this exercise, **Natural Breaks** will be used. This method selects breaks that best group similar values and maximizes the differences among classes. Reclassify the ***crw_idw_temp*** raster dataset using the following procedure.

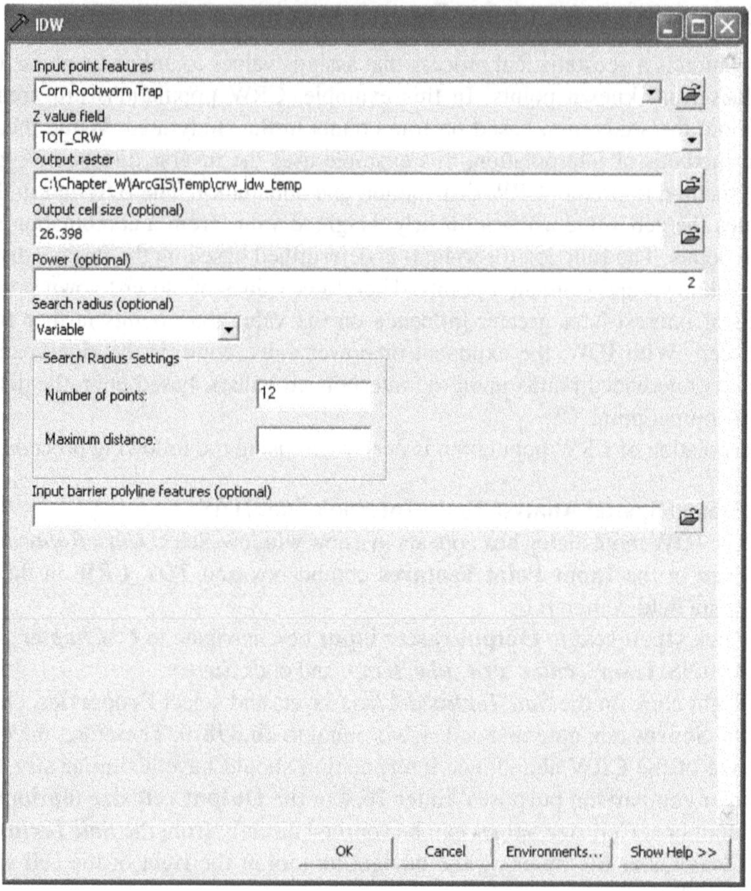

FIGURE 11.6 ArcMap™ Spatial Analyst tools, IDW input dialog box. (The ArcGIS® graphical user interface is the intellectual property of ESRI and is reproduced herein by permission. Copyright © 1999–2009 ESRI. All rights reserved.)

1. Expand **Spatial Analyst Tools>Reclass** >, and then select **Reclassify**.
2. Select ***crw_idw_temp*** in the **Input Raster** combo box.
3. Leave the **Reclass field** combo box default value of **Value**.
4. Click **Classify**, and a **Classification** dialog box opens. In the combo box, select **Natural Breaks (Jenks)** for **Method** and ***5*** for the number of **Classes**. On a scratch pad, record break values shown at the right side of the **Classification** box. After reclassification occurs, these values will be reclassed as bin values 1–5. Values for bins 1–5 are 2–69, 70–120, 121–196, 197–315, and 316–690, respectively. Click **OK**; when completed, a **Reclassify** dialog box appears. Click **Close**.
5. Click the **Open** file symbol next to the **Output raster** input box, navigate to ***C:\Chapter_11\ArcGIS\Rasters***, enter **crw_idw_class** as a name for the new raster, and click **Save** that will return to the **Reclassify** dialog box (Figure 11.8). Click **OK**. When completed, a **Reclassify** dialog box appears. Click **Close**.

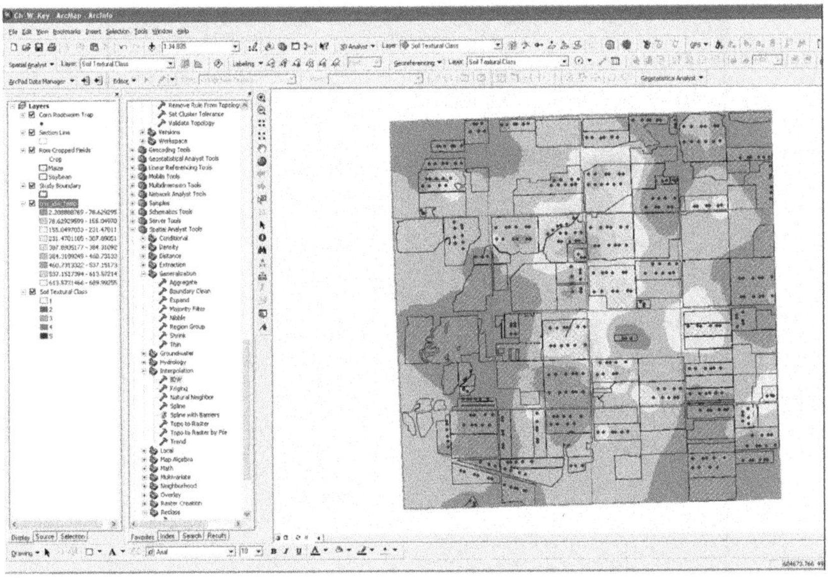

FIGURE 11.7 Interpolation layer of CRW population in the ArcMap™ environment. (The ArcGIS® graphical user interface is the intellectual property of ESRI and is reproduced herein by permission. Copyright © 1999–2009 ESRI. All rights reserved.)

The new raster is based on the same data as the first output; however, the data have been regrouped into 5 rather than 10 classifications, and the symbols differ. Remove the ***crw_idw_temp*** layer from the map, leaving the newly created ***crw_idw_class*** raster file as a map layer.

11.5.3 Comparing CRW Population with Soil Texture

Datasets can be created as "continuous" data, meaning that points at a particular location have an exact value or are "classified" meaning that a point within a classification will have a value that falls within the class range. In this exercise, CRW population data are "classified" with abundance values grouped into five classes. Soils data are grouped by USDA textural classes based on sand, silt, and clay content. The five soil classifications used were silty clay (SiC), silty clay loam (SiCL), silt loam (SiL), loam (L), and sandy loam (SaL). The SiC soils have the highest clay and lowest sand content whereas the SaL soils have the lowest clay and the highest sand content. Based on these classifications, each cell within a layer has a single value representing a soil texture in the soil layer or CRW population class in the density layer.

Methods presented here examine the relationship between the numbers of CRW adults captured and soil texture at each location. Detailed methods for examining statistical significance of this relationship using chi square analysis are presented in Beckler et al.[4]

Cross tabulation, using zonal statistical analysis, provides a comparison of a cell in one layer to a cell in another layer. Performing zonal statistical analysis provides

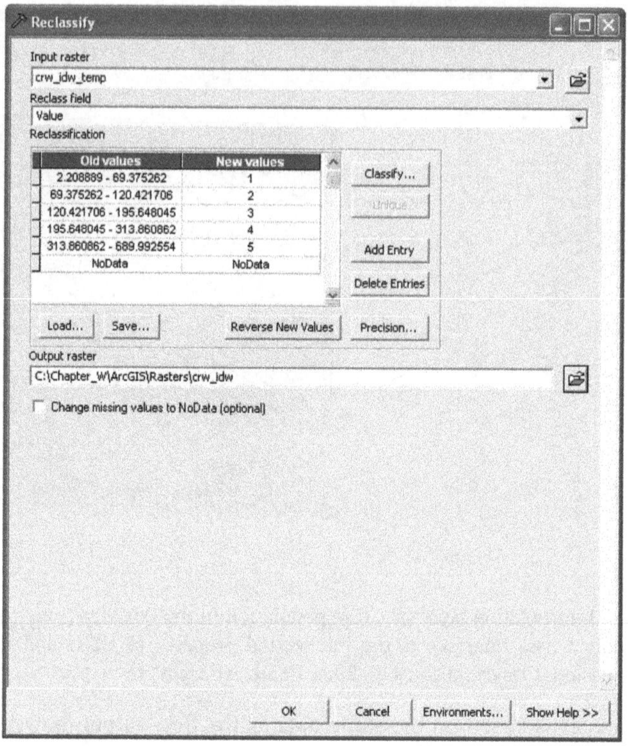

FIGURE 11.8 Reclassify tool in ArcMap™ Spatial Analyst tools. (The ArcGIS® graphical user interface is the intellectual property of ESRI and is reproduced herein by permission. Copyright © 1999–2009 ESRI. All rights reserved.)

an indication of the relationship between relative CRW abundance and soil texture. The following procedure results in a comparison table.

1. Open ArcTools by clicking on the red toolbox symbol on the menu bar.
2. Expand **Spatial Analyst Tools>Zonal>**, and then select **Zonal Statistics as Table** to open the dialog box (Figure 11.9).
3. Select ***Soil Textural Class*** in the **Input raster or feature zone** data combo box.
4. Select ***T_NAME*** in the **Zone field** combo box.
5. Select ***crw_idw_class*** in the **Input value raster** combo box. Click **Add**.
6. Click **Open** next to **Output table** input box, navigate to ***C:\Chapter_11\Tab_Data***, enter ***soil_crw_zstats*** in the **Name** input box, click **Save**, and click **OK**. Click **Close** when the **Zonal Statistics as Table** dialog box appears after the computation.
7. Click the **Source** tab at the bottom of the **Table of Contents**; note that the newly created table now appears.
8. Right click on ***soil_crw_zstats***, and select **Open** to open the table and view the results of the analysis (Figure 11.10).

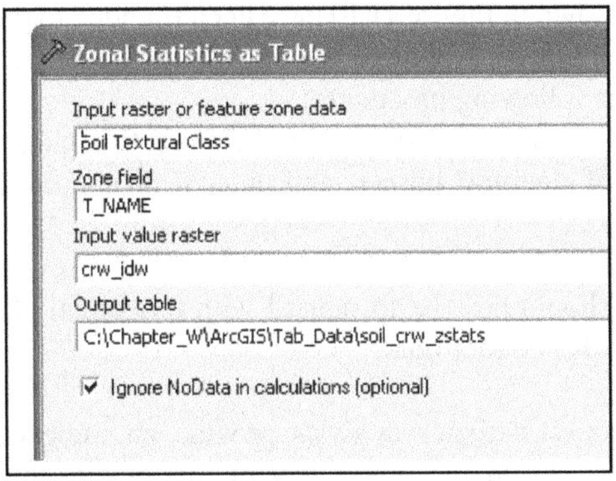

FIGURE 11.9 Zonal statistics as table dialog box. (The ArcGIS® graphical user interface is the intellectual property of ESRI and is reproduced herein by permission. Copyright © 1999–2009 ESRI. All rights reserved.)

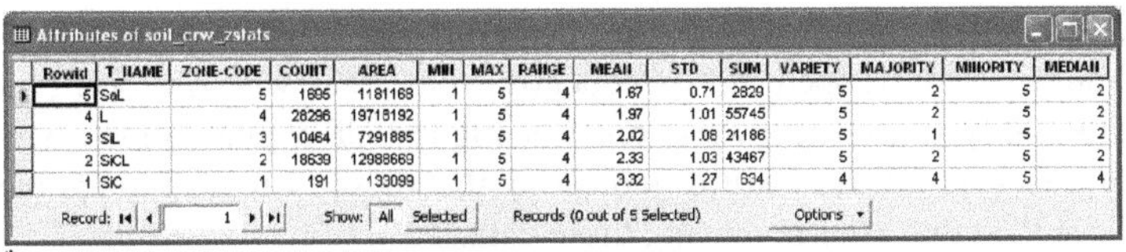

Attributes of soil_crw_zstats

Rowid	T_NAME	ZONE-CODE	COUNT	AREA	MIN	MAX	RANGE	MEAN	STD	SUM	VARIETY	MAJORITY	MINORITY	MEDIAN
5	SaL	5	1695	1181168	1	5	4	1.67	0.71	2829	5	2	5	2
4	L	4	28296	19718192	1	5	4	1.97	1.01	55745	5	2	5	2
3	SiL	3	10464	7291885	1	5	4	2.02	1.08	21186	5	1	5	2
2	SiCL	2	18639	12988669	1	5	4	2.33	1.03	43467	5	2	5	2
1	SiC	1	191	133099	1	5	4	3.32	1.27	634	4	4	5	4

Record: 1 Show: All Selected Records (0 out of 5 Selected) Options

*

Zonal statistics table—field definitions	
Rowid	Internal ID use by ArcMap™
T_NAME	Soil texture name abbreviation
ZONE-CODE	Internal ID for individual zones
COUNT	Number of cells occurring in each soil texture
AREA	Total area of cells for each soil texture (m^2)
MIN	Minimum CRW population class value
MAX	Maximum CRW population class value
RANGE	Range of CRW population class values
MEAN	Mean of CRW population class values
STD	Standard deviation of CRW population class values
SUM	Sum of CRW population class values
VARIETY	Number of CRW population class values
MAJORITY	CRW population class occurring in greatest number of cells
MINORITY	CRW population class occurring in least number of cells
MEDIAN	Median CRW population class value occurrence

* Some field values have been modified rounding to the nearest integer or two decimal places.

FIGURE 11.10 Results of zonal statistics analysis of soil texture versus CRW population class. (The ArcGIS® graphical user interface is the intellectual property of ESRI and is reproduced herein by permission. Copyright © 1999–2009 ESRI. All rights reserved.)

9. Table values shown in Figure 11.10 have been rounded to the nearest integer or two decimal places. Although not necessary, the field properties can be changed by the following procedure:
   a. Right click the field to modify, select **Properties**, click **Numeric**, select **Number of decimal places**, and enter ***0*** or appropriate value in the scroll box.
   b. Click **OK**.
   c. Numbers should now be formatted with 0 decimal places. Other field types can also be modified.

Results from the zonal statistics analysis provide an indication of comparative spatial distribution of soil texture and CRW populations in the study area. In the table, **COUNT** and **AREA** indicate that SiC soils occupy the least amount of area (191 cells) while loam (L)-textured soils occur most frequently (28,296 cells) in the study area followed by SiCL, SiL, and SaL. Most CRWs were captured in the dominant soil types, L and SiCL as indicated by the sum values of 55,604 and 43,348, respectively. The CRW abundance class occurring most frequently over all soil types was 2 (70–120) as indicated by the **MAJORITY** and **MEDIAN** values. The **MAJORITY** value for the SiC soil type, however, was 4, indicating higher numbers of CRW adults were trapped most frequently from areas with SiC soil types even though this soil type had the lowest area. The highest CRW population class (5) was found the least over all soil textures indicating that these extreme populations rarely occur. Close the table after reviewing these data.

Further analysis of these data may provide additional understanding of CRW adult population with soil texture. A cross tabulation of soil texture with CRW population class is performed in ArcMap™ using the **Tabulate Area** tool with the following procedure.

1. Expand **Spatial Analyst Tools>Zonal>**, and then select **Tabulate Area**, opening the tool dialog box (Figure 11.11).
2. Select ***Soil Textural Class*** in the **Input raster or feature zone data** combo box.
3. Select ***T_NAME*** in the **Zone field** combo box.
4. Select ***crw_idw_class*** in the **Input raster or feature class data** combo box, and click **Add**.
5. Select ***VALUE*** in the **Class field** combo box.
6. Click **Open** next to the **Output table** input box, navigate to ***C:\Chapter_11\Tab_Data***, enter ***soil_crw_tarea*** in the **Name** input box, click **Save**, and click **OK**. Click **Close** when the **Tabulate Area** dialog box appears after the calculations are complete.
7. Click the **Source** tab at the bottom of the **Table of Contents**, to view the newly created table.
8. Right click on ***soil_crw_tarea***, and select **Open** to open the table and view the results of the analysis (Figure 11.12).
9. If desired, round table values to the nearest m^2 using the previous procedure.

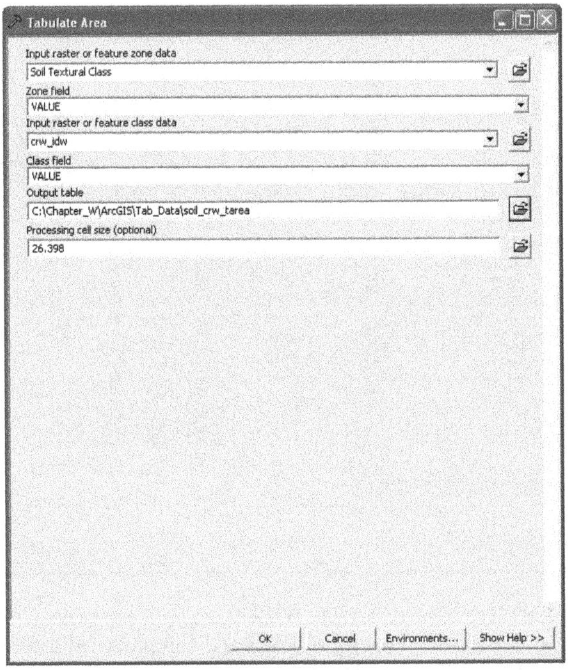

FIGURE 11.11 Tabulate area dialog box. (The ArcGIS® graphical user interface is the intellectual property of ESRI and is reproduced herein by permission. Copyright © 1999–2009 ESRI. All rights reserved.)

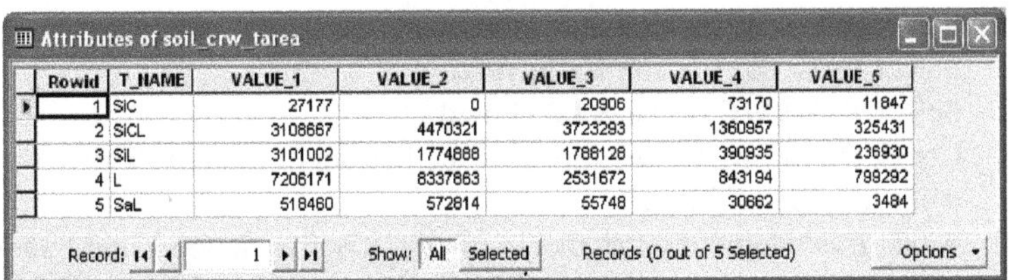

Attributes of soil_crw_tarea

Rowid	T_NAME	VALUE_1	VALUE_2	VALUE_3	VALUE_4	VALUE_5
1	SiC	27177	0	20906	73170	11847
2	SiCL	3108667	4470321	3723293	1360957	325431
3	SiL	3101002	1774888	1788128	390935	236930
4	L	7206171	8337863	2531672	843194	799292
5	SaL	518460	572814	55748	30662	3484

Record: 1 Show: All Selected Records (0 out of 5 Selected) Options

* Values in table have been rounded to the nearest m^2.

Tabulate area table — field definitions			
Table field	Raster value	CRW adult abundance	Total area (m^2)
VALUE_1	1	2–69	13,961,478
VALUE_2	2	69–120	15,155,886
VALUE_3	3	120–195	8,119,747
VALUE_4	4	195–313	2,698,917
VALUE_5	5	313–690	1,376,984

FIGURE 11.12 Results of tabulate area analysis of soil textural class versus CRW abundance class; all area values are in m^2. (The ArcGIS® graphical user interface is the intellectual property of ESRI and is reproduced herein by permission. Copyright © 1999–2009 ESRI. All rights reserved.)

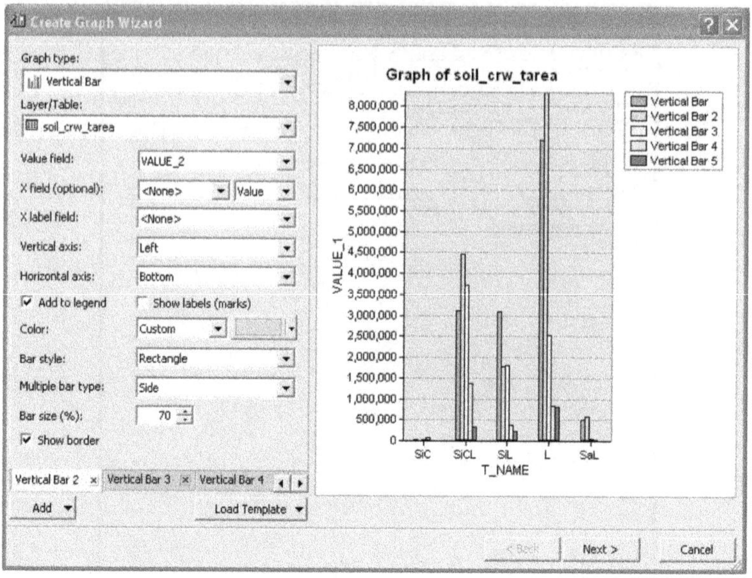

FIGURE 11.13 ArcMap™ Create Graph Wizard dialog box. (The ArcGIS® graphical user interface is the intellectual property of ESRI and is reproduced herein by permission. Copyright © 1999–2009 ESRI. All rights reserved.)

10. Create a bar graph of the data by clicking **Options** and selecting **Create Graph**; the **Create Graph Wizard** appears (Figure 11.13).
    a. Select **Vertical Bar** as **Graph type**.
    b. Select **soil_crw_tarea** as **Layer/table**.
    c. Select **VALUE_1** as the **Value field**.
    d. Select **T_NAME** as the **X label field**.
    e. Check the **Show Border** box near the bottom of the box. This will put a border around the bars.
    f. A bar graph depicting the area in m^2 of CRW population value 1 should appear in the graph window. Click **Add**; select **New Series** to add the remaining values, repeating the procedure above for each CRW population value.
    g. At the bottom of the left-hand box, the names for each series are present as Vertical Bar, Vertical Bar 1, etc. Change these names by clicking on the name and typing in the new name. Click **Vertical Bar**, rename it ***2–69***, and repeat the procedure for each **Vertical Bar** (1–4) renaming them ***70–120***, ***121–195***, ***196–313***, and ***314–690***, respectively. You may also customize the color of the bar for each series by click on **Color**, choosing **Custom** or **Palette** in the drop down box and then picking a color. Click **Next**.
    h. Enter ***Soil Texture vs. CRW Population*** in the **Title:** input box.

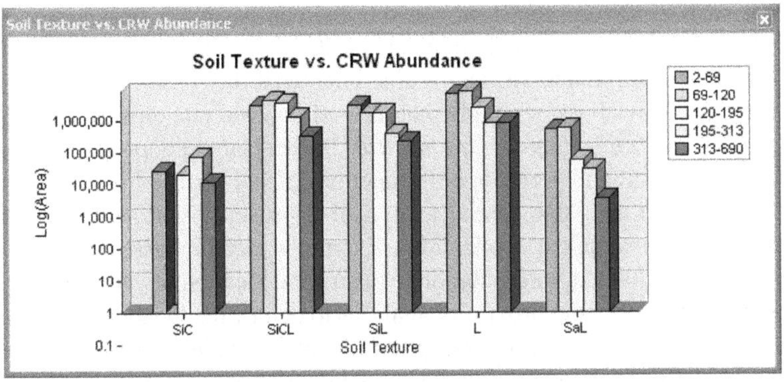

FIGURE 11.14 Bar graph of $\log_{10}$(area) of soil texture and CRW abundance classes. (The ArcGIS® graphical user interface is the intellectual property of ESRI and is reproduced herein by permission. Copyright © 1999–2009 ESRI. All rights reserved.)

i. Check the **Graph in 3D view**, if desired. In the **Axis Properties** box, for **Left** tab, check the **Logarithmic** and **Visible** boxes and enter ***Area*** in the **Title**: input box.
j. Click **Finish**. The newly created bar graph should appear similar to Figure 11.14.

Within the study area, CRW population classes 4 and 5 occupy the least amount of area but have the greatest occurrence in areas with SiC soils (Figure 11.14). CRW classes 2 and 3 occupied the greatest amount of area on SiCL soils. CRW classes 1 and 2 appear to dominate in areas where SiL, L, and SaL soils occur.

11.6 CONCLUSION

Interpolation is central to many ecological field studies and allows for the inference of values over an entire area, thus reducing sampling time and costs. In the case of irregularly distributed trap locations, weighted moving averages or IDW is a useful tool for data analysis as variable or fixed search radius to determine the interpolated value.[17] In general, the IDW algorithm allows for rapid calculations and generates quick contour plots for relatively smooth data values. We used these interpolated values to estimate the spatial distribution of adult CRW and their association with soil texture.

We calculated the Moran's *I* coefficient to measure spatial autocorrelation of measured CRW populations. The high value of the Moran's *I* coefficient for the sticky trap points indicated highly aggregated populations. Aggregated CRW populations have been reported within cornfields.[18–20] In this example, the aggregated spatial distribution of the CRW populations showed that physiographic factors such as soil texture can influence their distribution at a broader scale. This occurs perhaps because the microhabitat of a preferred oviposition site for female CRW is influenced by soil properties such as texture, moisture content, and compaction.[21] Based on our 1998 sample data, the distribution and abundance of CRW in the management site were highly associated with L and SiC loam soils. The variability in soil texture at our site was comparable to a laboratory study that showed survival rate of western CRW and southern CRW (*D. undecimpunctata howardi* Barber) larvae depended on the clay percentage of the soil and porosity, both functions of soil texture.[22]

There are many available spatial analytical approaches within a GIS that can be used in pest management applications. Examples include basic map overlays to determine suitable habitat, interpolation to determine location of varying concentration levels of attributes, and proximity analyses to determine locations of organisms in relation to buffers.[23,24] In this example, we applied spatial analytical techniques using ArcMap GIS for determining the spatial distribution of CRW and associated soil textural class. The GIS procedures were used to determine interactions of soil textural class with number of insects trapped. These results and additional spatial distribution analyses are the initial steps in improving management strategies for CRW and other agricultural pests. A more comprehensive analysis of all data collected during this study is presented in Beckler et al.[4]

ACKNOWLEDGMENTS

We thank farm cooperators of the United States Department of Agriculture, Agricultural Research Service Areawide Pest Management Program, Deb Hartman and several students for data collection, and Dave A. Beck for georeferenced land features. Partial funding and support were provided by the United State Department of Agriculture, Natural Resource Conservation Service; South Dakota State University; South Dakota Agricultural Experiment Station; South Dakota Corn Utilization Council; and Sustainable Agricultural Research and Education (SARE). Mention of a proprietary product does not constitute an endorsement or a recommendation by the USDA for its use.

REFERENCES

1. Sappington, T.W., Siegfried, B.D., and Guillemaud, T., Coordinated *Diabrotica* genetics research: Accelerating progress on an urgent insect pest problem, *Am. Entomol.*, 52, 90, 2006.
2. Gray, M.E., Sappington, T.W., Miller, N.J., Moeser, J., and Bohn, M.O., Adaptation and invasiveness of western corn rootworm: Intensifying research on a worsening pest, *Annu. Rev. Entomol.*, 54, 303, 2009.
3. Chiang, H.C., Bionomics of the northern and western corn rootworms, *Ann. Rev. Entomol.*, 18, 47, 1973.
4. Beckler, A.A., French, B.W., and Chandler, L.D., Using GIS in areawide pest management: A case study in South Dakota, *Trans. GIS*, 9, 109, 2005.
5. Beckler, A.A., French, B.W., and Chandler, L.D., Characterization of western corn rootworm (Coleoptera: Chrysomelidae) population dynamics in relation to landscape attributes, *Agric. Forest Entomol.*, 6, 129, 2004.
6. French, B.W., Beckler, A.A., and Chandler, L.D., Landscape features and spatial distribution of adult northern corn rootworm (Coleoptera: Chrysomelidae) in the South Dakota areawide management site, *J. Econ. Entomol.*, 97, 1943, 2004.

7. Hein, G.L. and Tollefson, J.J., Use of the Pherocon AM trap as a scouting tool for predicting damage by corn rootworm (Coleoptera: Chrysomelidae) larvae, *J. Econ. Entomol.*, 78, 200, 1985.
8. French, B.W., Chandler, L.D., and Riedell, W.E., Effectiveness of corn rootworm (Coleoptera: Chrysomelidae) areawide pest management in South Dakota, *J. Econ. Entomol.*, 100, 1542, 2007.
9. Chiang, H.C., Survival of northern corn rootworm eggs through one and two winters, *J. Econ. Entomol.*, 58, 470, 1965.
10. Krysan, J.L., Foster, D.E., Branson, T.F., Ostlie, K.R., and Cranshaw, W.S., Two years before the hatch: Rootworms adapt to crop rotation, *Bull. Entomol. Soc. Am.*, 32, 250, 1986.
11. Chou,Y., *Exploring Spatial Analysis in GIS* (1st edn.), OnWord Press, Santa Fe, NM, 1997.
12. ESRI® ArcMap™ 9.3, ESRI® ArcGIS™ 9.3, Copyright ©1999–2008, ESRI Inc.
13. Microsoft® Office Excel®, Microsoft Office Enterprise 2007, Copyright ©2006, Microsoft Corporation.
14. Vasiliev, I.R., Visualization of spatial dependence: An elementary view of spatial auto-correlation. In Arlinghaus, S.L. and Griffith, D.A. (eds.), *Practical Handbook of Spatial Statistics*, CRC Press, Boca Raton, FL, p. 17, 1996
15. Johnston, C.A., *Geographic Information Systems in Ecology*, Blackwell Science, Malden, MA, 1998.
16. Krajewski, S.A. and Gibbs, B.L., *Understanding Contouring: A Practical Guide to Spatial Estimation Using a Computer and Variogram Interpretation*, Gibbs Associates, Boulder, CO, 2001.
17. McCoy, J. and Johnston, K., *Using ArcGIS Spatial Analyst*, ESRI Press, Redlands, CA, 2002.
18. Steffey, K.L. and Tollefson, J.J. Spatial dispersion patterns of northern and western corn rootworm adults in Iowa cornfields, *Environ. Entomol.*, 11, 283, 1982.
19. Midgarden, D.G., Youngman, R., and Fleischer, S.J., Spatial analysis of counts of western corn rootworm (Coleoptera: Chrysomelidae) adults on yellow sticky traps in corn: Geostatistics and dispersion indices, *Environ. Entomol.*, 22, 1124, 1993.
20. Ellsbury, M.M., Woodson, W.D., Clay, S.A., Malo, D., Schumacher, J., Clay, D.E., and Carlson, C.G., Geostatistical characterization of the spatial distribution of adult corn rootworm (Coleoptera: Chrysomelidae) emergence, *Environ. Entomol.*, 27, 910, 1998.
21. Kirk, V.M., Calkins, C.O., and Post, F.J., Oviposition preferences of western corn rootworms for various soil surface conditions, *J. Econ. Entomol.*, 61, 1322, 1968.
22. Turpin, F.T. and Peters, D.C., Survival of southern and western corn rootworm larvae in relation to soil texture, *J. Econ. Entomol.*, 64, 1448, 1971.
23. Burrough, P.A., In *Principles of Geographical Information Systems for Land Resources Assessment*, Oxford University Press, Oxford, 1986.
24. Walker, P.A., Spatial modeling and population ecology. In Floyd, R.B., Sheppard, A.W., and De Barro, P.J. (eds.), *Frontiers of Population Ecology*, CSIRO Publishing, Melbourne, Victoria, Australia, p. 419, 1996.

12 Improving Surveillance for Invasive Plants: A GIS Toolbox for Surveillance Decision Support

Julian C. Fox and David Pullar

CONTENTS

12.1 Executive Summary....256
12.2 Introduction....256
12.3 Methods....258
12.3.1 Design Elements of the Toolbox....258
12.3.2 Modeling Seed Dispersal....259
12.3.3 Wind Dispersal Kernel....262
12.3.3.1 Modeling the Influence of Wind Direction and Strength....262
12.3.3.2 Modeling Terrain Influences Wind Dispersal....263
12.3.3.3 Modeling Dispersal along Roads and Rivers....264
12.3.3.4 Multiple Dispersal Events....265
12.3.4 Modeling Life History....265
12.3.5 Simulating Surveillance....266
12.3.6 Parameterization for Chilean Needle Grass....267
12.3.6.1 Potential Habitat for Invasion....267
12.3.6.2 Dispersal Parameters....267
12.3.6.3 Life History Parameters....268
12.3.6.4 Evaluating Surveillance....269
12.3.6.5 Evaluating Eradication....270
12.4 Results and Discussion....270
12.4.1 Evaluating Surveillance....270
12.4.2 Evaluating Eradication....273
12.4.3 Implications for Management of CNG....273
12.5 Summary....274
References....274

12.1 EXECUTIVE SUMMARY

Surveillance for newly detected invasive plants is expensive and is often done on an "ad hoc" basis due to limited resources. A novel way to improve surveillance efficiency is to concentrate surveillance in areas that are more likely to contain the weed, or are more susceptible to invasion, by replicating the various dispersal syndromes and plant life history factors that influence the spread of invasive plants. These areas can then be targeted for surveillance, potentially improving the success of containment and eradication efforts. Management and control actions such as eradication scenarios can then be applied to the simulated invasion and evaluated for effectiveness. This chapter presents a geographic information system (GIS) toolbox for surveillance decision support, discusses design elements of the toolbox, and demonstrates its application to Chilean needle grass (CNG) (*Nassella neesiana*), a highly invasive perennial tussock grass that has spread across agricultural areas of Australia. The toolbox consists of Python functions that replicate elements of invasive plant establishment, growth, and dispersal and also simulate surveillance and eradication strategies. These elements are placed in an iterative loop so the annual cycle of establishment, growth, dispersal, and management strategies can be modeled for multiple years. The toolbox is integrated with ArcGIS for direct application to known spatial information on invasive plant incursions. Simulation outputs are written to the GIS for spatial interrogation and map preparation for rapid communication to land managers. The toolbox has been designed as a decision support tool for managers and researchers of serious plant invasions and is being used in Australia for this purpose. It is available for free download at http://lir.gpa.uq.edu.au/weedtoolbox/. Tutorial information is also available at this Web site.

12.2 INTRODUCTION

Consider the scenario of a newly detected invasive plant incursion. If the invasive plant is targeted for containment or eradication, the first step in managing the incursion is to determine the spatial extent using surveillance around the known infestation.[1] Current surveillance methods are generally ad hoc, involving intensive survey in the vicinity of the known infestation, with sporadic efforts at greater distances. These surveillance efforts are invariably expensive and resource limited, so efforts that maximize the chance of detection are preferred. A novel approach to this problem is to simulate the various dispersal and life history mechanisms that influence incursion, spread, and establishment and concentrate surveillance efforts in areas most susceptible to invasion. The approach may improve efficiency, reduce wasted resources, and improve the success of management actions such as containment and eradication. Here, we describe an integrated landscape modeling tool for predicting incursion extent and spread, and simulating management actions using Python programming language and ArcGIS. The real value of simulation systems is their ability to simulate and compare different management strategies that would otherwise not be possible. Simulation models allow available data to be integrated in a manner that is comprehensive, explicit, and repeatable, which then allows a transparent assessment of the consequences of different management strategies.[2] With a set of simulation tools integrated within a GIS, it is hoped that practitioners can quickly calibrate a simulation of plant invasion for evaluation of different surveillance and management strategies.

Previous simulation models for plant invasions have emphasized the critical influence that dispersal, particularly the choice of kernel, has on simulation outputs.[3] In the context of seed dispersal, a kernel can be defined as a mathematical function that for each spatial location around a plant defines the probability distribution of seed dispersal. The shape of the kernel (particularly its tail), and whether it captures long-distance dispersal, have been shown to be pivotal to predictions of invasion and the inferred management recommendations.[4,5] The literature on modeling seed dispersal is diverse and expansive. Here we concentrate on wind-, water-, and road-mediated dispersal, the most important mechanisms of plant invasion in agricultural and forested landscapes in Australia.

Wind-mediated seed dispersal is an important mechanism of spread and colonization for invasive plants, and there has been a proliferation of models for wind dispersal in the last decade. Modeling this mechanism is challenging; it varies with strength, duration and direction of wind, meteorological conditions, the nature of the terrain, and seed morphology. Recently, complex models incorporating seed uplift processes, aerial transport processes, and atmosphere dynamics have been used to characterize long-distance dispersal.[6,7] Despite these advances, model complexity and the required parameters restrict models to specific and detailed instances of wind dispersal rather than application at the landscape level in a generalized simulation setting. The challenge is to simplify this complexity into simple but meaningful models of wind dispersal. For application at the landscape level, it is envisaged that a simple but flexible kernel will be parameterized for short- and long-distance dispersal and then weighted for wind strength, direction, and terrain influences.

Vegetation studies along watercourses and roads have indicated an enhanced presence of invasive plants. These conduits act as corridors for rapid dispersal and often provide suitable habitat for germination.[8] In a rare study where the water column of a stream was sampled for seeds, Boedeltje et al.[9] observed aquatic dispersal events of almost 1 km for species with limited normal dispersal. Similar results have been observed elsewhere, with experimentally released rhizomes retrieved 1.5 km downstream for an aquatic plant[10] and 90% of seeds for two floodplain trees being retrieved 1800 m downstream.[11] Studies of the material attached to vehicles (primarily mud attached to mudguards) indicated a surprising diversity of flora.[12] Distances associated with vehicular dispersal may be several orders of magnitude greater than other dispersal syndromes[12] and have been estimated to be, on average, 3–40 km, although longer dispersal distances of over several hundred kilometers have been reported. The challenge is to adequately characterize this highly stochastic mechanism in a relatively simple kernel.

As well as describing design elements of the toolbox, we demonstrate its application for CNG *N. neesiana*, a highly invasive perennial tussock grass that was introduced into Australia in the 1930s and has spread across agricultural areas of

New South Wales and Victoria.[13] CNG is named for its sharp pointed seeds that cause injury to stock and downgrade wool, skins, and hides. This grass also is comparatively unpalatable and can reduce grazing productivity by as much as 50%.[14] Environmentally, CNG eliminates biodiversity in native grasslands where it displaces native species and can form dense, continuous clumps.[15] For its economic and environmental impacts, CNG is regarded as one of the worst invasive plants in Australia and has been identified as a weed of national significance.[16]

The first record of CNG in Queensland was a specimen collected near Felton on the Darling Downs in 1998.[13] Since then, it has spread along roads in the Clifton Shire and along the Condamine river catchment. Due to its invasiveness, and potential for spread in the Darling Downs, CNG is a declared Class 1 pest under the Land Protection Act 2002 in Queensland and is the target of an active surveillance and eradication program. Current surveillance is implemented on an "ad hoc" basis using expert opinion to inform where the plant may invade. Managers of the incursion have responded positively to the explicit modeling of CNG and the possibility to inform and refine surveillance methods.

12.3 METHODS

12.3.1 Design Elements of the Toolbox

Components of invasive plant life history and spread along with surveillance and eradication scenarios are written as a series of Python functions. Python is an object-oriented programming language developed in the 1990s as a platform for nonprogrammers to quickly and easily write complex programs. It has several advantages over traditional high-level languages such as C, C++, and Java. Python programs are far quicker to write and easier to maintain due to the elegance and simplicity of the language. Python is open source software (http://www/python.org) and is portable to a wide range of hardware and software platforms. Programs can therefore be deployed unrestricted by licensing issues and preferred user platforms.

Python functions are placed in a recursive loop that can be run for the desired number of years and can integrate directly to ArcGIS products such as ArcMap and ArcCatalog. Input parameters and spatial data for the toolbox are controlled through a graphical user interface in ArcToolbox. The integrated GIS and simulation system has several advantages. Simulations can be applied to real incursions in real landscapes inclusive of terrain, hydrological, and wind models, and geographic layers describing road placement, vegetation, and animal distributions. GIS packages such as ArcMap have many functions and utilities useful for preparing, managing, interrogating, summarizing, and presenting geographic information. This functionality can be used for preparing input data, for analyzing simulation outputs, and for communicating results using inbuilt map preparation tools. An output example of these functions is presented in Figure 12.1 showing the graphical user interface and a simulated plant invasion over 10 years. Figure 12.1a shows the early stages of the invasion, Figure 12.1b the invasion after 5 years, and Figure 12.1c the invasion after 10 years.

12.3.2 Modeling Seed Dispersal

For an invasion simulation system, we require a general dispersal kernel that can be parameterized for a suite of dispersal agents under different conditions operating at a landscape scale. Therefore, a model that has been shown to fit actual seed dispersion data and has the flexibility to assume a variety of shapes is preferred. Given these criteria, the generalized model of Clark et al.[17] was selected. This generalized dispersal kernel has been adopted in several recent studies[18–21] and specifically in the

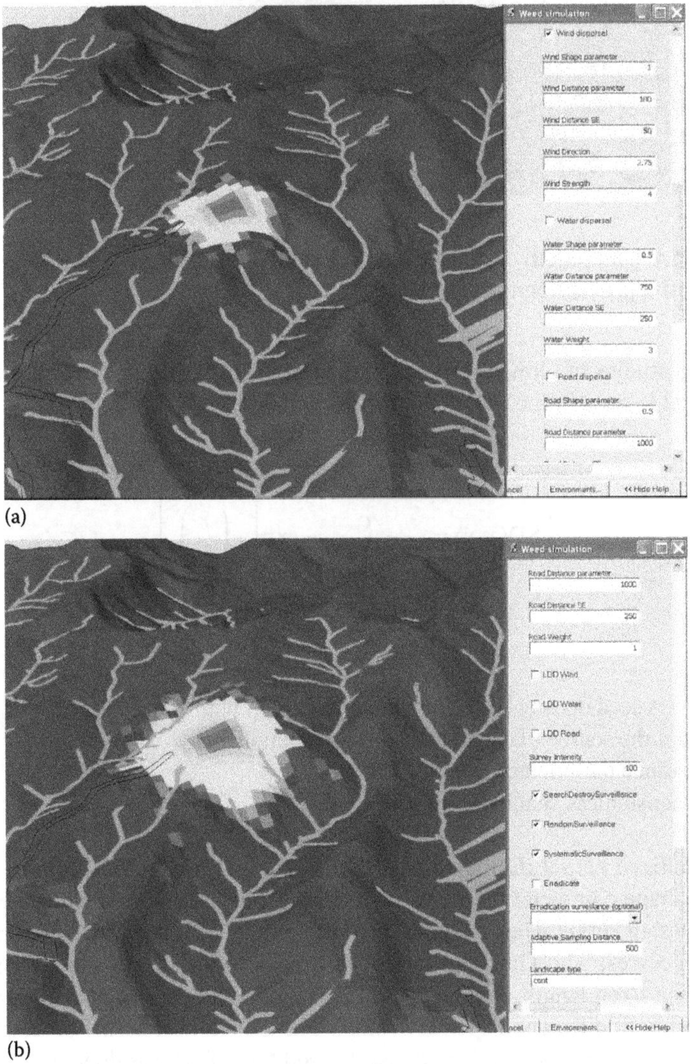

FIGURE 12.1 Graphical user interface of the surveillance decision support toolbox and an example incursion: (a) the early stages of the invasion, (b) the invasion after 5 years *(continued)*

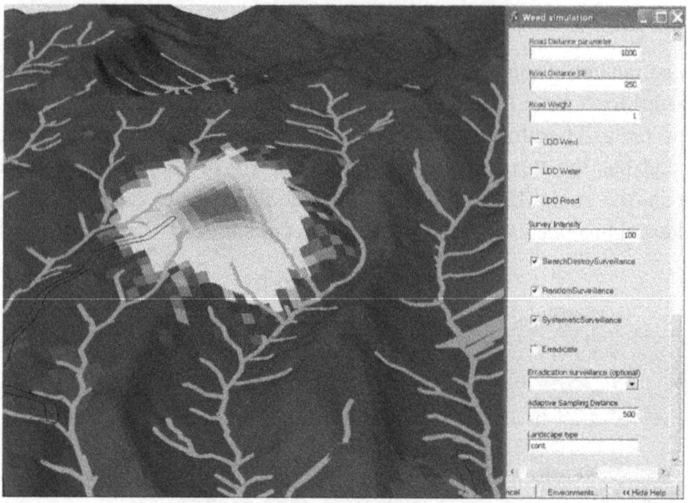

(c)

FIGURE 12.1 (continued) (c) the invasion after 10 years. The depicted invasion is for a highly invasive plant unconstrained by habitat.

simulation studies of Grevstad[22] and Clark et al.[17] This equation describes a family of dispersal kernels in the functional form (Equation 12.1):

$$SD(x,y) = \frac{c}{2\pi\alpha^2\Gamma(2/c)}\exp\left[-\left(\frac{r}{\alpha}\right)^c\right] \tag{12.1}$$

where

- $SD(x, y)$ is seed density per square meter for a cell with coordinates x, y at distance r from the source (the classical gamma function[23])
- α is distance parameter
- c is dimensionless shape parameter

The generalized kernel (Equation 12.1) is flexible and results in a fat-tailed kernel when the shape parameter (c) is less than 1, an exponential kernel when c equals 1, and a Gaussian kernel when c equals 2.[24] Different kernels suppose different mechanistic processes. The Gaussian kernel results from seeds moving by Brownian motion for a fixed length of time.[25] Different parameterizations of the kernel have been found to fit different dispersal vectors. The generalized kernel of Clark et al.[17] was adapted to distribute seeds in a raster-based GIS. This was done by calculating the dispersal into a raster cell based on the Euclidian distance of the cell centroid from the source and the size of the cell. Clark et al.[17] model calculates the seed deposition for a 1 × 1 cell unit; therefore, we need to scale deposition to the cell size as described in Equation 12.2. To sum all seeds for a cell size with centroid distance r from the source

$$\text{Seeds} \cdot (\text{cell size}^2) \cdot \frac{c}{2\pi\alpha^2\Gamma(2/c)} \exp\left[-\left(\frac{r}{\alpha}\right)^2\right] \tag{12.2}$$

Using the solutions for the gamma function (Γ) and Equation 12.2, the dispersal kernel can be estimated as a fat-tailed (Equation 12.3), exponential (Equation 12.4), or Gaussian (Equation 12.5) kernel:

$$\text{Seeds} \cdot (\text{cell size}^2) \cdot \frac{1}{24\pi\alpha^2} \exp\left[-\left(\frac{r}{\alpha}\right)^{1/2}\right] \tag{12.3}$$

$$\text{Seeds} \cdot (\text{cell size}^2) \cdot \frac{1}{2\pi\alpha^2} \exp\left[-\left(\frac{r}{\alpha}\right)\right] \tag{12.4}$$

$$\text{Seeds} \cdot (\text{cell size}^2) \cdot \frac{1}{\pi\alpha^2} \exp\left[-\left(\frac{r}{\alpha}\right)^2\right] \tag{12.5}$$

Graphically, the three kernels are shown in Figure 12.2 with a fixed distance parameter of 15. In this example, the total number of seeds dispersed is equivalent for the three kernels; however, the Gaussian kernel distributes most seeds close to the parent plant, whereas the fat-tailed kernel distributes seeds further from the parent.

The parameterized kernel is used to distribute seeds from each source cell to the surrounding cells on an annual basis. In the programming, this occurs in reverse, with dispersal calculated from each source cell into the analysis cell based on the Euclidian distance, and all dispersal events into the analysis cell are summed.

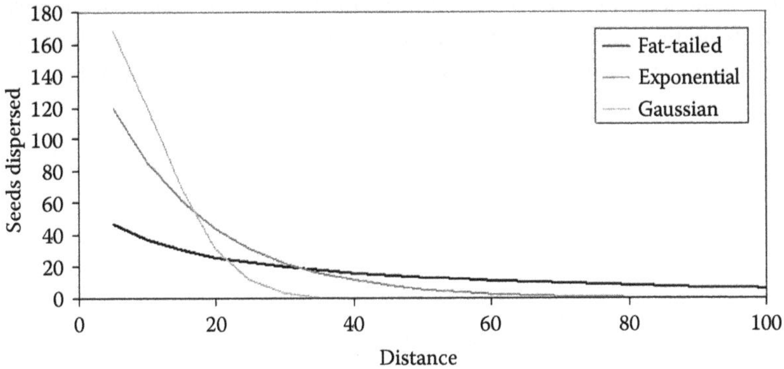

FIGURE 12.2 Comparison of dispersal kernels.

In its most basic form, dispersal will occur concentrically around the source cell. Additional information on the dispersal event such as wind direction and strength and terrain effects are incorporated using weighting functions. The application of these weights results in anisotropic dispersal pattern; however, the average of all weights for a particular influence is always one, thus ensuring that the overall magnitude of dispersal from each source cell is unchanged. For example, a weight for the influence of wind results in values less than 1 for areas upwind of the source and values greater than 1 for areas downwind of the source. Thus, when the function is applied, more seeds are dispersed to locations downwind than upwind.

Dispersal is a highly stochastic or random event, and the deterministic kernel needs to be modified to reflect this. To generate more realistic stochastic dispersal events, the distance parameter can be randomly sampled from a designated distribution. This is similar to the 2Dt mixture kernel of Clark et al.[17] Distance parameters can be sampled from different distributions dependent on the dispersal syndrome. Stochastic short-dispersal events can be realized by sampling a normal distribution with a mean equal to the average expected dispersal distance and a standard deviation to control the stochastic variation. For incorporating rare long-distance dispersal events a lognormal distribution can be sampled. Using a skewed long-tailed lognormal distribution, rare long-distance dispersal events will be represented when large distance parameters are sampled.

At the commencement of the dispersal function, the chosen distribution is sampled for each source cell, and sampled values are stored in an array. During implementation of the kernel, dispersal from each source cell will use the sampled value in the array. Therefore, each time the dispersal function is called, a new set of random values will be sampled from the distribution. Distributions are truncated to ensure that the sampled distance parameters are positive.

12.3.3 Wind Dispersal Kernel

The analysis of Clark et al.[17] suggests that an exponential kernel (Equation 12.4) provides the best fit to the observed wind-mediated seed dispersal. This is a good starting point for a simple model of wind dispersal.

12.3.3.1 Modeling the Influence of Wind Direction and Strength

A Python function was written to determine the bearing, in radians from north, for dispersal from the source cell to the analysis cell. The prevailing wind direction was then subtracted from this bearing and subjected to the following formula:

$$WindWt = WindStrgth\left(\left(1-\left(\frac{(1-\cos(CellDir - WindDir))}{2}\right)^{0.75}\right)-0.5\right) \quad (12.6)$$

where

WindWt is the weight applied to the kernel
WindStrgth makes the ellipse more oblique
CellDir is the bearing from the source cell to the analysis cell
WindDir is the bearing of the prevailing wind

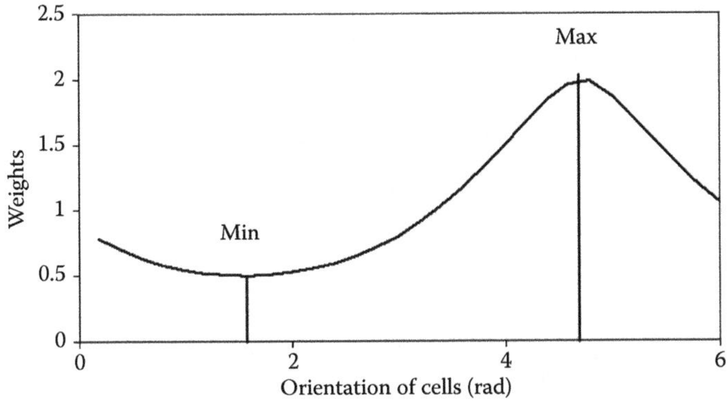

FIGURE 12.3 *WindWt* for a wind direction of 4.712 rad (270°) with a *WindStrgth* of 4.

All Python functions can be accessed from the following Web site: http://lir.gpa.uq.edu.au/weedtoolbox/

The function, shown in Equation 12.6, results in values greater than 1 when cell direction and wind direction are in agreement and less than 1 when they differ. The weights average to 1 across all directions, thus the total dispersal is not altered due to the weight function. The variable *WindStrgth* makes the ellipse more oblique and concentrates dispersal in the downwind of the prevailing wind. Figure 12.3 demonstrates the calculated *WindWt* values for a wind direction of 4.712 rad (270°) with a *WindStrgth* of 4. By examining Figure 12.3, you notice that when the orientation of cells is against the wind (1.571 rad), the magnitude of dispersal is halved, whereas when the orientation is aligned with the wind the magnitude of dispersal is doubled.

12.3.3.2 Modeling Terrain Influences Wind Dispersal

A Python function was written that determines the angle due to difference in terrain (elevation) between the analysis cell and the cell from which dispersal is occurring. This function is intended to capture the influence of terrain on wind dispersal. The angle is determined as described in the following equation:

$$\text{Angle} = a\tan\left(\frac{elev_{\text{a}} - elev_{\text{s}}}{Dis}\right) \tag{12.7}$$

where

$elev_{\text{a}}$ and $elev_{\text{s}}$ are elevation of the analysis and source cells, respectively
Dis is the Euclidian distance between analysis and source cells

The terrain weight is then calculated by dividing the Angle by π/2 (90°) to give a unitless ratio between 0 and 1. For upslope dispersal, the quantity is subtracted from 1 to provide a unitless ratio between 0 and 1:

$$TerrWght = \left[1 - \frac{Angle}{1.571}\right]^4 \tag{12.8}$$

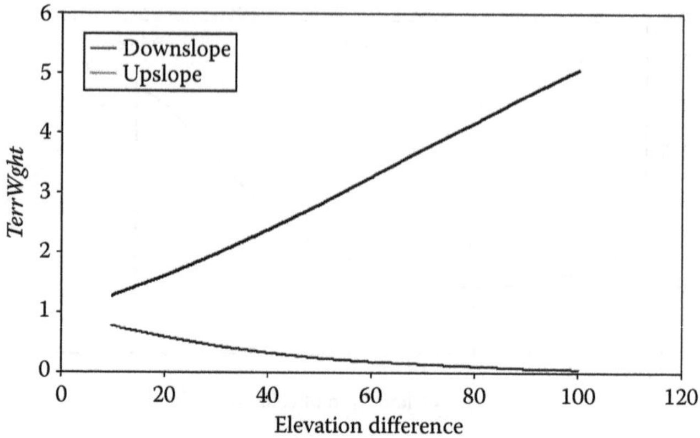

FIGURE 12.4 *TerrWght* values for upslope and downslope wind dispersal for a constant distance between cells of 100 m but subject to increasing elevation difference.

For downslope dispersal, the quantity is added to 1 to provide a unitless ratio greater than 1:

$$TerrWght = \left[1 + \frac{Angle}{1.571}\right]^4 \tag{12.9}$$

In both upslope and downslope dispersal, a power factor of 4 is applied to increase the influence of terrain on dispersal.

Figure 12.4 demonstrates calculated *TerrWght* values for upslope and downslope wind dispersal for a constant distance between cells of 100 m, but subject to increasing elevation difference.

12.3.3.3 Modeling Dispersal along Roads and Rivers

Initially, the dispersal kernel is modified so that dispersal occurs along a linear path rather than in all directions. Clark[26] describes a linear dispersal kernel as

$$\frac{c}{2\alpha\Gamma(1/c)}\exp\left[-\left(\frac{r}{\alpha}\right)^c\right] \tag{12.10}$$

where

α is distance parameter
c is dimensionless shape parameter

Based on the available literature, a fat-tailed kernel was selected for dispersal along rivers and roads as dispersal events further from the parent plant are better represented by the fatter tail of the distribution. Using the solution to the gamma function, the fat-tailed kernel is

$$\text{Seeds} \cdot \text{cell\_length} \cdot \frac{0.5}{2\alpha} \exp\left[-\left(\frac{r}{\alpha}\right)^{1/2}\right] \tag{12.11}$$

Quantifying dispersal based on the Euclidian distance between two points rather than the actual distance along a road or watercourse is inexact. An improvement to using Euclidian distance to describe this dispersal is to use network theory and apply the kernel along the stream or road represented as a network.[27] This method has been shown to be a more accurate method for seed dispersal along streams and roads.

For dispersal along watercourses the hydrology tools in the ArcGIS Toolbox are used to estimate stream location and calculate distances along streams from a digital elevation model (DEM). By deciding on a cutoff, a flow accumulation layer can be used to define potential watercourses, and flow length estimates dispersal distances along watercourses. Linear dispersal along the river network should be used between two points on a watercourse when the source cell is located upstream relative to the analysis cell.

For dispersal along roads, railroad tracks, etc., a GIS layer that includes the pathways is used. Cells that intersect with roads and paths are then assigned a weighting function based on their usage rates. Higher dispersal weight is assigned for roads that are used often versus lower dispersal weight for roads that are rarely used. A buffering algorithm can be used to determine cells in proximity to roads. Linear dispersal along the network occurs between two points on a road when the usage rate is the same (i.e., the same road).

12.3.3.4 Multiple Dispersal Events

Studies have indicated that polychory, i.e., seed dispersal, occurs by several mechanisms such as combinations of wind-, water-, and animal-mediated dispersal, which is common in the dispersal biology of plants.[9,25,28] In the instance where seeds are subject to several alternative dispersal modes, seeds available for dispersal at the source can be proportioned and dispersed according to each kernel. Seeds arriving at a source from several different modes are additive and provide a total for a particular species.[25,29] This is realized in simulations as follows: initially, seeds available for dispersal are divided among the applicable mechanisms. Each mechanism then disperses its quota of seeds. The final dispersal into each cell is the sum of dispersal due to each mechanism.

12.3.4 Modeling Life History

The life history component of the simulation model drives plant germination, growth, fecundity, and mortality. A stage matrix approach was used to track the movement of individuals through the various life stages, with the survival, growth, and reproduction of each life stage summarized in the columns of a stage matrix. This is consistent with the raster-based approach, as the number of individuals in each life stage can be summarized at the raster cell level, and simple matrix multiplication can be used to project the population into the next time step. The stage matrix approach is the underpinning of population viability analysis, a tool used for assessing populations under threat in conservation biology.[30] This approach also is used widely in population models for invasive plants.[31]

An example matrix (**A**) is shown in Equation 12.12 for a hypothetical perennial herbaceous or woody plant with high fecundity (1500) and short seed longevity (0.25):

$$\mathbf{A} = \begin{vmatrix} 0.25 & 0 & 1500 \\ 0.0375 & 0 & 14 \\ 0 & 0.1 & 0.5 \end{vmatrix} \qquad (12.12)$$

The three columns represent three basic life stages, seed, seedling, and adult, whereas the rows represent fecundity and survivorship for seeds, seedlings, and adults into the next time period. Plant life history can either be modeled deterministically using matrix **A**, or demographic stochasticity can be incorporated by sampling matrix **A** from a standard deviation matrix of the same dimension.

12.3.5 Simulating Surveillance

Surveillance strategies are performed annually after dispersal and population growth. Incursions detected by a particular strategy are preserved from year to year, and cells with detected incursions are removed from the pool of potential surveillance points in subsequent years. Therefore, the proportion of the detected adults to the actual number of adults will increase from year to year if the population is not expanding. If the invasion colonizes new cells quicker than they are surveyed, then the proportion of the population detected will decrease.

Several surveillance strategies are compared including systematic-, random-, habitat-based, seek and destroy, as well as adaptive versions of each. Systematic surveillance uses a grid to systematically survey an entire area at grid intersections. Random surveillance is implemented by randomly selecting cells from the entire population. Both these strategies are unbiased, in that every cell in the study area has an equal probability of being selected for surveillance. Seek-and-destroy surveillance is a biased surveillance strategy that is often applied in the management of invasive plants. It is applied using the knowledge of how the plant disperses to optimize chances of finding it and is biased as it concentrates the sampling on particular areas around the incursion. For example, if a plant is believed to disperse along roads, then surveillance efforts are concentrated along roads. Another biased surveillance strategy is habitat-based surveillance, where surveillance is restricted to areas of suitable habitat for plant germination and growth.

Adaptive cluster sampling[32] uses information on detected incursions to concentrate the sampling effort around the detected patch. Usually a search radius is applied around a detected incursion. The adaptive design is applied to all of the above sampling methods.

12.3.6 Parameterization for Chilean Needle Grass

12.3.6.1 Potential Habitat for Invasion

CNG grows in temperate regions with annual rainfall between 500 and 800 mm on a wide range of soils and prefers grassy woodlands and disturbed habitats such as pastures, roadsides, and stream banks.[13] On the Darling Downs, habitat susceptible to invasion includes grassy areas on roadsides, stream banks, and drainage channels. CNG has been observed to spread linearly on roads, rivers, and drainage channels and then invade adjacent paddocks or grasslands through stock and wildlife movement (Grant Beutel, pers. comm.). Paddocks under a grazing regime are susceptible to invasion, but CNG does not survive under cultivation. Given these conditions for habitat susceptible to invasion, a rule set was constructed and applied to GIS layers for land use, vegetation, roads, streams, and drainage channels to create a GIS layer of habitat susceptible to invasion. Given the size of the incursion (spanning 35 km^2), and computational considerations, a cell size of 50 m was selected. Under the assumption that each square meter could support two plants, each 50 m cell of potential CNG habitat was assigned a carrying capacity of 5000 plants.

12.3.6.2 Dispersal Parameters

12.3.6.2.1 Wind Dispersal

CNG seeds are poorly adapted for dispersal by wind and have been observed to travel only 5 m from the parent plant. Based on this, an exponential kernel with a distance parameter of 5 m with a standard deviation of 5 m was selected to characterize wind dispersal. The distance parameter was sampled from the normal distribution, as wind dispersal for CNG operates only over short distances, and longer-distance events are unlikely.

12.3.6.2.2 Dispersal along Watercourses

Despite efforts to sample streams for dispersing material,[9,24] general quantitative estimates of aquatic dispersal are lacking. This is due to the variability associated with the season and duration of seed release, stream velocity and deposition characteristics, frequency of flood events, and seed buoyancy and survival. Despite the inherent variability of aquatic dispersal, we know that dispersal will occur linearly in a downstream direction, and by observing the past spread of an invasive plant along a watercourse we can estimate a dispersal kernel for the process. Floodwaters have been observed to play a significant role in transporting the buoyant CNG seeds. For CNG, by matching the rate of spread along the Condamine river to that predicted by different distance parameters, we can identify the parameter that best characterizes the syndrome. Based on these data, a distance parameter on a fat-tailed kernel of 750 m with a standard deviation of 250 m was identified. The distance parameter was sampled from the lognormal distribution to incorporate rare long-distance dispersal events.

Stream flow on the Darling Downs is intermittent for all but the largest streams. To incorporate intermittent flow in stream dispersal models, a random draw was applied to second-order streams, and dispersal only occurred every second year when the stream actually flowed.

12.3.6.2.3 Dispersal along Roads

Vehicular dispersal along roads is particularly effective for species with small, persistent seeds that can adhere and survive in soil deposited in wheel arches and mud guards.[12] This is the case for CNG, and its rapid spread along roadsides in the Clifton area indicates that dispersal along roads is the most important dispersal vector. It is hypothesized that CNG is spread along roadsides by Shire Council slashers, and this has been confirmed by work in Victoria (C. Grech, pers. comm.). The kernel for road dispersal was parameterized by assuming that the incursion was initiated in 1998 at the Clifton Show grounds and simulating spread in 1997 for different distance parameters. A distance parameter of 1 km with a standard deviation of 250 m resulted in a predicted current extent that matched the actual current extent. Similar to water dispersal, the distance parameter was sampled from the lognormal distribution to characterize rare long-distance dispersal events. The parameterization of dispersal kernels for CNG is summarized in Table 12.1.

12.3.6.3 Life History Parameters

The studies of Gardener et al.[33,34] examined the ecology and life history of CNG and provide information for parameterizing the life history component of the simulation. Three life stages of seed, juvenile, and adult are assumed for CNG. Gardener et al.[33,34] estimated fecundity on a square meter basis, and, under the assumption that in a

TABLE 12.1
Summary of Dispersal Kernels for CNG

	Kernel	α	SD on α	α Distribution
Wind	Exponential: $\frac{1}{2\pi\alpha^2}\exp\left[-\left(\frac{r}{\alpha}\right)\right]$	5	5	Normal
Water	Fat-tailed linear: $\frac{0.5}{2\alpha}\exp\left[-\left(\frac{r}{\alpha}\right)^{1/2}\right]$	750	250	Lognormal
Roads	Fat-tailed linear: $\frac{0.5}{2\alpha}\exp\left[-\left(\frac{r}{\alpha}\right)^{1/2}\right]$	1000	250	Lognormal

TABLE 12.2
Life History Parameters for CNG

0.75	250	750
0.025	0	0
0	0.7	0.9

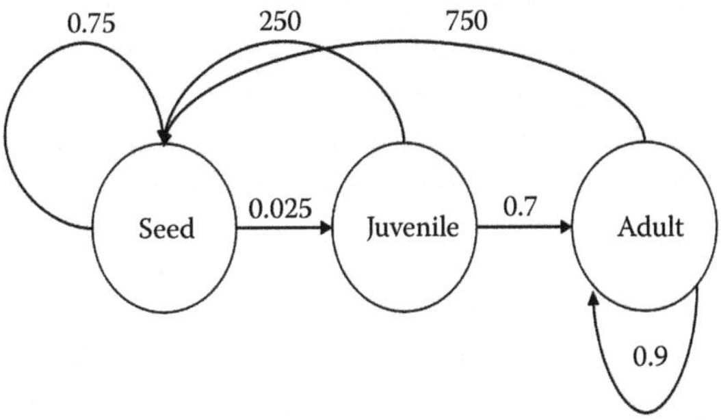

FIGURE 12.5 Life cycle diagram for CNG.

dense infestation 20 plants occupy a square meter, each plant has an annual seed production of 750 seeds. An estimated 2.5% of seeds germinate and emerge as juvenile plants that can produce about 250 seeds in their first year. Juvenile survivorship is good with 70% of plants surviving to become adults. Adult survivorship is very good with an estimated 90% of plants surviving into the next year. These parameters are the basis of a stage matrix for CNG as shown in Table 12.2. The stage matrix can be depicted as a life cycle diagram in Figure 12.5.

TABLE 12.3
Standard Deviation Matrix for CNG

0.1	100	500
0.01	0	0
0	0.1	0.1

Populations on the Darling Downs are at the edge of their range in terms of the higher summer temperatures they experience, and some mortality has been observed during especially hot summers. Seed production is also dependent on the arrival of spring rains, and in drier years when the rains have failed, plants have failed to produce seeds (Grant Beutel, pers. comm.). To incorporate these stochastic influences on demographic parameters, a standard deviation matrix is used as described in Table 12.3.

12.3.6.4 Evaluating Surveillance

Initially, several surveillance methods were applied to CNG: random, systematic, seek and destroy, and adaptive versions of each. An adaptive seek-and-destroy method was identified as superior for CNG in terms of the proportion of the projected population that is intersected over 10 years. This method involved surveillance within 2 km of existing incursions along the dispersal pathways of roads and streams and in cells with suitable habitat. Adaptive seek-and-destroy surveillance was evaluated against predicted CNG spread for varying surveillance intensities. Based on current surveillance intensities, the surveillance was conducted after dispersal and life history processes at three intensities: 500, 1000, and 5000 ha per annum. Currently, annual surveillance for CNG varies between 3000 and 5000 ha per annum. Three different detection probability levels were also applied: 100% (perfect), 75%, or 50% probability of detection. CNG is only detectable during the flowering stage, and this is the time when most surveillance and eradication activity occurs.

Surveillance was evaluated by accumulating detected incursion cells each year and calculating the proportion of the population in these detected cells against the total population size. Once an incursion cell has been detected it is stored and removed from the pool of potential surveillance cells in subsequent years. Because surveillance is adaptive, cells in close spatial proximity to incursion cells are targeted in subsequent years.

12.3.6.5 Evaluating Eradication

An eradication program was implemented at each time step following surveillance by randomly selecting from the pool of incursion cells detected during surveillance at three intensities: 500, 1000, and 5000 ha per annum. Successful eradication of juvenile and adult plants was assumed for selected cells, though seeds were unaffected.

12.4 RESULTS AND DISCUSSION

Using the parameterization described earlier, the simulation was run for 10 years, and this was replicated 100 times to generate a mean projection and a standard deviation. Projected population growth is shown geographically in Figure 12.6 for the Condamine river.

Figure 12.6a shows the current invasion (invaded cells are shown in darker color), Figure 12.6b shows the invasion after 5 years, and Figure 12.6c shows the invasion after 10 years. Projected population growth is depicted graphically in Figure 12.7. It can be observed that the population can be expected to grow from about 400,000 adults to over 3 million adults during the 10 year time step in the absence of any control measures.

12.4.1 Evaluating Surveillance

Assuming that as of 2007, the surveyed population extent reflects the actual extent, Figure 12.8 depicts the proportion of detected CNG incursion under different surveillance intensities of 500, 1000, and 5000 ha per annum.

It can be observed that at intensities of 500 and 1000 ha per annum, we fail to detect an increasing proportion of the CNG population, until in the 10th year only 10% of the population is detected. Clearly, these intensities are too low to keep track of the highly invasive CNG as seeds disperse to new areas and plants become established. At a survey intensity of 5000 ha per annum a higher proportion of the population is detected through the 10 year projection, with 75% of all incursion cells detected at 10 years.

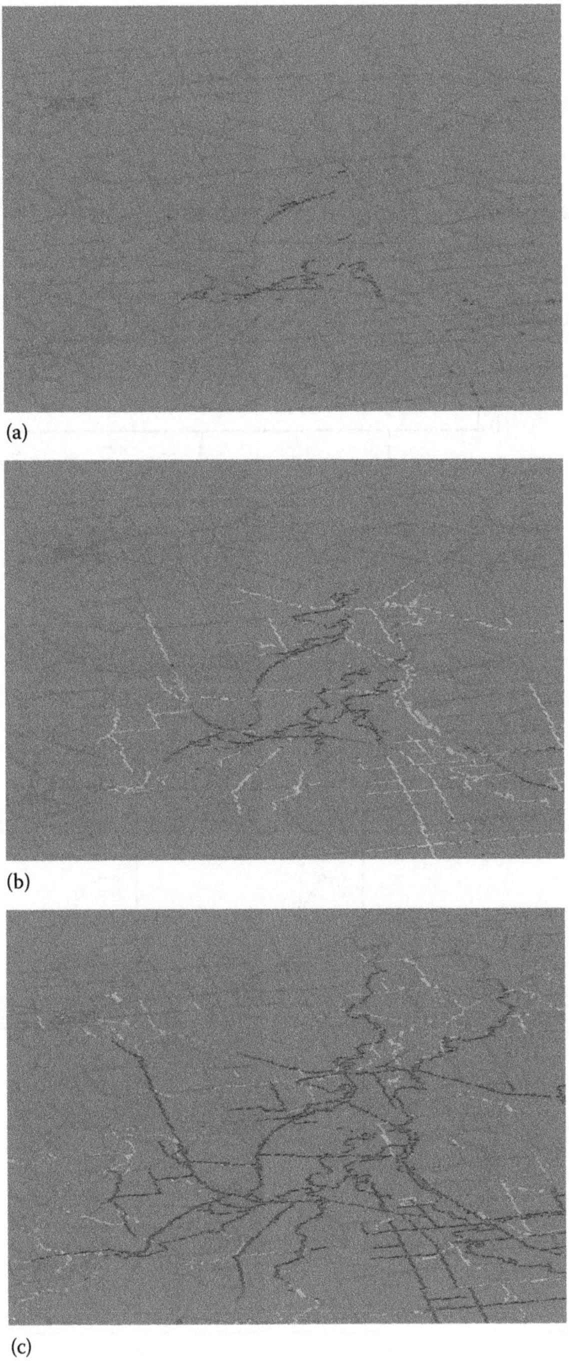

(a) (b) (c)

FIGURE 12.6 Projected population growth for the Condamine river: (a) the current invasion (invaded cells are shown in darker color), (b) the invasion after 5 years, and (c) the invasion after 10 years.

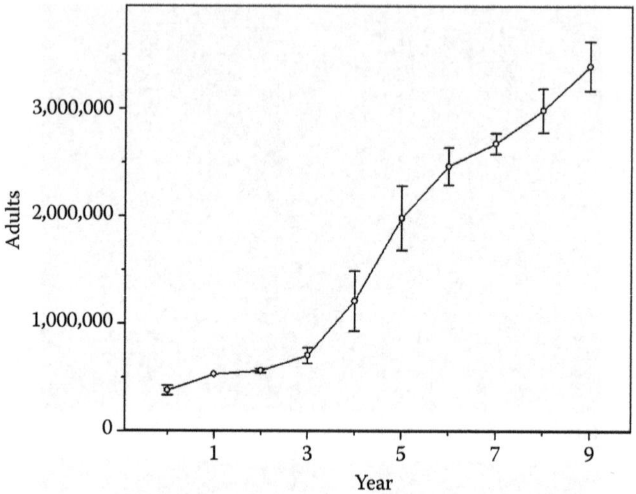

FIGURE 12.7 Projected population growth for CNG from simulations over 10 years. Simulations were replicated 100 times to generate a mean projection and a standard deviation.

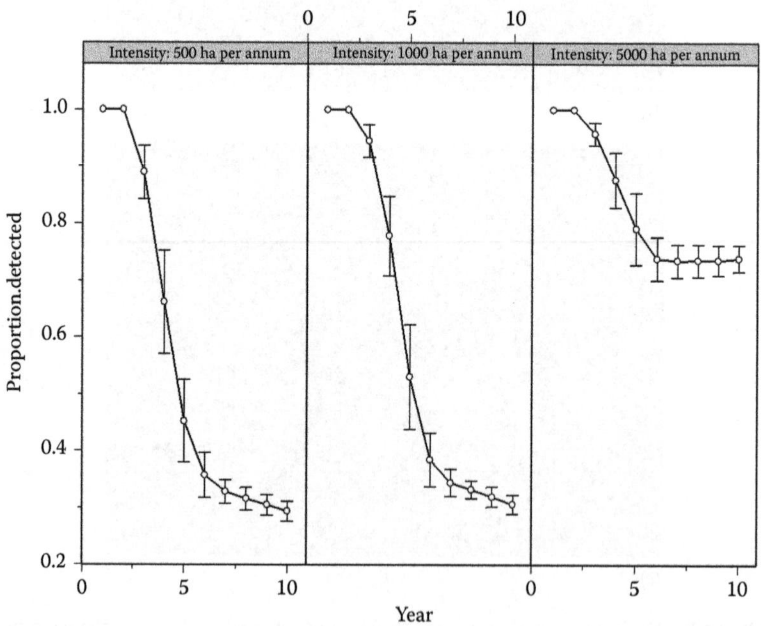

FIGURE 12.8 Proportion of detected CNG incursion under different surveillance intensities of 500, 1000, and 5000 ha per annum assuming perfect detection.

12.4.2 Evaluating Eradication

Following surveillance, the pool of detected incursion cells were randomly sampled and eradicated at the rate of 500, 1000, and 5000 ha per annum. To clarify, at year 0 all incursion cells are known (assumed to be known) and during each year dispersal and population projection results in an expanding incursion. At the end of each year, surveillance using an adaptive seek-and-destroy method is implemented at the prescribed intensities. Surveillance will result in an expanding pool of detected incursion cells, which will be less than the total number of incursion cells. Eradication is implemented at this point by sampling the pool of incursion cells at the prescribed intensity and resetting adults and juveniles to 0. Results from different eradication intensities are depicted in Figure 12.9. It can be observed that eradication at the rate of 5000 ha per annum results in the population staying below 1 million adult plants after 10 years. Eradication at lower intensities has only a small impact relative to taking no action as depicted in Figure 12.7.

12.4.3 Implications for Management of CNG

Results from the simulations indicate that an extensive eradication program is required to control CNG and that current surveillance and eradication levels (2000–5000 ha per annum) may need to be increased if the invasion is to be negated. The difficulty of surveillance and eradication in this instance is caused by a high population growth rate, large seed bank, and rapid means of dispersal that results in a highly invasive plant that is very hard to control. This has been observed in Victoria

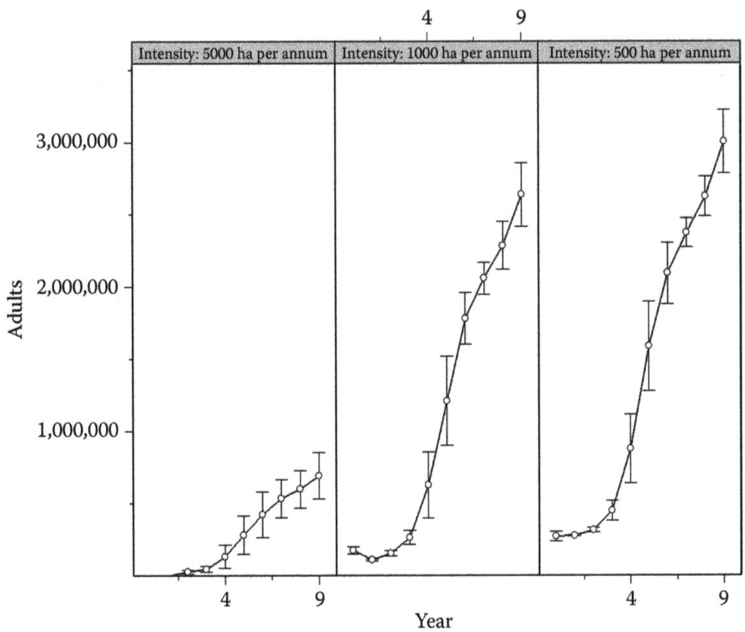

FIGURE 12.9 Project population growth of CNG subject to different eradication intensities.

and New South Wales, where control efforts have met with little success, and CNG is now widely dispersed through agricultural areas. Although it is sobering to learn that current control efforts may be ineffective, it is better to learn this now and perhaps modify tactics for CNG and other newly invasive weed. However, as with any simulation, results are bound by numerous assumptions and abstractions required to make the simulation function. The CNG simulation is parameterized using data from northern New South Wales, and there is anecdotal evidence that the Queensland population may be behaving differently due to the warmer summer on the Darling Downs. Despite this difference, simulations indicate that CNG is highly invasive and may be very difficult to eradicate.

12.5 SUMMARY

Through a case study for CNG, it was demonstrated that the surveillance support toolbox is useful for exploring the many aspects that function when examining a new invasive species. The toolbox was used to model invasion spread and population growth, identify areas susceptible to invasion, help define a surveillance method that is appropriate for the invasive plant in question as well as surveillance intensity, and evaluate the feasibility and intensity of eradication efforts.

REFERENCES

1. Panetta, F.D. and Lawes, R., Evaluation of weed eradication programs: The delimitation of extent, *Diversity and Distributions*, 11, 435, 2005.
2. Higgins, S.I., Richardson, D.M., and Cowling, R.M., Using a dynamic landscape model for planning the management of alien plant invasions, *Ecological Applications*, 10, 1833, 2000.
3. Higgins, S.I., Richardson, D.M., and Cowling, R.M., Validation of a spatial simulation model of a spreading alien plant population, *Journal of Applied Ecology*, 38, 571, 2001.
4. Clark, J.S., Lewis, M., McLachlan, J.S., and HilleRisLambers, J., Estimating population spread: What can we forecast and how well? *Ecology*, 84, 1979, 2003.
5. Buckley, Y.M., Brockerhoff, E., Langer, L., Ledgard, N., North, H., and Rees, M., Slowing down a pine invasion despite uncertainty in demography and dispersal, *Journal of Applied Ecology*, 42, 1020, 2005.
6. Nathan, R., Katul, G.G., Horn, H.S., Thomas, S.M., Oren, R., Avissar, R., Pacala, S.W., and Levin, S.A., Mechanisms of long-distance dispersal of seeds by wind, *Nature*, 418, 409, 2002.
7. Katul, G.G., Porporato, A., Nathan, R., Siqueira, M., Soons, M.B., Poggi, D., Horn, H.S., and Levin, S.A., Mechanistic analytical models for long-distance seed dispersal by wind, *American Naturalist*, 166, 368, 2005.
8. Parendes, L.A. and Jones, J.A., Role of light availability and dispersal in exotic plant invasion along roads and streams in the H.J. Andrews Experimental Forest, Oregon, *Conservation Biology*, 14, 64, 2000.
9. Boedeltje, G., Bakker, J.P., Bekker, R.M., Van Groenendael, J.M., and Soesbergen, M., Plant dispersal in a lowland stream in relation to occurrence and three specific life-history traits of the species in the species pool, *Journal of Ecology*, 91, 855, 2003.
10. Johansson, M.E. and Nilsson, C., Hydrochory, population-dynamics and distribution of the clonal aquatic plant Ranunculus-Lingua, *Journal of Ecology*, 81, 81, 1993.

11. Schneider, R.L. and Sharitz, R.R., Hydrochory and regeneration in a Bald Cypress Water Tupelo swamp forest, *Ecology*, 69, 1055, 1988.
12. Hodkinson, D.J. and Thompson, K., Plant dispersal: The role of man, *Journal of Applied Ecology*, 34, 1484, 1997.
13. Csurhes, S., *Weed Risk Assessment: Chilean Needle Grass*, Unpublished report. Queensland Department of Natural Resources and Mines, Brisbane, Queensland, 2005.
14. Smithyman, D., McLaren, D., Iaconis, L., Storrie, A., Carter, R., and Thorp, K., *Weed Management Guide—Chilean Needle Grass*, CRC for Australian Weed Management, Adelaide, Australia, 2003.
15. Hocking, C., Land management of Nassella areas—Implications for conservation, *Plant Protection Quarterly*, 13, 86, 1998.
16. McNaught, I., Thackway, R., Brown, L., and Parsons, M., *A Field Manual for Surveying and Mapping Nationally Significant Weeds*, Bureau of Rural Sciences, Canberra, Australia, 2006.
17. Clark, J.S., Silman, M., Kern, R., Macklin, E., and HilleRisLambers, J., Seed dispersal near and far: Patterns across temperate and tropical forests, *Ecology*, 80, 1475, 1999.
18. Dick, C.W., Etchelecu, G., and Austerlitz, F., Pollen dispersal of tropical trees (*Dinizia excelsa*: Fabaceae) by native insects and African honeybees in pristine and fragmented Amazonian rainforest, *Molecular Ecology*, 12, 753, 2003.
19. Austerlitz, F., Dick, C.W., Dutech, C., Klein, E.K., Oddou-Muratorio, S., Smouse, P.E., and Sork, V.L., Using genetic markers to estimate the pollen dispersal curve. *Molecular Ecology*, 13, 937, 2004.
20. Robledo-Arnuncio, J.J. and Gil, L., Patterns of pollen dispersal in a small population of *Pinus sylvestris* L. revealed by total-exclusion paternity analysis, *Heredity*, 94, 13, 2005.
21. Robledo-Arnuncio, J.J. and Austerlitz, F., Pollen dispersal in spatially aggregated populations, *American Naturalist*, 168, 500, 2006.
22. Grevstad, F.S., Simulating control strategies for a spatially structured weed invasion: *Spartina alterniflora* (Loisel) in Pacific Coast estuaries, *Biological Invasions*, 7, 665, 2005.
23. Abramowitz, M. and Stegun, I.A., *Handbook of Mathematical Functions with Formulas, Graphs and Mathematical Tables*, Applied Mathematics Series, 55. National Bureau of Standards, Washington, DC, 1964.
24. Nathan, R., Perry, G., Cronin, J.T., Strand, A.E., and Cain, M.L., Methods for estimating long-distance dispersal, *Oikos*, 103, 261, 2003.
25. Bullock, J.M., Shea, K., and Skarpaas, O., Measuring plant dispersal: An introduction to field methods and experimental design, *Plant Ecology*, 186, 217, 2006.
26. Clark, J.S., Why trees migrate so fast: Confronting theory with dispersal biology and the paleorecord, *American Naturalist*, 152, 204, 1998.
27. Lawes, R.A. and McAllister, R.R.J., Using networks to understand source and sink relationships to manage weeds in riparian zone. In *15th Australian Weeds Conference: Managing Weeds in a Changing Climate* (eds. Preston, C., Watts, J.H., Crossman, N.D.), pp. 466–469, Adelaide Convention Centre, Adelaide, 2006.
28. Nathan, R. and Muller-Landau, H.C., Spatial patterns of seed dispersal, their determinants and consequences for recruitment, *Trends in Ecology & Evolution*, 15, 278–285, 2000.
29. Higgins, S.I., Nathan, R., and Cain, M.L., Are long-distance dispersal events in plants usually caused by nonstandard means of dispersal? *Ecology*, 84, 1945, 2003.
30. Burgman, M.A., Ferson, S., and Akcakaya, H.R., *Risk Assessment in Conservation Biology*, Chapman & Hall, London, 1993.
31. Buckley, Y.M., Briese, D.T., and Rees, M., Demography and management of the invasive plant species *Hypericum perforatum*. II. Construction and use of an individual-based model to predict population dynamics and the effects of management strategies, *Journal of Applied Ecology*, 40, 494, 2003.

32. Thompson, S.K., *Sampling*, 2nd edn., John Wiley & Sons, New York, 2002.
33. Gardener, M.R., Whalley, R.D.B., and Sindel, B.M., Ecology of *Nassella neesiana*, Chilean needle grass, in pastures on the northern tablelands of New South Wales. I. Seed production and dispersal, *Australian Journal of Agricultural Research*, 54, 613, 2003.
34. Gardener, M.R., Whalley, R.D.B., and Sindel, B.M., Ecology of *Nassella neesiana*, Chilean needle grass, in pastures on the northern tablelands of New South Wales. II. Seedbank dynamics, seed germination, and seedling recruitment, *Australian Journal of Agricultural Research*, 54, 621, 2003.

13 Tracking Invasive Weed Species in Rangeland Using Probability Functions to Identify Site-Specific Boundaries: A Case Study Using Yellow Starthistle (*Centaurea solstitialis* L.)

Lawrence W. Lass, Timothy S. Prather, Bahman Shafii, and William J. Price

CONTENTS

13.1 Executive Summary 278
13.2 Introduction 278
  13.2.1 Background 278
13.3 Methods 279
  13.3.1 Model of Development 279
  13.3.2 Productivity Model Components 281
  13.3.3 Spatial Network Models 283
13.4 Case Study 285
  13.4.1 IDRISI Software 285
    13.4.1.1 Additional Software Required for the Exercise 285
  13.4.2 Topographic Correlates of the Site (Slope, Aspect, and Sunlight Difference between Spring and Summer "Sundiff") 285
    13.4.2.1 Preliminary Steps 286
    13.4.2.2 Steps for Importing and Calculating Slope, Aspect, and Sun Angle Differencing 287

13.4.3 Developing Vegetation Indices (NDVI and TSAVI1) from Landsat Images 289
13.4.3.1 Atmospheric Correction 289
13.4.3.2 Georectification 292
13.4.3.3 Vegetation Index 293
13.4.4 Productivity Modeling with the Logit Regression Module 295
13.4.4.1 Specifying a Sampling Scheme 296
13.4.5 Network Modeling 296
13.5 Conclusions 298
References 298

13.1 EXECUTIVE SUMMARY

Rangeland and forest productivity, and species diversity, are impacted by invasive species. The lack of intensive cropping or rotational regimes to break pest cycles contributes to the persistence and spread of invasive species across landscapes. Efficient deployment of resources to achieve landscape level management objectives is difficult without some idea of the location and rate of spread. The objective of this chapter is to illustrate that spatial data collection at one site can be used to aid in the prediction of plant movement at another through the integration of plant productivity models and spatial network models. In this context, the plant productivity model provides the likelihood of an invasive species occurrence while the spatial model reflects the environmental conditions and competitive barriers that determine plant movement. Modeling plant movement across actual landscapes is demonstrated using a yellow starthistle (*Centaurea solstitialis* L.) plant movement project with computations, and processing of the spatial data is carried out using the IDRISI software package from Clark Labs.

13.2 INTRODUCTION

The spread of introduced plants depends on their ability to establish and disperse.[1] Species with high growth rates that tolerate disturbance are likely to capture resources that may allow them to dominate.[2] Expansion to other plant communities then may depend on competitive interactions along specific gradients that may be characterized as disturbance, resource, or competitive gradients.[3] Identifying environmental gradients important to a species' ability to increase its distribution should allow construction of predictive models to describe the plant's movement across a given area.

13.2.1 BACKGROUND

Plant species diversity in forested communities increased along a productivity gradient in Kentucky,[4] but in other studies, plant diversity was higher at intermediate sections of a productivity gradient[2,5–7] suggesting fewer species compete and survive as biomass increases. Plant community structure varies with biotic and abiotic factors that include topography and soil organic matter.[8] Since plant community structure and diversity are explained in part by plant productivity and topography, landscape models that explain a plant species distribution should include variables that reflect productivity and topography. Variables that reflect plant productivity and topography include vegetation indices; topographic variables of slope, aspect, and elevation; soil type; and sun angle. Our modeling approach for predicting yellow starthistle occurrence and movement utilizes productivity and topography variables.

Yellow starthistle is an invasive Eurasian weed and presently infests over 2 million ha in the Western United States. It was first introduced to the state of California in the mid-1800s as a contaminant of alfalfa (*Medicago sativa* L.) seed.[9] Yellow starthistle was identified in 23 of the 48 contiguous states in 1985 but is presently found in 41 of the 50 states.[10,11] Yellow starthistle has been present for more than 40 years in many locations in the northwestern states of Idaho, Oregon, and Washington.[12–14] In Idaho, yellow starthistle took advantage of favorable conditions for expansion, and populations increased from 10 to 150,000 ha in less than 50 years.

The most serious invasion sites have occurred on marginal rangeland and non-crop land, but cultivated lands, including those for dry land grain, set aside, Conservation Reserve Program (CRP), grass and legume seed crops, and irrigated pastures are also susceptible.[15] This invader is opportunistic and fills in areas where the previous vegetation has been dominated by weedy annual grasses such as bromes (*Bromus* sp.), medusahead rye (*Taeniatherum caput-medusae*), or fescues (*Festuca* sp.). Colonies of yellow starthistle at many sites often exceed more than 300 plants/m^2 and compose more than 90% of the vegetative cover.[15] Due to toxins in the plant, horses feeding on a diet of yellow starthistle develop brain lesions that lead to "chewing disease" (equine nigropallidal encephalomalacia), a fatal nervous disorder.[16,17]

Successful control and management strategies for yellow starthistle require the application of a multilayered prevention, containment, plant community restoration, and mechanical, chemical, and biological control options. A successful program must not just control the starthistle but also improve the site in order to slow invasion and avoid costly follow-up treatments.[12–14] For example, areas at risk to rapid movement should have priority for chemical control of adjacent infestations and also for restoration activities.

Critical to all these processes is the ability to predict the spread of yellow starthistle, thereby allowing managers to adjust their activities to either favor spread of desirable species or discourage the spread of yellow starthistle. Models that incorporate dispersal have been useful for adjusting management actions for the control of invasive plant species.[18]

13.3 METHODS

13.3.1 Model of Development

The first step in the process is developing a productivity model to identify an ecological gradient where the plant will thrive and reproduce or diminish due to environmental constraints, limited resources, or competition. The goal of this component in the exercise is to define a map of productivity or the likelihood of occurrence of the invasive species. Data are first collected from locations documented with Global Positioning System (GPS) or paper maps indicating where the weed is present or absent. This should be conducted for the area of interest or an area of similar environmental context. Surveyed sites should be representative of the target area of interest in terms of vegetation, topography, and climate (temperature and rainfall patterns). For weed species, many federal, state, and county agencies may monitor existing infestations with fairly accurate mapping tools and are often willing to share this information or have it posted on the Internet. It is recommended that the data are validated with site visits as known infestations may have been treated or undocumented infestations may be found outside the survey area. As initial data, however, these data provide a useful resource for model development.

Once on-the-ground presence–absence occurrence data have been collected, the data should be summarized into proportions relative to variables representing a relevant environmental gradient. For example, it was noted that yellow starthistle followed topographic trends present in the canyonlands of north Idaho. Hence, the topographic variables of slope and aspect were utilized as environmental correlates. For other situations, the environmental correlates may differ depending on the gradient being represented and could include such measurements as soil moisture levels, light intensities, or soil depths. With the predictors in hand, categories or bins based on these variables are created such that multiple observations of the binary presence–absence response variable are available for each binned correlate combination. From these multiple observations, the proportions of the invasive species present at each correlate combination can be developed. For yellow starthistle, 10 slope and 16 aspect bins were created resulting in a 10 × 16 matrix of slope–aspect categories. Each cell within the matrix was then given a proportion of yellow starthistle presence, $\mathbf{p}_{ij}$, where i and j represent the ith and jth levels of slope and aspect, respectively.

The second step is to develop the model. Model development may take one of two paths. In the first path, the model may arise from theoretical expectations of the inherent system. This approach requires considerable knowledge of the invasive species and its interactions with the surrounding environment. In the second path, the model is based on empirical data where parametric model forms are tested, through trial and error, until an adequate functional form is identified. While this approach has the advantage of not requiring extensive knowledge of the environmental interactions and biology of the invasive species, it does not lend itself to supporting or testing of related theoretical expectations. For the goals expressed here, however, either path is applicable, providing that the resulting model provides accurate predictions.

Regardless of the modeling approach, estimation of the model must meet the standard statistical requirements of regression analysis. These require, among other things, that the response variable, in this case, the $\mathbf{p}_{ij}$'s, follow a normal distribution with constant variance. This requirement is often met through a transformation of the data. For yellow starthistle proportions, the transformation used was a logit function given by

$$\text{logit}_{ij} = \ln\left[\frac{\mathbf{p}_{ij}}{(1-\mathbf{p}_{ij})}\right] \tag{13.1}$$

In addition, the form and behavior of the environmental correlates should be considered. Aspect, for example, is a cyclic variable repeating over the interval 0°–360°. A common method of circumventing this cyclic nature is to transform such data to a polar coordinate system that describes both the distance and angle from a fixed point.

In this case study, the polar coordinate system was applied to the yellow starthistle aspect data prior to modeling.

Either linear or nonlinear parametric forms may be applicable to the likelihood of occurrence response. Yellow starthistle was found to be well represented by a linear multiple regression model.[19–21] Before completion of the model, the data should meet standard statistical requirements of significant parameter estimates, adequate residual structures, and in the case of nonlinear models, minimal parameter correlations. Once an adequate occurrence model has been developed, the spatial network modeling phase can begin.

The independent variables used in the development of productivity models should reflect conditions and factors that influence plant development and potential to reproduce. Typically, information on the environmental and climatic factors is not available or difficult to obtain without coarse interpretation due to large distances between observation stations. Alternatively, topographic factors, such as slope and aspect, and environmental conditions such as sunlight, as well as competitive correlates such as vegetation indices related to plant community biomass, have been used to model plant survival and seed movement. These independent variables help define what makes the invaded site different from areas not being invaded. In other words, if you are using the productivity model with Brazilian pepper tree (*Schinus terebinthifolius*) in the Florida Everglades, slope and aspect may not be as important as depth to water table and associated vegetation. The secret to success when using this method is to identify a group of independent variables that explain why the invader is successfully spreading. Our work with yellow starthistle started with topographic variables (slope, aspect, and elevation) because most of Idaho's starthistle patches are growing in south- and western-facing slopes of drier annual grass rangeland. A single model, however, could not describe yellow starthistle's productivity when growing plants were in forest and pasture sites. Instead of developing multiple models for each land use pattern, a more biological-based approach was developed. A hybrid model that uses both topographic and vegetative characteristics of the site is currently being used to model yellow starthistle productivity at a site and predict its movement from the site.

13.3.2 Productivity Model Components

For yellow starthistle, the relevant environmental characteristics that have been correlated to presence/absence include topography, solar radiation, and vegetative productivity. The data corresponding to these were slope, aspect, and elevation; seasonal sun angle difference; and vegetation indices. These are described below and in the demonstration exercise in more detail.

The slope, aspect, and elevation data use USGS National Elevation Dataset (NED) (1/3 arc s) with a spatial resolution of 10 × 10 m. There may be other sources for elevation data on a regional or local scale that have greater accuracy and finer spatial resolution. Consider purchasing Lidar data if spatial resolution and data accuracy are an issue, but processing this type of data is outside the objectives of this demonstration. Slope data are grouped into 5° intervals and aspect are grouped into 22.5° intervals to reduce any smoothing effects of the filtering process. The average slope value for yellow starthistle found in a 1981 Idaho survey was 31° ± 8°, and the average aspect value was 237° ± 89°. Aspect data may be rotated from 0° to 50° in 10° increments to align with the solar radiation effect. Additionally, aspect data also may be weighted and categorized to reflect solar radiation; for yellow starthistle, this resulted in the highest value of the aspect class at 237° and lowest value at 45° (see exercises on topographic parameters of the site).

Evaporation demand is greater in full sun, and, in Idaho, these southwest-facing slopes tend to be populated with nonnative annual grasses, providing an opportunity for yellow starthistle invasion. Seasonal sun angle difference (*sundiff*) data subtract June's estimate of the amount of the hill still shaded based on slope and aspect and the sun angle for June 12 (56.6°) at sun azimuth of 237° from March's hill shade using a sun angle from March 12 (23.8°) and 237° sun azimuth. The 237° sun azimuth represented the average aspect value for yellow starthistle growing in Idaho. The sun angle difference data, therefore, show areas always receiving full sunlight or partial shade when plants are developing (see exercises on topographic parameters of the site). More sun does not equal more yellow starthistle in Idaho but dries soil faster and reduces competition from native plants.

The Normalized Difference Vegetation Index (NDVI) and Transformed Soil Adjusted Vegetation Index (TSAVI1) are calculated from red and near-infrared bands of Landsat 5 data. Images are atmospherically corrected and georeferenced using National Agricultural Imaging Program (NAIP) images for reference, and the spatial resolution is reduced to 10 × 10 m using nearest neighbor analysis prior to calculating the vegetation indices given as

$$\text{NDVI} = \frac{([\text{NIR}]-[\text{Red}])}{([\text{NIR}]+[\text{Red}])} \tag{13.2}$$

$$\text{TSAVI1} = \frac{([a]\times(([\text{NIR}]-[a])\times([\text{Red}]-[b])))}{(([\text{Red}]+[a])\times([\text{NIR}]-([a]\times[b])))} \tag{13.3}$$

where

[NIR] is the near-infrared band
[Red] is the visible red band
[*a*] is the regression slope rate of the soil reflectance model
[*b*] is the regression intercept of the soil reflectance model

Red and NIR units of measure are mW $\text{cm}^2\ \text{sr}^{-1}\ \mu\text{m}^{-1}$. NDVI and TSAVI1 are unit-less relative ratios ranging between 0 and 1.

The NDVI and TSAVI1 values are scaled to match vegetation training sites. NDVI estimates the percent green vegetation cover or healthy vegetation while the TSAVI1 indicates the amount of living vegetation present (biomass) adjusted for soil background. Hence, both indices provide information on competitive ability (see exercises on vegetation index data development).

Other datasets may be useful when developing productivity models. Defining what is unique to a site being invaded will often lead to locating other sites. Consider adding model components such as fire history, ownership, construction disturbance (building permits and new roads), prevailing wind direction and speed, and distance from roads. Resources for additional data are often located through Federal and State GIS Clearing Houses. A list of these sources may be obtained by searching on "GIS Resources State" or the list developed and maintained by the University of Oregon located at http://libweb.uoregon.edu/map/map_section/map_Statedatasets.html. Past experience has shown these data to be useful when refining the predictive productivity model to predict areas susceptible to first invasion (see exercises on other data development).

13.3.3 Spatial Network Models

Movement of a plant species across a specific landscape requires the input of site-specific data to accurately represent how a species may move within that landscape. Spatial data residing in a geographic information system (GIS) serve this process well. The algorithm within GIS software, commonly referred to as network analysis, will be applied to plant dispersal modeling.

Network algorithms commonly are implemented in GIS software packages such as IDRISI®, a raster-based GIS,[22] or Arc Info®, a vector-based GIS.[23] These network routines model the cost of movement along a predefined grid of cells for a raster-based GIS or line segments for a vector-based GIS. Cost in terms of travel rate is determined as a predefined force and friction through each cell or segment. The path of greater or least resistance can be determined from the summation of the costs linking cells or segments along a path from point A to B.[22,24,25]

Cost assessment reflects both active and passive resistance of an object through the network. An active resistance model (dispersal) assumes that cost values greater than 1.0 are forces (dispersal is assisted), and values less than 1.0 are frictions (dispersal is impeded).[2] A practical analogy to this model would be that of walking with (force) or against (friction) a wind. Here, the wind is producing an active resistance.[25] The alternative model, passive resistance, assumes that cost values greater than 1.0 result in friction while values less than 1.0 represent forces. An analogy in this case would be one of walking up (friction) or down (force) a hill. In this example, gravity produces a passive form of resistance. In either scenario, cost values equal to 1.0 produce a neutral level of resistance.

An additional component of the network model is a directional effect. In the simplest form, network models that apply friction equally in all directions (isotropic) result in series of rings similar to growth rings of a cross section tree trunk (Figure 13.1). Rarely does this happen in nature, and, typically, friction varies as the movement direction changes (anisotropic) causing slower or faster movement along some of the paths. Anisotropic networks may still have a series of ringed movement patterns but often exhibit lobes (Figure 13.2). In such systems, the friction/force information is provided with both magnitude and direction. Within the algorithm, the effective friction (EF) is given as a power function: $EF = (FR)^f$, where FR is the potential full magnitude of the friction and $f = \cos(a)^K$ represents the difference between the direction of movement and the direction of the applied friction, a, and where K is a parameter controlling the sensitivity to the difference angle, a.[22] A value of $K = 0$ produces an isotropic model while larger K values lead to a higher degree of directionality in the model. Recalling the hill-walking analogy, when one is walking directly up slope, the

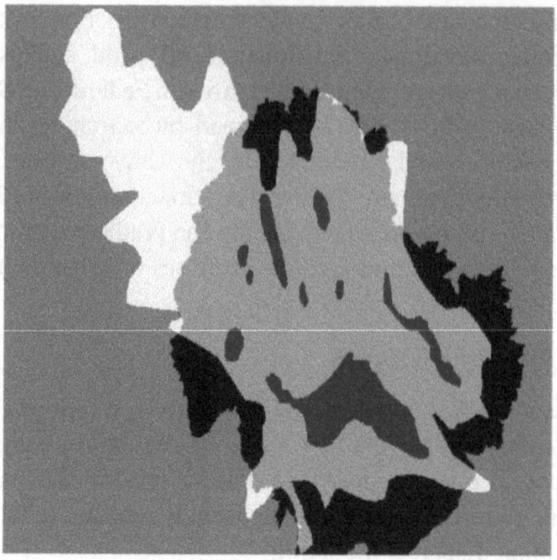

FIGURE 13.1 Isotropic model of yellow starthistle (1981 infestation dark gray in center islands, light gray is predicted areas, black is over predicted areas black, and white is unpredicted areas based on 1987 infestation.

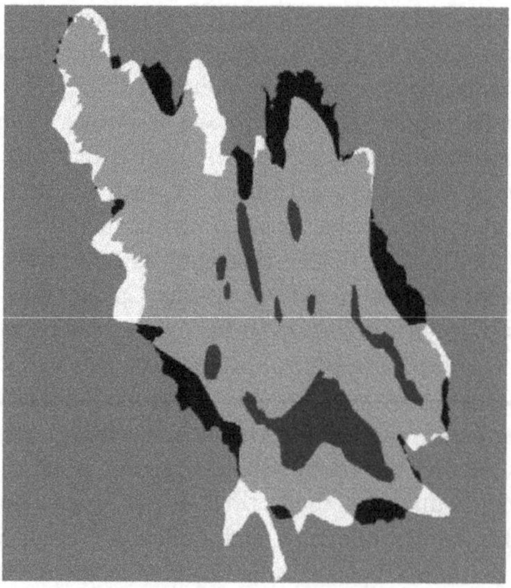

FIGURE 13.2 Anisotropic model of yellow starthistle (1981 infestation dark gray in center islands, light gray is predicted areas, black is over predicted areas black, and white in unpredicted areas based on 1987 infestation.

friction is felt at 100%, while walking along the slope produces no resistance or 0% friction. Walking down hill produces an *EF* of –100%, that is, a force.

A network is superimposed on a landscape with the results of the productivity model (topographical variables, vegetation indices, etc). The establishment and dispersal models estimate the time the invasive plant infestation would take to grow across the network according to the characteristics encountered at that location. To apply the network model to species movement, the productivity model showing likelihood of occurrence predictions will play the role of frictions, whereas relevant topographic correlates can provide the directional attributes of the movement model.

13.4 CASE STUDY

The raw data for this project are available on the web and also included on the CD accompanying the book located in the Chapter 13 folder. The project will develop a predictive model for the spread of yellow starthistle near White Bird, Idaho, based on the distribution of yellow starthistle near Kamiah, Idaho (80km north of White Bird). The Kamiah data are used to develop the productivity models estimating the likelihood of occurrence and network models while the White Bird data will be used to validate the models. The Kamiah data come from an aerial survey conducted in 1981 and repeated in 1986 with additional ground validation in 1995. The White Bird data are from USFS and Idaho County ground surveys from 2006. The initial steps of the demonstration exercise will develop a data cube where data layers stack on each other. The productivity model will use the data layers to estimate the likelihood of occurrence for each 10 × 10 m pixel. The network model will predict the rate of invasion based on the productivity model.

13.4.1 IDRISI Software

Before you start the exercises you will need to download the **IDRISI** software from the Clark Labs site at http://www.clarklabs.org/ and look under the Contact Us pull-down menu. **IDRISI** comes with a 15 day free trial that may be extended to 6 months by mailing in the coupon located at the end of this chapter. **IDRISI** is the backbone remote sensing software for our research because it is affordable for students and land managers and offers unique applications and solutions to natural resources and agricultural management problems. The algorithms discussed in these exercises are also available in most of the commercial remote sensing software (ERDAS and Envi). We have found that ESRI Arc/Info products offer limited algorithms for logistic regression and network modeling. If you are new to **IDRISI**, we advise that you start with the tutorials located under the Help menu to aid in your navigation through the software. We are also assuming you have some understanding of GIS and the associated data types. If you need a refresher course on the basics, refer to Wikipedia at http://en.wikipedia.org/wiki/Geographic_information_system.

13.4.1.1 Additional Software Required for the Exercise

Google Earth at http://earth.google.com will be used for developing the vegetation index when atmospherically correcting the Landsat images.

13.4.2 Topographic Correlates of the Site (Slope, Aspect, and Sunlight Difference between Spring and Summer "Sundiff")

Soil temperature and soil moisture influence the development of plants, yet are difficult to measure over a large area. We would expect areas receiving more direct sunlight to have warmer and drier soil than areas where slope and aspect limit the amount of

BOX 13.1 DEVELOPMENT OF TOPOGRAPHIC COMPONENTS

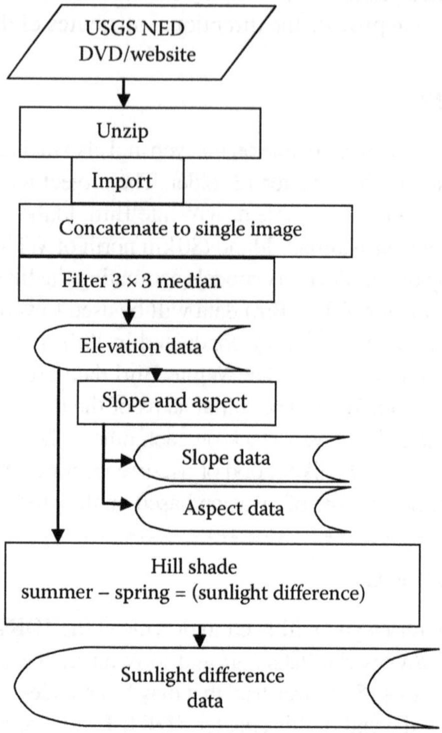

incoming direct sunlight. For example, south-western-facing slopes typically green up faster in the spring and dry out earlier in the summer than northeast slopes.

We use elevation data originated from USGS NED with a spatial resolution of 1/3 arc s (Box 13.1). USGS provides elevation in 2, 1, 1/3, and 1/9 arc s and roughly equivalent to 60, 30, 10, and 3 m resolution between nodes. The raw elevation data are available from the USGS Seamless map server at http://seamless.usgs.gov/. Use the **define download area by coordinates** options to identify the specific area and manually ***enter*** the **latitude coordinates** in *2*° intervals and **longitude** in ***1.5***° intervals (i.e., 47 to 45 and −117.5 to −116 for this demonstration exercise). While the server will generate the appropriate files for download, they will not be the right file type or format. The user will need to select **Modify Data Request** (top of menu). To do this, ***uncheck*** the **1 arc s box. Check** the **National Elevation Data (NED) (1/3 arc s)** option, change the format to ***geo tiff***, and then select the button box for **Save Changes and Return**. You are ready to start the downloading process. Tip: during the save process, it will be useful to ***name the zip file*** for the upper left coordinate (i.e., ned_4700_11700 for 47.00 and −117.00).

13.4.2.1 Preliminary Steps

To start this process, we examine the general phenological stages of the weed. Select a starting date in the spring when most yellow starthistle seeds have germinated. For the demonstration exercise, we will use March 12 for the postgermination phase because it simulates yellow starthistle growing conditions for northern Idaho. The ending date is predetermined as near mid-summer when the sun is approaching its highest point. In our case, we chose a time when most of the plants were bolting and prior to flowering (June 12, about 9 days before mid-summer).

Sun elevation and azimuth are calculated using several sources on the web. A good choice is http://www.geocities.com/senol_gulgonul/sun/. **Calculate and record** the sun altitude angle at **237°** azimuth for **March 12** and **June 12** at ***−116.5*** longitude and ***46.0*** latitude.

13.4.2.2 Steps for Importing and Calculating Slope, Aspect, and Sun Angle Differencing

Step 1: Download data from USGS

Go to the **USGS Seamless map server** at http://seamless.usgs.gov/. The site has an online tutorial and is a good starting point if you are new to this site. Download the **NED (1/3 arc s)** data in **geotiff** format. For the exercise, use the **bounding coordinates** of ***−117.5 to −116 longitude*** *and* ***45 to 47 latitude,*** and download all four files generated.

Step 2: Unzip, import, and concatenate

Unzip the NED into a working directory on your computer. Open **IDRISI**, and identify this working directory by right clicking over the **Project** button on right side of screen to select a **new** project. **Import** the geotiff NED files to **IDRISI** format using the following set of commands from the menu: **File, Import, Government formats**, and **GEOTIFF/TIFF**. Select the ***files*** to be imported and click **OK**. When all files have been converted, use the **CONCAT** command found under the **Reformat** menu to make single **NED file** covering the study area.

Step 3: Re-projection

The current spatial units of the NED data from the USGS seamless map server are in degrees latitude and longitude. For our purpose, they need to be projected to **UTM Zone 11 N** to match other data found in the exercise. This is done using the **projection module** found under the **Reformat** menu to re-project the data. Copy the **reference parameter** from the ***STUDYSITE*** file located in the **processed** data directory.

If you are developing your own study area, it is recommended you set the parameters on increments of 10 m, so that the x and y coordinates round up or down to achieve a 10 m interval. The size of the study area and data availability will ultimately define the spatial resolution, but the spatial resolution of the elevation data is a good starting point. The development of a ***STUDYSITE*** file defines the ***datum***, ***projection***, ***dimensions***, ***spatial resolution***, and ***extent study area***, so all further data development matches the footprint. You will want to think of the data being developed as a cube of data layers that stack on each other.

Step 4: Process slope and aspect data

Prior to generating the slope and aspect data, use the **Filter module** located under the **Image Processing—Enhancement** menus to remove any re-projection artifacts (flat spots) in the NED data. A ***3 by 3 matrix*** and ***median filter type*** often works well for **filter settings**, but a 5 × 5 matrix may be appropriate when the flat spot is wide.

Other types of filters may work better for flatter topography. Without this step, the slope, aspect, and sundiff data will have visible striping.

Open the **slope module** under the **GIS analysis menu, surface analysis box, and topographic variables menu. Select** probes both slope and aspect at the same time. Enter ***input*** *(click on* ***box by name to pull down directory****)* and ***output file names***. Select **degrees** for slope units and a scale factor of 1. Enter a ***filename*** and then click the ***OK*** box.

Display the data and view it for errors. Zoom in sufficiently to see individual pixels. Examine the range of values. Slope should range from 0 to 90, and aspect should range from −1 to 360.

Group the slope data into 5° intervals using the **reclass module** under the **GIS analysis menu and database query** button. Enter names of the ***input file*** and ***output file***. Select **equal-interval** reclass, specify ***5°*** for the **class width**, and define the **output minimum** as ***0*** and **maximum** as ***100***. Click the **no** box when asked if you want the data to be in the **integer** format. Use the same procedure for **aspect**, but use an interval of ***22.5°***. The grouped intervals help offset any error that may have been introduced when the NED data were filtered.

Although we commonly use 5° for slope and 22.5° for aspect when grouping the data, other sequential methods such as log curves, smaller or larger groups, and even rotating the data to match the heating of the sun may be appropriate. We often rotate aspect classes such that southwest becomes class 8 and northeast becomes class 1 (Table 13.1). Use the **Assign** module to make a weighted aspect image to match Table 13.1.

Step 5: Sundiff—Sunlight difference between spring and summer

Sun angle differencing data will subtract June's estimate of hill shading based on slope and aspect and the sun angle for June 12 (56.6°) at sun azimuth of 237° from March's shading using a sun angle from March 12 (23.8°) and 237° sun azimuth for a difference of 32.8°.

First calculate the **March hill shade** dataset using the same module as slope and aspect with the **hill shade box checked**. The numeric results represent a relative illumination and are not quantified by the amount of incoming sunlight (cloudy and sunny days) nor other shading material (trees, shrubs, and tall grass). Values should

TABLE 13.1
Directional Rotation to Improve Aspect Data Grouping

Original Direction	Group Number	New Group Number
N	1	2
NE	2	1
E	3	3
SE	4	4
S	5	6
SW	6	8
W	7	7
NW	8	5

TABLE 13.2
Reclass Parameters for Hill Shade Data

Assign a New Value of	To All Values from	To Just Less than
0	−1000	0
1	1	10,000

range from 0 to 1 with 1 in full sunlight (100%). Values outside the normal range are usually on the edge of the data where calculation errors are common. Use the **Reclass** module found under the **GIS analysis** pull-down menu and in **Database query** menu to set all values between 0 and above 1 based on Table 13.2.

Repeat these steps for the **June hill shade image**. Then, use the **image calculator** to subtract the March hill shade image from the June hill shade image to estimate **sundiff**, and apply the equation:

$$((1-(\text{June image}-\text{March image}))\times 100) \tag{13.4}$$

The resulting image should show larger numbers for areas receiving full sunlight all spring and lower number for areas only receiving partial sun.

After a sundiff image has been calculated, **view a histogram** of the data with the intervals set to ***1*** in numeric output. Use the ***weed training dataset*** for Kamiah (in CD Chapter 13 folder dataset) as the **mask**, and repeat histogram again. Examine the data for consistency where sundiff values of the weed fall into a bell-shaped curve of a single peak. We often reclassify the sundiff into 2% or 5% intervals to provide a larger number of data points within a class for the modeling, but in many cases, it is not necessary. To **reclassify** the ***sundiff data***, see the slope and aspect section above.

13.4.3 Developing Vegetation Indices (NDVI and TSAVI1) from Landsat Images

This is a three-part process of atmospheric correction, rectification, and vegetation index calculation (Box 13.2)

Normally, the first part of the process is **importing** the raw Landsat imagery into **IDRISI**, but it has been done to save you this step. You can find the data for bands 1, 2, 3, and 4 with the header files under in the CD Chapter 13/***landsat data directory***. If you have your own data, consult the **IDRISI** Help menu under **satellite data import** for specific file type and parameter setting (see **Box 13.2**)

13.4.3.1 Atmospheric Correction

Atmospheric correction removes haze and other atmospheric effects from remotely sensed imagery. Images acquired prior to 2000 may not have information on optical thickness from the NASA NEO project, so it should be estimated as ***0.6*** for an average clear day. Although older images are of historic importance, images acquired more than 5 years ago may not reflect current conditions.

BOX 13.2 DEVELOPING TSAVI1 AND NDVI FROM LANDSAT IMAGES

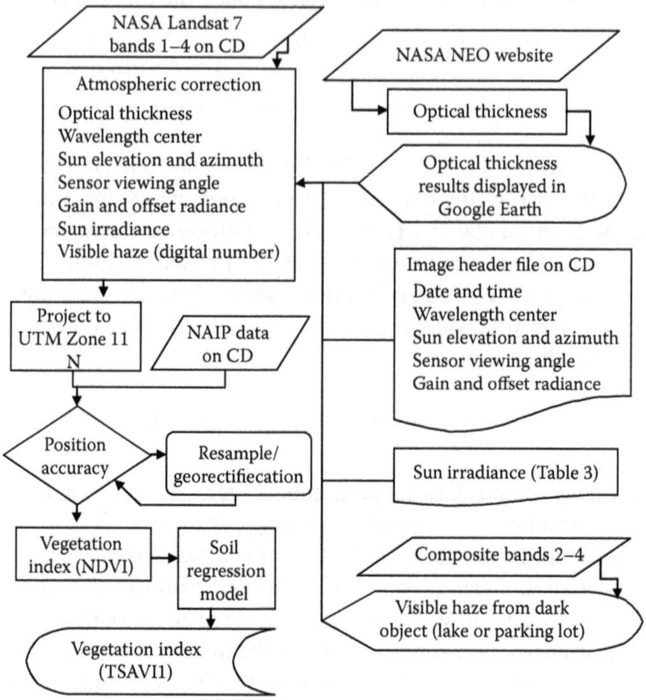

Steps for atmospheric correction

Step 1: Determine the optical properties of the atmosphere when the image was acquired
Optical thickness is measured with the MODIS satellite and a series of ground-based Lidar units. **Open** the Landsat image ***header file*** with a **text editor** (file names vary, but names with extensions ***h***** or ***metadata.txt*** are good places to start). Within the header file, the **date** and **time** of acquisition are often recorded. Sometimes, the **sun azimuth** and **elevation** may also be listed. If the metadata are complete, the **radiometric correction** is also listed. For Landsat 7, the file name extension is ***h1***.

Using the date and time, determine the optical thickness from http://neo.sci.gsfc.nasa.gov. Select **Search Parameters,** enter **date range** of **±*30*** days to the image acquisition date, and select **search NEO**. Select the closest matching date for aerosol optical thickness MODIS under **Matching Data Sets**, and click **Select**.

Select **open** in **Google Earth** to view results. Use the **scene center** to "fly to" the correct location. If the center numbers are very large, they may need to be scaled by 1000 to get the correct latitude and longitude. You also may need to zoom out to 150–180 km to view the full Landsat footprint.

The **optical thickness scale bar** is along the top of the map. Areas with an optical thickness of 0.6 are clear, and you should see the normal Google Earth layer below.

Select the **representative optical thickness value** for your area of interest on the Landsat image. For our exercise, it will be between ***47*** and ***46*** latitude and ***–117.5*** and ***–116*** longitude. Typical values range from ***0.2*** to ***0.4*** for clear dry days.

Step 2: Identify the sensor parameters used for acquiring the images
Determine the wavelength center, sun elevation and azimuth, sensor viewing angle, and gain and offset radiance values from the metadata file. Center wavelengths may need to be calculated by averaging the minimum and maximum wavelengths if the center values are not listed. The units for wavelength should be in microns (0.830).

Gain and offset radiance are also called Gains/Bias. Multiplying the highest digital number (DN) in the image by the offset value should yield a number between 10 and 30. If the result is larger, the gain and offset are in W m^2 sr^{-1} μm^{-1} units, and the gain and offset values need to be reduced by a factor of 10 (i.e., divide the DN by 10) to obtain mW cm^2 sr^{-1} μm^{-1}. See detailed notes in the **ATMOSC** help file of **IDRISI**.

The **viewing angle** for Landsat and SPOT images is ***zero***, but some sensors allow forward viewing and may have a nonzero value. Sun elevation and azimuth should also be in the metadata file. It may also be calculated separately using several sources available on the web (http://www.geocities.com/senol_gulgonul/sun/).

Step 3: Calculate the amount of sunlight (sun irradiance) and visible haze
Sun irradiance may be estimated for each band based on center wavelength value from Table 13.3. Table 13.3 used the mean distance between sun and earth and is acceptable for most work. For those requiring a more accurate estimate of sun irradiance, an accounting for seasonal distance differences will be necessary. Advanced users may want to consult the **IDRISI ATMOSC** help file to correct for changes in earth-to-sun distance.

Create a false color composite of bands ***2***, ***3***, and ***4***. Select **display**, and then the **display launcher open file list** button (box with three dots) by double clicking. **Select *band 2***. Change the **autoscale options** to ***equal intervals*** and the **palette file** to ***grayscale***. Select **OK**. In the **composer** box, select ***band 2*** and click the **blue** button box. Add ***bands 3*** and ***4*** as layers using the **add layer** box in the **composer** box. Select **add layer**.

Select **raster image** and open **file list** button (box with three dots) by double clicking. Select ***band 3*** and **OK**. Change the **autoscale** options to ***equal intervals*** and the **palette file** to ***grayscale***. Select **OK**. In the **composer** box, select ***band 3*** and click the **green** button box.

Repeat for ***band 4***, but use **red** button box to get the final false color composite image.

Step 4: Determining the DN haze value
The haze value will be subtracted from the images in the correction process to account for visible haze. Deep lakes generally have very low reflectance and are useful for determining haze values. A large empty parking lot or other feature with low reflectance will also work. **Highlight *band 4*** (red) in the **composer** box, and select **Cursor Inquiry** button (box with arrow and question mark in the main menu). Find the lowest DN value of the large deep lake or other nonreflective feature for band 4. Sample the nonreflective surface 20+ times looking for the lowest value. Record this number.

Repeat for the other bands by highlighting the ***band name*** in the **composer** box and selecting a nonreflective surface. Record these numbers. Typically, the DN haze value will be between 4 and 50 but may be more depending on the type of sensor.

TABLE 13.3
Estimated Solar Irradiance (SI) above Atmosphere at Mean Earth–Sun Distance

Wavelength	SI	Wavelength	SI	Wavelength	SI
0.310	0.063	0.900	0.092	1.540	0.028
0.350	0.097	0.940	0.084	1.580	0.026
0.370	0.113	0.980	0.077	1.620	0.024
0.390	0.108	1.020	0.071	1.660	0.022
0.410	0.169	1.060	0.065	1.720	0.020
0.430	0.200	1.100	0.059	1.900	0.014
0.500	0.191	1.140	0.056	2.100	0.010
0.540	0.187	1.180	0.052	2.300	0.007
0.580	0.182	1.220	0.048	2.500	0.005
0.620	0.169	1.260	0.045	2.700	0.004
0.660	0.155	1.300	0.041	2.900	0.003
0.700	0.143	1.340	0.038	3.100	0.002
0.740	0.130	1.380	0.036	3.300	0.002
0.780	0.119	1.420	0.033	3.500	0.001
0.820	0.109	1.460	0.031	3.700	0.001
0.860	0.098	1.500	0.029	3.900	0.001

Source: Adapted from ASTM G173 found at http://rredc.nrel.gov/solar/spectra/am1.5/ASTMG173/ASTMG173.html

Note: Wavelength units are microns, and solar irradiance units are mW cm^2 sr^{-1} μm^{-1}.

If there are many bands (+10), consider making a mask layer and **Extracting** the DN values. See **IDRISI Tutorial** to do this.

Step 5: Atmospherically correcting the scene

Open the **ATMOSC** module in **IDRISI** under **Image Processing**, **Restoration** menus. Select **full model button, input image, and output image**. Using values from Steps 1 to 4 for each band fills the missing information. Select **OK**. Consider using a macro to run a batch if several bands need to be corrected. See the **IDRISI ATMOSC Help file** for more information.

13.4.3.2 Georectification

Georectification or resampling is the process of moving the data of the grid system to register with the ground coordinates. Landsat imagery generally has some positioning errors. This can be assessed by overlaying the data with ground control points. If error is minor, you may not need to resample the images and can just project the data into UTM zone 11N directly.

Step 1: Re-project data

Landsat imagery comes in several projections, so check the header file for the type of projection. The most common types are **ALBERS** and **UTM**. In our exercise, the Landsat data are in **ALBERS**. Project the atmospherically corrected Landsat data into the study site projection as outlined in *Step 3* of the *Topographic Data Development* (**13.4.2.2**) section. Repeat the projection process for each band.

Step 2: Verify positional accuracy

Display a true color composite of bands 1, 2, and 3 using the instructions outlined in the **IDRISI** tutorial. Add the ***NAIP*** imagery data from the CD under the directory ***Chapt_13/naip*** as another display layer, and adjust the transparency by selecting **Image format** in the display box and **visibility**. Increase the transparency of the NAIP data, so the Landsat data are visible. Compare the position of roads and buildings between the satellite imagery and NAIP images.

If the position of the Landsat image needs to be adjusted, read the tutorial in the **IDRISI** help menu to get an understanding of the process involved. Select the **Resample module** in the **Reformat pull-down menu**. Enter the ***Landsat 3 band name*** from atmospheric correction as the **input reference** name and the ***NAIP file*** as the **output reference** image. Tie points (i.e., the same location on both images) will be used to adjust the image. Zoom in and find the same locations on both images to select as tie points. Start at the upper right corner, and work clockwise around the whole Landsat image. The maximum number of tie points is 256, but, typically, you will only need to use ***75 to 100*** points.

Select the **quadratic function**, and evaluate root mean square (RMS) error. To reduce large residual errors (e.g., start with an RMS value of ***+50*** and end with ***+30***), add more tie points between the images, select the **quadratic function**, and evaluate the new RMS value. Save the resulting ***GCP file*** periodically, and when complete, add ***input image*** and ***output image names*** and identify the **reference parameters** from the study area image. Click **OK**, and repeat for the other atmospherically corrected bands of Landsat images.

NAIP image data have become a standard for background maps. Data are provided on the CD for the exercise, but you may want to import your own. NAIP is highly compressed using ***MrSid*** software and will need to be uncompressed and converted to a format that **IDRISI** can read. Lizard Tech, the makers of MrSid, has free decoding software (mrsidgeodecode_win.exe) located at http://www.lizardtech.com. If you are using Windows XP, you can start **cmd** in **run** menu of the **Start Menu**. Use the commands ***cd, cd..***, and ***cd xxx*** to change directories where ***xxx*** is the directory name containing the **NAIP** file. Extract with **mrsidgeodecode_win** using the command line ***mrsidgeodecode_win.exe –I nnn.sid –o nnn.jpg –of jpg –wf –s 2*** (or ***3***) where ***nnn*** is the file name and ***2*** (or ***3***) is the amount of scaling (2 = 4 and 3 = 8 m spatial resolution). The **mrsidgeodecode–help** command will list the full help menu. If you are using Windows Vista or you want to avoid ***cmd***, you can create a shortcut for the *mrsidgeodecode_win.exe* and modify the shortcut properties to include the command line. After extraction, the images may be imported directly into **IDRISI** using the **IDRISI to JPG** module located in the **File pull down Menu** and **Import**.

13.4.3.3 Vegetation Index

After georectification and atmospheric correction, we will calculate the vegetation index. **IDRISI** offers 19 vegetation index models falling into either the slope-based or distance-based models. For more information on vegetation index models, refer to the electronic **IDRISI** manual under Help section. Each model has advantages and disadvantages; however, most applications start with the NDVI.

13.4.3.3.1 Developing an NDVI Layer

First, open **IDRISI**, and select ***image processing > transformation > vegindex***. Select **NDVI**, and input the ***red* (*band 3*)** and ***nir* (*band 4*)** images. Type an ***output file name*** and click **OK**. The results should have a scale of ***–1.0*** to ***1.0***, and the display color should range from brown = –1.0 to green = 1.0 with zero ≈ no green vegetation.

Add the weeds vector layer to the displayed NDVI by selecting ***add layer > vector***, and select the ***file name* (*Kamiah_YST 1981*)**. Select the **white palette** before proceeding. Zoom into a weed patch. Try changing the properties of the NDVI image by selecting the **layer properties** and changing the **display min** to ***–0.01*** in order to show areas with green vegetation. Change the display min to ***–0.20*** to include areas with matured annual grasses or ***–0.30*** to include bare ground. Reduce the display maximum to ***0.2*** to show areas with limited green plants or increase to ***0.4*** to show more green plants. Determine what the minimum and maximum display thresholds are for a weed species to be found growing there.

13.4.3.3.2 Developing a TSAVI1 Layer

Advanced users may select other models for their ability to reduce extreme values that potentially mask information such as high leaf area in productive sites and base soils in dry climates. In rangeland, it is anticipated that there may be considerable soil background between plants; hence, we will use the **Transformed Soil-adjusted Vegetation Index 1** (TSAVI1) to remove the soil background effects.

The soil adjustment of TSAVI1 index is based on linear regression of the NIR band as a function of the red band. The development of the linear regression starts with the creation of a soil value profile. The soil locations are isolated with a soil mask image, and, finally, the regression module is run.

Step 1: Creating a soil value profile

Using the NDVI image, identify two or three areas with bare soil (i.e., areas that typically have negative values). Using the **on screen digitizing module** (represented as a circle with crosshairs in main menu bar), digitize 50–100 points that are representative of bare soil areas (see **IDRISI** tutorials if you need help using the digitizing module). **Save** the resulting ***vector point file***. Convert the **vector point data** to **raster** using the **Rastervector module** under the **Reformat** pull-down menu. Run the **Extract module** located under the **GIS analysis menu and database query heading** to create a ***values file*** for the red and infrared images with the rasterized bare soil layer as the **Feature definition image**. Select ***all summary types*** and press ***OK*** to calculate.

Step 2: Create soil mask image

Divide the **range value** from the **extraction output** by ***4***, or add and subtract the **standard deviation value** from the mean to calculate a weighted maximum and minimum soil values, respectively. Group data of the red band ***3*** by setting the **non-soil** to ***0*** and the **soil** to ***1*** using the **Reclass module** and the ***weighted maximum and minimum soil values*** as the **thresholds.**

Step 3: Calculate the regression intercept and slope values

Using the **regression module** under **GIS Analysis—Statistics**, enter the **dependent variable** as ***infrared band 4***, the **independent variable** as ***red band 3***, and the ***soil mask*** (created in *Step 2* above). Typically, soils will have an intercept of the soil line ranging from –0.1 to 0.1 and a slope of the soil line ranging from 1.0 to 1.25.

Step 4: Calculate the TSAVI1 values

Open the **image processing—transformation** and **vegindex module**. Select **TSAVI1**, and input the ***file names*** for the georectified and atmospherically corrected red (band 3) and NIR (band 4) images. Type in an ***output file name***. Enter the **intercept** and **slope** values from the regression model results (*Step 3* above), and click ***OK***. The results should be scaled from 0 to 1.0, and color should range from brown = 0 (no vegetation) to green >0.5 to 1.0. Sometimes, the regression model does not fully account for bare soil, and the results will show small negative values. In this case, a reclassification is required to collapse all negative values to zero using the **RECLASS** module. When using the **RECLASS** module, remember not to convert the results to an integer. The final step for preparing the **TSAVI1** data is to use the **SCALAR** module under the **GIS analysis menu** to convert the data to a percentage by multiplying by ***100***. At this point, you can convert the data to an integer format using the **Convert** module under the **Reformat** pull-down menu.

Recheck data prior to starting the next section

Before starting the productivity modeling section, take some time to examine the data using the **EXTRACT** and **HISTOGRAM** commands. Generate the mean and standard deviation of each independent variable (i.e., slope, aspect, sun angle, and TSAVI1) that may be used in the **Logistic regression** modeling to insure that there are the expected variability and data consistency. You are really looking to make sure the data do not fall into one class or are all represented as by a couple of classes. Compare the grouped data against the raw data to see whether the reclassification process caused a loss of variability. In addition, the **histogram module** may be useful for looking at the data distribution within the Kamiah 1995 weed-training area. If the histogram is bell shaped, the model may be improved by squaring or even cubing the independent variable to better fit the shape of the curve. Use **Image calculator module** to square or cube the data.

13.4.4 Productivity Modeling with the Logit Regression Module

In **IDRISI**, the **logistic regression** module uses a binomial logistic regression to calculate a prediction image. Input the ***file name*** for the **dependent** variable developed as training data. Training data are the ***1981 Kamiah data*** on the CD in the Chapter 13 folder, discussed earlier in this chapter. Enter the **independent** variable images either individually or as a ***Raster Group File*** (*.rgf*). (The Raster Group Files may be created using **IDRISI Explorer**.) The maximum number of independent variables **IDRISI** can handle is 20. We usually start the analysis with single independent variable working up to more complex models with interactions. For this exercise, test each of the independent variables developed as well as combinations of the variables. Examine the **Relative Operating Characteristics** (**ROC**) statistic at the bottom of the output. A perfect fit has a value of 1, and a random fit will have a value of 0.5. The higher the ROC, the better the combination of variables fit the data. Which variables provided the best fit? The demonstration exercise requires two models to complete the network section where the results of the model with TSAVI1 are used for the friction component and the results of the model without TSAVI1 are used for the direction component (see Network Modeling code located in Section 13.4.5). The other independent variable used in the models will be selected by you based on other ROCs listed in the model output as part of the learning process in this exercise. Not all independent variables developed above will have a positive affect when added to the logistic model.

13.4.4.1 Specifying a Sampling Scheme

If there are a large number of training sites, proper sampling can reduce spatial interdependence as well as processing time and demand on computer memory. The sampling method can either be stratified or systematic, and the proportion of selected pixels can be set. We will often use a mask file of the training sites surrounded by a 500 or 1000m buffered area to select pixels of interest for the analysis. The mask image is a byte binary format file with 1's in all cells that should be considered and 0's elsewhere and is generated using the **Buffer module** located under the **GIS analysis** pull-down menu and the **Distance operators** menu.

Specify an ***output name*** for the predicted **dependent** variable and a ***name*** for the **regression residuals. IDRISI** will also write, in text format, a numeric regression and test results file in the working directory with the same name. Set your file naming structure in a series format for easy renaming and tracking future changes to the model because many combinations of independent variables will want to be tested.

Press ***OK*** to start. (Tip: perform a ***ctrl-alt-del***, and select **performance** button under **task manager** to show processor status.) If the processor slows (flat lines) near 100%, this is expected and good; however if it begins to stall near 5%, something is wrong, and you may need to reduce the percent pixels sampled. Also in **IDRISI**, the button box in lower left screen may be double clicked to show the progression of the regression. If you would like to try all combinations of independent variables, it might be a good time to investigate the **Macro Modeling** tool under the **Modeling** button within **IDRISI** or try the **new prediction** button in the **Logistic regression** module.

13.4.5 Network Modeling

For this final part of this demonstration exercise, an anisotropic cost surface will be computed to estimate the spread of the weed (Box 13.3). The exercise will (1) use the ***Kamiah 1981 data*** to estimate the movement of yellow starthistle over a 14 year period, (2) assess the network model accuracy, and (3) apply the network model to a new yellow starthistle infestation in the Riggins area to assess future spread.

The **VARCOST** module incorporates friction with different strengths (productivity model with TSAVI1) and direction (of friction). An area with zero friction is an area where the weed is likely to grow and would have high productivity values in the logit model. So "Step one" is to generate a friction direction image based on aspect values so the direction of movement corresponds with the friction. The direction image mask in created by **RECLASS** of the aspect image where a value of 180 is assigned to all values between 0° and 180°; –180 is assigned to all values from 180° to 360° and 0 to all values between –1° and 0°. Merge the direction image mask with the original aspect image by adding the two images in the **OVERLAY** module. In this case, the direction image estimates the direction of spread. Advanced users also may include an isotropic friction surface representing a secondary friction encountered moving through a network. An example would be a productivity model with the vegetation component or weed management where friction is altered, acting independently of the direction of movement.

BOX 13.3 NETWORK MODELING

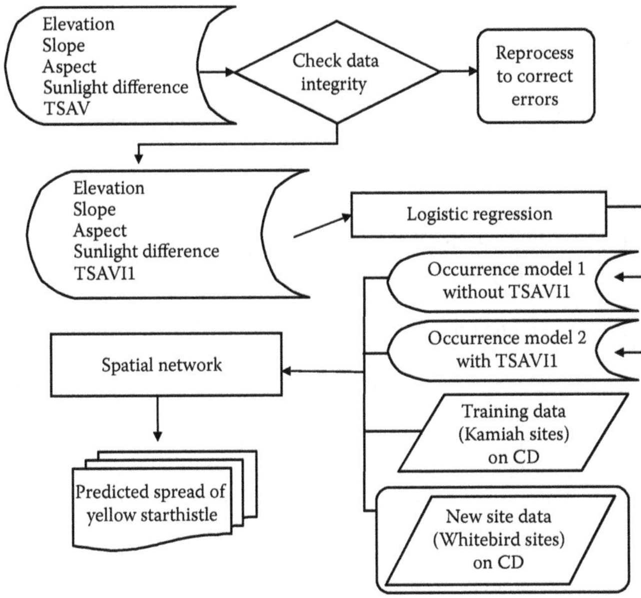

Invert the productivity model with TSAVI1 image using **image** calculator to make the **friction image** if you have not done so already.

Open **VARCOST** module under the **GIS analysis** pull-down menu and **Distance Operators** selection. Select the **source image** as the ***Kamiah 1981 survey data***, the **friction images** as the ***inverted productivity model with TSAVI1***, and the **direction image** as the direction of friction image. Enter an ***output image file name***. Define the **cosine function** as *2* (default) and click ***OK***.

Overlay the ***Kamiah 1995 survey data*** as **vector data** with a red line symbol by using **add layer** in the **image composer**. Change the **layer properties** of the new network image by reducing the **max display** to closely match the perimeter of the Kamiah 1995 survey data. If overlay of the Kamiah 1995 survey data match to predicted movement is poor try reducing the **cosine function** to 0.5 and run again.

Open **Contour** under the **GIS analysis** pull-down menu and **surface analysis** and **feature extraction** submenu. Input the ***file name*** of the image generated with **VARCOST** and the ***output file name***. Specify the **minimum contour** as *zero* and the **maximum contour** as the ***maximum value*** found when adjusting the **max display properties** to match those in Kamiah 1995 survey data. Set the **contour interval** as the ***maximum display value/14 years*** (1995 survey date–1981 survey date).[22]

Run **VARCOST** again using ***White Bird survey data*** as the **source** image. Develop contour lines based on the contour interval from Kamiah data.

Are some infestations expanding faster than other infestations? Which populations should be controlled first?

13.5 CONCLUSIONS

Prediction of yellow starthistle dispersal has important economic and managerial advantages. It would allow strategic planning efforts to focus on areas at highest risk and identification of defensible boundaries for optimizing limited management resources. The network model provided a reasonable assessment of dispersal of yellow starthistle given the known infestation in the specified area, and validated well in an independently larger area.

REFERENCES

1. Bakker, J.P., Poschlod, P., Strykstra, R.J., Bekker, R.M., and Thompson, K., Seed banks and seed dispersal: Important topics in restoration ecology. *Acta Botanica Neerlandica* 45, 461, 1996.
2. Grime, J.P., The humped-back model: A response to Oksanen. *Journal of Ecology* 85, 97, 1997.
3. Keddy, P.A., Population ecology on an environmental gradient: *Cakile edentula* on a sand dune. *Oecologia* 52, 348, 1982.
4. Adkison, G.P. and Gleeson, S.K., Forest understory vegetation along a productivity gradient. *Journal of the Torrey Botanical Society* 131, 32, 2004.
5. Grytnes, J.A., Fine-scale vascular plant species richness in different alpine vegetation types: Relationships with biomass and cover. *Journal of Vegetation Science* 11, 87, 2000.
6. Wisheu, L.C. and Keddy, P.A., Species richness-standing crop relationships along four lakeshore gradients: Constraints on the general model. *Canadian Journal Botany* 67, 1609, 1989.
7. Theodose, T.A. and Bowman, W.D., Nutrient availability, plant abundance and species diversity in two alpine tundra communities. *Ecology* 78, 1861, 1997.
8. Fu, B.J., Liu, S.L., Ma, K.M., and Zhu, Y.G., Relationships between soil characteristics, topography and plant diversity in a heterogeneous deciduous broad-leaved forest near Beijing, China. *Plant and Soil* 261, 47, 2004.
9. Benefield, C.B., DiTomaso, J.M., Kyser, G.B., and Tschohi, A., Reproductive biology of yellow starthistle: Maximizing late-season control. *Weed Science* 49, 83, 2001.
10. Maddox, D.M., Mayfield, A., and Poritz, N.H., Distribution of yellow starthistle (*Centaurea solstitialis*) and Russian knapweed (*Centaurea repens*). *Weed Science* 33, 315, 1985.
11. U.S.D.A., *U.S.D.A. Plants Database*. http://plants.usda.gov/, 2009.
12. Callihan, R.H., Sheley, R.L., and Thill, D.C., Yellow starthistle: Identification and control. Moscow, ID: University of Idaho, College of Agriculture. Current Information Series No. 634, 4pp., 1982.
13. Callihan, R.H., Smith, L., McCaffrey, J.P., and Michalson, E., Yellow starthistle management for small acreages. Moscow, ID: University of Idaho, Cooperative Extension System Bulletin. Current Information Series No. 1025, 8pp., 1995.
14. Roché, C.T. and Roché, B.F. Jr., Distribution and amount of four knapweed (*Centaurea* L.) species in eastern Washington. *Northwest Science* 62, 242, 1988.
15. Lass, L.W., McCaffrey, J.P., Thill, D.C., and Callihan, R.H., Yellow starthistle biology and management in pasture and rangeland. University of Idaho Bulletin Number 805, 18pp., 1999.
16. Cordy, D.R., Nigropallidal encephalomalacia (chewing disease) in horses on rations high in yellow star thistle. *Proceedings of the American Veterinary Medical Association* 91, 149, 1954.
17. Checke, P.R. and Shull, L.R., Other plant toxins and poisonous plants. In: *Natural Toxicants in Feeds and Poisonous Plants*. Westport, CT: The AVI Publishing Co., Chap. 11, p. 358, 1985.
18. Higgins, S.I. and Richardson, D.M., Predicting plant migration rates in a changing world: The role of long-distance dispersal. *The American Naturalist* 153, 464, 1999.

19. Shafii, B., Price, W.J., Prather, T.S., Lass, L.W., and Thill, D.C., Predicting the likelihood of yellow starthistle occurrence using landscape characteristics. *Weed Science* 51, 784, 2003.
20. Shafii, B., Price, W.J., Prather, T.S., Lass, L.W., and Thill, D.C., Using landscape characteristics as prior information for Bayesian classification of yellow starthistle. *Weed Science* 52, 948, 2004.
21. Shafii, B., Prather, T.S., Price, W.J., Lass, L.W., and Howard, D., Modeling dispersal of yellow starthistle in the canyon grasslands of North Central Idaho. In: *Proceedings of the Eighteenth Annual Kansas State University Conference on Applied Statistics in Agriculture [CDROM]*, April 30–May 2, 2006, Manhattan, KS, 2007.
22. Clark Labs. *IDRISI GIS Software*. Worcester, MA: Clark Labs, Clark University, 2004.
23. [ESRI] Software, version 8.2, *Four Parts: ArcINFO Work Station, ArcMap, ArcCatalog, and ArcToolbox*. New York: Environment Systems Research Institute, 2002.
24. Eastman, R.J., Kyem, P.A.K., Toledano, J., and Waigen, J., Explorations in geographic information system technology. In: *GIS and Decision Making*, vol. 4. Geneva, Switzerland: UNITAR, 1993.
25. Tian, F., Shafii, B., Williams, C.J., Prather, T.S., Price, W.J., and Lass, L.W., Prediction of yellow starthistle survival and movement over time and space. In: *Proceedings of the Sixteenth Annual Kansas State University Conference on Applied Statistics in Agriculture*, Kansas State University, Manhattan, KS, p. 74, 2004.

14 Using GIS to Map and Manage Weeds in Field Crops

Mary S. Gumz and Stephen C. Weller

CONTENTS

14.1 Executive Summary 301
14.2 Introduction 302
14.3 Materials and Methods 304
14.3.1 Uploading Images into ERDAS 306
14.3.2 Unsupervised Image Classification and Accuracy Assessment 306
14.3.3 Supervised Image Classification in ERDAS and Accuracy Assessment 309
14.4 Results 314
14.4.1 Crop Health Assessment 314
14.4.2 Weed Detection by Supervised Classification 315
14.4.3 On-Farm Use of GIS-Based Weed Mapping 315
14.5 Conclusions 316
Acknowledgments 316
References 317

14.1 EXECUTIVE SUMMARY

Peppermint (*Mentha piperita*) and spearmint (*Mentha spicata* and *Mentha cardiaca*) are high-value, essential oil crops in Indiana, Michigan, and Wisconsin and the Pacific northwestern states of Washington, Oregon, Idaho, Montana, and Northern California. Although the mints are profitable alternatives to corn and soybeans, mint production efficiency must improve in order to allow industry survival against foreign produced oils and synthetic flavorings. Weed control is the major input cost in mint production, and tools to increase efficiency are necessary. Remote sensing-based site-specific weed management (SSWM) offers potential for decreasing weed control costs through simplified weed detection and control from accurate site-specific weed and herbicide application maps. This research showed the practicality of remote sensing for weed detection in the mints. Unsupervised classification of multispectral images, which did not require reference field data, allowed for the characterization of mint crop health so that weak areas in the stand where weeds could potentially invade were easily detected. Supervised classification, which required field reference data (i.e., ground truthed observations) to train the classification, allowed for identification of weed-free crop areas as well as identification of individual weed species. Based on the results presented here, multispectral remote sensing can provide the weed location data necessary to develop GIS-based site-specific weed management programs for peppermint.

14.2 INTRODUCTION

Peppermint and spearmint (mint) are grown as essential oil crops in the United States in Indiana, Michigan, Wisconsin, and the Pacific Northwest. Mint oil is used as a flavoring in gum, candies, and oral hygiene products. Mint is grown as a short-term perennial crop in a 3–5 year rotation in the Midwestern United States. In the first year of production (row mint), mint stolons are planted in rows. The mint foliage is cut as a hay crop, and the hay is processed through steam distillation to harvest the essential oil. After mowing, mint regrows and is fall plowed for winter protection and then reemerges the following spring in a more even pattern across the field (meadow mint) where the process of harvesting is repeated. Spearmint is often double cut each year. First year, meadow mint is vigorous and develops a complete crop canopy more quickly than row mint. Meadow mint is harvested from late June through July while row mint is harvested 3–4 weeks later.[1]

Mint is a high-value crop; however, continued economic success for farmers growing this crop requires development of more efficient production methods that lower producer costs while maximizing yield. Mint is an at-risk crop due to increased competitive pressure from imported foreign mint oil, increased use of synthetic flavorings in value added products, and increasing field production costs.[2] Weeds are the worst pest control problem in mint production, and their management requires high-cost inputs.[2] Uncontrolled weed infestations can significantly decrease mint oil yield and result in contamination of the distilled oil even at populations below economic threshold levels for yield loss.[3] Weeds are difficult to control in mint since the meadow crop cannot be cultivated and manual weed removal is cost prohibitive. Herbicides are necessary to attain acceptable weed control; however, growers have only a few chemical options, and they are presently applied broadcast over the entire field. Vigorous stands of mint compete well with weeds; however, mint stands vary greatly in their vigor, and in most fields, there are areas where the less vigorous mint is more prone to weed infestations. Weed infestation patterns of this type would lend themselves to the use of SSWM, if such technology was dependable and affordable. SSWM uses spatial information to locate and treat specific weed infestations in a field and has potential in allowing growers to target herbicide applications to weed infested areas of the field, saving time and money in herbicide costs.[4] SSWM would allow for efficient spot application of effective postemergence herbicides, such as bromoxynil, which are not widely used due to phytotoxicity issues associated with broadcast applications.[2]

Since weed management in production fields is a major cost for growers, any improvement in weed control efficiency should allow higher profits. The development of practical and cost-effective remote sensing and SSWM technologies would address these needs; however, such technology needs to be developed.

The first step in development of effective SSWM programs is to be able to acquire detailed information about spatial distribution of weeds in a field.[4] Traditionally, the only means to accurately determine site-specific weed composition within a field was through intensive human-based field scouting or grid sampling. Recent advances in remote-sensing technology may prove a valuable asset in implementing site-specific weed management and reducing herbicide use in mint. Agricultural remote sensing uses an aerial (airplane or satellite-based) image of a crop field that can then be analyzed to identify variations in crop conditions, weed infestations, or other field anomalies.

Differences in spectral reflectance of different plant species allow identification of crops and weeds in an analyzed image. By spectrally analyzing different pixels in an image, an analyst can determine what the image is showing and where it is located in the field. Multispectral remote sensing measures spectral reflectance in 3–4 broad spectral bands (typically a green, a red, and a near-infrared [NIR] band).[5] There are many methods to analyze the information in remotely sensed images including supervised classification,[6] discriminant analysis,[7] spectral vegetation indices,[8] and wavelet transform.[9] Goel et al.[8] used ratios of the magnitude of spectral reflectance in the red and NIR portions of the spectrum to differentiate velvetleaf (*Abutilon theophrasti*) from corn in multispectral images. Koger et al.[7] differentiated grass mixtures (barnyardgrass [*Echinochloa crus-galli*], browntop millet [*Brachiaria ramosa*], and large crabgrass [*Digitaria sanguinalis*]) from soybeans in multispectral images using discriminant analysis of reflectance values. Koger et al.[7,9] also used waveform analysis of hyperspectral imagery to map infestations of barnyardgrass, large crabgrass, and pitted morningglory (*Ipomoea lacunosa*) in soybean fields. Hyperspectral remote sensing has been useful in mapping several rangeland weeds, including baby's breath (*Gypsophila paniculata*),[10] leafy spurge (*Euphorbia esula*),[11] and spotted knapweed (*Centaurea maculata*)[12] by analyzing color differences between the target weeds and background vegetation.

The research cited above suggests that remote sensing has potential to assist in production management in mint. This is based on the fact that mint plants have unique characteristics and production practices making them ideally suited for remote-sensing and site-specific weed control. Production fields tend to have patchy weed infestation patterns, the crop is a low-growing perennial crop, which seldom exceeds 1 m in height, which results in weeds, when present, often overtopping the crop. Problem weed species in mint, such as foxtails (*Setaria* spp.), *Amaranthus* spp., and white cockle (*Silene alba*)[2] differ greatly in color from each other and offer a sharp contrast to the uniform dark green to purplish green of a mint plant.

Remote-sensing images in mint should be able to identify variation in the crop stand, which, when analyzed, could direct growers to field areas where weed infestation is more likely to occur and may need preemergent and/or postemergent herbicide applications, resulting in savings in time and negating the need for in-field weed scouting. Another advantage of SSWM is that mint is harvested early in the summer (mid-June to early August) and allowed to regrow in late summer into the fall. In order to maintain a healthy crop stand in subsequent years, weeds must be monitored and managed after harvest when the crop is starting to regrow and is least competitive with emerging or existing weeds. Late summer weeds compete with the crop and if left uncontrolled produce many seeds. This situation leads to a perpetual renewal of the soil weed seed bank and continuous weed problems in subsequent years. Since mint is a high-value, high-input crop, potential increases in revenues and decreases in input costs with effective SSWM make development of these technologies economically feasible.[4]

The objective of this research was to determine if multispectral remote sensing was effective for detecting weed infestations in mint. The utility of multispectral imagery was evaluated in two different applications. The first application examined the use of imagery to create a crop health map of a weed-free peppermint field in order to identify low crop density areas. These types of data can help direct growers' weed management time and resources to those areas, since weeds are generally more prevalent in weak crop areas. The second application evaluated if late season imagery (obtained after harvest) could (1) identify weedy and non-weedy areas of the peppermint field and (2) allow for the differentiation of problem weed species. Information interpreted from a remotely sensed image is beneficial as it gives a grower a GPS-referenced map of a field showing anomalous areas, which can then be used as the basis for crop management decisions.[13]

14.3 MATERIALS AND METHODS

Experiments were conducted in rain-fed meadow peppermint in 2004. Aerial images were obtained of a 14 ha field of second year "Murray Mitcham" peppermint (South Bend) located west of South Bend, Indiana. The two predominant soil types were Antung muck in the north half of the field and Maumee mucky loamy fine sand in the south half. The field was managed by the grower during the season with standard production practices for fertility, pest control, and harvesting practices. The aerial image datasets for this chapter's case study are located on the CD in the Chapter 14 file folder. Copy these folders onto the C drive of the computer before continuing this case study.

Aerial multispectral images were acquired at South Bend on June 4 (Figure 14.1) and September 10, 2004 (Figure 14.2) by Precision Aviation, Rantoul, Illinois, using a multispectral camera flown at an altitude of 1800 m. Band centers were Band 1 (near [NIR]): 735–865 nm; Band 2 (red): 620–700 nm; and Band 3 (green): 510–590 nm. The spatial resolution at nadir was 0.71 m.

Ground reference observations were made on a 20 m grid pattern and mapped using ArcPad 6.0.2 on a handheld computer equipped with a compact flash-type GPS receiver. Differential GPS signal correction was provided via wide area augmentation system (WAAS). Field observations cataloged included weed species present, severity of weed infestations, vigor of crop stand, and the presence of disease or insects, nutrient deficiencies, or moisture stress. At the sampling points, crop vigor was visually rated on a 0–5 scale (0 = 75%–100% crop cover; 1 = 50%–75% crop cover; 2 = 25%–50% crop cover; 3 = 25%–50% crop cover; 4 = 1%–25% crop cover; and 5 = bare soil [no crop]). Weed infestation severity was visually rated on a 0–5 scale (0 = no weeds; 1 = one weed plant; 2 = 25% weed cover; 3 = 50% weed cover; 4 = 75% weed cover; and 5 = 100% weed cover). In addition to the grid observations, patches of weed infestations were mapped as polygons in the field. Weed patches

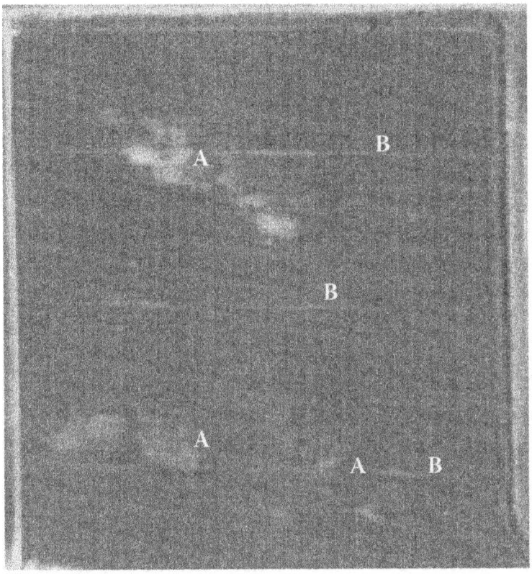

FIGURE 14.1 Grayscale infrared image (Bands 3, 2, 1) of a peppermint field near South Bend, Indiana, taken June 4, 2004. The image, taken 6 weeks prior to harvest, shows variation in the fitness of the peppermint stand. Healthy vegetation appears dark; bare soil appears white. Overall field health is good, but there are bare areas visible due to poor drainage (A) and dead furrows (B).

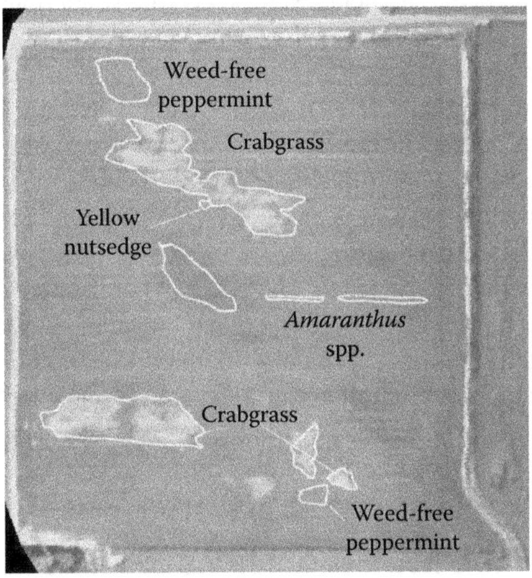

FIGURE 14.2 Grayscale infrared image (Bands 3, 2, 1) of a peppermint field near South Bend, Indiana, taken September 10, 2004, showing postharvest weed regrowth. Field scouting notes superimposed on the image show examples of locations where yellow nutsedge, mixed Amaranthus species, dead crabgrass, and weed-free occurred. Late season crabgrass infestations occurred where the crop was bare or thin and was not able to outcompete weeds.

consisted of areas of weed infestation where the canopy cover of the weeds was greater than that of the peppermint.

14.3.1 Uploading Images into ERDAS

Images on the CD were clipped from the original imagery for demonstration purposes. Images can be uploaded into ERDAS Imagine software from the Chapter 14 folder. (*Note:* ERDAS Imagine 9.3 was used for data analysis in this chapter; newer updated versions may be available.)

1. Open **ERDAS IMAGINE** software.
2. Click the **Viewer Tool** bar, and click the **Open File** icon.
3. On the **Select Layers To Add** tab
   a. Highlight ***file name (Chapter_14\images\mint_6_4.img).***
   b. Click on the **Raster Options** tab, select **pseudo-color**, and in the layer box, select ***3***. Click **OK** to close the **Raster Options** box.
4. Click **OK** in the **Select Layers To Add** box. The image will project in the **Viewer** area.

14.3.2 Unsupervised Image Classification and Accuracy Assessment

An unsupervised classification of peppermint crop health using the June 4 image was performed originally using ERDAS Imagine 8.6 image analysis software. Peppermint, at the time of imaging, ranged in height from 30 to 40cm. Early season weed control was excellent, and fields were weed-free when images were acquired.

In unsupervised classification, the spectral analysis program groups spectrally similar pixels into an arbitrary number of groups specified by the analyst. For ***n*** groups, the program calculates ***n*** centers equally spaced in central space and places pixels into the nearest group. The new spectral mean of each group is calculated and pixels reassigned into the class with the spectral mean closest to its own value. Unsupervised classification is an iterative process, and in each iteration, pixels may remain in their current group or are moved to a new group with a spectral mean closer to the value of the pixel. The process is repeated until a convergence factor, where a certain percentage of pixels remain unchanged, is reached.[14] In unsupervised classification, the input parameters defined by the operator are number of groups, maximum number of iterations, and the convergence factor. After all pixels are assigned to a group, the analyst labels the groups, based on information from ground referencing.

After uploading the image, to conduct unsupervised classification,

1. Select the **Classifier Icon** from the ERDAS main menu bar.
2. Select **Unsupervised Classification**.
3. Click the file icon next to the **Input Raster File**, and select ***Chapter_14\images\mint_6_4.img***. After the name is entered, click **OK**.
4. Uncheck the **Output Signature Set**.

5. Check the box next to **Output Cluster Layer**, click the file icon, and specify an ***output file name*** and location (e.g., ***Chapter_14\images\6_4_clstr8.img***).
6. Under **Clustering Options**, choose **Initialize from Statistics**, and enter ***8*** for **Number of Classes.**
7. Under **Processing Options**, enter ***25*** for Maximum Iterations and enter ***0.980*** for **Convergence Threshold.**
8. Click **OK** to process the image. When processing is complete, click **OK.** Then, use the image-loading procedure using **Viewer** tool to open the new file created above, and see the results of the classification.

After the unsupervised classification, the eight classes were labeled as one of four crop health categories: healthy, moderate, thin, and bare peppermint, based on the GPS-referenced ground scouting done in the field. While eight classes were originally chosen, the analysis revealed that these could be combined to four in order to give the analyst better control over the break points between categories. Therefore, repeat the procedure above selecting ***4*** as the number in the **Number of Classes** box, and change the ***output file name*** to a new file. After processing, upload the processed data by using the **Viewer** tool, and compare the two figures in adjacent **Imagine** image viewers.

The class distribution can now be examined by selecting **Raster** on the menu toolbar and choosing **Attributes.** Note that there are five attributes listed, although the image was clustered using four classes. One of the classes is labeled "**unclassified**," these are the pixels that did not fall into any of the classifier bins, and it is suggested that these pixels have a white color in the **Color** cell column. It is suggested that for the other attributes, the analyst changes the color in the **Color** cell column to sharply contrasting colors. The number of pixels that are in each attribute bin can be seen in the **Histogram** column. Ground-referencing data from the field can be used to help identify what features are represented by Classes 1, 2, 3, and 4. Class names can be changed by editing names in the **Class name** column. Based on the scouting data, Class 1, which contains 0 or few pixels, is bare soil. Classes 2, 3, and 4 can be renamed in a similar manner as 0%–25%, 25%–75%, and 75%–100% cover, respectively. Figure 14.3 shows the unsupervised classification map using four break points between classes.

Accuracy of the unsupervised classification within the field borders was assessed by comparing the generated classification to crop health ratings at sampling points manually assessed in the field. While not all the data needed to conduct an accuracy assessment for this image are given, Table 14.1 does present the accuracy assessment of the original dataset using the user's (also referred to as reference), producer's (also referred to as reliability), and overall accuracy. The user's or reference accuracy is defined as the probability of a pixel classified on the map that actually represents the category on the ground. The user's accuracy is calculated by dividing the correct pixels in a category (e.g., 68 correct pixels in the healthy crop category) by the total number (row total) of pixels classified within the category (total number of pixels in healthy category = 71 [3 pixels that should have been healthy were classified as moderate]). The user's or reference accuracy for the healthy category is then calculated as 68/71 or 96%.

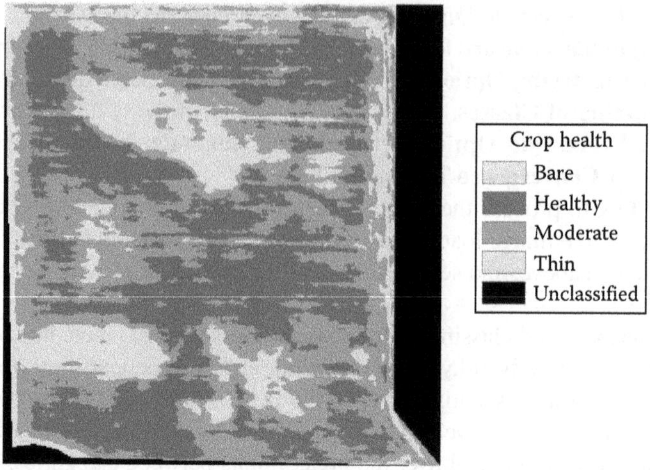

FIGURE 14.3 Unsupervised classification (gray scale) of multispectral image from Figure 14.1 showing four computer-generated crop health classes (healthy [>75% soil crop cover], moderate [25%–75% soil crop cover], thin [<25% soil crop cover], or bare [no crop]). Healthy areas, shown in dark gray, correspond to the darkest areas of the original field image (Figure 14.1). The image shows large areas are aggregated into similar crop health categories. Unclassified areas, in black, are areas outside the field boundary.

TABLE 14.1

Accuracy Assessment of an Unsupervised Classification of a Multispectral Image of a Peppermint Field Taken June 4, 2004, Showing the Predicted Crop Health Class of 93 Reference Points in the Field

		Predicted Group Membership					
		Healthy	Moderate	Thin	Bare	Row Total	Reference Accuracy (%)
Actual group membership	Healthy	68	3	0	0	71	96
	Moderate	2	8	0	0	10	80
	Thin	0	2	6	2	10	60
	Bare	0	0	0	4	4	100
	Column total	70	13	6	6		
	Reliability accuracy (%)	97	67	100	80		

Notes: Predicted crop health classes are shown in columns and actual crop health classes are shown in rows. Reliability accuracy of each class is the proportion of incorrect pixels excluded from a class. Reference accuracy of each class is the proportion of correct pixels included in each class. Overall accuracy was 92%. The Kappa statistic, which measures the probability that a classification is more accurate than a random assignment of pixels, was 81%.

The producer's or reliability accuracy gives an idea of how well a certain area is classified by the classification system. Producer's accuracy is computed by dividing the number of correct pixels (e.g., group membership in the moderate group is 8) by the column total (the total predicted group membership is 13 [some pixels representing healthy and thin areas were both classified in moderate category]; therefore, the reliability accuracy = 8/13 or 67%). The overall accuracy is calculated by dividing the total number of correctly identified pixels (i.e., 68 [healthy] + 8 [moderate] + 6 [thin] + 4 [bare soil] = 86) by the total number of classified pixels (95) (i.e., 86/95 = 92% accuracy).

14.3.3 Supervised Image Classification in ERDAS and Accuracy Assessment

When supervised classification is used, the analyst assigns pixels from areas representing known field conditions (such as healthy peppermint) to a spectral class, based on ground-referencing data. These are termed "**training pixels**" because they provide the spectral characteristics for the classifying algorithm to use in categorizing the remaining pixels in the image. Separate "**test pixels**" of known field areas are also chosen to test the validity of the classifying algorithm. The original supervised classification of the peppermint field from the September 10 image is shown in Figure 14.4. The field had postharvest regrowth of peppermint, and several weed species were present. The classification was done originally using ERDAS Imagine 8.6, and healthy peppermint and five other classifications were distinguished. The classification in this chapter was conducted in ERDAS 9.3, and

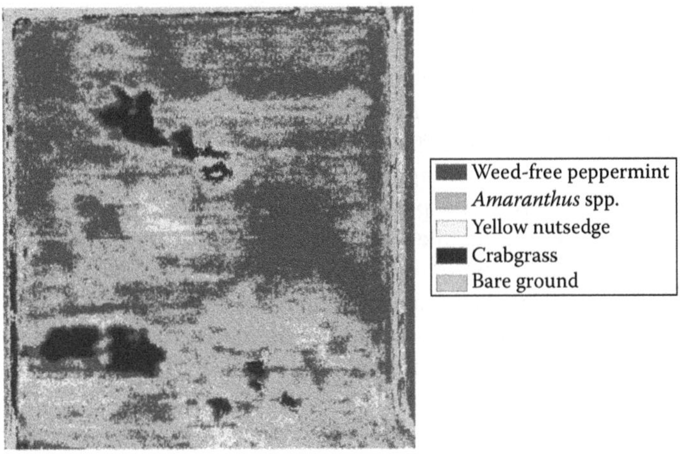

FIGURE 14.4 Original supervised classification of the multispectral image of peppermint field taken September 10 during crop and weed regrowth following harvest. Supervised classification placed image pixels into five categories: weed-free peppermint, mixed *Amaranthus* species, yellow nutsedge, crabgrass, and bare soil with 96% accuracy. Areas of weed infestation correspond with thin or bare areas of the stand identified in the June 4 image (Figures 14.1 and 14.2).

FIGURE 14.5 September 10 image uploaded into ERDAS.

to simplify the process for demonstration, the number of classification groups has been reduced to four from the original work. To conduct a supervised classification,

1. Upload the ***September 10 (Chapter_14\images\mint_9_10.img)*** image from the C drive into ERDAS as described above. On the **Select Layers To Add** tab,
   a. Highlight ***file name (Chapter_14\images\mint_9_10.img).*** and then
   b. Click on the **Raster Options** tab, select **pseudo-color**, and in the layer box, select ***3*** and check the boxes next to **Clear Display** and **Fit to Frame**. Click **OK** to add the image to the viewer (Figure 14.5).
2. In the **Viewer** window, select **File** and select **Open** and **Vector Layer**. Change **Files of Type** to **Shapefile** (***.shp**). Navigate to ***Chapter_14\ground_truth\weed_831.shp***, which was uploaded onto the C drive from the CD. Click **OK**. Note that polygons appear in the viewer, and these polygons will be used to create a signature file for image classification (Figure 14.6).
3. Click **Vector** in the **Viewer** menu bar, and select **Symbology** to open the **Symbology editor** where polygons can be symbolized based on attribute data. In the **Symbology editor**, select **Automatic** from the menu bar, and choose **Unique Value** from the option list. Select **Weed** in the value field and click **OK**. Four classes appear.
4. Click the bar under **Symbol** next to ***Amaranthus***, and select **Outlined Yellow**. Repeat this procedure for the other classes choosing unique colors for each class (bare soil, healthy peppermint, crabgrass). Click **Apply** but leave the **Symbology editor** open. Polygons will now have unique color outlines for each class (Figure 14.7).
5. On the main ERDAS menu bar, click **Classifier** and select **Signature Editor**. In the **Viewer** menu, select **AOI** (area of interest) and then choose

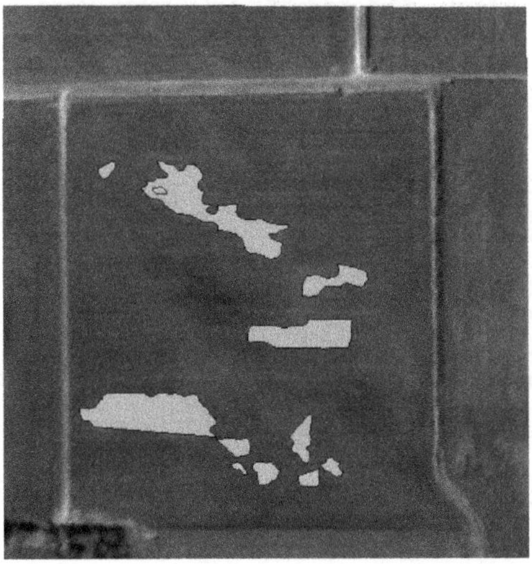

FIGURE 14.6 Polygons that appear on the image in viewer after shapefiles are uploaded.

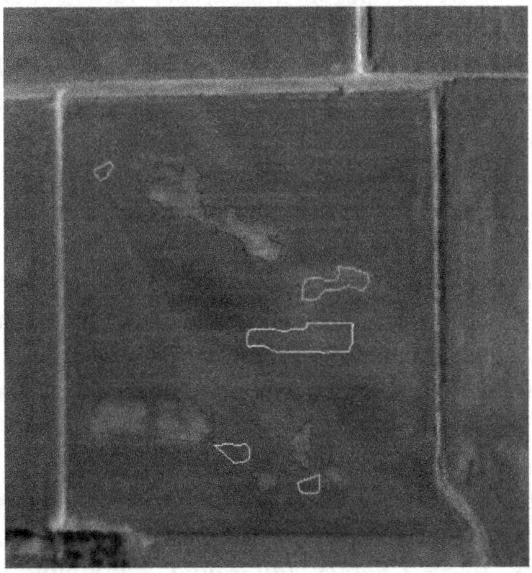

FIGURE 14.7 Outlines of polygons for classified data. Each type of polygon (e.g., bare soil, *Amaranthus*, etc.) should be outlined in a unique color for easy differentiation.

Tools. The **Polygon Tool** in the **AOI** dialog box will be used to select appropriate areas for classification. Click the zoom-in (+) icon in the **Viewer** tool bar, and navigate to an area close to the ***bare soil*** polygon (Figure 14.8).

Note that this polygon around a dark cluster of cells is slightly offset from the lighter cells that suggest bare soil. As the analyst, you must decide if the polygon surrounds an appropriate set of pixels for classification or if the polygon should be moved to surround a different cluster of cells, as the classification relies on good input information to accurately describe the

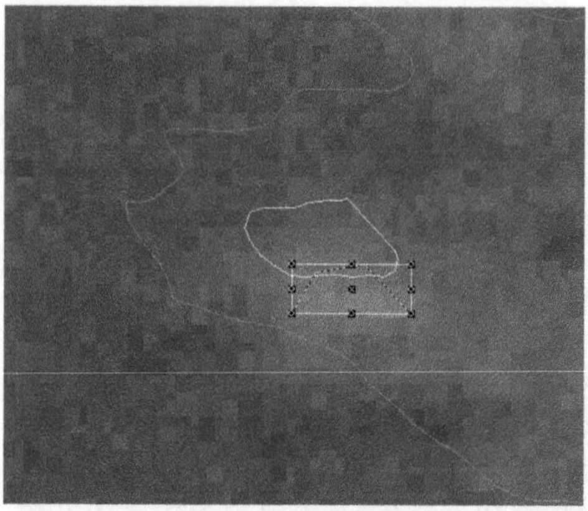

FIGURE 14.8 Close-up view of pixels in the original bare soil polygon and a new polygon created by the analyst to surround light-colored pixels.

rest of the image. In this case, it was decided to draw a new polygon around the lighter cells. To do this, click on the **Polygon** tool in the **AOI** dialog box and drag to the desired shape, clicking at each vertex, double click to close the polygon (if the polygon is not closed, the entire image will be selected).

6. In the **Signature** editor, click the **Create New Signature from AOI** icon, and a new signature class appears. Click the cell below **Signature Name** and enter ***bare soil***.
7. In the **Viewer** frame, click the zoom-out (-) icon. Locate the next polygon type (e.g., peppermint) and zoom-in and create another **AOI** using the **polygon** tool in the approximate region of the reference polygon. In the signature editor, **Create New Signature from AOI**, click in a new cell, and add the name for that polygon. Repeat these procedures until the four class signatures have been created.
8. In the **Signature Editor** menu bar select **File**, click **Save**, navigate to ***Chapter_14\Ground_Truth***, and save the file as ***mint_9_10.sig***. Close the **Signature**, **Symbology**, and **AOI** editors.
9. Click the **Classifier** icon and select **Supervised Classification**. In the **Input Raster File**, select the ***Chapter_14\images\mint_9_10.img***. In the **Input Signature File**, select ***Chapter_14\Ground_Truth\mint_9_10.sig***. In the **Classified File**, enter ***C:\Chapter_14\Images\mint_9_10_sc.img.*** Click **OK**. Set the **Nonparametric Rule** to **None** and the **Parametric Rule** to **Maximum Likelihood**. Click **OK**. After the image classification is finished, click **OK**.
10. Click on the **Viewer** icon to open a new viewer window. Load the ***mint_9_10_sc*** imagery, as previously described.
11. Click on **Raster** in the **Viewer** menu bar, and select **Attributes**. Change each class symbol to an appropriate color by clicking on the bar below **Color**. The image should resemble that of Figure 14.9.

In the original paper, five classes (weed-free peppermint, mixed *Amaranthus* species, dead crabgrass, yellow nutsedge [*Cyperus esculentus*], and bare ground) chosen

FIGURE 14.9 Supervised classification based on four classes. *Note:* Differences may occur for each conducted analysis based on pixels selected for classification.

for the South Bend image were based on the predominant weed species in the field (compare Figure 14.4 and your final figure, shown here as Figure 14.9). **Test pixels** from the original dataset were then assigned to the category with the highest probability of membership.[5] Test pixels, those not used in creating the class means and covariance, are used to verify the classifier's accuracy. The classification accuracy was determined by comparing the predicted class membership of the test pixels to their actual class membership as determined from ground referencing. In supervised classification, all pixels in the image are classified, resulting in a georeferenced map of weed-free and weedy areas of the field. While the data to conduct the accuracy are not provided, the general method to conduct an **accuracy assessment** of the **Supervised Classification** is as follows:

1. Select **Accuracy Assessment** from the **Classification** dialog box.
2. Select **Edit—Create/Add Random Points**. Set search count to ***95***. Click **OK** to generate points. The generated points will show up on the image. Only pixels that were not used as training pixels can be used in the accuracy classification.
3. Using the GPS-referenced field observations, enter the ***class*** in the **Reference Column**. In the **Accuracy Assessment** viewer, select **Edit—Show Class Value**. The class each point was assigned to by the supervised classification will appear in the **Class Column**.
4. In the **Accuracy Assessment** viewer, select **Report—Options**. Select the checkboxes next to **Kappa Statistic**, **Error Matrix**, and Accuracy Totals. Select **Report—Accuracy Report** and **Report—Cell Report**. The accuracy report can either be viewed in the **Imagine Text Editor** or **saved** as a text file. The results from the accuracy assessment of the original data are presented in Table 14.2.

TABLE 14.2
Accuracy Assessment of a Supervised Classification of a Multispectral Image of a Meadow Peppermint Field Taken September 10, 2004, Showing the Predicted Crop or Weed Class of 331 Test Pixels in the Field (Weed-Free Peppermint Regrowth, Yellow Nutsedge, Mixed *Amaranthus* spp., or Bare Ground)

		Predicted Group Membership					
		Weed Free	Yellow Nutsedge	*Amaranthus* spp.	Bare Ground	Row Total	Reference Accuracy (%)
Actual group membership	Weed-free	226	0	9	0	235	96
	Yellow nutsedge	0	10	0	0	10	100
	Amaranthus spp.	4	1	24	0	29	83
	Bare Ground	0	0	0	57	57	100
	Column total	230	11	33	57	331	
	Reliability accuracy (%)	68	91	70	100		

Notes: Predicted classes are shown in columns and actual classes are shown in rows. Reliability accuracy of each class is the proportion of incorrect pixels excluded from a class. Reference accuracy of each class is the proportion of correct pixels included in each class. Overall accuracy was 96%. The Kappa statistic, which measures the probability that a classification is more accurate than a random assignment of pixels, was 95%.

14.4 RESULTS

14.4.1 Crop Health Assessment

The multispectral image obtained on June 4, 2004 allowed visualization of variations in meadow (Figure 14.1) peppermint growth. Healthy vegetation appears dark (dark red in color image on CD), and bare soils appear light (green in color image) in the adjusted gray scale image. Highly reflective surfaces such as gravel roads surrounding the field appear light blue in the color image on the CD. The overhead view of the whole field captured by the aerial images showed variations in field vigor, such as patches of thin peppermint surrounded by healthy peppermint or vice versa, not readily apparent from the horizontal vantage point of the ground scouting observations. The images vary slightly in color from each other due to atmospheric differences (e.g., differences in sun angle, time of day, or humidity) between the two observation dates.

The June 4 meadow peppermint image had low spatial variation (Figure 14.1) as healthy peppermint was grouped into large areas (dark) and poor growth or bare ground areas were discrete patches (light colored). Unsupervised classification

(Figure 14.2) placed pixel values from the multispectral image into four discrete crop health categories, healthy, moderate, or thin peppermint or bare ground. There is also an unclassified category for the outside edges of the image. Accuracy of the classification within the field borders was assessed by comparing the computer classification to crop health ratings at sampling points manually assessed in the field (Table 14.1).

For each class, the accuracy matrix calculates two accuracies: (1) a reliability accuracy, which is a measure of the ability of the classification to identify a given class and exclude pixels from other classes, and (2) a reference accuracy, which measures the ability of the classification to include all pixels from a given class into their correct class (as noted from ground scouting notes). Overall accuracy, the proportion of correctly identified pixels, was 92% using this method. In addition to assessing class and overall accuracy, the Kappa statistic, which measures the probability that a classification is more accurate than a random assignment of pixels,[5] was 81%. Misclassified pixels were placed in one class above or below its actual class (e.g., a moderate crop health pixel misclassified as thin crop health). Gross misclassifications, such as a thin or bare pixel being classed as healthy, did not occur.

14.4.2 Weed Detection by Supervised Classification

Postharvest multispectral images were taken of the South Bend meadow peppermint field on September 10, 2004 (Figure 14.3) to allow determination of the extent of weed regrowth following harvest. Ground-referencing notes taken at the same time as the images were superimposed over the aerial images and used to determine where actual weed patches occurred. Meadow peppermint was harvested on July 20, and after harvest, considerable weed regrowth occurred, particularly in areas where the peppermint stand had been poor earlier in the season. Weed infestations included mixed *Amaranthus* species (primarily *Amaranthus powellii* and *Amaranthus hybridus*), large crabgrass, and yellow nutsedge. Both peppermint and weeds were 20–30 cm tall at the time of image acquisition. On August 31, quizalofop was applied by the grower to control crabgrass, and at the time of image acquisition, the patches of dead crabgrass in the field were easily discernible. The field was classified originally based on five classes (yellow nutsedge, dead crabgrass, *Amaranthus* spp., bare soil, and weed-free peppermint regrowth) by supervised classification to yield a classification map (Figure 14.9) with >96% overall accuracy (Table 14.2) with a Kappa statistic of 95%.

14.4.3 On-Farm Use of GIS-Based Weed Mapping

Since many Midwest peppermint producers also farm large acreages of agronomic crops,[2] a reduction in time spent on weed management in peppermint would be of benefit to these growers. Unsupervised classification maps could also lead to better use patterns of herbicides such as terbacil, the most widely used preemergence herbicide in peppermint.[2] Terbacil is very effective, but most growers try to limit its use, if possible, in the latter stages of a peppermint rotation to minimize potential damaging soil carryover to following rotational crops. For example, a grower in the third year of a peppermint rotation who wanted to minimize terbacil applications in order to prevent carryover to rotational crops could use the maps shown in Figure 14.2 to determine herbicide placement. Terbacil could then be applied only to areas with weaker crop stands but greater weed pressure, while areas of strong, vigorous, and more competitive peppermint could be left untreated since crop competition would provide weed suppression. This "spray–no spray" decision aid for peppermint production is similar to that employed by cotton growers who base their rate of defoliant on an unsupervised classification of crop biomass.[15]

A weakness of the unsupervised classification technique, however, is the need for a weed-free field image at the time of the analysis. Unsupervised classification clusters pixels showing similar amounts of green vegetation. Weeds present in the image would not be differentiated from a crop in an unsupervised classification. Supervised classification can be used to identify problem weed species in a current production field from an in-season image that would offer growers assistance in selecting the most effective postemergence herbicide for each field. An example of the usefulness of supervised classification would be the placement of bromoxynil in a field. While bromoxynil is labeled for postemergence application in peppermint and is effective in controlling several *Amaranthus* species, it is not widely used because of potential crop phytotoxicity[2] and, thus, is not an herbicide well suited for broadcast application across an entire peppermint field. Maps with areas that depict weed infestations, such as those shown in Figures 14.2 through 14.4, would provide information to growers about where in the field to apply bromoxynil (i.e., only to areas infested with *Amaranthus* species).

14.5 CONCLUSIONS

The overall results show that multispectral images can be used to locate weed infestations in peppermint by either unsupervised classification of crop vigor early in the season that identifies areas where weeds are likely to occur or supervised classification that categorizes weed species into different groups. Both methods are useful; however, unsupervised classification of crop health does not require extensive ground referencing and can quickly and accurately direct a peppermint producer to the areas where weed management resources would be most effective. In this experiment, a map of crop health developed by unsupervised classification of multispectral images had an accuracy of 90%. This result was possible without inputting any ground reference information (although ground reference data were used in the accuracy assessment). Unsupervised classification can render a map of potential weed infestation (based on crop vigor) that can then be used as the framework for a GIS-based system of weed management.

ACKNOWLEDGMENTS

Funding was provided by a USDA Block Grant and the Indiana Mint Market Development and Research Council.

REFERENCES

1. Weller, S.C., Green, R., Janssen, C., and Whitford, F., Mint production and pest management in Indiana, Purdue Pesticide Programs, Bulletin PPP-103, 16, 2000.
2. Pest Management Strategic Plan for the Indiana, Wisconsin, and Michigan Mint Industries (PMSP), NSF Center for IPM, 2002, Web page: http://pestdata.ncsu.edu/pmsp/pdf/MidwestMintpmsp.pdf/ (verified February 16, 2007).
3. Heap, I., The effect of weeds on mint oil yield and quality. Special Report, Department of Crop and Soil Science, Oregon State University, Corvallis, OR, p. 27, 1996.
4. Swinton, S.M., Economics of site-specific weed management, *Weed Sci.*, 53, 259, 2005.
5. Lillesand, T.M., Kiefer, R.W., and Chipman, J.W., *Remote Sensing and Image Interpretation*, 5th edn., John Wiley & Sons, New York, p. 724, 2004.
6. Gibson, K.D., Dirks, R., Medlin, C.R., and Johnston, L., Detection of weed species in soybean using multispectral digital images, *Weed Technol.*, 18, 742, 2004.
7. Koger, C.H., Shaw, D.R., Watson, C.E., and Reddy, R.N., Detecting late-season weed infestation in soybean (*Glycine max*), *Weed Technol.*, 17, 696, 2003.
8. Goel, P.K., Prasher, S.O., Patel, R.M., Smith, D.L., and DiTommaso, A., Use of airborne multi-spectral imagery for weed detection in field crops, *Trans. ASAE*, 45, 443, 2002.
9. Koger, C.H., Bruce, L.M., Shaw, D.R., and Reddy, K.N., Wavelet analysis of hyperspectral reflectance data for detecting pitted morningglory (*Ipomoea lacunosa*) in soybean (*Glycine max*), *Remote Sens. Environ.*, 86, 108, 2003.
10. Lass, L.W., Prather, T.S., Glenn, N.F., Weber, K.T., Mundt, J.T., and Pettingill, J., A review of remote sensing of invasive weeds and example of the early detection of spotted knapweed (*Centaurea maculosa*) and babysbreath (*Gypsophila paniculata*) with a hyperspectral sensor, *Weed Sci.*, 53, 242, 2005.
11. Williams, A.P. and Hunt, E.R., Estimation of leafy spurge cover from hyperspectral imagery using mixture tuned matched filtering, *Remote Sens. Environ.*, 82, 446, 2002.
12. Lass, L.W., Thill, D.C., Shafil, B., and Prather, T.S., Detecting spotted knapweed (*Centaurea maculosa*) with hyperspectral remote sensing technology, *Weed Technol.*, 16, 426, 2002.
13. Gao, J., Integration of GPS with remote sensing and GIS: Reality and prospect, *Photogramm. Eng. Remote Sens.*, 68, 447, 2002.
14. Jensen, J.R., Unsupervised classification. In: *Introductory Digital Image Processing: A Remote Sensing Perspective*, Prentice-Hall, Inc., Upper Saddle River, NJ, p. 331, 1996.
15. Plant, R.E., Munk, D.S., Roberts, B.R., Vargas, R.L., Rains, D.W., Travis, R.L., and Hutmacher, R.B., Relationships between remotely sensed reflectance data and cotton growth and yield, *Trans. ASAE*, 43, 535, 2000.

15 Adapting Geostatistics to Analyze Spatial and Temporal Trends in Weed Populations

Nathalie Colbach and Frank F. Forcella

CONTENTS

15.1 Executive Summary..320
15.2 Introduction..320
15.3 Analysis Steps..321
15.4 Data Collection..322
15.4.1 Experimental Field..322
15.4.2 Sampling Grid..322
15.5 Exploratory Data Analysis..323
15.5.1 Objective..323
15.5.2 Method..324
15.5.3 Results..325
15.6 Data Transformation..330
15.6.1 Objective..330
15.6.2 Method..330
15.6.3 Results..330
15.7 Detrending Data..330
15.7.1 Objective..330
15.7.2 Median Polishing..330
15.7.3 Estimating Trend with Linear Regressions..331
15.8 Empirical Semivariograms..334
15.8.1 Objective..334
15.8.2 Method..335
15.8.3 Results..335
15.9 Semivariogram Model Fitting..335
15.9.1 Objective..335
15.9.2 Method..340
15.9.3 Results..340
15.10 Analysis of Variogram Parameters..346
15.10.1 Objective..346

15.10.2 Method....346
15.10.3 Results....354
15.11 Kriging....354
15.11.1 Objective....354
15.11.2 Method....356
15.11.3 Results....356
15.12 Cross-Semivariograms and Cokriging....356
15.12.1 Objective....356
15.12.2 Cross-Semivariograms....360
15.12.3 Cokriging....360
15.13 Error Analysis....363
15.13.1 Prediction of Weed Means....363
15.13.2 Prediction of Weed Locations....363
15.14 Summary: Using Geostatistical Information for Decision Making....369
Glossary....369
References....370

15.1 EXECUTIVE SUMMARY

Geostatistics were originally developed for the mining industry to estimate the location, abundance, and quality of ore over large areas from soil samples to optimize future mining efforts. These methods have been adapted for many different situations. In this chapter, geostatistics are used to examine the weed distribution inside a single production field, variations of distribution over time, skewed data distributions, and correlations with species traits. A geostatistical study starts with selecting a sampling plan that "catches" the spatial relationships among the variables of interest. Exploratory data analysis then examines data distributions and checks whether the prerequisites for a geostatistical analysis are fulfilled. If necessary, data are transformed and detrended to meet these prerequisites. Then, empirical semivariograms are calculated and used to (1) explain small-scale spatial trends (e.g., weed patch shapes and progress in time as a function of species dispersal and germination traits), (2) determine the variances for unsampled distances to allow prediction of values in unsampled points and maps to be plotted, using kriging, and (3) reduce estimation errors at unsampled points. Cross-semivariograns and cokriging describe covariation of variables in space, and these relationships are used to estimate a sparsely sampled primary variable with the help of an extensively sampled secondary variable. Here, these methods were adapted to predict weed maps with past observations and variograms. Last, error analysis evaluates how close predicted weed maps are to observations and the risk of spraying insufficiently or unnecessarily when basing herbicide spraying in precision agriculture on weed maps predicted with past observations using cokriging.

15.2 INTRODUCTION

Geostatistics, originally developed for the mining industry, have been adapted to natural resource management, climatology, and ecology. The application of geostatistics to ecology and related disciplines is hindered by extremely skewed data resulting from aggregation in vegetation. For instance, weeds occur in patches[1–9] because they tend to cluster where conditions, such as nutrient and soil moisture availability, are favorable for growth, persistence of propagule banks, and limited short distance dispersal of seed and propagules.[4,10]

In addition, the relevant working scale of a geostatistical application in an agricultural application can be quite different from the original large-scale mining applications. For many weed species, their abundance in a given field can mostly be explained by the past field history,[11] without any relationship to neighboring fields. Moreover, weed densities and locations, unlike ore deposits, vary over time. Indeed, fields are mostly infested by annual species,[11] which emerge, grow, reproduce, and die inside a single crop season. The success of individual species depends on weather as well as the crop sown by the farmer and on the choice of control techniques (e.g., herbicide and tillage) that interfere with the weed life cycle.[12,13] Last, agronomists and ecologists want not only to predict the future of weed populations but also to understand and explain these dynamics. The objective of this chapter is to demonstrate how geostatistics can be used to (1) describe small-scale spatial trends in weed populations and create weed contour maps, (2) relate the observed spatial variability to the weeds biological characteristics, (3) describe the development of the spatial trends over time and to predict the location of future weed patches, and (4) evaluate the usefulness of actual and predicted weed maps for site-specific herbicide spraying.

15.3 ANALYSIS STEPS

A geostatistical study starts with data collection, choosing an adequate sampling plan to "catch" the spatial relationships to be studied (Section 15.4). Exploratory data analysis then looks at data distributions and checks whether the prerequisites (i.e., normal or at least unskewed distribution, independence of mean and variance) for a geostatistical analysis are fulfilled (Section 15.5). If necessary, data are transformed to meet these prerequisites (Section 15.6). The next step checks for trends, i.e., large-scale spatial relationships, in data and tries to remove these with additional data transformations (Section 15.7). Only then are calculated empirical semivariograms to describe small-scale spatial trends, i.e., the variance of a variable between locations as a function of the distance between these points (Section 15.8). Fitting models to these data (Section 15.9) makes possible the estimation of variogram parameters, which can then be correlated to other data to explain small-scale spatial trends (Section 15.10). Semivariogram models are also used to estimate variance for unsampled distances and, thus, to plot maps with kriging (Section 15.11). Cross-semivariograns and cokriging (Section 15.12) describe the variance between two variables as a function of the distance between the sampling points and then use these relationships to estimate a sparsely sampled primary variable with the help of an extensively sampled secondary variable. Last, error analysis compares measured values with predicted values (Section 15.13).

Several software programs were used in the present work. Simple data analyses were carried out with Excel or SAS.[14] All the SAS program samples shown here were written for a UNIX environment (see an SAS manual,[14] for specific SAS script instructions). SAS was always run in the same directory where the data file was located; data files were obtained by saving the relevant excel sheets as text files (***prn*** files, using blanks as separators). In the boxes that contain SAS code, procedure names, options, etc. are shown in black, variable names, etc. chosen by the user are in ***bold italics*** (in SAS code, these will be blue) and comments are in *italics* (in SAS code, these will be green). SAS also was used for calculating empirical semivariograms, fitting variogram fittings, and parameter analyses. GSLIB[15] (DOS version) was used for calculating empirical semivariograms and cross-semivariograms, kriging, and cokriging and maps. GSLIB is a free program that can be downloaded in several versions including DOS and a newer Windows (WinGslib version 1.5.6) version from the Web site http://www.gslib.com/. (*Note:* Program routines given in boxes are for GSLIB DOS version; however, these same parameters are asked for and used as inputs into the appropriate boxes for the WinGslib version.)

15.4 DATA COLLECTION

15.4.1 EXPERIMENTAL FIELD

The details can be found in Colbach et al.[16] with only the main points repeated here. A field survey of weed seedling populations was conducted from 1993 to 1997 at the Swan Lake Research Farm, Stevens Co., Minnesota. The field was 54 m wide (east–west) and 244 m long (north–south). Each year in mid-May, the field was planted with soybean, using a no-till planter. Rows were spaced 0.76 m apart and oriented north to south axis of the field. Weeds were treated with postemergence herbicides and interrow cultivation. An additional application of glyphosate was applied to control perennial weeds, either before planting or after harvest.

15.4.2 SAMPLING GRID

The sampling grid must be adapted to capture the spatial relationships that are to be analyzed. The minimum sampling distance must be carefully chosen, with this distance considerably lower than the distance at which spatial relationships are expected. Usually, regular sampling grids are preferred to avoid subconscious choices of the assessor (e.g., a preference for large and numerous plants) when placing quadrats in random sampling plans. In this study, weed seedlings were identified and counted in permanent 0.1 m^2 quadrats that covered both the crop-row and the interrow areas. Assessments were carried out once a year (except in 1995 when the data were not collected) prior to postemergence herbicide applications. Weed density by species was obtained at the same locations each year. Beginning at the field margins, samples were collected at 410 locations located on regular sampling grid with 10 rows and 41 columns (Figure 15.1).

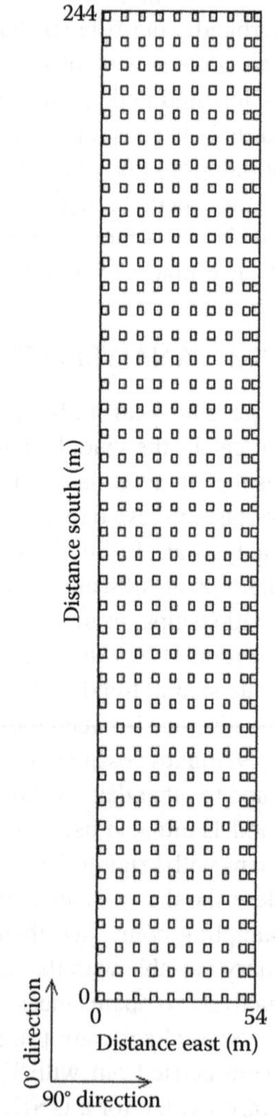

FIGURE 15.1 Sampling grid at the Swan Lake experiment station.

TABLE 15.1
Extract of Data File Found on CD in Chapter 15 Folder Labeled *Chapter_15_data.xls* for the Worksheet *Data1993*

xlocation	ylocation	agrre	amare	ascsy	cheal	cirar	setvi	sinar
6.1	0	50	10	0	30	0	50	0
6.1	6.1	10	0	0	0	0	20	0
6.1	12.2	0	10	0	0	0	0	0
6.1	18.3	0	10	0	20	0	0	0
6.1	24.4	0	0	0	0	0	20	0
6.1	30.5	0	0	0	0	0	10	0
6.1	36.6	0	10	0	0	0	60	0
6.1	42.7	0	0	0	0	0	10	0
6.1	48.8	0	0	0	0	0	20	0
6.1	54.9	0	0	0	0	0	20	0
[...]								

Note: The file contains several other worksheets that have weed density data by species and x and y locations from the Swan Lake Research Farm, Stevens Co., Minnesota.

Distances between grid points were 6.1 m in both the x and y directions, except between the last two sample rows that were separated by 3.05 m. Seven plant species were identified and analyzed in detail. These included redroot pigweed (*Amaranthus retroflexus* L.), common milkweed (*Asclepias syriaca* L.), wild mustard [*Brassica kaber* (DC.) L.C. Wheeler (=*Sinapis arvensis* L.)], common lambsquarters (*Chenopodium album* L.), Canada thistle (*Cirsium arvense* (L.) Scop.), quackgrass [*Elytrigia repens* (L.) Nevski], and green foxtail [*Setaria viridis* (L.) Beauv.].

In this study, x and y always indicate spatial coordinates (m), and z, the dependent variable, describes weed density (plants/m^2). Table 15.1 shows an extract of the ***Chapter_15_data.xls*** data file for the ***Data1993*** worksheet (found on the CD in Chapter 15 file) containing the weed densities counted on the sampling grid. Figure 15.2 shows an example of densities counted for *S. viridis* in 1993 with the sampling grid shown in Figure 15.1.

15.5 EXPLORATORY DATA ANALYSIS

15.5.1 Objective

The conditions for geostatistical analysis are that the values be normally distributed and independent. Quite often, this is not the case, especially when looking at vegetation data such as weeds. Weed densities tend to be skewed, with a large number of sampling points having zero or very small values (Figure 15.3). One of the first data analysis steps is to determine the frequency distribution and, if necessary, transform the data to better fit a normal distribution. This step involves calculating the mean, variance, skewness, kurtosis, and a test for normality. In addition, mean and variances

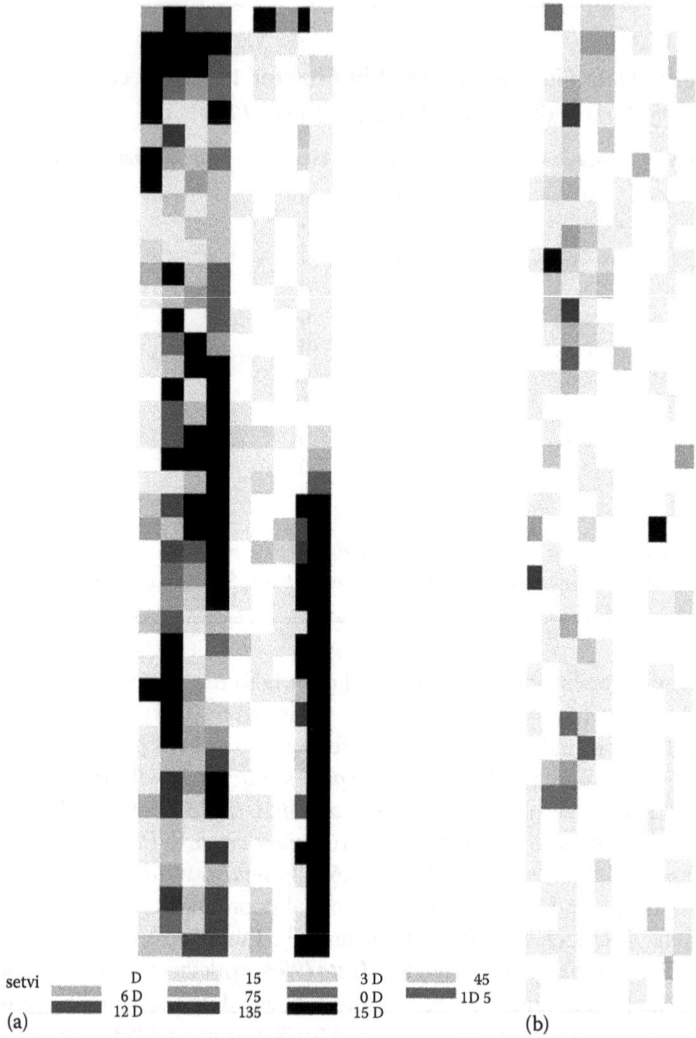

FIGURE 15.2 Weed densities counted in the sampling quadrats at the Swan Lake experimental field. Example of *S. viridis* distributions in 1993 (a) and 1997 (b). Graph obtained with **PROC GMAP** of **SAS** using data from ***Data1993*** and ***Data1997*** worksheets for ***Chapter_15_data.xls*** (see Box 15.3).

were calculated along rows and columns, and a linear regression was used to check for a possible correlation between $\log_e$(variance) and $\log_e$(mean)[18] (Box 15.2).

15.5.2 Method

There are a range of different software applications that achieve these calculations and tests. In Excel, the statistical functions, **mean**, **standard error**, and **skew**, can be found in the drop-down menu Insert/Functions. In the Excel file available on the

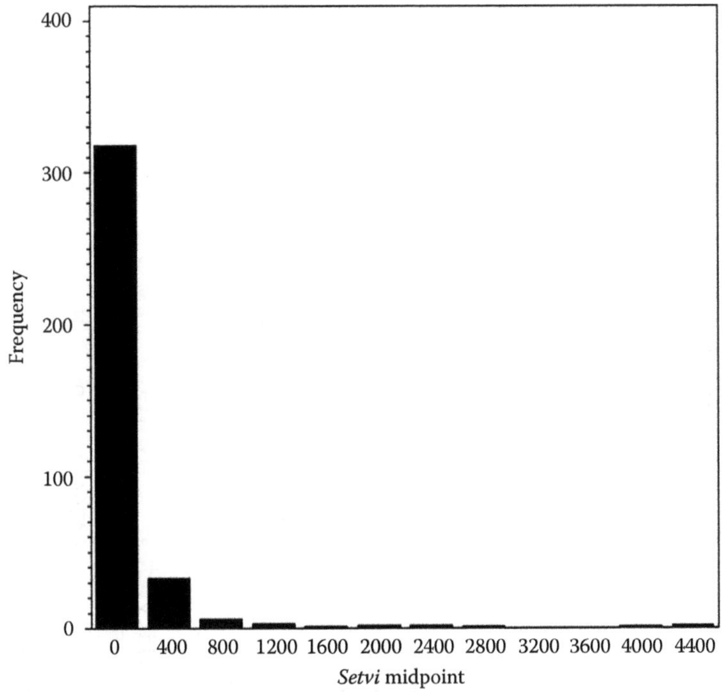

FIGURE 15.3 Distribution of *S. viridis* densities counted on the Swan Lake experimental field in 1993. Graph produced with **PROC GCHART** of **SAS**.

CD for Chapter 15 with the title ***Chapter_15_data.xls*** data, some of these variables have been calculated and placed on top of each data sheet. In SAS, these variables can all be calculated with the **PROC UNIVARIATE** function (Box 15.1). (*Note:* remember that the ***xls*** files were converted to ***prn*** files for upload into SAS, and, if using other file formats, use appropriate infile or datafile variables; see Section 15.3.) Box 15.1 shows part of the output produced by this function for ***SETVI***. To examine the correlation among data points in both the x and y directions, the mean–variance correlation can be tested in SAS (see Box 15.2). First, the data for mean and variance are power transformed in each direction (Box and Cox test[17]), using the natural log [i.e., ***logmean*** = ***log$_e$(mean) or logvar*** = ***log$_e$(variance)]***, and then the results are regressed using model ***logvar = constant + slope × logmean + error*** (Box 15.3).

15.5.3 Results

Basic statistics (Table 15.2) indicate a wide variation in mean plant densities, ranging from 0 for *Asclepias syriaca* to over 150 plants/m^2 for *S. viridis*, and high values for standard deviations of the means. In the present example, none of the density distributions were normal, and all were highly skewed. In addition, variance and mean were always significantly correlated (P values <0.001) (see parameter estimate in output [Box 15.2] for dependent variable ***logvar***).

BOX 15.1

SAS program (top) for calculating descriptive statistics for weed species and extract of output for ***SETVI*** variable (bottom) of the ***Data1993*** worksheet of ***Chapter_15_data.xls***, descriptive statistics for the other weeds are shown in Table 15.2.

```
data table1;
infile 'Data1993.prn' firstobs = 6;
input xlocation ylocation agrre amare ascsy cheal cirar setvi sinar;
proc univariate normal;
var agrre amare ascsy cheal cirar setvi sinar;
run; quit;
```

```
                         The UNIVARIATE Procedure
                            Variable: setvi

                                Moments
N                       369    Sum Weights                369
Mean             151.273713    Sum Observations         55820
Std Deviation    490.041317    Variance            240140.493
Skewness         6.43733769    Kurtosis            47.2596874
Uncorrected SS     96815800    Corrected SS        88371701.4
Coeff Variation  323.943472    Std Error Mean       25.510532
[...]
                          Tests for Normality
          Test                  --Statistic---     -----p Value-----
          Shapiro-Wilk          W      0.304304    Pr < W      <0.0001
          Kolmogorov-Smirnov    D      0.378776    Pr > D      <0.0100
          Cramer-von Mises      W-Sq   17.66014    Pr > W-Sq   <0.0050
          Anderson-Darling      A-Sq   86.49038    Pr > A-Sq   <0.0050
[...]
```

BOX 15.2

SAS program (top) and extract of output (bottom) for analyzing mean–variance dependency of the ***SETVI*** variable. This test consists in regressing $\log_e$(variance) against $\log_e$(mean); if the slope is significant (i.e., if `Pr > |t|` is lower than .05), the dependent variable y should be transformed as $y^{slope/2}$ (Box and Cox test[17]).

```
data table1;
infile 'Data1993.prn' firstobs = 6;
input xlocation ylocation agrre amare ascsy cheal
  cirar setvi sinar;

y = setvi;
proc sort; by xlocation;
proc means noprint;
var y; by xlocation;
output out=tablex mean = mean var = variance;

data table1; set table1;
proc sort; by ylocation;
proc means noprint;
var setvi; by ylocation;
output out=tabley mean = mean var = variance;

data table2; set tablex tabley;
logmean = log(mean);
logvar = log(variance);
proc reg;
model logvar = logmean;
run; quit;
```

```
                    The REG Procedure
                      Model: MODEL1
               Dependent Variable: logvar
[...]
                   Parameter Estimates

              Parameter   Standard
Variable    DF   Estimate     Error   t Value   Pr > |t|

Intercept    1    1.66556   0.61979      2.69     0.0312
logmean      1    1.78683   0.14511     12.31     <.0001
```

BOX 15.3

SAS program for **PROC GMAP** used to produce the output of Figure 15.2.

```
*---distance between sampling points---;
%let length = 6.1;

*---reading data file with observed weed densities---;
data table1;
infile 'Data1993.prn' firstobs = 6;
input xlocation ylocation agrre amare ascsy cheal
  cirar setvi sinar;
x = xlocation;
y = ylocation;
plot= _N_ ; *creating a plot name from line number _N_ ;

*---create coordinate table for proc gmap---;
*for each (x,y) location, four coordinates (x+length/2,
y+length/2),    (x+length/2,y-length/2),    (x-length/2,
y-length/2) and (x-length/2,y+length/2) are created
where length = sampling distance;
data table3; set table1;
x = x+&length/2;
y = y+&length/2;
keep x y plot;
data table4; set table1;
x = x+&length/2;
y = y-&length/2;
keep x y plot;
data table5; set table1;
x = x-&length/2;
y = y-&length/2;
keep x y plot;
data table6; set table1;
x = x-&length/2;
y = y+&length/2;
keep x y plot;
data table3; set table3 table4 table5 table6;
keep x y plot;
proc sort; by plot;

*---draw map---;
goptions reset=all cback=white colors=(white grayee
graydd graycc graybb grayaa gray77 gray66 gray56
gray44 black); *color options for background and map;
```

BOX 15.3 (continued)

```
pattern value=msolid;  *each basic spatial unit is to
be filled;
proc gmap data=table1 map=table3;
*table1  contains  the  weed  densities,  table3  the
coordinates;
id plot;        *plot is the identification of the basic
spatial unit;
choro setvi/coutline=same   cempty=white   ctext=black
midpoints= 0 to 150 by 15;
*setvi is the variable to represent on the map, the
color outline of each plot is the same as the color
used for filling, empty plots are drawn with white
lines, the legend is written in black and densities
are represented from 0 to 150 by 15;
run; quit;
```

TABLE 15.2
Exploratory Data Analysis for Weed Species of the *Data1993* Worksheet of *Chapter_15_data.xls* Calculated Using PROC UNIVARIATE and PROC GLM of SAS (for SAS Code, Refer to Boxes 15.1 and 15.2)

Weed Species	Bayer Code	Mean (Plants/m²)	Standard Deviation (Plants/m²)	Skewness[a]	Test for Normality[b]	Mean–Variance Correlation[c]
Elytrigia repens	agrre	4.6	42.9	15.86	<0.0001	<0.0001
Amaranthus retroflexus	amare	10.5	19.8	2.88	<0.0001	0.0011
Asclepias syriaca	ascsy	0.0	0.0			
Chenopodium album	cheal	2.7	10.2	5.23	<0.0001	<0.0001
Cirsium arvense	cirar	4.6	13.0	3.95	<0.0001	<0.0001
Setaria viridis	setvi	151.2	490.0	6.44	<0.0001	<0.0001
Brassica kaber	sinar	0.8	3.0	3.93	<0.0001	0.0010

[a] The more different the skewness value (calculated with **PROC UNIVARIATE** of SAS, see Box 15.1) is from zero, the more skewed (asymmetrical) the data distribution.

[b] Probability of error when stating that the distribution is not normal (Shapiro–Wilk test, see Box 15.1).

[c] *P* statistic for correlation between $\log_e$(variance) and $\log_e$(mean) calculated using **PROC REG** of SAS (see example in Box 15.2).

15.6 DATA TRANSFORMATION

15.6.1 Objective

The above analysis indicated that the data were highly skewed, variances and means were correlated, and the dependent variable z had to be transformed before analysis. The objective of this step was to identify a suitable transformation that decorrelates the variance and mean values and converts the data to a normal, or less skewed, distribution. One approach frequently used to obtain a normal distribution is to remove outliers. Unfortunately, in the case of weed densities, skewed distributions are not just due to a single point (Figure 15.3). It is therefore necessary to find variable transformations that do not remove large portions of the observed data.

15.6.2 Method

Box and Cox[17] proposed a transformation to decorrelate variances and means. If the linear regression $\boldsymbol{log_e}(variance) = a + b\ \boldsymbol{log_e}(mean)$ is significant, then, a suggested transformation for the dependent variable z is $z^{1-b/2}$. Another interesting transformation that may be tried is $\boldsymbol{log_e}(z + k)$ transformation where k is a constant. The constant, k, is needed because of the many zero values. In addition, a suitable chosen value for k can achieve normality or decrease skewness.[18] Other transformations tested on the Swan Lake data included z^k, e^{kz}, and $\boldsymbol{log_e}(\boldsymbol{log_e}(z + k))$.

15.6.3 Results

The best results (i.e., lowest skewness and decorrelation of mean–variance) were obtained using the $\boldsymbol{log_e}(z + 1)$ transformation, even though none of the transformed variables showed a normal data distribution (Table 15.3).

15.7 DETRENDING DATA

15.7.1 Objective

Detrending datasets aims to remove large-scale spatial trends. In the present case study, soil depth and texture could, for instance, vary in the experimental field and lead to a large-scale variation in weed densities.

15.7.2 Median Polishing

Even with transformation, a mean–variance correlation was still significant in the dataset. To reduce this correlation, one should look for a large-scale spatial trend in the data and apply a detrending transformation such as median polishing:

$$z' = z - row\ median - column\ median + overall\ median \tag{15.1}$$

This transformation can only be used for gridded data and only works for additive trends. It is not adequate if the trend comprises an interaction between row and

TABLE 15.3
Variable Transformation Tested for the *S. viridis* (SETVI) Densities Observed at Swan Lake in 1993

Variable Transformation	Mean ($Plants/m^2$)	Standard Deviation ($Plants/m^2$)	Skewness[a]	Test for Normality[b]	Mean–Variance Correlation[c]
None	151.2	490.0	6.44	<0.0001	<0.0001
$y^{1-1.78/2}$	1.10	0.742	−0.627	<0.0001	0.2170
$\log_e(y + 0.1)$	2.22	3.14	−0.441	<0.0001	0.3432
$\mathbf{\log_e(y + 1)}$	**2.91**	**2.22**	**0.053**	**<0.0001**	**0.7779**
$\log_e(y + 10)$	3.75	1.40	0.92	<0.0001	0.0258
$\exp(y)$	10^{258}				<0.0001
$\log_e(\log_e(y + 1) +1)$	1.13	0.759	−0.629	<0.0001	0.2146

Notes: Statistics were calculated with **PROC UNIVARIATE** and **PROC GLM** of SAS. The best transformation is indicated in **bold**.

[a] The more different the skewness value (calculated with **PROC UNIVARIATE** of SAS, Box 15.1) is from zero, the more skewed (asymmetrical) the data distribution.

[b] Probability of error when stating that the distribution is not normal (Shapiro–Wilk test with **PROC UNIVARIATE** of SAS, see Box 15.1).

[c] *P* for correlation between $\log_e$(variance) and $\log_e$(mean) calculated with **PROC REG** of SAS (see Box 15.2).

column variables. Other transformations $z' = f(z, row, column)$ should then be tested. Moreover, median polishing does not work well if the mean plant density is low, for example, 1 $plant/m^2$. In that case, row, column, and overall median values are nil, and Equation 15.1 will not modify variable z. If this occurs, the correlation of mean and variance is then not due to a large-scale trend, but simply to the extreme patchiness of the weed population.

With the 1993 Swan Lake date, median polishing was only possible for the more frequent species such as AMARE. In this instance, the transformed variable z' presented independent means and variances ($\boldsymbol{P} < 0.05$), and the density distribution was not skewed (0.16 vs. 0.67 for log-transformed data). Despite this apparent improvement of data distribution, median polishing was not satisfactory because it introduced artifacts into the data, i.e., weed patches are now present where there were none before, etc. This was probably because the seedling densities were so low.

15.7.3 Estimating Trend with Linear Regressions

We tried other detrending transformations by estimating the large-scale trend by fitting a linear regression to the $\boldsymbol{\log_e}(z + 1)$ data (Box 15.4):

$$\boldsymbol{\log_e}(z{+}1) = constant + \alpha \cdot row + \beta \cdot column + \gamma \cdot row^2 + \delta \cdot column^2 + \zeta \cdot row \cdot column + error \quad (15.2)$$

BOX 15.4

SAS program (top) and output (bottom) for determining a large-scale trend in densities using *Setaria viridis* as an example. Trends for other species can be explored by changing the input variable (***SETVI***).

```
data table1;
infile 'Data1993.prn' firstobs = 6;
input xlocation ylocation agrre amare ascsy cheal cirar setvi sinar;
logsetvi=log (setvi+1);
proc glm;
model logsetvi =xlocation ylocation xlocation*xlocation ylocation*ylocation xlocation*ylocation
/solution;
run; quit;
```

```
Dependent Variable: logsetvi
                                        Sum of
Source                   DF           Squares     Mean Square     F Value    Pr > F
Model                     5        285.126316       57.025263       13.50    <.0001
Error                   363       1533.399489        4.224241
Corrected Total         368       1818.525805

                  R-Square       Coeff Var       Root MSE    setvi Mean
                  0.156790        70.43131       2.055296      2.918157
```

Source	DF	Type I SS	Mean Square	F Value	Pr > F
xlocation	1	48.1963485	48.1963485	11.41	0.0008
ylocation	1	14.7703157	14.7703157	3.50	0.0623
xlocation*xlocation	1	83.9781075	83.9781075	19.88	<.0001
ylocation*ylocation	1	1.0143972	1.0143972	0.24	0.6244
xlocation*ylocation	1	137.1671471	137.1671471	32.47	<.0001

Source	DF	Type III SS	Mean Square	F Value	Pr > F
xlocation	1	34.6634823	34.6634823	8.21	0.0044
ylocation	1	12.5222972	12.5222972	2.96	0.0860
xlocation*xlocation	1	83.9781075	83.9781075	19.88	<.0001
ylocation*ylocation	1	1.0143972	1.0143972	0.24	0.6244
xlocation*ylocation	1	137.1671471	137.1671471	32.47	<.0001

Parameter	Estimate	Standard Error	t Value	Pr > |t|
Intercept	3.612001877	0.62487591	5.78	<.0001
xlocation	−0.101280327	0.03535602	−2.86	0.0044
ylocation	0.011190925	0.00649977	1.72	0.0860
xlocation*xlocation	0.002458564	0.00055141	4.46	<.0001
ylocation*ylocation	0.000011263	0.00002298	0.49	0.6244
xlocation*ylocation	−0.000554058	0.00009723	−5.70	<.0001

The final model using only those explanative variables with alpha ≤0.05 then constitutes the large-scale trend that should be retracted from the z variable. In the case of the *S. viridis* data from 1993, the final regression model (15.2) was

$$trend = 88.06 - 19.48 \cdot x\ location + 2.34 \cdot y\ location + 0.64 \cdot x\ location \cdot x\ location - 0.10 \cdot x\ location \cdot y\ location$$

The detrended variable then becomes

$$z' = \boldsymbol{log_e}(z + 1) - trend \qquad (15.3)$$

The R^2 of this final model was very low (0.156) (Box 15.4), indicating that there was no large-scale trend. In addition, the resulting detrended z' variable presented worse distribution characteristics (skewness = −0.097, $\boldsymbol{P}$ for mean–variance correlation = 0.0002) than the untrended $\boldsymbol{log_e}(z + 1)$ variable. Consequently, weed densities were only transformed with $\boldsymbol{log_e}(z + 1)$ for the subsequent analysis. As the distance over which the variograms were calculated was limited and as ordinary kriging (which does not assume a stable mean over the whole field) was used, the resulting bias would be negligible.

15.8 EMPIRICAL SEMIVARIOGRAMS

15.8.1 Objective

Semivariograms describe small-scale spatial trends, i.e., the variance between locations as a function of the distance between these points. Semivariograms were developed for each species and year, using the following equation:

$$\gamma_h = \frac{1}{2N_h} \sum (z_{i+h} - z_i)^2 \qquad (15.4)$$

where

γ_h is the empirical semivariance for the distance $\boldsymbol{h}$
N_h is the number of points separated by the distance $\boldsymbol{h}$
z_i is the weed density at location $\boldsymbol{i}$

This statistic is then plotted for each separation distance $\boldsymbol{h}$ (termed an empirical semivariogram) and characterizes the spatial variability of weed densities as a function of distance among locations. Separate empirical semivariograms were established in four directions: along the rows (i.e., 0° or north-south direction), across the rows (i.e., 90° or east-west direction), as well as along the diagonals (i.e., 45° and 135°) in order to check for anisotropy. For each of these four directions, only points located at an angle of $\pm\boldsymbol{a}$° relative to the nominal direction (i.e., 0°, 45°, 90°, or 135°) were used for the semivariogram. In the present study, the lowest possible angle was used to optimize the discrimination of the analyzed directions.

The maximum distance between points used for the semivariogram can be limited. Here, distance was limited to 50 m for both directions, as the greatest possible distance for both x and y axes (the field was only 54 m wide).

15.8.2 METHOD

In the original paper by Colbach et al.,[16] the semivariograms were calculated with **GSLIB (DOS** version) using the **GAMV** function, which is convenient for irregular data as the sampling grid in this research was not entirely regular. To run **GAMV**, parameterize the ***GAMV.par*** file located in ***GSlib/DOSEXEC/GAMV*** (see, e.g., Box 15.5) and double click on **GAMV.EXE** located in **GSlib/DOSEXEC/BIN** and enter the parameter file name (i.e., ***GAMV.par*** with the relevant path if applicable). The output will be located in the directory specified in the ***GAMV.par*** file. Alternatively, **gamv** function in the drop-down menu of **Variogram** of **WinGslib** could be used (Figure 15.4). On the **Files** page, **Input** and **Output** Files are defined, as well as the **number of lags**, **unit lag separation distance**, and **lags tolerance**. On the **Variograms** page, data for the **azimuth** (angles and associated information) and **Tail/Head** data are selected for the variograms. After data entry, click the **triangular "play"** button to run the program. The **"page"** button can then be clicked to see the output. Here, we also used **PROC VARIOGRAM** of SAS to calculate variograms, followed by **PROC GPLOT** to draw the graph (Box 15.6).

15.8.3 RESULTS

Figure 15.5 shows an example of the empirical semivariograms calculated along rows (0° direction), columns (90°), and the two diagonals (45° and 135°) for SETVI in 2003. For both directions, variance was high at the lowest sampling distance (i.e., 6.1 m); this variance at low distances close to zero is called *nugget*. Variance increased more or less with increasing distance between analyzed data points. In the case of the 135° direction, variance reaches a threshold value (called *sill*) at approximately 25 m while in the 0° direction, variance continued to increase up to 40 m. These distances are called *ranges*, and a variation in range with direction points to geometric anisotropy. Final variance was much lower in the 0° direction than in any other direction, pointing to a marked zonal anisotropy.

The conclusions that can be drawn from a graphical analysis of empirical semivariograms are though limited. Thus, the next step fits models to the empirical semivariogram.

15.9 SEMIVARIOGRAM MODEL FITTING

15.9.1 OBJECTIVE

Empirical semivariograms allow description of spatial correlations. Fitting models to these data makes possible the estimation of (1) variance for unsampled distances, which is necessary to plot maps, and (2) variogram parameters, which can then be correlated to other data.

BOX 15.5

Data file extract from ***setvi199394.dat*** (top) and parameter file ***GAMV.par*** (bottom) for running **GAMV.EXE** of **GSLIB** for calculating the 1993 semivariogram for *S. viridis* and the cross-semivariograms of 1993 with 1994 densities. The ***setvi199394.dat*** sheet was created manually by writing the six first lines and then by copying the relevant columns from the ***Data1993.xls*** and ***Data1994.xls*** datasheets. Both years of data were computed for four directions (0°, 45°, 90°, and 135°) and an angle tolerance **(atol)** of ***0.1***° (e.g., only data points within 45° ± 0.1° are considered for the 45° direction); ***8*** distance classes (**number of lags option**), **distance** by ***6.1*** m and **lag separation distance** option with a tolerance of ***3.05*** m were considered, thus using pairs of data points distanced up to 8·6.1 = 48.8 m in the 0° and 90° directions.

```
Setvi densities for 1993 and 1994
4
Xlocation
Ylocation
Setvi1993
Setvi1994
6.1    0      3.931825633    3.713572067
6.1    6.1    3.044522438    0
6.1    12.2   0              2.397895273
6.1    18.3   0              0
6.1    24.4   3.044522438    0
6.1    30.5   2.397895273    0
[...]
```

```
                              Parameters for GAMV
                              ********************

START OF PARAMETERS:
c:/swanlake/setvi199394.dat        \file with data
1  2  0                            \columns for X, Y, Z coordinates (here, no z)
2  3  4                            \number of variables,column numbers
-98   1.0e21                       \trimming limits
c:/swanlake/setvi1993.sem          \file for variogram output
8                                  \number of lags
6.1                                \lag separation distance
3.05                               \lag tolerance
4                                  \number of directions
0.0  0.1  0.1  0.0  0.10  0.10     \azm,atol,bandh,dip,dtol,bandv
45.0  0.1  0.1  0.0  0.10  0.10    \azm,atol,bandh,dip,dtol,bandv
90.0  0.1  0.1  0.0  0.10  0.10    \azm,atol,bandh,dip,dtol,bandv
135.0  0.1  0.1  0.0  0.10  0.10   \azm,atol,bandh,dip,dtol,bandv
0                                  \standardize sills? (0=no, 1=yes)
2                                  \number of variograms
1  1  1  1                         \tail=1st var., head=1st var., variogram type
1  2  2  2                         \tail=1st var., head=2nd var, variogram type

type 1 = traditional semivariogram
     2 = traditional cross semivariogram
     3 = covariance
     4 = correlogram
     5 = general relative semivariogram
     6 = pairwise relative semivariogram
     7 = semivariogram of logarithms
     8 = semimadogram
     9 = indicator semivariogram - continuous
    10 = indicator semivariogram - categorical
```

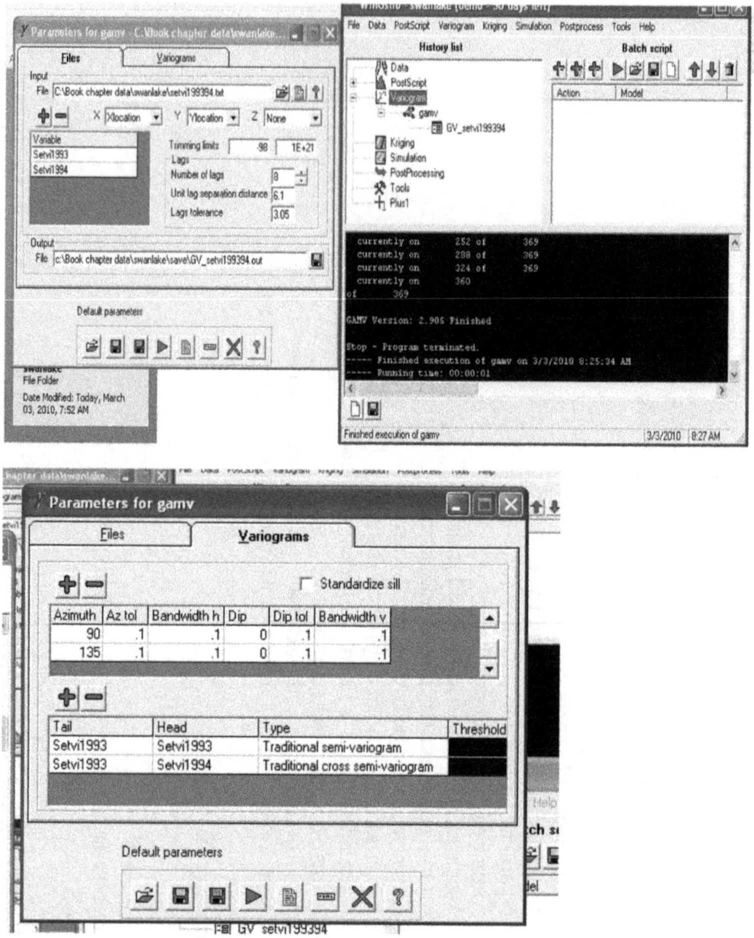

FIGURE 15.4 Using **WinGslib** to process **Variogram** information using **gamv** function for parameters given in Box 15.5 (DOS code for **GSLIB**).

Various contrasting models can be fitted to empirical variograms, for example, linear, power, spherical, exponential, or Gaussian models. The latter can be problematic as its derivative at distance = 0 is zero. In the present case, the linear and power models were inadequate because they do not allow for a sill value (i.e., variance reaching a threshold for large distances). Spherical (vs. exponential) models present the advantage of having a "real" range (i.e., the distance when variance reaches its sill). Here, a nested spherical model was used, i.e., a sum of two spherical models and a nugget value, with ranges depending on the directions.

$$\text{Model 1:} \quad \begin{array}{ll} \text{if } h < a_1 & \gamma_1(h) = c_1 \cdot \left[1.5 \cdot \dfrac{h}{a_1} - 0.5 \cdot \left(\dfrac{h}{a_1} \right)^3 \right] \\ \text{if } h \geq a_1 & \gamma_1(h) = c_1 \end{array}$$

BOX 15.6

SAS program for calculating (**PROC VARIOGRAM**) and drawing (**PROC GPLOT**) an empirical semivariogram. Output is shown in Figure 15.5. The information used for computing the semivariograms is described in Box 15.5.

```
data table1;
infile 'Data1993.prn' firstobs = 6;
input xlocation ylocation agrre amare ascsy cheal cirar
  setvi sinar;

*---data transformation---;
logsetvi=log(setvi+1);

*---calculating empirical semivariogram---;
proc variogram outVar=tabVariogram;
compute   angletolerance=0.1 lagdistance=6.1   ndirec-
tions=4 maxlags=8;
coordinates xcoord=xlocation ycoord=ylocation;
directions 0 (0.1) 45(0.1) 90 (0.1) 135(0.1);
var logsetvi;

*---drawing graph of empirical semivariogram---;
goptions reset=all cback=white colors=(black red blue
green orange);*
symbol1 v=dot;
axis1 order=(0 to 10 by 1);
axis2 order=(0 to 55 by 5);
proc gplot data=tabVariogram;
plot variog*distance=angle /vaxis=axis1 haxis = axis2;

run; quit;
```

* Color graph is on the CD accompanying this book.

Model 2:

$$\text{if } h < a_2 \qquad \gamma_2(h) = c_2 \cdot \left[1.5 \cdot \frac{h}{a_2} - 0.5 \cdot \left(\frac{h}{a_2}\right)^3\right]$$

$$\text{if } h \geq a_2 \qquad \gamma_2(h) = c_2$$

$$\text{Total model: } \gamma(h) = c_0 + \gamma_1(h) + \gamma_2(h) \tag{15.5}$$

where

c_0 is the nugget (representing small-scale variation that cannot be described with the present sampling scheme)

c_1 and c_2 are the contributions of the first and second spatial structures to the total variance (sill)

a_1 and a_2 are the ranges (with different values for the 0° and 90° directions)

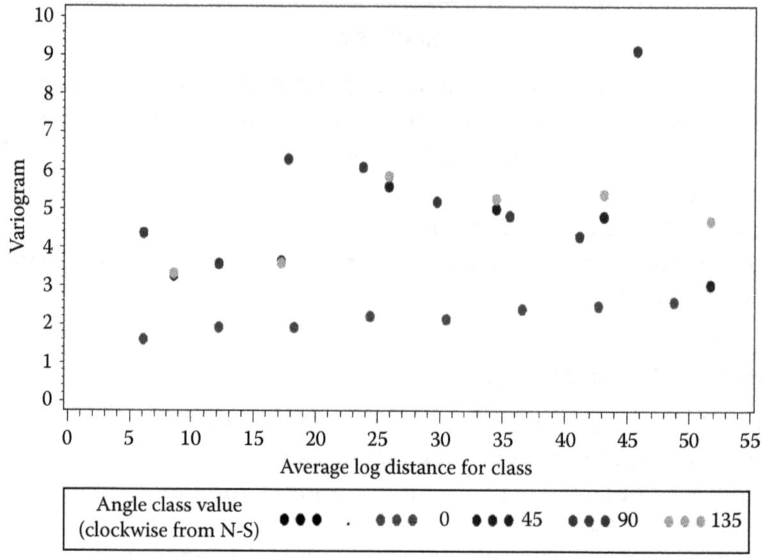

FIGURE 15.5 Example of empirical semivariogram calculated with Equation 15.4 for *S. viridis* in 1993 with **PROC VARIOGRAM** of **SAS** (Box 15.6) (black circles appear in the legend because of empty lines in the SAS data table, which are considered as a fifth direction"." by **PROC GPLOT**).

This model was fit to the empirical semivariogram for each species and year using an iterative least-squares procedure. Points with fewer than 50 pairs were excluded because they were considered unreliable.[19–21] Values for the ranges (a_1, a_2), contributions (c_1, c_2), and nugget (c_0) were estimated using weighted least squares based on number of pairs, N_h.

Equation 15.5 has the advantage of covering a large range of possible situations. The most complicated situation occurs when all parameters are significantly different. A nested model is necessary when sills differ between directions (i.e., zonal anisotropy). If, however, the sills are identical and only the ranges differ, the analyzed variable presents a geometric anisotropy. Equation 15.5 can then be reduced to a single model with one sill, irrespective of direction. Only the ranges then vary with the direction.

15.9.2 Method

PROC NLIN of SAS was used to fit Equation 15.5 and estimate its parameters (Box 15.7).

15.9.3 Results

Figure 15.6 shows an example of fitting a nested spherical model to an empirical semivariogram. For direction 0, the final variance (i.e., sill1) is lower than for

BOX 15.7

SAS program extract (top, must follow program shown in Box 15.6) and output extract (bottom, see also Figure 15.6) for fitting a variogram model to the empirical semiovariogram. Explanations are given as comments in program (*comment;). Model fitting is often difficult and requires bounding the parameters, using likely values seen of graphs. In the present example, bounds were used for ***nugget, c1*** and ***a1_0***.

```
data tabVariogram; set tabVariogram;
*to eliminate distances with less than 50 pairs;
if count >= 50;
proc nlin eformat method=gauss;
*initial parameter values to set off iteration (determined on graphs);
parms    nugget= 1.4
         c1 = 1.3
         c2 = 3.2
         a1 _ 0 = 50
         a2 _ 90= 17;
*two parameters are fixed as they are difficult to estimate because of insufficient sam-
pling at distances close to zero and at larger distances;
a1 _ 90= 0;
a2 _ 0 = 400000;
h = distance;    *renaming distance;
pi = 3.141592;
angle = angle*(pi/180); *transforming from degree to radian;
```

(continued)

BOX 15.7 (continued)

```
*first model;
range1= ( a1_90**2 * (sin(angle) )**2 + a1_0**2 * (cos(angle) )**2)**0.5;
if h < range1
then semi1 = c1 * ( 1.5 * (h/range1) - 0.5 * (h/range1)**3 );
else semi1 = c1;

*second model;
range2 = (a2_90**2 * (sin(angle) )**2 + a2_0**2 * (cos(angle) )**2 )**0.5;
if h < range2
then semi2 = c2 * ( 1.5 * (h/range2) - 0.5 * (h/range2)**3 );
else semi2 = c2;
*fixing limits on parameter values;
*logical limits;   *limits determined on graphs in case of bad fits;
bounds nugget >= 0;   bounds nugget <= 1.4 ;
bounds c1 >= 0;       bounds c1 <= 1.4 ;
bounds c2 >= 0;
bounds a1_0 >= 0;     bounds a1_0 >= 45;
bounds a2_90 >= 0;

*sum of partial models;
```

```
model variog = nugget + semi1 + semi2;

*weighing least squares by the inverse of the number of pairs;
_weight_ = 1/count;

*preparing a data table with the predicted values;
output out=new p=pred;

*graph of predicted and observed variance vs. distance;;
data new; set new;
if angle = 0 or angle = 90; *to simplify graph;
goptions reset=all cback=white colors=(black );
symbol1 v=dot i=none c=red ;
symbol2 v=dot i=none c=green;
symbol3 v=none i=join c=red ;
symbol4 v=none i=join c=green;
axis1 order=(0 to 7 by 1);
axis2 order=(0 to 55 by 5);
proc gplot;
plot variog *distance=angle/vaxis=axis1 haxis = axis2;
plot2 pred*distance = angle/vaxis=axis1 haxis = axis2;
run; quit;
```

(continued)

BOX 15.7 (continued)

The NLIN Procedure

Source	DF	Sum of Squares	Mean Square	F Value	Approx Pr > F
Regression	2	2.5877	1.2939	-9.21	
Residual	25	0.3707	0.0148		
Uncorrected Total	27	2.9584			
Corrected Total	26	0.2342			

Parameter	Estimate	Approx Std Error	Approximate 95% Confidence Limits		Label
nugget	1.4000	0	1.4000	1.4000	
c1	1.4000	0	1.4000	1.4000	
c2	2.1777	0.6746	0.7883	3.5671	
a1 _ 0	45.0000	0	45.0000	45.0000	
a2 _ 90	20.1176	26.3363	-34.1227	74.3579	
Bound4	0.1406	0.0368	0.0649	0.2163	c1 <= 1.4
Bound6	0.1495	0.0407	0.0657	0.2333	nugget <= 1.4
Bound1	0.000624	0.000368	0.00013	0.00138	45 <= a1 _ 0

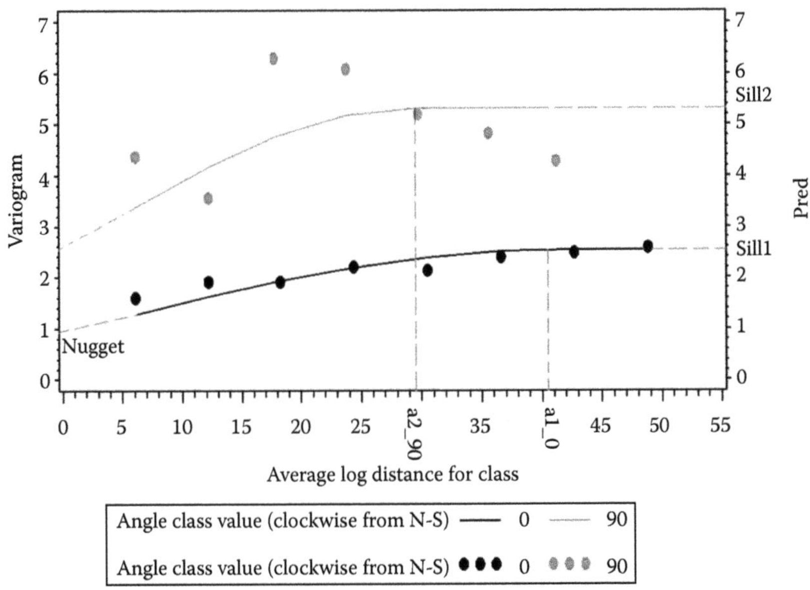

FIGURE 15.6 Example of fitting a nested spherical model (——) using Equation 15.5 to an empirical semivariogram for directions 0° and 90° for *S. viridis* in 1993 with **PROG NLIN** (Box 15.7). Lines and text were added manually to the **SAS** output graph.

direction 90 (i.e., sill2), pointing to lower variability in weed densities along vs. across crop rows. The nugget (i.e., the variance at zero distance) did not depend on the direction and is considered as "white noise," i.e., small-scale variability that cannot be described with the present sampling scheme. The contribution c_1 of the first model is the difference between sill1 and the nugget; it was reached after distance a1_0 = 40 m in the 0° direction, whereas it was present at all distances in the 90° direction as a1_90 was nil. In any other direction, the c_1 contribution depended on the value of range1 varying with the angle (see Box 15.7). The contribution c_2 of the second model is the difference between sill2 and the nugget; it was reached after distance a2_90 = 29 m in the 90° direction, whereas its effect was negligible in the 0° as the a2_0 was infinite. In any other direction, the c_2 contribution depended on the value of range2 varying again with the angle (see Box 15.7). Model fitting was not always easy as variance did not increase smoothly (Figure 15.5). In addition, with nonlinear regressions, the result can vary considerably with the initial values used for the parameters or the method (DUD vs. GAUSS), and sometimes, it is necessary to manually limit the possible ranges for parameters (with the bounds option) to achieve not only convergence during **NLIN** iteration but also visually satisfactory results, i.e., fitted lines vs. observations.

In the field, this directional effect (anisotropy) was evident as elliptical weed patches that were longest in the direction of the crop rows (Figure 15.2). The most likely reason for the difference in ranges (geometric anisotropy) is that weed seeds and other propagules are moved in the direction of crop rows by agricultural implements, such as tillage and harvesting equipment. Other factors, such as water and gravity, also may play a role in creating this distribution. A possible explanation for the difference in total variation (zonal anisotropy) would be the variation in performance of field implements (e.g., planter, cultivator, and combine). The differing speeds, depth adjustments, etc. that occur during north–south passes of these tools across the field may have contributed to an east–west heterogeneity in weed growth conditions.

15.10 ANALYSIS OF VARIOGRAM PARAMETERS

15.10.1 Objective

The analysis of variogram parameters is pertinent when several variables are investigated in a site to examine the differences in weed densities and species. The simplest way is to explain variogram parameters as a function of a year and a weed species effect:

$$\textit{Variogram parameter} = \textit{constant} + \textit{year effect} + \textit{species effect} + \textit{error} \qquad (15.6)$$

This linear model was tested separately for each of the five parameters, i.e., the nugget, and the two contributions. The two ranges were analyzed together, adding a direction effect to model Equation 15.6.

The most interesting way is to look for relationships between the different variogram parameters and a series of explanatory variables that discriminate the datasets. Here, two types of discriminating variables are pertinent: environmental variables that explain the year effect and life traits that explain the differences among weed species. We concentrated on species characteristics by looking at the effect of pre- and postharvest seed production on spatial variability. The model used was as follows:

$$\begin{aligned} \textit{Variogram parameter} &= \textit{constant} \\ &+ \textit{year effect} \\ &+ \beta \times \textit{numbers of seeds dispersed before harvest} \\ &+ c' \times \textit{numbers of seeds dispersed during harvest} \\ &+ \textit{error} \end{aligned} \qquad (15.7)$$

where β and χ are the parameters associated to the covariates "seeds dispersed before harvest" and "seeds dispersed after harvest," respectively. Seed production data were not collected during the field trials but were adapted from the means of the two values reported by Forcella et al.[22] for seed production by annual species. Other models looked at the effects of plant densities or germination behavior. In all cases, only explanative variables significant at $\boldsymbol{P} = 0.01$ were kept in the final model.

15.10.2 Method

Analyses were carried out with **PROC GLM** of SAS. Programs are given in Box 15.8 (Equation 15.6) and Box 15.9 (Equation 15.7).

BOX 15.8

SAS program (top) and output extract (bottom) for analyses of variance of semivariogram parameters.

```
*---reading data file---;
data table1;
infile 'variogramParameters.prn' firstobs=2;
input year species$ nugget c1 c2 a1_0 a2_90;

*---aggregating 2 ranges into a single range variable---;
data table2; set table1;
range = a1_0;
direction = 0;
keep year species direction range;
data table3; set table1;
range = a2_90;
direction = 90;
keep year species direction range;
data table2; set table2 table3;

*---analysis of variance---;
proc glm data = table1;
class year species;
```

(continued)

BOX 15.8 (continued)

```
model nugget c1 c2 = year species;
means year/lsd lines;
means species/lsd lines;

proc glm data = table2;
class year species direction;
model range = direction year species;
means year/lsd lines;
means species/lsd lines;
means direction/lsd lines;

run; quit;
```

```
Dependent Variable: range

                                         Sum of
     Source                  DF         Squares      Mean Square    F Value   Pr > F
     Model                   10     5445.966816       544.596682       6.33   <.0001
     Error                   33     2840.190050        86.066365
     Corrected Total         43     8286.156866

               R-Square     Coeff Var      Root MSE     range Mean
               0.657237      57.68332      9.277196       16.08298
[...]
```

```
     Source                     DF     Type III SS      Mean Square    F Value    Pr > F
     direction                   1     2119.878510      2119.878510      24.63    <.0001
     year                        3      231.823975        77.274658       0.90    0.4526
     species                     6     2873.895250       478.982542       5.57    0.0004
[...]
                                 t Tests (LSD) for range

NOTE: This test controls the Type I comparisonwise error rate, not the experimentwise
error rate.

                         Alpha                                   0.05
                         Error Degrees of Freedom                  33
                         Error Mean Square                   86.06637
                         Critical Value of t                  2.03452
                         Least Significant Difference          12.008
                         Harmonic Mean of Cell Sizes         4.941176

                           NOTE: Cell sizes are not equal.

                Means with the same letter are not significantly different.

                     t Grouping            Mean         N     species

                         A               30.125         8     setvi
                     B   A               25.550         2     sinar
```

(continued)

BOX 15.8 (continued)

```
        B   A   C         20.188        4     agrre
        B   D   C         17.574        8     cirar
            D   C         11.411        8     amare
            D              8.099        8     cheal
            D              6.355        6     ascsy

                   t Tests (LSD) for range

NOTE: This test controls the Type I comparisonwise error rate, not the experimentwise
error rate.

        Alpha                                     0.05
        Error Degrees of Freedom                    33
        Error Mean Square                     86.06637
        Critical Value of t                    2.03452
        Least Significant Difference            5.6909

Means with the same letter are not significantly different.

      t Grouping       Mean      N    direction
               A     23.024     22    0
               B      9.142     22    90
```

BOX 15.9

SAS program (top) and output extract (bottom) for analyzing semivariogram parameters as a function of species traits.

```
data table1;
infile 'weednest.mod' firstobs=2;
input year variable$ stage$ species$ transfo r2 nugget c1 c2 a1max a2min ;

*transformation to obtain relative variations and improve species comparision;
sill = nugget + c1 + c2;
nugget = nugget/sill;
c1 = c1/sill;
c2 = c2/sill;

*definition of seed production per plant;
if species = 'setvi' then seedprod = 43.2;
if species = 'amare' then seedprod = 61.2;
if species = 'cheal' then seedprod = 105.2;
if species = 'sinar' then seedprod = 117.8;

*proportion of seed shed after to harvest;
if species = 'setvi' then rec = 0.27;
if species = 'amare' then rec = 0.26;
if species = 'cheal' then rec = 0.40;
if species = 'sinar' then rec = 0.16;
```

(*continued*)

BOX 15.9 (continued)

```
*seed amount shed before and after to harvest;
seedpost = rec*seedprod;
seedpre = (1-rec)*seedprod;

data table2; set table1;
range = a1max;
direction = 0;
keep year species direction range seedpost seedpre;
data table3; set table1;
range = a2min;
direction =90;
keep year species direction range seedpost seedpre;

proc glm data = table1;
class year species;
model nugget c1 c2 = year seedpost seedpre/solution;

data table2; set table2 table3;
proc glm data = table2;
class year species direction;
model range = direction year seedpost seedpre/solution;

run; quit;
```

Dependent Variable: range

Source	DF	Sum of Squares	Mean Square	F Value	Pr > F
Model	6	2608.511848	434.751975	3.77	0.0121
Error	19	2189.420768	115.232672		
Corrected Total	25	4797.932616			

R-Square	Coeff Var	Root MSE	range Mean
0.543674	62.26256	10.73465	17.24094

[...]

Source	DF	Type III SS	Mean Square	F Value	Pr > F
direction	1	1037.526211	1037.526211	9.00	0.0074
year	3	360.956840	120.318947	1.04	0.3958
seedpost	1	771.477725	771.477725	6.69	0.0181
seedpre	1	0.091953	0.091953	0.00	0.9778

Parameter		Estimate		Standard Error	t Value	Pr > |t|
Intercept		20.17335864	B	7.40120899	2.73	0.0134
direction	0	12.63405782	B	4.21047543	3.00	0.0074
direction	90	0.00000000	B	.	.	.
year	93	8.91374000	B	6.19765203	1.44	0.1666
year	94	2.71061308	B	6.16775976	0.44	0.6653
year	96	−1.07820041	B	6.19765203	−0.17	0.8637
year	97	0.00000000	B	.	.	.
seedpost		−0.52907114		0.20447499	−2.59	0.0181
seedpre		0.00423835		0.15003857	0.03	0.9778

15.10.3 Results

The output extract in Box 15.8 indicates that the range was not significantly influenced by the survey year ($Pr > F = 0.4525$). The most important effect (i.e., highest F value) was due to the direction, with significantly higher ranges in the 0° direction for all species. Ranges varied significantly among species, from 6 m for ASCSY to 30 m for SETVI.

This species effect was then further explained by covariables describing seed dispersal and germination behavior. The results of linear model Equation 15.7 indicate that an increase in seed dispersal before harvest resulted in a decrease in spatial variability in the direction of the crop rows (c_1) but increased unexplained variability (nugget) (Table 15.4). Indeed, the sampling grid used in this work was not fine enough to model the effect on patch shape of preharvest seed dispersal, i.e., seeds falling close to their source plants. The simultaneous decrease in the magnitude of spatial variability between directions (c_2) suggests that preharvest seed dispersal is spatially uniform in all directions. This is not surprising in fields planted with small-stature crops, like *Glycine max*, where weed seed heads are usually positioned above the crop canopy. Even if seeds were moved slightly further in one direction by wind gusts or animals, these movements probably would have resulted in dispersal more random than that of combines and would have contributed little to changes in directional spatial structure.

An increase in weed seed dispersal during harvest resulted in an increase in spatial variability in the direction of crop rows but had no effect on differences in spatial variability along the rows vs. across rows. Moreover, the range parameter tended to decrease as more seed is dispersed during harvest. This decrease is not easily explained but could be the result of an overlap with existing weed patches and greater dispersal distance as influenced by the combine harvester. Weed densities in the patch intersection would not be correlated to densities in the centers of existing patches.

15.11 KRIGING

15.11.1 Objective

The next step in the analysis process is kriging to estimate seedling densities (or other variables) on unsampled locations and plot maps. Kriging is an interpolation technique that estimates the value of an attribute, z, at unsampled locations in the field based on available data at neighboring locations as well as semivariogram model parameters. Basically, there are two types of kriging, ordinary and simple kriging.

Simple kriging assumes a stable mean for the whole field that must be incorporated into the kriging analysis. This method is not convenient in the present case as the analyzed weed variables were often skewed, with mean–variance dependence, even after variable transformation.

Ordinary kriging, on the other hand, uses a variable mean. This kriging method is more adequate for analyzing weed densities due to the dataset values skewness and mean–variance correlation.

TABLE 15.4
Effect of Direction, Year, and Seed Rain Timing on the Parameters of the Semivariogram Models

	Direction Effect		Year Effect		Seed Dispersal			
					Before Harvest		During Harvest	
Variogram Parameter	$P(H = 0)$	Regression Parameter	$P(H = 0)$	Regression Parameter	$P(H = 0)$	Regression Parameter	$P(H = 0)$	Regression Parameter
Nugget	Not tested		0.4840		0.0015	0.0120	0.0019	−0.0192
c_1	Not tested		0.0833		0.0056	−0.00707	0.0001	0.0218
c_2	Not tested		0.5293		0.0766	−0.00495	0.5464	
Range	0.0074	0° 6.6 90° −6.6	0.3958		0.9778		0.0181	−0.529*

* Regression parameters were only shown for significant variables.

Notes: Results of linear model (Equation 15.7) with PROG GLM (see Box 15.9). Regression parameters were only shown for significant variables.

When estimating variable *z* at an unsampled location, kriging software programs use the measured *z* values of the neighboring locations located within a specified search radius. This radius should be smaller than the semivariogram range. Indeed, locations farther than the range are only randomly correlated, and this information, therefore, will not increase the quality of the estimation on the unsampled point.

15.11.2 Method

Various software programs exist for kriging and drawing maps, though with varying degrees of complexity. In the original work,[16] we used the **KB2D** and **KB3D** functions of **GSLIB** for bi-dimensional and three-dimensional kriging, respectively. These functions accept sums of any number of variogram models, including combinations of different models (e.g., a spherical + power model). The data resulting from kriging can then be processed with yet another function and/or software to draw maps, for instance, **PIXELPLT** of **GSLIB**, which produces a postscript file. Here, we used **PROC KRIGE2D** of **SAS** for kriging followed by **PROC GMAP** for drawing the map (Box 15.10).

15.11.3 Results

Maps estimated with ordinary kriging for *S. viridis* plant densities in 1993 with the **SAS** program (Box 15.10) are presented in Figure 15.7. When compared to the original map showing the sampled densities (Figure 15.2), it clearly appears that the weed patches are located in the same parts of the field in both maps, but the kriged map has "smoothed" the raw map into a continuous and more detailed representation.

15.12 CROSS-SEMIVARIOGRAMS AND COKRIGING

15.12.1 Objective

Cross-semivariograns are based on a similar principle as semivariograms, but instead of looking at correlations between locations for a given variable, they describe the variance between two variables as a function of the distance between the locations where the variables were measured. Quite often, the variables to be correlated are a biological (e.g., crop yield) and an environmental variable (e.g., soil nitrogen content). Cross-semivariogram models are then fitted and used for cokriging. While kriging only uses measurements of one variable to estimate this same variable, cokriging is usually used to estimate a sparsely sampled primary variable (e.g., soil nitrogen) with the help of an extensively sampled secondary variable (e.g., weed density). The idea is to estimate a variable that is expensive to sample with the help of an easy-to-measure variable. Cokriging requires the semivariograms for the primary and secondary variable, their cross-semivariogram, and their respective means.

BOX 15.10

SAS program for kriging and drawing a density map for ***S. viridis*** in 1993; the output graph is shown in Figure 15.7. Kriging is ordinary, using only observations located in a radius of 20 m for estimations (radius = 20); the observed variable is ***setvi***. The variogram is a nested model of two spherical models (form option), using the parameter values determined with the program of Box 15.7. The nugget option indicates the nugget irrespective of direction. The scale option lists the contributions, and the range option lists the maximum ranges for the 90° and 0° directions (see angle option). The ratio option lists the ratio of minimum range/maximum range for the two directions, here 1/45 (Instead of 0/45 to avoid nil ranges that are not accepted by **KRIGE2D**) and 20.1/400000.

```
*---defining kriging step as macro-variable---;
%let length = 1 ;

*---reading data file---;
data table1;
infile 'Data1993.prn' firstobs = 6;
input xlocation ylocation agrre amare ascsy cheal
  cirar setvi sinar;

*---transforming weed densities---;
logsetvi=log(setvi+1);

*---kriging with nested anisotropic spherical model---;
proc krige2D data=table1 outest=table2;
coordinates xcoord=xlocation ycoord=ylocation;
grid X=0 to 54 by 1 y=0 to 244 by &length; *grid of coordi-
nates for estimating densities;
predict radius=20 var=logsetvi ;
model form=(spherical spherical) nugget = 0.92
range=(40.9 400000)
ratio=(0.0244 0.0000732) scale=(1.60 2.7) angle=(0 0);

*---retransform weed densities---;
data table2; set table2;
estimate = exp(estimate)-1; *estimate is variable cre-
ated by proc krige2D;

plot = _N_; *creating a plot name from line number
_N_;

*---create coordinate table for proc gmap---;
*for each (x,y) location, four coordinates (x+length/2,
y+length/2),(x+length/2,y-length/2), (x-length/2,
y-length/2) and (x-length/2,y+length/2) are create where
length = kriging step.
```

(continued)

BOX 15.10 (continued)

```
gxc and gxy are variables created by proc krige2D;
data table3; set table2;
x = gxc+&length/2;
y = gyc+&length/2;
keep x y plot;
data table4; set table2;
x = gxc+&length/2;
y = gyc-&length/2;
keep x y plot;
data table5; set table2;
x = gxc-&length/2;
y = gyc-&length/2;
keep x y plot;
data table6; set table2;
x = gxc-&length/2;
y = gyc+&length/2;
keep x y plot;
data table3; set table3 table4 table5 table6;
keep x y plot;
proc sort; by plot;
*---draw map---;
goptions reset=all cback=white colors=(white grayee
graydd graycc graybb grayaa gray77 gray66 gray56
gray44 black ); *color options for background and map;
pattern value=msolid; *each basic spatial unit is to
be filled;
proc gmap data=table2 map= table3; *table2 contains
the weed densities, table3 the coordinates;
id plot;                  *plot is the identification of
the basic spatial unit;
choro estimate/coutline=same cempty=white ctext=black
midpoints = 0 to 150 by 15;
*estimate is the variable to represent on the map, the
color outline of each plot is the same as the color
used for filling, empty plots are drawn with white
lines, the legend is written in black and densities
are represented from 0 to 150 by 15;
run; quit;
```

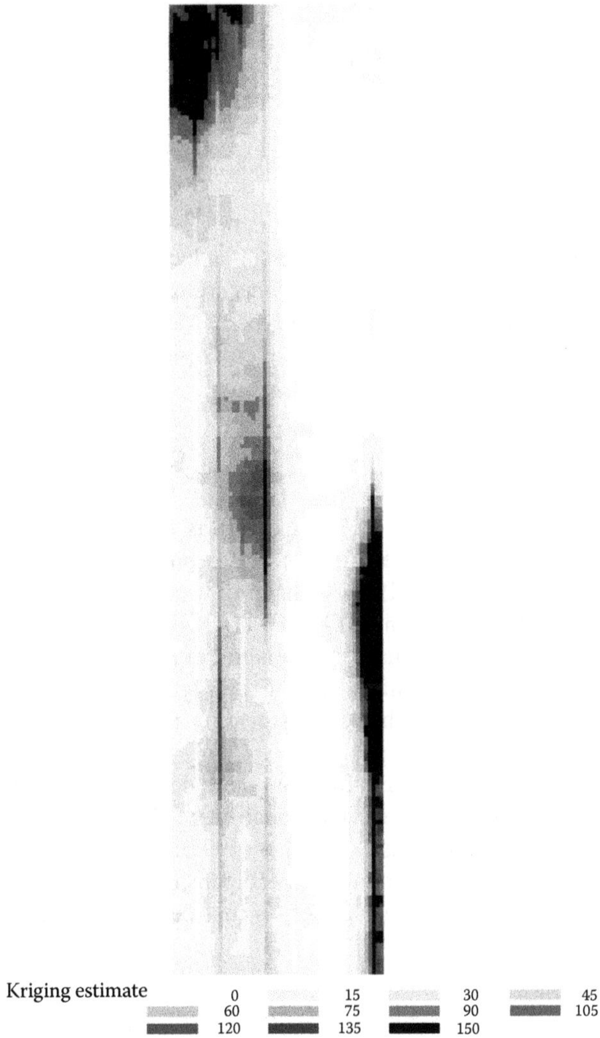

FIGURE 15.7 Maps of *S. viridis* seedling densities for 1993 at the Swan Lake Research Farm based on kriging with **SAS** program shown in Box 15.10.

In this case study, both methods were used somewhat differently. Cross-semivariograms were used to look for correlations between weed densities of a given species of years j and $j + 1$. The aim was to test whether cokriging could use the sampled data of year j to predict the weed distribution of year $j + 1$. Neither the semivariogram of year $j + 1$ nor the cross-semivariogram for years j and $j + 1$ could be known in advance. Therefore, we used the semivariogram and the mean density of year j also for year $j + 1$ as well as the mean of past cross-semivariograms for all pairs of years i and $i + 1$ with $i \leq j$.

15.12.2 Cross-Semivariograms

Empirical cross-semivariograms were calculated as follows:

$$\gamma_h = \frac{1}{2 \cdot N_h} \sum (z_{i+h,j} - z_{i,j})(z_{i+h,j+1} - z_{i,j+1}) \tag{15.8}$$

where $z_{i,j}$ was the weed density of a given species at location i for year j. In the original work,[16] the empirical cross-semivariograms were calculated with the **GAMV** function of **GSLIB** (Box 15.5). The same type of variogram models as for semivariograms can then be fitted to the empirical cross-semivariogram, using the same methods (see Section 15.9).

15.12.3 Cokriging

The semivariogram and mean of 1996 as well as the cross-semivariogram of 1993 and 1994 (Table 15.5) were then used to predict the weed map of the 1997 from the observations of 1996. Cokriging was carried out with the three-dimensional **COKB3D** function of **GSLIB** (DOS version). These calculations are beyond the scope of this chapter although cokriging parameters are given in Box 15.11. The resulting maps (shown in Figure 15.8) were drawn with **PIXELPLT** (Box 15.12) after back transforming the output file produced by **COKB3D** with $\exp(z) - 1$.

Cokriging with the 1996 observations and mean for predicting 1997 weed densities predicted patch location correctly though weed densities were grossly overestimated. Using the actual density mean of 1997, rather than the 1996 mean, considerably increased the prediction quality (Figure 15.8 vs. Figure 15.2). This is a feasible option, as estimating a mean density is considerably faster and cheaper than estimating patch locations.

TABLE 15.5
Variogram Models Used for Predicting the 1997 Map of *S. viridis* from 1996 Observations Using Cokriging

			Variogram Parameters						
			Contribution		Range (m)				
Variogram Type	Year	Nugget	c_1	c_2	a1_90	a1_0	a2_90	a2_0
Semivariogram	1996	1.38	1.20	2.82	0	49.47	34.12	400,000
Cross-semivariogram	mean of 1993/1994 and 1996/1997	0.169	0.587	1.86	0.995	43.0	25.4	5×10^{18}

BOX 15.11

COKB3D.PAR file parameterized for cokriging 1997 ***S. viridis*** densities from 1996 observations (column 3 of ***setvi1996.dat***) while disregarding 1997 observations (a dummy variable located in column 4 of ***setvi1996.dat*** was called instead and then eliminated), using the mean densities actually observed in 1996 (i.e., $4.7 = \mathbf{log_e}(109 + 1)$) and 1997 (i.e., $2.55 = \mathbf{log_e}(11.32 + 1)$), respectively, the 1996 semivariogram for both the main (semivariogram 1 1) and the secondary variable (2 2) and a mean cross-semivariogram (1 2). See Table 15.5 for variogram models.

```
Parameters for COKB3D
               ****************************************

START OF PARAMETERS:
setvi1996.dat              \file with data
2                          \  number of variables primary+other
1  2  0  4  3              \columns for X,Y,Z and variables dummy variable and setvi1996
-98   1.0e21               \trimming limits: lower limit fixed to eliminate all values for
                            dummy variable
0                          \co-located cokriging? (0=no, 1=yes)
setvi1996.dat              \  file with gridded covariate
3                          \  column for covariate = setvi 1996 densities
1                          \debugging level: 0,1,2,3
cokb3d.dbg                 \file for debugging output
set9697.out                \file for output
54   0.5  1.0              \number of x units,xmin,xsize
244 0.5  1.0               \number of y units,ymin,ysize
1    0.5  1.0              \nz,zmn,zsiz
1    1    1                \x, y, and z block discretization
```

(continued)

BOX 15.11 (continued)

```
1  12  8                       \min primary,max primary,max all sec
20.0  20.0  20.0               \maximum search radii: primary
20.0  20.0  20.0               \maximum search radii: all secondary
 0.0   0.0   0.0               \angles for search ellipsoid
1                              \kriging type (0=SK, 1=OK, 2=OK-trad)
2.55 4.7 0.0 0.0               \means for setvi1996 1997
1    1                         \semivariogram for "i" and "j" (here 1996)
2  1.38                        \  nb of structures, nugget effect
1  1.20 0.0  0.0  0.0          \  it,cc,ang1,ang2,ang3
   49.47 0.00 1.0              \  a_hmax, a_hmin, a_vert
1  2.82 0.0 0.0  0.0           \  it,cc,ang1,ang2,ang3
   400000 34.12  1.0           \  a_hmax, a_hmin, a_vert
2    2                         \semivariogram for "i" and "j" (here 1996)
2  1.38                        \  nb of structures, nugget effect
1  1.20 0.0  0.0  0.0          \  it,cc,ang1,ang2,ang3
   49.47 0.00 1.0              \  a_hmax, a_hmin, a_vert
1  2.82 0.0 0.0  0.0           \  it,cc,ang1,ang2,ang3
   400000 34.12  1.0           \  a_hmax, a_hmin, a_vert
1    2                         \semivariogram for "i" and "j" (here mean cross-variogram)
2  1.6999E-01                  \  nst, nugget effect
1  5.8703E-01 0.0  0.0  0.0    \  it,cc,ang1,ang2,ang3
   4.3086E+01 9.954E-01 1      \  a_hmax, a_hmin, a_vert
1  1.863E+00  0.0  0.0  0.0    \  it,cc,ang1,ang2,ang3
   5.49E+18 2.5417E+01 1       \  a_hmax, a_hmin, a_vert
```

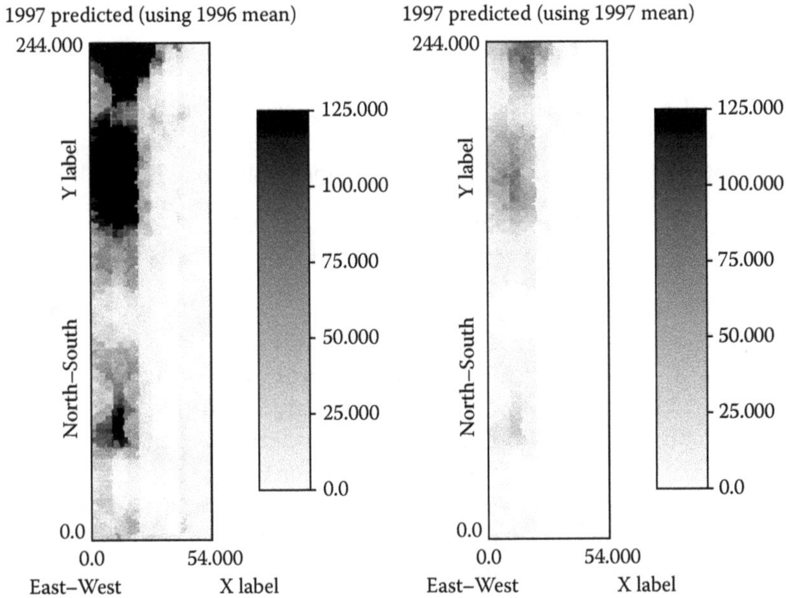

FIGURE 15.8 Maps of *S. viridis* seedling densities for 1997 based on cokriging from the weed densities sampled in 1996, using the 1996 mean density (a) or the 1997 mean density (b).

15.13 ERROR ANALYSIS

15.13.1 Prediction of Weed Means

The mean *S. viridis* density observed in 1997 (i.e., 11.92) can be compared to the mean density calculated from the cokriged maps to determine the prediction error of the mean weed density. When cokriging with the 1996 mean, the predicted 1997 mean was 48.38, which is considerably higher than the actual mean density counted in 1997 (error = +306% = [(48.38 − 11.92)/11.92]*100). This was due to the higher mean density in 1996 (i.e., 109). If the actual 1997 mean was used for cokriging, then the predicted 1997 was reduced to 5.11, with an error of −57% (=[(5.11 − 11.92)/11.92]*100).

15.13.2 Prediction of Weed Locations

Box 15.13 shows an SAS program for comparing kriged or cokriged weed densities to observed densities to calculate mean residual error and mean prediction error for the weed densities. Mean residual error is as follows:

$$\text{MRE} = \frac{\sum (z_i - \hat{z}_i)}{n} \qquad (15.9)$$

where

z_i are observed values (with mean $\overline{z_i}$)
$\hat{z}_i$ is the predicted values
n is the number of values

BOX 15.12

PIXELPLT.PAR file parameterized for drawing map from output of ***COKB3D.EXE*** with ***PIXELPLT.EXE***. The content of set9697 was backtransformed with exp(z) – 1 before drawing the graph. Output is shown in Figure 15.8.

```
                  Parameters for PIXELPLT
                  ************************

START OF PARAMETERS:
set9697.out          \file with gridded data
1                    \  column number for variable
-1.0e21  1.0e21      \  data trimming limits
set9697.ps           \file with PostScript output
1                    \realization number
54   0.5  1          \number of x units,xmin,xsize
244  0.5  1          \number of y units,ymin,ysize
1  0.0  1.0          \nz,zmn,zsiz
1                    \slice orientation: 1=XY, 2=XZ, 3=YZ
1                    \slice number
1997 predicted (Using real 1997 mean)          \Title
East-West            \X label
North-South          \Y label
0                    \0=arithmetic, 1=log scaling
0                    \0=gray scale, 1=color scale
0                    \0=continuous, 1=categorical
0.0  125.0  25.0     \continuous: min, max, increm.
7                    \categorical: number of categories
1  9  Code_One       \category(), code(), name()
2  3  Code_Two
3  2  Code_Three
4  1  Code_Four
5  8  Code_Five      \category(), code(), name()
6  6  Code_Six
7  7  Code_Seven

Color Codes for Categorical Variable Plotting:
      1=red, 2=orange, 3=yellow, 4=light green, 5=green,
      6=light  blue,  7=dark  blue,  8=violet,  9=white,
      10=black, 11=purple, 12=brown, 13=pink, 14=inter-
      mediate green, 15=gray
```

BOX 15.13

SAS program (top) and output (bottom) for calculating prediction errors and decision errors when using cokriged weed maps for estimating weed patches and making spraying decisions. ***Data1997.prn*** contains the observed weed densities; ***set-9697realMean.out*** is an output file produced by ***COKB3D*** of ***GSLIB***

```
*---reading data file with observations---;
data observation;
infile 'Data1997.prn' firstobs = 6;
input xlocation ylocation agrre amare ascsy cheal cirar setvi sinar;
x = floor(xlocation);
y = floor(ylocation);
obs = setvi;
proc sort; by y x;

*---reading cokriging output---;
data prediction;
infile 'set9697realMean.out' firstobs = 5;
input sim var;
*sim = exp(sim)-1;    *only necessary if the content of the out file has not yet been back-
transformed manually;

*---creating coordinates for cokriging output---;
data coord;
do x = 0 to 54 by 1; output; end;
data coord; set coord;
```

(*continued*)

BOX 15.13 (continued)

```
do y = 0 to 244 by 1; output; end;
proc sort; by y x;

*---merge all data tables---;
data prediction; merge coord prediction;
data table1; merge observation prediction; by y x;
if obs ne '.'; *to eliminate lines with missing values;
if sim ne '.';

*---mean residual error---;
error = sim - obs;
e2 = error **2;
proc univariate;
var error ;

*---mean prediction error---;
proc means noprint;
var e2 ;
output out=new SUM = se2;

data new; set new;
racMSEP = (se2/_FREQ_)**0.5;
proc print; var racMSEP;

*---cases with inadequate spraying decision---;
data table1; set table1;
```

```
threshold = 27;*maximum acceptable weed density;
tminus= 0;*variable for summing cases without spraying though the threshold was exceeded;
tplus= 0;*variable for summing cases with spraying though the threshold was not exceeded;
if (obs > threshold and sim > threshold) or (obs <= threshold and sim <= threshold)
then do; *these are the cases with adequate spraying; end;
else do;
  if obs > threshold and sim <= threshold
  then tminus = 1;
  else tplus = 1;
end;
Proc means; var tminus tplus;
run; quit;
```

```
                    The UNIVARIATE Procedure
                       Variable:  error
                           Moments

N                         400    Sum Weights               400
Mean               -5.6266984    Sum Observations   -2250.6794
Std Deviation      25.6789766    Variance           659.409839

Skewness           -2.6487863    Kurtosis           13.9112055
Uncorrected SS      275768.42    Corrected SS       263104.526
Coeff Variation    -456.37734    Std Error Mean     1.28394883
```

(continued)

BOX 15.13 (continued)

```
                 Basic Statistical Measures
             Location                    Variability
         Mean     -5.62670        Std Deviation          25.67898
         Median   -0.33846        Variance               659.40984
         Mode     -0.81350        Range                  265.57028
                                  Interquartile Range    12.76573

NOTE: The mode displayed is the smallest of 2 modes with a count of 2.

                        Tests for Location: Mu0=0

      Test            -Statistic-    -----p Value-----

      Student's t     t  -4.38234    Pr > |t|     <.0001
      Sign            M       -32    Pr >= |M|    0.0016
      Signed Rank     S     -7870    Pr >= |S|    0.0006
[...]
             Obs       racMSEP
               1       26.2568
[...]
                       The MEANS Procedure

Variable      N          Mean        Std Dev        Minimum        Maximum
--------------------------------------------------------------------------
tminus      400     0.1125000      0.3163763              0      1.0000000
tplus       400     0.0500000      0.2182179              0      1.0000000
--------------------------------------------------------------------------
```

Mean prediction error is the root-square of the mean-squared error of prediction.[23,24]

$$\text{rMSEP} = \sqrt{\frac{\sum (z_i - \hat{z}_i)^2}{n}} \qquad (15.10)$$

In the example where cokriging was carried out with the actual 1997 mean (Box 15.13), MRE = −5.62 plants/m^2 (significantly different from zero with Pr > |*t*| < .0001) and Skewness = −2.64, pointing to a tendency of underestimating densities. Mean error prediction was 25.9 plants/m^2, which is approximately double the observed mean (i.e., 11.92).

If cokriging was done with the 1996 mean, MRE increased to 40.3 (significantly different from zero with Pr > |*t*| < .0001) and Skewness to 2.79, showing that weed densities were systematically overestimated. rMSEP was multiplied by nearly four (98.8 plants/m^2).

15.14 SUMMARY: USING GEOSTATISTICAL INFORMATION FOR DECISION MAKING

The program of Box 15.13 also calculates the error frequency if herbicide spraying in precision agriculture was based on cokriged maps. In this example where the 1997 weed map was cokriged from 1996 observations and the 1997 means, 11% of the field was not sprayed though its weed density exceeded the maximum acceptable threshold of 27 plants/m^2, whereas 4% was sprayed even though its density was sufficiently low; the remaining 85% was managed correctly, i.e., areas were treated when densities exceeded the threshold and left untreated when densities were low enough.

When the 1997 map was cokriged using the 1996 density, the decision error was different: 7% was erroneously left untreated, 30% was unnecessarily sprayed, and only 63% was managed correctly. The comparison between these two methods indicates that data used for making maps can significantly impact final outcomes.

GLOSSARY

Anisotropy: It is present when spatial autocorrelation of a process changes with direction.

Anisotropy (geometric): It occurs when the range of the semivariogram changes with direction while the sill remains constant.

Anisotropy (zonal): It occurs when the sill of the semivariogram changes with direction while the range remains constant.

Autocorrelation: $\rho_h = \dfrac{1-\gamma_h}{\sigma \cdot \sigma'}$, where γ_h is the autocorrelation and γ_h is the empirical semivariance for distance h, σ_j and σ_{j+t} the standard deviations of variables z_i and z_i', respectively.

Cokriging: It estimates a sparsely sampled variable, z', using the sampled data of this same variable and that of an extensively sampled variable, z, as well as the semivariograms of the variables z and z' and their cross-semivariogram.

Contribution: Difference between sill and nugget (if any).

Cross-semivariogram (empirical): $\gamma_h = \frac{1}{2N_h}\sum (z_{i+h} - z_i)(z'_{i+h} - z'_i)$, where γ_h is the empirical cross-semivariance for the distance h, N_h the number of points separated by the distance h, and z_i and z'_i the data values of two variables measured at location i.

Kriging: Linear interpolation method that allows estimation of variable z_i at unsampled locations, using a weighted linear combination of available samples and a modeled semivariogram.

Nugget: It represents microscale variation that cannot be described with sampling plan used or measurement error.

Range: The distance (in any) at which data are no longer autocorrelated.

Semivariance (empirical): $\gamma_h = \frac{1}{2 \cdot N_h}\sum (z_{i+h} - z_i)^2$, where γ_h is the empirical semivariance for the distance h, N_h the number of points separated by the distance h, and z_i a data value measured at location i. The semivariogram provides a description of how the data are related (correlated) with distance.

Sill: Value of semivariance γ_h for distance larger than range.

REFERENCES

1. Bigwood, D.W. and Inouye, D.W., Spatial pattern analysis of seed banks: An improved method and optimized sampling, *Ecology*, 69, 497, 1988.
2. Marshall, E.J.P., Field-scale estimates of grass weed populations in arable land, *Weed Research*, 28, 191, 1988.
3. Van Groenendael, J.M., Patchy distribution of weeds and some implications for modelling population dynamics: A short literature review, *Weed Research*, 28, 437, 1988.
4. Thornton, P.K., Fawcett, R.H., Dent, J.B., and Perkins, T.J., Spatial weed distribution and economic thresholds for weed control, *Crop Protection*, 9, 337, 1990.
5. Nordmeyer, H. and Niemann, P., Möglichkeiten der gezielten Teilflächenbehandlung mit Herbiziden auf der Grundlage von unkrautverteilung und Bodenvariabilität, *Zeitschrift für Pflanzenkrankheiten und Pflanzenschutz*, 13, 539, 1992.
6. Wiles, L.J., Oliver, G.W., York, A.C., Gold, H.J., and Wilkerson, G.G., Spatial distribution of broadleaf weeds in North Carolina soybean (*Glycine max*) fields, *Weed Science*, 40, 554, 1992.
7. Halstead, S.J., Gross, K.L., and Renner, K.A., Geostatistical analysis of the weed seed bank, *Proceedings of the North Central Weed Science Society*, 45, 123, 1990.
8. Mortensen, D.A., Johnson, G.A., and Young, L.J., Weed distribution in agricultural fields. In: *Soil Specific Crop Management* (P.C. Robert, R.H. Rust, and W.E. Larson, eds), American Society of Agronomy, Madison, WI, p. 113, 1993.
9. Johnson, G.A., Mortensen, D.A., and Martin, A.R., A simulation of herbicide use based on weed spatial distribution, *Weed Research*, 35, 197, 1995.

10. Streibig, J.C., Gottschau, A., Dennis, B., Haas, H., and Polgaard, P., Soil properties affecting weed distribution, in *7th International Symposium on Weed Biology, Ecology, and Systematics*, Paris, vol. 7, p. 147, 1984.
11. Fried, G., Norton, L.R., and Reboud, X., Environmental and management factors determining weed species composition and diversity in France, *Agriculture, Ecosystems & Environment*, 128, 68, 2008.
12. Colbach, N., Chauvel, B., Gauvrit, C., and Munier-Jolain, N.M., Construction and evaluation of ALOMYSYS, modelling the effects of cropping systems on the blackgrass life-cycle. From seedling to seed production, *Ecological Modelling*, 201, 283, 2007.
13. Chikowo, R., Faloya, V., Petit, S., and Munier-Jolain, N., Integrated weed management systems allow reduced reliance on herbicides and long term weed control, *Agriculture, Ecosystems & Environment*, 132, 237, 2009.
14. SAS Institute Inc., *SAS/STAT User's Guide, Version 6*, SAS Institute Inc., Cary, NC, 1989.
15. Deutsch, C.V. and Journel, A.G., *GSLIB: Geostatistical Software Library and User's Guide*, Oxford University Press, New York, 369pp., 1998.
16. Colbach, N., Forcella, F., and Johnson, G. A., Temporal trends in spatial variability of weed populations in continuous no-till soybean, *Weed Science*, 48, 366, 2000.
17. Box, G.E.P., Hunter, W.G., and Hunter, J.S., *Statistics for Experimenters: An Introduction to Design, Data Analysis, and Model Building*, Wiley, New York, 1978.
18. Colbach, N., Duby, C., Cavelier, A., and Meynard, J.M., Influence of cropping systems on foot and root diseases of winter wheat: Fitting of a statistical model, *European Journal of Agronomy*, 6, 61, 1997.
19. Journel, A.G. and Huijbregts, C., *Mining Geostatistics*, Academic Press, New York, 1978.
20. Hamlett, J.M., Horton, R., and Cressie, N.A.C., Resistant and exploratory techniques for use in semivariogram analyses, *Journal of Soil Science Society of America*, 50, 868, 1986.
21. Cressie, N., *Statistics for Spatial Data*, John Wiley & Sons, New York, 1991.
22. Forcella, F., Peterson, D.H., and Barbour, J.C., Timing and measurement of weed seed shed in corn (*Zea mays*), *Weed Technology*, 10, 535, 1996.
23. Wallach, D. and Goffinet, B., Mean squared error of prediction in models for studying ecological and agronomic systems, *Biometrics*, 43, 561, 1987.
24. Wallach, D. and Goffinet, B., Mean squared error of prediction as a criterion for evaluating and comparing system models, *Ecological Modelling*, 44, 299, 1989.

16 Using GIS to Investigate Weed Shifts after Two Cycles of a Corn/Soybean Rotation

Kurtis D. Reitsma and Sharon A. Clay

CONTENTS

16.1 Executive Summary.......... 374
16.2 Introduction.......... 374
16.3 Materials and Methods.......... 375
16.3.1 Minimum Recommended System Requirements to Reproduce These Analyses.......... 375
16.3.2 Field Methods.......... 375
16.3.3 Analyses Method Overview.......... 375
16.3.4 Aggregating Data in MS Excel.......... 376
16.3.4.1 Aggregating Weed Densities by Year.......... 376
16.3.4.2 Weed Density and Species Change Calculations.......... 377
16.3.4.3 Data for Estimating Direction Distribution.......... 378
16.3.5 ArcMap™.......... 380
16.3.5.1 Creating Layers Using ArcMap™.......... 380
16.3.5.2 Creating Data Subsets.......... 383
16.3.5.3 Data Exploration in ArcMap™.......... 384
16.4 Results.......... 387
16.4.1 Spatial Data Exploration.......... 387
16.4.2 Creating Interpretive Maps.......... 392
16.4.2.1 Spatial Data Interpolation Using an Ordinary Kriging Method.......... 393
16.4.2.2 Spatial Data Interpolation Using an Inverse Distance Weighting Method.......... 397
16.5 Conclusions.......... 402
Acknowledgments.......... 402
References.......... 403

16.1 EXECUTIVE SUMMARY

Genetic engineering has led to the development of crops tolerant of broad-spectrum herbicides. The adoption of these crops has reduced the amount and types of herbicides used for weed management. As with any weed management strategy, shifts in weed densities and species may occur. Field maps depicting changes in weed species or temporal/spatial can be used to assess the effectiveness of the strategy or evolvement of new problems. This case study investigates grass and broadleaf weed densities and species shifts after two rotations (4 years) of glyphosate tolerant corn (*Zea mays* L.) and soybean (*Glycine max* [L.] Merr.) in a production field. Point data collected from the field for 2 years are provided on the accompanying CD in Chapter 16 file folder. This case study will show how to summarize the data in MS® Excel and do spatial analyses using geostatistical methods in ArcMap™ (ESRI® Geographic Information System ver. 9.3). Estimates of the spatial extent of grass and broadleaf species in year 1 (Yr1) and year 4 (Yr4) and the shift in species over years are interpolated from point data and extrapolated to unsampled areas.

16.2 INTRODUCTION

Weeds occur in patches across almost all production fields.[1–9] Patchiness can result from spatial variation of soil types and microclimate factors, such as temperature and water content. Herbicide effectiveness can be influenced by spatial variation due to some of the same site-specific factors that influence weed distribution, as well as distribution of weed species more or less tolerant to the chosen control technique. In addition, better growth and establishment of a susceptible weed species may occur in favorable field microclimates, and, consequently, the herbicide may be less efficacious than if applied to the same species in less favorable environments. Weed patchiness makes it difficult to develop systems to accurately sample and model weed distributions on a field-scale level.[10]

Herbicides used for selective weed control were introduced in the late 1940s. They were adopted rapidly into cropping systems due to their effectiveness, relatively low cost, and ease of application. Rotation of glyphosate tolerant crops is one of the latest management innovations in production fields. Shifts in weed species have been documented when glyphosate is the dominant weed control method.[11] Understanding spatial and temporal variations in weed density and species composition leads to knowledge of the strengths and weaknesses of a given system that, in turn, can lead to better management strategies.

Weed density maps can be developed from point data using methods such as kriging, inverse distance weighting (IDW), or other geostatistical methods.[10,12] Interpolation estimates values at locations not sampled based on neighboring measured points resulting in a contiguous surface depicting spatial variation of the parameter of interest. These analyses assume that values measured at closely sampled points are more closely related (i.e., small variance in means) to each other than those that are located at further sampling distances (i.e., larger variance in means).[10,12,13] The objectives of this case study are to evaluate spatial weed density distribution and changes in weed density and dominant species type (weed free, grass, and broadleaf) data at the start and end of two glyphosate resistant corn/soybean crop rotations (4 years) in a 65 ha field based on point data. Once the basic steps are mastered, the information and procedures provided can be used to examine the spatial analysis of individual weed species, perennial versus annual weed types, or other inquiries of interest for each year or their change over years.

16.3 MATERIALS AND METHODS

16.3.1 Minimum Recommended System Requirements to Reproduce These Analyses

IBM-PC Pentium® 4 or higher processor, 1 Gb RAM, and 10 Gb free harddrive space. Operating System: Microsoft Windows Vista (Ultimate, Enterprise, Business, and Home Premium), Windows 2000, or Windows XP (Home Edition or Professional). Installed Software: MS Office Professional or MS Excel, MS Access (ver. 2003), ESRI® ArcGIS™ ArcMap™ (ver. 9.3) with Spatial Analyst™, and Geostatistical Analyst™ extensions.

16.3.2 Field Methods

The field location and global positioning system (GPS) coordinates of sampling points have been modified to insure landowner anonymity. Conventional corn and soybean were rotated in this field prior to 1998, with glyphosate tolerant soybean and corn introduced into the rotation in 1998. The weed control strategy consisted of herbicide burn-down applications that included glyphosate in early spring and postemergence glyphosate applications in late May each year. Application rates were consistent with product label instructions.

Weed densities by species were collected in 0.1 m^2 quadrats across a 65 ha field in mid-June 1998 and 2001, about 2 weeks after the last glyphosate application each year. The sampled points were located on a 30 m × 15 m irregular grid pattern, with about 1200 points sampled each year. Sampling points were flagged and then georeferenced using a GPS that collected data in decimal degree, North American Datum, 1983 coordinate system. Nine broadleaf and six grass (although green and yellow foxtail counts were combined for total "foxtail" density) species accounted for more than 95% of the total counts across the field (Table 16.1).

The MS® Excel file (***weed_yr1yr4.xls***) located in Chapter 16 file folder of the CD contains two worksheets, Yr1 and Yr4, with the density data for each species by location. The points are assumed to have the same coordinates each year. This case study will examine the spatial distribution of total grass and broadleaf density at the beginning and end of the study and the temporal change in dominant species shift that occurred across years.

16.3.3 Analyses Method Overview

Copy the folder ***Weed_Map*** from the CD to the root directory of **C:** on your computer. A completed MS® Excel (***Weed_data_key.xls***) spreadsheet is provided in this folder, although instructions on how to aggregate using the raw data from ***Weed_Yr1Yr4.xls*** are presented.

TABLE 16.1
Weed Species by Classification and Data Table Designation

Broadleaf Species Common Name	Scientific Name	Designation	Grass Species Common Name	Scientific Name	Designation
Common cocklebur	*Xanthium strumarium*	B1	Woolly cupgrass	*Eriochloa villosa*	G1
Common sunflower	*Helianthus annuus*	B2	Large crabgrass	*Digitaria sanguinalis*	G2
Common waterhemp	*Amaranthus rudis*	B3	Foxtail– Green and Yellow	*Setaria viridis/S. pumila*	G3
Redroot pigweed	*A. retroflexus*	B4	Barnyardgrass	*Echinochloa crus-galli*	G4
Common ragweed	*Ambrosia artemisiifolia*	B5	Witchgrass	*Panicum capillare*	G5
Canada thistle	*Cirsium arvense*	B6			
Field bindweed	*Convolvulus arvensis*	B7			
Horseweed	*Conyza canadensis*	B8			
Common lambsquarters	*Chenopodium album*	B9			

Summarized datasets will be imported into ArcMap™ for analysis, interpolation (using kriging and IDW), and map layer creation. Validation of the models, to determine how well the interpolation predicts values between measured points, will be accomplished by jackknifing, that is, dividing the dataset into a "training" (**trn**) subset for interpolation and "testing" (**tst**) subset for validation.

16.3.4 Aggregating Data in MS Excel

16.3.4.1 Aggregating Weed Densities by Year

1. Open ***Weed_Yr1Yr4.xls***. In order to maintain the integrity of the original dataset, click **File** in the menu bar, select **Save As**, navigate to ***C:\Weed_Map\Excel***, and type in a unique file name such as ***Weed_Data.xls*** in the **File Name** input box, as ***Weed_data_key.xls*** contains the completed worksheet.
2. In ***Weed_Data.xls***, click **Insert** in the menu bar and select **Worksheet**. A new worksheet is created named **Sheet1**; double click the tab of **Sheet1** until a block appears, and rename it ***Weed_Sum***.

3. Coordinate values are the same for both years and will be only referenced to Yr1. In the Yr1 worksheet, select and copy cells A1–C1 and paste into corresponding cells in ***Weed_Sum*** to have the headers ***Latitude***, ***Longitude***, and ***Stake***.
4. Copy stake and coordinate values into ***Weed_Sum*** by entering = ***Yr1!A2***, = ***Yr1!B2***, and = ***Yr1!C2*** into cells A2, B2, and C2, respectively. Copy this information into cells A3–C1203.

Label ***Weed_Sum*** cells D1, E1, F1, G1, H1, and I1 ***Yr1_br***, ***Yr1_gr***, ***Yr1_tot***, ***Yr4_br***, ***Yr4_gr***, and ***Yr4_tot***, respectively. Use Yr1 and Yr4 worksheets as reference, and calculate these values from the worksheets using the following steps:

1. In cell D2, enter = ***SUM(Yr1!D2:M2)***, and in cell E2, enter = ***SUM(Yr1!N2:R2)***, to sum Yr1 broadleaf and grass densities.
2. In cell F2, enter = ***SUM(D2:E2)*** for total weed density in Yr1.
3. In cell G2, enter = ***SUM(Yr_4!D2:M2)***, and in cell H2, enter = ***SUM(Yr_4!N2:R2)***, to sum Yr4 broadleaf and grass densities.
4. In cell I2, enter = ***SUM(G2:H2)*** for total weed density in Yr4.
5. Copy formulas in cells D2–I2 to fill remaining rows D3–I1203.

16.3.4.2 Weed Density and Species Change Calculations

Change in weed density indicates weed control strategy effectiveness and spatial differences may provide information for making strategy changes. Label ***Weed_Sum*** cells J1, K1, and L1 ***Br_chg***, ***Gr_chg***, and ***Tot_chg***, respectively. To calculate the density change, values of broadleaf, grass, or total density for Yr1 are subtracted from similar values in Yr4 (=***Yr4 row*** – ***Yr1 row***) by

1. In cell J2 enter = ***G2-D2***.
2. Copy and paste this formula into cells K2 and L2 to calculate grass and total change, respectively. The formulas = ***H2-E2*** and = ***I2-F2*** should appear in cell K2 and L2, respectively. Copy to cells D3–F1203.

A positive number indicates that weed density increased, a negative number indicates that the density decreased, and a 0 indicates weed-free areas if values both years are 0.

The dominant species in Yr1 is compared to the dominant species in Yr4 to determine if changes in weed species dominance occurred during the study. Three possibilities for dominance occur at each point, that is, broadleaf, grass, or no dominant species. After examining the dataset, there are no points where broadleaf = grass, so if a 0 occurs, the area is weed free. When Yr1 and Yr4 data are combined, there are nine distinct possibilities for weed shifts at each point (Table 16.2). These data will be interpolated in ArcMap™ and will be assigned numeric values; –1 = Broadleaves, 0 = Weed free, and 1 = Grass. Label ***Weed_Sum*** cells M1, N1, O1, P1, and Q1 ***Yr1_dom_cod***, ***Yr4_dom_cod***, ***Yr1_dom_type***, ***Yr4_dom_type***, and ***Yr1_Yr4_shift***,

TABLE 16.2
Dominant Species Possibilities Yr1 and Yr4 and Possible Shifts

Yr1 Dominant	Yr4 Dominant	Shift Possibilities
Broadleaf	Broadleaf	Broadleaf to broadleaf
		Broadleaf to weed free
		Broadleaf to grass
Weed free	Weed free	Weed free to broadleaf
		Weed free to weed free
		Weed free to grass
Grass	Grass	Grass to broadleaf
		Grass to weed free
		Grass to grass

respectively. Conditional statements in MS® Excel will assign values by evaluating a stated condition using the "If, Then, Else" statement. This is accomplished by

1. Enter = ***IF(D2>E2, −1,IF(G2>D2,1,0))*** in cell M2 and = ***IF(G2>H2, −1,IF(H2>G2,1,0))*** in cell N2 that codes the dominant weed type for Yrl and Yr4.
2. Enter=***IF(M2=−1,"Broadleaf,"IF(M2=0,"WeedFree,""Grass")),=IF(N2=−1, "Broadleaf,"IF(N2=0, "Weed Free," "Grass"))***, and =***O2&"–"&P2*** into cells O2, P2, and Q2, respectively. Copy cells M2–Q2 and paste into cells M3–Q1203.

The numeric codes will be used when **interpolating** and **intersecting** Yrl and Yr4 dominance maps. The text values are assigned to easily identify the original and final dominant weed type for quality assurance.

16.3.4.3 Data for Estimating Direction Distribution

Spatial data may exhibit a directional trend. Data that have no directional trend are termed "anisotropic" whereas data with a directional trend are termed "isotropic." In ArcMap™, the calculated standard deviational ellipsis geometry for an attribute is used to determine if directionality occurs. If the data are isotropic, geometric parameters of the standard deviational ellipsis are calculated and results used in geostatistical analysis for better interpolation. Negative numbers cannot be used in this calculation, and, therefore, the data must be examined and transformed to eliminate negative values but keep values at the same relative scale. The transformation equation used is as follows:

$$W_{elp} = \frac{W_i - (\text{Min}\,W_{chg})}{\text{Max}\,W_{chg} - \text{Min}\,W_{chg}} \tag{16.1}$$

where

W_{elp} is the transformed values for calculating the geometry of the standard deviation ellipse

W is the minimum ($MinW_{chg}$) or maximum ($MaxW_{chg}$) values for the broadleaf, grass, or total density change

This equation subtracts the minimum value of the column from each column value and divides by the maximum value – minimum value.

1. Label ***Weed_Sum*** cells R1, S1, T1, U1, and V1 ***Br_ch_elp***, ***Gr_ch_elp***, ***Tot_ch_elp***, ***Yr1_dom_elp***, and ***Yr1_dom_elp***, respectively
2. In cell R2, type = ***(J2-(MIN(J$2:J$1203)))/(MAX(J$2:J$1203)-(MIN(J$2:J$1203)))***
3. In cell S2, type = ***(K2-(MIN(K$2:K$1203)))/(MAX(K$2:K$1203)-(MIN(K$2:K$1203)))***
4. In cell T2, type = ***(L2-(MIN(L$2:L$1203)))/(MAX(L$2:L$1203)-(MIN(L$2:L$1203)))***
5. In cell U2, type = ***(M2-(MIN(M$2:M$1203)))/(MAX(M$2:M$1203)-(MIN(M$2:M$1203)))***
6. In cell V2, type = ***(N2-(MIN(N$2:N$1203)))/(MAX(N$2:N$1203)-(MIN(N$2:N$1203)))***
7. Copy the formulas in cells R2–V2, and paste into cells R3–V1203.

Save the finished MS® Excel spreadsheet (Figure 16.1).

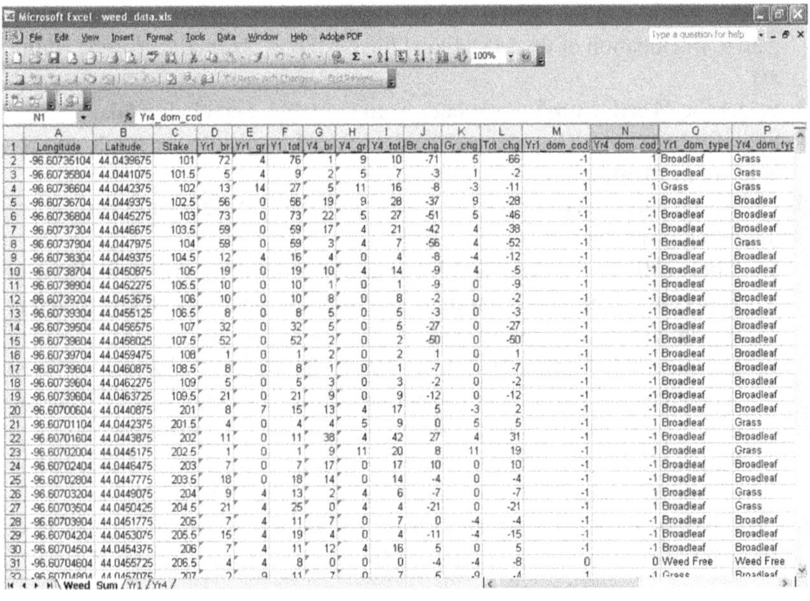

	A	B	C	D	E	F	G	H	I	J	K	L	M	N	O	P
1	Longitude	Latitude	Stake	Yr1_br	Yr1_gr	Y1_tot	Y4_br	Y4_gr	Y4_tot	Br_chg	Gr_chg	Tot_chg	Yr1_dom_cod	Yr4_dom_cod	Yr1_dom_type	Yr4_dom_typ
2	-96.60735104	44.0439675	101	72	4	76	1	9	10	-71	5	-66	-1	1	Broadleaf	Grass
3	-96.60735804	44.0441075	101.5	5	4	9	2	5	7	-3	1	-2	-1	1	Broadleaf	Grass
4	-96.60736604	44.0442375	102	13	14	27	5	11	16	-8	-3	-11	1	1	Grass	Grass
5	-96.60736704	44.0449375	102.5	56	0	56	19	9	28	-37	9	-28	-1	-1	Broadleaf	Broadleaf
6	-96.60736804	44.0445275	103	73	0	73	22	5	27	-51	5	-46	-1	-1	Broadleaf	Broadleaf
7	-96.60737304	44.0446675	103.5	59	0	59	17	4	21	-42	4	-38	-1	-1	Broadleaf	Broadleaf
8	-96.60737904	44.0447975	104	59	0	59	3	4	7	-56	4	-52	-1	1	Broadleaf	Grass
9	-96.60738304	44.0449375	104.5	12	4	16	4	0	4	-8	-4	-12	-1	-1	Broadleaf	Broadleaf
10	-96.60738704	44.0450875	105	19	0	19	10	4	14	-9	4	-5	-1	-1	Broadleaf	Broadleaf
11	-96.60738904	44.0452275	105.5	10	0	10	1	0	1	-9	0	-9	-1	-1	Broadleaf	Broadleaf
12	-96.60739204	44.0453675	106	10	0	10	8	0	8	-2	0	-2	-1	-1	Broadleaf	Broadleaf
13	-96.60739304	44.0455125	106.5	8	0	8	5	0	5	-3	0	-3	-1	-1	Broadleaf	Broadleaf
14	-96.60739504	44.0456575	107	32	0	32	5	0	5	-27	0	-27	-1	-1	Broadleaf	Broadleaf
15	-96.60739604	44.0458025	107.5	52	0	52	2	0	2	-50	0	-50	-1	-1	Broadleaf	Broadleaf
16	-96.60739704	44.0459475	108	1	0	1	2	0	2	1	0	1	-1	-1	Broadleaf	Broadleaf
17	-96.60739604	44.0460875	108.5	8	0	8	1	0	1	-7	0	-7	-1	-1	Broadleaf	Broadleaf
18	-96.60739604	44.0462275	109	5	0	5	3	0	3	-2	0	-2	-1	-1	Broadleaf	Broadleaf
19	-96.60739604	44.0463725	109.5	21	0	21	9	0	9	-12	0	-12	-1	-1	Broadleaf	Broadleaf
20	-96.60700604	44.0440875	201	8	7	15	13	4	17	5	-3	2	-1	-1	Broadleaf	Broadleaf
21	-96.60701104	44.0442375	201.5	4	0	4	4	5	9	0	5	5	-1	1	Broadleaf	Grass
22	-96.60701604	44.0443875	202	11	0	11	38	4	42	27	4	31	-1	-1	Broadleaf	Broadleaf
23	-96.60702004	44.0445175	202.5	1	0	1	9	11	20	8	11	19	-1	1	Broadleaf	Grass
24	-96.60702404	44.0446475	203	7	0	7	17	0	17	10	0	10	-1	-1	Broadleaf	Broadleaf
25	-96.60702804	44.0447775	203.5	18	0	18	14	0	14	-4	0	-4	-1	-1	Broadleaf	Broadleaf
26	-96.60703204	44.0449075	204	9	4	13	2	4	6	-7	0	-7	-1	1	Broadleaf	Grass
27	-96.60703504	44.0450425	204.5	21	4	25	0	4	4	-21	0	-21	-1	1	Broadleaf	Grass
28	-96.60703904	44.0451775	205	7	4	11	7	0	7	0	-4	-4	-1	-1	Broadleaf	Broadleaf
29	-96.60704204	44.0453075	205.5	15	4	19	4	0	4	-11	-4	-15	-1	-1	Broadleaf	Broadleaf
30	-96.60704504	44.0454375	206	7	4	11	12	4	16	5	0	5	-1	-1	Broadleaf	Broadleaf
31	-96.60704604	44.0455725	206.5	4	4	8	0	0	0	-4	-4	-8	0	0	Weed Free	Weed Free
32	-96.60704904	44.0457075	207	2	9	11	7	0	7	5	-9	-4	1	-1	Grass	Broadleaf

FIGURE 16.1 Excel® spreadsheet with summarized, aggregated data by GPS location for Yr1 and Yr4, change in density, and ellipse calculations for change columns.

Excel® spreadsheet with summarized, aggregated data by GPS location for Yr1 and Yr4, change in density, and ellipse calculations for change columns.

16.3.5 ArcMap™

The 1203 point dataset will be interpolated and modeled using two methods, kriging and IDW, to examine spatial weed density distribution at Yr1 and Yr4 and the temporal and spatial change in density and species dominance. Model validation will be done by jackknifing, a type of validation that involves removal of a subset of points from the original dataset to act as a testing (or validation) set, to examine the precision of the selected interpolation method for predicted values from the measured points. A tool in ArcMap™ that randomly selects points to create training and testing datasets was used and these sets stored in a geodatabase. To obtain results presented in this chapter, the training and testing datasets that were generated and saved in the Chapter 16 file folder on the CD must be used. However, if sets are generated by the user, results, while similar, may differ from those in this chapter.

16.3.5.1 Creating Layers Using ArcMap™

The tabular geographic coordinates and weed density data are imported from the MS® Excel worksheet as event themes. Shapefiles are created during the projection procedure with data defined in a common coordinate system using meters as the distance unit. Open ArcMap™, and select **Create a new map**. Event themes are created as follows:

1. From the main toolbar menu options, select **Tools>Add XY Data**.
2. Click the folder at the right of the table selection drop-down box, and navigate to the location of the MS® Excel file ***C:\Weed_Map\Excel\weed_data.xls***.
3. Select the **Weed_Sum$** worksheet from the table, and click **Add**.
4. Specify the X and Y coordinate fields, select **Longitude** for the X coordinate, and **Latitude** for the Y coordinate.
5. Define the coordinate system consistent with the GPS data collection by the following procedure:
   a. Click **Edit** in the **Add XY Data** dialogue box (Figure 16.2), **Spatial Reference Properties** dialogue box appears, and click **Select** to **select a predefined coordinate system**.
   b. Double click **Geographic Coordinate Systems**, double click **North America**, and select **North American Datum 1983.prj (NAD-83)** (the coordinate system used for data collection). Click **Add** and return to **Spatial Reference Properties**. Note: the spatial reference dialogue box is now populated, defining the spatial reference of the data.
6. Click **Apply** and **OK** to return to **Add XY Data** dialogue box. Click **OK**. The data are processed and displayed in the view frame (Figure 16.3).

If a check appears in the box next to the data frame in the **Display** frame, the data are shown in the **View** frame to the right. Uncheck boxes to remove layers from

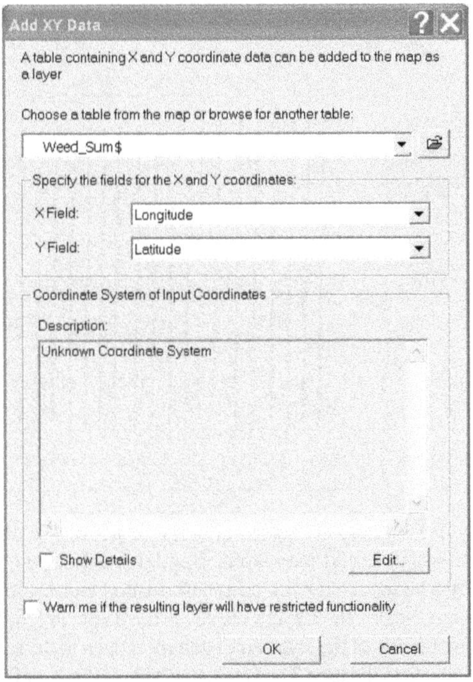

FIGURE 16.2 Display of the ArcMap™ screen when adding X and Y data to a new project. To define the coordinate system consistent with GPS data collection, in this dialogue box click edit and follow the instructions given in the text. (The ArcGIS® graphical user interface is the intellectual property of ESRI and is reproduced herein by permission. Copyright © 1999–2009 ESRI. All rights reserved.)

the frame. Each layer is an event theme that can be displayed and used; however, conversion of event themes to shapefiles allows the fullest extent of data manipulation. Layer conversion to a shapefile with projection to **North America, Continental** projection, and **USA Contiguous Albers Equal Area Conic** is done as follows:

1. Expand **Data Management Tools** in **ArcToolbox** features list.
2. Expand **Projections and Transformations>Feature**, and select **Project** by double clicking, which will open the **Project** dialogue box.
3. Select the event theme (or shapefile) **Weed_sum$ Events** in the **Input Dataset or Feature Class** drop-down box. Note that the **Input Coordinate System** is grayed out, but GCS_North_America_83 should appear (Figure 16.4).
4. Click the folder icon to the right of the **Output Dataset or Feature Class**, navigate to **C:\Weed_Map\Shapefiles\USA_Albers\, type the new shapefile name (e.g., *Weeds_Data.shp*)** in the **Name** input box, and click **Save** to save the projected shapefile in the ***USA_Albers*** folder.
5. Click the icon to the right of the **Output Coordinate System** input box to open the **Spatial Reference Properties** dialogue box.

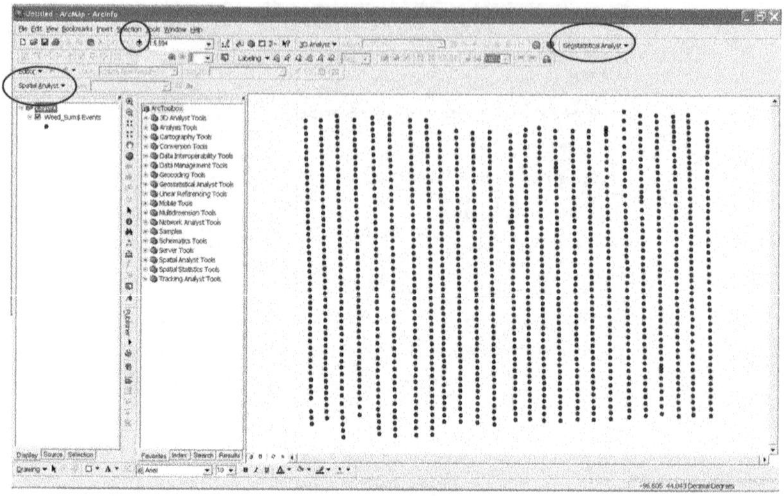

FIGURE 16.3 Display of the ArcMap™ screen after the first X and Y data are added from the Excel® dataset. Note the + sign to add map layers, **Spatial Analyst** button, **Editor** button, and the **Geostatistical Analyst** button along with other information on the top portion of the screen. The area to the far left that contains the **Layers** display is the **Table of Contents** where new layer names are displayed after addition of the layer. **ArcToolbox** is open in the middle of the screen and shows the many tools available to the user. The area to the right is the view frame where the created layers are displayed. (The ArcGIS® graphical user interface is the intellectual property of ESRI and is reproduced herein by permission. Copyright © 1999–2009 ESRI. All rights reserved.)

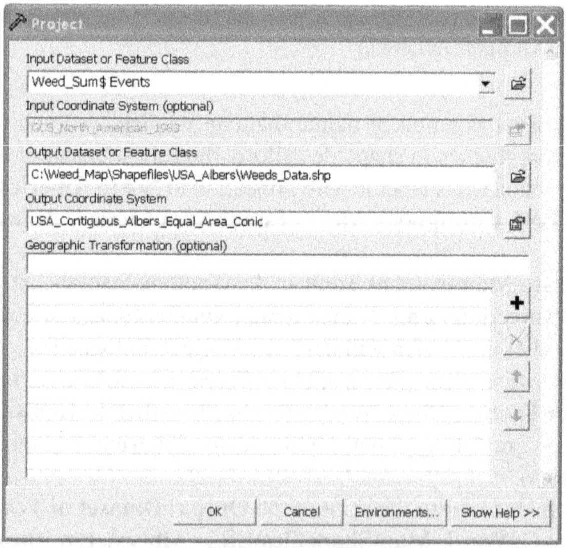

FIGURE 16.4 The **Input Dataset** is defined along with the **Input Coordinate System** used to collect the data. The **Output Dataset** is defined and next the **Output Coordinate System** will be defined. (The ArcGIS® graphical user interface is the intellectual property of ESRI and is reproduced herein by permission. Copyright © 1999–2009 ESRI. All rights reserved.)

6. Click **Select**, the coordinate system browser will open, navigate to the projection file by selecting **Projected Coordinate Systems>Continental>North America>USA Contiguous Albers Equal Area Conic.prj**, and click **Add**. Details of the projection will appear in the **Spatial Reference Properties** dialogue box. Note that distance units are defined as "meters" by default. Click **OK** to close the **Spatial Reference Properties** dialogue box.
7. A project box will appear after the conversion has occurred. Click **Close**. Click **OK** to create the projected shapefile, and note that the newly created shapefile ***Weeds_Data*** appears in the Display frame. However, the map view is defined as **Geographic, NAD-83**, as the map view defaults to the projection of the first layer added. Proper display of the data requires that either the map view be redefined or a new data frame be created with the projected layer added.
8. Create a new data frame by selecting **Insert>New Data Frame** from the main bar drop-down menu. A new data frame will appear in the Table of Contents. Add the ***Weeds_Data*** by right clicking on the **New Data Frame>Add Data** and selecting ***Weeds_Data***. The data layer can be renamed as *Albers* (or other appropriate name) by left clicking on the name and renaming. You can toggle between data frames by right clicking on the frame and selecting **Activate**.

16.3.5.2 Creating Data Subsets

The **Geostatistical Analyst** extension of ArcMap™ has a tool that randomly selects data from the original dataset and creates two datasets that will be used for training and validation. To create data subsets,

1. Activate **Geostatistical Analyst** tool bar by selecting **View** from the top tool bar, and click **Toolbars>Geostatistical Analyst** from the list. If **Geostatistical Analyst** is not listed, you must install it or seek help from your system administrator.
2. Click **Geostatistical Analyst>Create Subsets**.
3. Select ***Weeds_Data*** in **Input Layer**: drop-down box, and click **Next**.
4. Slide the **Subsets Percent/Samples** bar to the right so that **Training** = 70%/841 and **Testing** = 30%/361 percent/samples, respectively.
5. Click on the folder next to **Output geodatabase**, navigate to ***C:\Weed_Map\GeoDatabase***, enter ***Weeds.gdb*** in the **Name** box, and click **Save**.
6. Click the **Training**: Input box and rename the training dataset ***Weeds_Data_trn***, then rename the testing dataset ***Weeds_Data_tst*** in the same manner, and click **Finish**.
7. When asked if you want to add the layers, click **Yes**.

In summary, the procedure above created a geodatabase containing training and testing datasets using the subsets tool. (*Note:* To obtain exact results provided in this chapter, the training and testing datasets located in Chapter 16 file folder of the CD must be used, as the **Create subsets** tool generates random sets each time the commands are performed.) A geodatabase is an efficient way to store geographic datasets, managed in either a file folder or relational database. Geodatabases may contain tables, feature classes, and rasters.

Import the ***Weeds_Data*** layer into the geodatabase by the following steps.

1. Select **Tools** from the top tool bar, and click **ArcCatalog** that is specifically designed to manage geographic datasets.
2. Navigate to ***C:\Weed_Map\GeoDatabase***, and double click ***Weeds.gbd***. Note that the training and testing datasets appear as feature class data.
3. Right click in the **Contents** window, and select **Import>Feature Class... (single)**.
4. Click the folder icon next to the **Input Features** input box, navigate to ***C:\Weed_Map\Shapefiles\USA_Albers***, and select ***Weeds_Data.shp***.
5. Output location should be ***C:\Weed_Map\GeoDatabase\Weeds.gdb***. In the **Output Feature Class** input box, enter ***Weed_Sum_all*** to enter the entire dataset. Click **OK** and **Close** the dialog box. Note that the feature class with the complete dataset is now stored in the geodatabase with a file size of 692 KB versus the shapefile that required 1.31 MB for storage. Repeat these steps to import the fld_bnd.shp (field boundary shapefile) feature class. Remove ***Weeds_data.shp***, and add ***Weed_Sum_all*** to the Albers data frame.

16.3.5.3 Data Exploration in ArcMap™

Initial exploration of data can provide information on how data are distributed over a normal curve and if there are any spatial or nonspatial data trends. The **Histogram** option in **Geostatistical Analyst** creates a bar graph of frequency distribution of the observations in the dataset as well as provides values for the minimum, maximum, mean, median, standard deviation, skewness, kurtosis, and first and third quantiles. Values for the entire dataset are presented; however, subsets could be created and these properties calculated in a similar manner.

1. Click **Geostatistical Analyst>Explore Data Option>Histogram** to open the dialog box.
2. At the bottom **Histogram** dialog box, select ***Weeds_Sum_all*** in the **Data Source** box, and select ***Yr1_gr*** in the **Attribute** box. Click **OK** and the histogram screen appears (Figure 16.5). Each attribute can be explored by selecting it in the **Attribute** box. Attribute distribution statistics are summarized in Table 16.3. Note that by clicking on one of the histogram's bars, the points associated with the criteria are highlighted on the map. To clear these points, right click on the map, and select **Clear Selected Features**. If you change the attribute, but do not clear the last attributes features, the highlighted points from the last selection will be distributed across the new histogram.

Mean and median values indicate the average and most frequent value of the attribute, respectively. The standard deviation (square root of the variance) gives

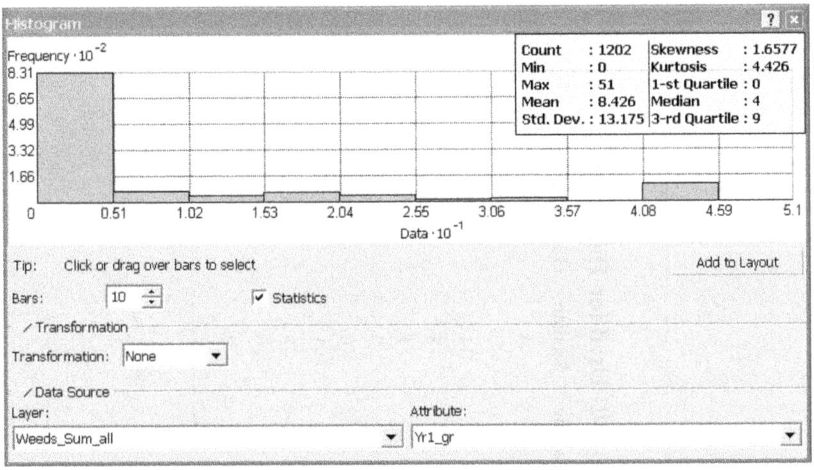

FIGURE 16.5 Histogram data generated from **Geostatistical Analyst**. Note that the **Data Source Layer** is ***Weeds_data***, and the **Attribute layer** is ***Yr1_gr*** (grass density from the year 1 test dataset). Both the **layer** and **attribute layer** can be changed to examine other histograms. The number of bars (currently 10) or transformation of data also can be changed on this page. (The ArcGIS® graphical user interface is the intellectual property of ESRI and is reproduced herein by permission. Copyright © 1999–2009 ESRI. All rights reserved.)

an indication of how closely values are clustered around the mean value with lower values indicating small variation and large values indicating greater variation. For example, Yr4 data for broadleaf, grass, and total density had lower standard deviations (range from 3.9 to 8.4) compared with Yrl data (range from 13.2 to 24.9) (Table 16.3). These trends indicate a more uniform density in Yr4 compared with Yrl. The first and third quantiles are equal to the cumulative proportion of the 0.25 and 0.75 percentile, respectively, i.e., if the data were arranged in increasing order, 25% of the values would be below the first quantile and above the third quantile.

The skewness coefficient measures the symmetry of data distribution with a symmetrical distribution having a value of 0. Positive skewness coefficients indicate a higher frequency of values greater than the median and the mean is larger than the median value. Negative skewness coefficients indicate a higher frequency of values smaller than the median and the mean is less than median value. The skewness coefficients for grass, broadleaf, and total densities in both years are greater than 0, indicating that mean values and a higher frequency of values are greater than the median. However, skewness values for grass and total change over years were somewhat negative whereas broadleaf change was positive. This indicates that the grass species and total density were less changed than the broadleaf density.

Kurtosis is a measure of the "heaviness" of the distribution tails. The kurtosis of a normal distribution is equal to 3. "Platykurtic" distributions have kurtosis values less than 3, which indicate thin tails. "Leptokurtic" distributions have kurtosis values greater than 3, which indicate thick tails and a large portion of the variance due to infrequent but extreme values. Kurtosis values for Yrl are closer to 3

TABLE 16.3
Histogram Distribution Characteristics Based on Entire Dataset (*Weed_Sum_all*) Used for This Chapter

Layer/Attribute	Min	Max	Mean	Std Dev	Skewness	Kurtosis	First Quantile	Median	Third Quantile
Yr1_gr	0	51	8.4	13.2	1.65	4.4	0	4	9
Yr1_br	0	102	13.9	20.7	1.57	4.5	0	3	19
Yr1_tot	0	131	22.3	24.9	1.10	3.4	2	11	42
Yr4_gr	0	46	4.9	7.7	2.54	10.2	0	4	5
Yr4_br	0	53	2.4	3.9	4.35	37.9	0	1	3
Yr4_tot	0	69	7.2	8.4	2.40	10.7	2	5	9
gr_chg	–46	42	–3.6	13.7	–0.08	4.8	–5	0	4
br_chg	–102	49	–11.5	20.7	–1.5	4.6	–16	–1	0
Tot_chg	–131	61	–15.1	25.3	–0.09	3.8	–32	–5	2
Yr1_Dom_Cod	–1	1	–0.16	0.9	0.3	1.3	–1	0	1
Yr4_Dom_Cod	–1	1	0.10	0.9	–0.2	1.2	–1	0	1

(range from 3.4 to 4.5), indicating a more normal distribution of points compared to Yr4 (range from 10.2 to 37.9), indicating a more patchy weed distribution. Note that only a few extreme values (>26 plants/m^2) out of the more than 1200 values in the dataset resulted in a very high kurtosis value for the ***Yr4_Br***. Mean values for the dominant weed type (Broadleaf = −1, Weed free = 0, and Grass = 1) in Yr1 (***Yr1_Dom_Cod***) were slightly negative, indicating a higher occurrence frequency of broadleaf species compared to grasses. By Yr4 (***Yr4_Dom_Cod***), the dominant type shifted to grass species as indicated by a positive mean. These data suggest that a weed shift has occurred during the rotation span. The next step is to explore trends in data and determine where and what weed shifts occurred in the field.

16.4 RESULTS

16.4.1 Spatial Data Exploration

Defining the neighborhood: Data collected in close proximity and in similar areas form a neighborhood. **Calculate Distance Band from Neighborhood Count** utility in **Spatial Statistics Tools** of ArcMap™ is used to calculate distance bands, returning minimum, maximum, and average values for the distance among sampling points. This information provides a better understanding of the spatial distribution of the data. These data are used to calculate the Moran's I value that examines autocorrelation among points.

1. Open **Arc Toolbox**.
2. Expand **Spatial Statistics Tools>Utilities.**
3. Double click **Calculate Distance Band from Neighborhood Count.**
4. Select ***Weeds_Data_trn*** in the **Input Features** box.
5. In this example, 25 neighbors will be used for interpolation, although other numbers may be used. Enter ***25*** in the **Number of Neighbors to Include** box (Figure 16.6).
6. In the **Distance Method** box, select **Euclidean Distance**. Click **OK**.

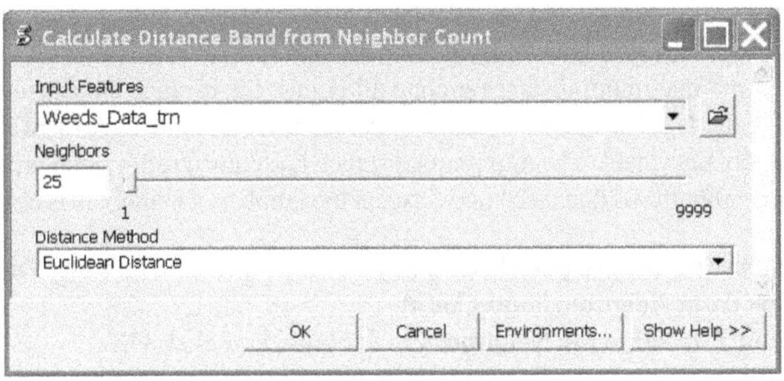

FIGURE 16.6 In **ArcToolbox** utilities, this screen appears when **Spatial Analysis Tools>Utilities>Calculate Distance Band from Neighbor Count** is selected. The **Input Features** box is filled by selecting the file, type ***25*** in the **Neighbors**, and use **Euclidean Distance** in the **Distance Method** box. (The ArcGIS® graphical user interface is the intellectual property of ESRI and is reproduced herein by permission. Copyright © 1999–2009 ESRI. All rights reserved.)

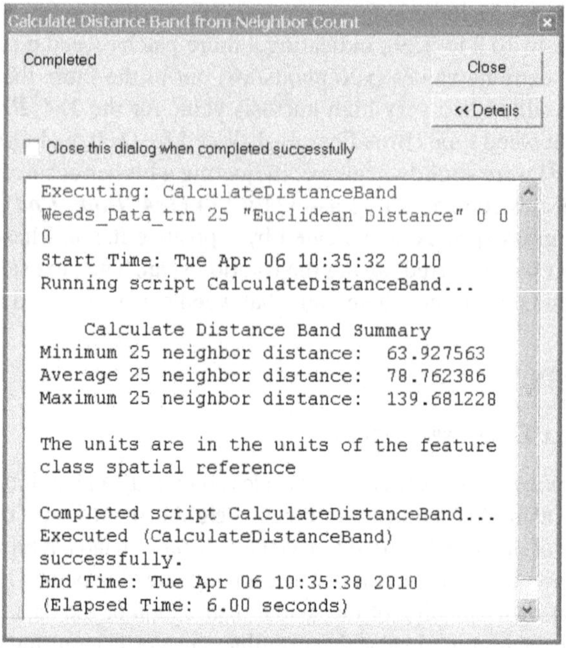
Calculate Distance Band from Neighbor Count
Completed
Close
<< Details
Close this dialog when completed successfully
Executing: CalculateDistanceBand Weeds_Data_trn 25 "Euclidean Distance" 0 0 0
Start Time: Tue Apr 06 10:35:32 2010
Running script CalculateDistanceBand...
Calculate Distance Band Summary
Minimum 25 neighbor distance: 63.927563
Average 25 neighbor distance: 78.762386
Maximum 25 neighbor distance: 139.681228
The units are in the units of the feature class spatial reference
Completed script CalculateDistanceBand...
Executed (CalculateDistanceBand) successfully.
End Time: Tue Apr 06 10:35:38 2010
(Elapsed Time: 6.00 seconds)

FIGURE 16.7 The dialog box that appears after **OK** is clicked when **Spatial Analysis Tools>Utilities>Calculate Distance Band from Neighbor Count** is calculated. Scroll the box to find the **Distance Band Summary** data for minimum, average, and maximum 25 neighbor distance. (The ArcGIS® graphical user interface is the intellectual property of ESRI and is reproduced herein by permission. Copyright © 1999–2009 ESRI. All rights reserved.)

7. Scroll the utility box that appears when the calculation is finished. Using 25 neighbors, the minimum, mean, and maximum distance values are 64, 79, and 145 m, respectively (Figure 16.7).

A value for lag distance should be calculated as lag distance influences the empirical semivariogram. If the grid is in an irregular pattern, a rule of thumb is to set the lag size equal to half the maximum distance among all points. Short-range autocorrelation may be masked if the lag distance is too large, whereas if lag is too small, there may be an excess of empty bin values. The maximum distance is obtained using same procedure for calculating neighborhood distance bands, except the number of neighbors is changed to ***1***.

1. Open **ArcToolbox>Spatial Statistics Tools>Utilities>Calculate Distance Bands from Neighborhood Count**.
2. Type in ***1*** in **Value for Neighbors to Include**, and click **OK**.
3. Scroll the utility box. The maximum distance for the 1 neighbor distance was 47 m. Therefore, an appropriate lag distance is 23.5 m (47/2).

Determining autocorrelation among points: The null hypothesis for a test of autocorrelation is that the spatial pattern is random with respect to point values. The data will show a clustered (positive autocorrelation) or dispersed pattern (negative autocorrelation) if neighboring points have similar or dissimilar values, respectively. Moran's I is just one of several statistical tests used to indicate spatial autocorrelation. Moran's I values range from −1 (dispersed) to +1 (clustered) with a zero value indicating a random spatial pattern. In addition to Moran's I value, a Z-score is also calculated. A Z-score that falls between +1.96 and −1.96 indicates that the null hypothesis (random distribution) is accepted. A very low or very high Z-score indicates that a random distribution is unlikely, the null hypothesis is rejected, and spatial autocorrelation is present. Moran's I statistic is calculated by the following procedure:

1. Open **Arc Toolbox**, and expand **Spatial Statistics Tools>Analyzing Patterns**.
2. Double click **Spatial Autocorrelation** to open the dialog box.
3. Select ***Weeds_data_trn*** in **Input Feature Class** box.
4. Select an attribute such as ***Tot_chg*** in **Input Field** box.
5. Check **Display Output Graphically** box.
6. Default values of **Inverse Distance** for **Conceptualization of Spatial Relationships** and **Euclidean Distance** for **Distance Method** should be left in these boxes.
7. Type ***145*** in the **Distance Band or Threshold Distance** box as this is the maximum distance calculated previously.
8. Leave **Weights Matrix** box blank. Click **OK**.

This analysis returns values for Moran's I of 0.31 and a Z-score of 48.9 (Figure 16.8). The ***clustered*** box is highlighted, indicating spatial autocorrelation that is supported by Moran's I value >0 and the Z-score >1.96. Repeat this procedure for each attribute. Based on these analyses, data for each attribute are clustered with varying degrees of spatial autocorrelation.

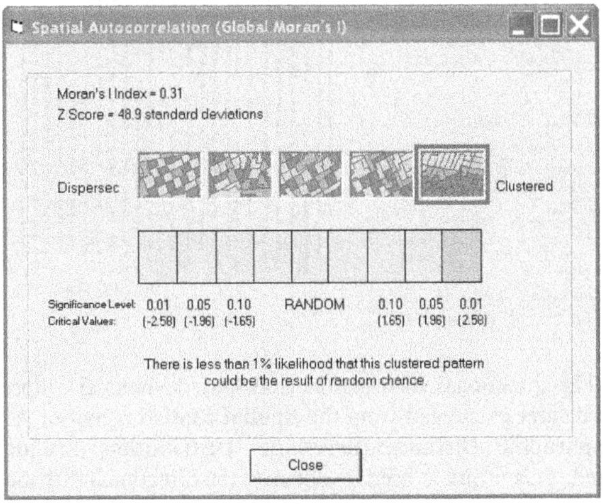

FIGURE 16.8 Example of results from the calculation of Moran's I using **Spatial Statistics Tools>Analyzing Patterns>Spatial Autocorrelation(Morans I)** in **ArcToolbox**. (The ArcGIS® graphical user interface is the intellectual property of ESRI and is reproduced herein by permission. Copyright © 1999–2009 ESRI. All rights reserved.)

Determining directional distribution: Field characteristics (e.g., soil type, field slope, or aspect) and management (e.g., tillage, planting, or harvest direction) can influence the geometry of a weed patch. Data that exhibit a directional trend are termed "isotropic" whereas data that have no directional trend are termed "anisotropic." The **Directional Distribution** tool in ArcMap™ **Utilities** calculates the ellipse geometry of the standard deviation directional trend of the spatial distribution of the data. The directional trend data can be used to adjust the search pattern during interpolation. The geometry of the standard deviational ellipse for each attribute is calculated using the following procedure:

1. Open **Arc Toolbox>Spatial Statistics Tools>Measuring Geographic Distributions.**
2. Double click **Directional Distribution** to open the dialog box.
3. Select ***Weeds_data_trn*** in the **Input Feature Class** box.
4. In the **Output Ellipse Feature Class** box, click the folder icon to navigate to a location to save the shapefile that will be created, and, in this case, save as ***C:\Weed_Map\Shapefiles\USA_Albers\Direct_Dist\Yr1_gr_elp.shp***.
5. Select ***Yr1_gr*** in the **Weight Field** box, leave **Case Field** blank, and click **OK**. A shapefile is created with the geometry of the ellipse in the attribute table (Figure 16.9).

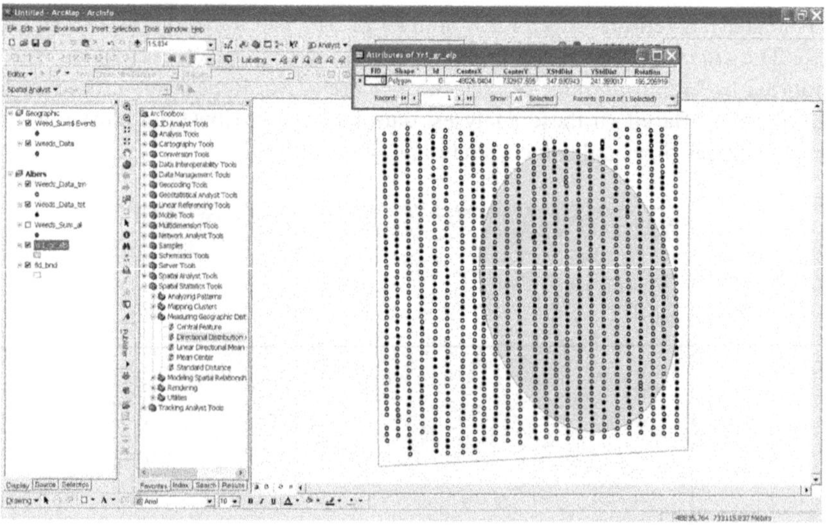

FIGURE 16.9 The directional distribution (standard deviational ellipse) of ***Yr1_gr*** in ***Weeds_data_trn*** dataset calculated from the **Spatial Statistics** tool of **ArcToolbox** when **Measuring Geophraphic Distance>Directional Distribution (Standard deviational ellipse)** tool is used. This utility is used to calculate the directional distribution of the attributes, and a new layer is generated each time this utility is run. For any ***change in density*** data, the input must come from the ***change ellipse*** attributes that were calculated in Excel®, as negative numbers cannot be used in this utility. (The ArcGIS® graphical user interface is the intellectual property of ESRI and is reproduced herein by permission. Copyright © 1999–2009 ESRI. All rights reserved.)

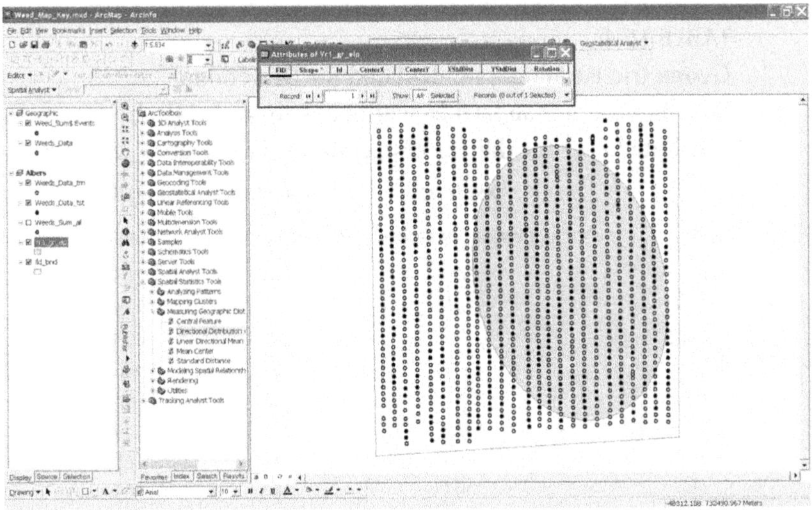

FIGURE 16.10 Right click on the layer of interest, left click on **Open Attribute Table**, and the information on the attribute of interest can be examined. In this case, the results of the ***Yr1_gr_ellipse*** calculation taken from the ***Weeds_data_trn*** dataset are displayed. Note that the Center of X and Y, standard distributions of X and Y, and the rotation of the ellipse are presented. (The ArcGIS® graphical user interface is the intellectual property of ESRI and is reproduced herein by permission. Copyright © 1999–2009 ESRI. All rights reserved.)

Attribute tables for shapefiles can be accessed by the following procedure:

1. Right click on the attribute layer of interest.
2. Scroll to **Open Attribute Table** and left click.
3. The attribute table contains ***CenterX*** (x-coordinate of the center of the ellipse), ***CenterY*** (y-coordinate of the center of the ellipse), ***XStdDist*** (standard distance in the x direction), ***YStdDist*** (standard distance in the y direction), and ***Rotation*** (rotation of the ellipse) (Figure 16.10). Repeat this procedure for other attributes. Since negative values cannot be used to calculate the geometry of the standard deviational ellipse, the transformed ellipse values (***Br_chg_elp***; ***Gr_chg_elp***; ***Tot_chg_elp***; ***Yr1_dom_el***; ***Yr4_dom_el***) were calculated in MS Excel® and should be used as input values for the associated attributes. The values for each attribute ellipsis are presented in Table 16.4 and will be used later in this exercise. All ellipsis layers may be removed from the table of contents if desired.

Up to this point, the data have been summarized, aggregated, and analyzed for Yr1 and Yr4 of the corn–soybean rotation. Recall that the questions relevant to this study are

1. Did weed management strategies change weed densities over time?
2. Did weed type shifts occur over the 4 years?
3. Are there spatial changes in weed densities or shifts in weed type?

TABLE 16.4
Geometric Parameters Calculated for Directional Distribution of Standard Deviational Ellipses for Each Attribute Layer for the Training Dataset (*Weed_data_trn*)

Layer	XStdDist (Major Semiaxis)	YStdDist (Minor Semiaxis)	Rotation (Angle)
Yr1_gr	348	241	165
Yr1_br	249	318	79
Yr1_tot	319	288	123
Yr4_gr	329	296	146
Yr4_br	251	358	46
Yr4_tot	315	343	79
gr_chg	300	318	60
br_chg	307	314	29
Tot_chg	306	315	45
Yr1_Dom_Cod	324	287	177
Yr4_Dom_Cod	333	290	152

Note: This analysis was performed using **Spatial Statistics Tools>Measuring Geographic Distributions>Directional Distribution (Standard Deviational Ellipse) in ArcToolbox.**

The first two questions can be answered with the results of the above data analyses. Mean broadleaf, grass, and total weed density declined across the field (Table 16.3), although the broadleaf density declined to a greater extent than the grass density. Based on only the field mean, grass species may be targeted for control across the entire field. However, spatial analysis of the data will create map layers to indicate where problem areas exist. These maps can be used to take a prescriptive approach for modifying current weed control strategies.

16.4.2 Creating Interpretive Maps

Ordinary kriging will be used to interpolate the initial, final, and change in weed population density. Inverse distance weighting (IDW) will be used to examine spatial distribution of weed type dominance. Data from the previous analyses will be used as input parameters to refine interpolation prediction models.

Spatial Analyst, **3D Analysis**, and **Geostatistical Analyst** are extensions in ArcMap™ used for interpolation. **Geostatistical Analyst**, used in this case study, has the greatest control of input parameters, although results using several analysis methods can be compared.

Spatial variation of a kriged dataset is quantified by the semivariogram, estimated by the sample variogram, and computed from the input point dataset. The value of the sample semivariogram for a separation distance of h (lag) is the average squared difference in the z-value (weed attribute) between pairs of sample points separated by h. The sample semivariogram is mathematically expressed as

$$\lambda(h) = \frac{1}{2n}\sum_{i=1}^{n}\left[Z(xi) - Z(xi+h)\right]^{z} \tag{16.2}$$

where n is the number of pairs of sample points separated by the distance h.

Trend removal decomposes the data into a deterministic trend component and an autocorrelated random component. In this case, the geometry of the data is somewhat circular in pattern (see geometry above), and, therefore, the third-order trend removal function (see below) is selected. After trend removal, residuals are interpolated by the selected method, the trend is added back, final predictions are calculated for unsampled points, and the output surface is created.

16.4.2.1 Spatial Data Interpolation Using an Ordinary Kriging Method

Add the **Geostatistical Analyst** button to the tool bar (if not present), and click **OK**. Scroll to **Geostatistical Wizard**, and click to open the dialog box.

1. Select **Kriging** in the **Geostatistical Wizard** dialog box.
2. Select the desired data set for kriging by selecting ***Weeds_data_trn*** in **Input Data** box and ***Yr1_gr*** in **Attribute** box.
3. Click the **Validation** tab.
4. Select ***Weeds_data_tst*** in **Input Data** box and ***Yr1_gr*** in **Attribute** box. Click **Next**.
5. Expand **Ordinary Kriging>Prediction Map**.
6. Select **Third** in **Order of Trend Removal** box. Click **Next**.
7. Click on the **Neighborhood** slide bar to change the neighborhood to 10% of local.
8. Click **Advanced** options.
9. Change the value for **Include at Least** to *2*.
10. Click the **eight sector button** (the one with the most "slices") to indicate a queen's neighborhood.
11. In the **Ellipse** section, enter the values for **rotation**, **XStdDist, and YStdDist** corresponding to the attribute obtained from the standard deviational ellipse analysis calculated above (Table 16.4).
12. Click the **Optimize Weight Distance** button.
13. Select **Neighbors** in the **Preview Type** box. Note that the ellipse closely resembles the ellipse that was calculated in directional distribution analysis. Click **Next**.
14. Select **Spherical** for the model type (default value).
15. Enter ***23.5*** in the **Lag Size** box (calculated above). Click **Next**.
16. Uncheck the **Default** box in the ellipse input, and enter ***25*** in the **Neighbors to include** box.
17. Enter the parameters found in Table 16.4 to define the ellipse. Click **Next**, and the cross validation prediction model appears (step 5 of 6) (Figure 16.11). Click **Next**.

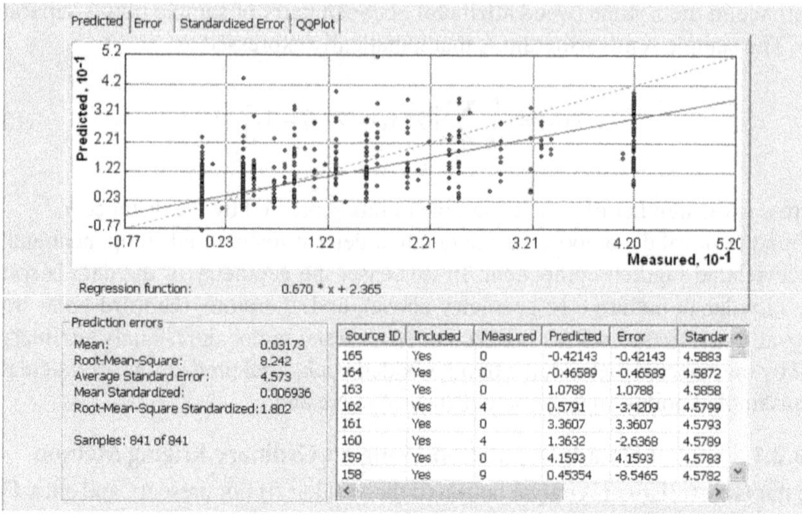

FIGURE 16.11 Measured versus predicted error in cross validation of using the training dataset for ***Yr1_gr*** using the **kriging** function in **Geostatistical Wizard** of **Geostatistical Analyst.** The number of neighbors, neighborhood designation, ellipse data for the attribute of interest, and other information (Table 16.4) were entered on earlier prompt screens. Enter ***25*** in the **Neighbors to Include** box and ***2*** in the **Include at Least** box (this value is used for the points close to the field edge). (The ArcGIS® graphical user interface is the intellectual property of ESRI and is reproduced herein by permission. Copyright © 1999–2009 ESRI. All rights reserved.)

18. The next screen (step 6 of 6) provides the fit or validation of the model and compares the training data to the testing data; save this validation if further statistical analysis is desired. Click **Finish**.
19. The next screen provides a method summary and the option to save it. The method summary is attached to the Geostatistical Analyst (GA) layer, so saving as a separate file is not necessary. Click **OK**. A GA is created by kriging the point data using the input parameters.
20. Right click on the newly created GA layer, select **Properties>General** tab, and enter ***Yr1_Gr_GA*** in the **Layer Name** box. Click the **Extent** tab, and select **the rectangular extent of Fld_Bnd_Alb** from the **Set extent to**: box.
21. While still in **Properties**, click **Symbology**. This allows the user to customize the map output. Check **Filled Contour**. Click **Classify** to show the classification menu. The **Classes** value indicates how many contour breaks will be shown on the map, and **Method** allows for **equal interval, geometric interval, manual**, or **quantile**. If **manual** is selected, the user can input desired values into the **Breaks** area. In this example, five classes were selected in the manual method, and breaks were chosen to show grass densities of 10 (range of 2–10), 20 (range of 11–20), 30 (range of 21–30), and **max** (in this case, the greatest density was 42, so range is 30–42). (Suggested class bin values for various attributes are given in Table 16.5.) Click **OK>Apply**. This will update the completed map.

TABLE 16.5
Suggested Legend Bin Values for Weed Attributes

Yr1_br_GA	Y1_gr_GA	Yr1_tot_GA	Br_chg_GA	Gr_chg_GA	Tot_chg_GA
Min	Min	Min	Min	Min	Min
10	10	10	−80	−30	−100
20	20	20	−60	−20	−80
40	30	40	−40	−10	−60
60	Max	60	−20	0	−40
80		80	−10	10	−20
Max		100	0	20	−10
		Max	10	30	0
			20	Max	10
			40		20
			Max		40
					Max

Yr4_br_GA	Yr4_gr_GA	Yr4_tot_GA	98_DOM_TYPE	01_DOM_TYPE
Min	Min	Min	−1 = Broadleaf weeds	−1 = Broadleaf weeds
5	5	10	0 = Zero count area	0 = Zero count area
10	10	20	1 = Grass weeds	1 = Grass weeds
15	15	30		
20	20	40		
30	30	50		
40	Max	Max		
Max				

22. The contour color scheme can be modified by clicking on **Color ramp** and choosing the colors you wish to show. In this case, a light gray to dark gray ramp was selected with dark gray indicating areas with highest grass densities (Figure 16.12). Compare map layers with statistical information obtained from the mean values. The ***Yr1_gr*** field mean value was 8.4/0.1 m^2, indicating acceptable grass control. The map, however, shows areas with very high densities where additional control may have been beneficial.
23. Repeat this same **Geostatistical Analyst** procedure for other attributes of interest.

The standard error map of prediction associated with the above interpolation(s) is created. The standard error of prediction estimates the standard deviation, derived from a particular sample used to compute the estimate with greater standard errors indicating more errors associated with predicted values than were estimated by the interpolation method. This layer indicates how well the interpolation method

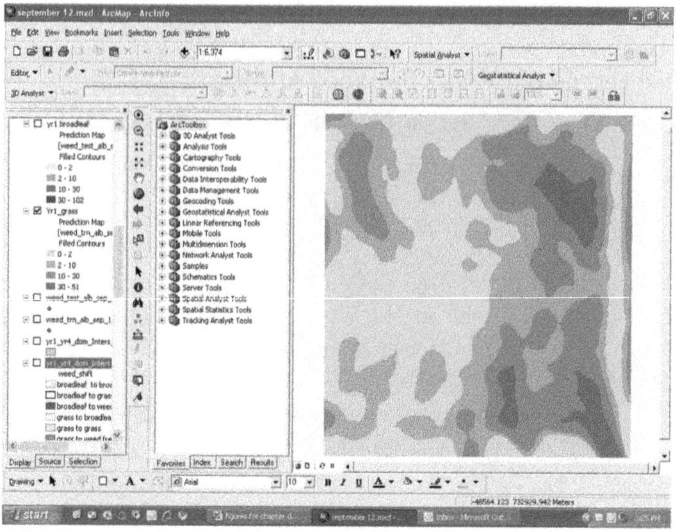

FIGURE 16.12 Results of kriging for ***Yr1_gr*** data. The darker colors represent higher weed densities with the darkest shade representing 40 + grass plants in 0.1 m^2. Note that in the upper right quadrant of the map, an area of high grass density is present. This area corresponds to a wet area of the field with poor drainage. (The ArcGIS® graphical user interface is the intellectual property of ESRI and is reproduced herein by permission. Copyright © 1999–2009 ESRI. All rights reserved.)

predicted the change in grass weed density with respect to measured values in the validation dataset. To create a prediction standard error map:

1. Right click on the***Yr1_gr_GA*** layer. Select **Create prediction of standard error map**.
2. The GA layer ***Yr1_gr_GA_2*** appears at the top to the **Table of Contents** in ArcMap™, and **prediction standard error map** is noted. This layer has the same four class divisions as the original layer and is uniform in color. To examine the variability in the map:
3. Double click the layer to open **Properties>General** tab, and rename the map ***Yr1_gr_StdErr***.
4. Select **Extent>Set the extent to** box, and select **the rectangular extent of fld_bnd**.
5. Click **Symbology** tab. Note that there is very little variability in the standard error (4.57–4.6). Change the classification break intervals to smaller increments to visualize differences in variability, as described above. Click **Apply>OK**.
6. Repeat this procedure for all GA layers of interest.

Changing the classification break intervals to small increments shows that standard error changes across the field, with the smallest values in the field's center and larger values radiating outward. Increasing error near the interpolation boundary is referred to as the "edge effect" and occurs because no points beyond the edge are used to estimate outer values, and there are fewer neighboring points at the margins. One way to reduce the edge effect is to measure points beyond the interpolation boundary, but this is not practical when interpolating across fields with clear-cut boundaries.

16.4.2.2 Spatial Data Interpolation Using an Inverse Distance Weighting Method

The dominant weed species maps for Yrl to Yr4 and the weed species shift map will be created using the IDW method. The IDW interpolation determines a cell value using a linearly weighted value from a combination set of sample points. The function value (or weight) is determined by the inverse distance from the known point to the estimated point. For example, known closer points have greater influence on the value (i.e., a point 2 m away would have ½ influence) with further points having less or no influence on the function value (i.e., a point 50 m away would have a 1/50 influence on the weight). Results from IDW interpolation often produce map layers that have "target" or blocky shapes, rather than a smoothed surface. The dominant weed species at each sampling point was coded as Broadleaf = −1, Weed free = 0, or Grass =1 in the ***dom_cod*** fields. Interpolation using the IDW method is performed as follows:

1. Click **Geostatistical Analyst>Geostatistical Wizard** to open the dialog box.
2. Select **Inverse Distance Weighting.**
3. Select ***Weeds_data_trn*** in **Input Data** box and ***Yrl_dom_cod*** in **Attribute** box.
4. Click **Validation** tab, and select ***Weeds_data_tst*** in **Input Data** box and ***Yrl_dom_cod*** in **Attribute** box. Click **Next**.
5. In **Neighbors to include** type ***25*** and in **Include at Least** type ***2***.
6. Click the eight sector button to indicate a queen's neighborhood.
7. In the **Ellipse** section, enter values corresponding to the values for the attribute layer obtained from the standard deviational ellipse analysis (Table 16.4).
8. Click **Optimize power value.**
9. Select **Neighbors** in the **Preview Type** box. Click **Next>Next.**
10. Examine the prediction of the model and click **Finish**. This screen provides the method summary and is attached to the GA layer, so saving is not necessary. Click **OK**. A GA layer using IDW interpolation is created.
11. Double click the layer to open **Properties** dialog box.
    a. Select **General** tab, and rename the layer ***Yrl_dom_GA***.
    b. Select **Extent** tab>**Set the extent to.** Choose **the rectangular extent of fld_bnd.**
    c. Select **Symbology** tab. Select **Classify**, and change **Classes** to ***3***, **Method** to **Manual**, and **Values** to ***−0.25*** and ***0.25***. Click **OK**.
    d. Change label names by clicking on the label and typing the new name in the input box. Negative values represent broadleaf dominance, 0 represents weed-free areas, and positive values represent grass dominance. Click **Apply**. Note the changes in labeling in the **Table of Contents** area. The map has "bull's eye" type contours that are typical when using IDW interpolation.
12. Repeat the above procedure using ***Yr4_dom_cod*** attribute for the ***Weeds_data_trn*** and ***Weeds_data_tst*** for the training and validation set, respectively. After layer creation, rename the layer ***Yr4_dom_GA***, and set the extent, as described earlier.

Data interpolation using the **Geostatistical Analyst** extension results in a **Geostatistical Analyst Output Layer** (labeled **prediction map**), a raster type layer composed of square cells with assigned values calculated during interpolation. **Geostatistical Analyst Output Layers** can only be created by **Geostatistical Analyst**, but these layers cannot be used directly in other applications. Shapefiles (polygons) from the GA layers allow for further analysis and manipulation of spatial data. **Yr1_dom_GA** and **Yr4_dom_GA** need to be converted from the interpolated GA layers to filled contour shapefiles for further analysis. *Note:* make sure the break value "bins" are defined appropriately in **Symbology**, as the values become attributes of the new shapefile layer and are not easily changed after conversion. In **ArcToolbox**,

1. Expand **Geostatistical Analyst Tools>GA Layer to Contour**.
2. Select ***Yr1_dom_GA*** from the **Input Geostatistical Layer** box.
3. Select **Filled contour** from **Contour Type** box.
4. Click the folder icon next to **Output Feature Class**, and navigate to ***C:\Weed_Map\Temp***, enter ***Yr1_dom_type.shp*** in the **Name** box.
5. Select **Presentation** in the **Contour quality** box. Click **OK**. Note that the newly created shapefile (classes) is added to the **Table of Contents**.
6. To keep the map within the field boundaries, Click **ArcToolboxs>Analysis Tools>Extract**. Select **Clip**. Select ***Yr1_dom_type*** in the **Input Features** box, and select ***fld_bnd*** in **Clip Features** box.
7. Click the folder icon next to **Output Feature Class** box and navigate to ***C:\Weed_Map\Temp\file name*** and enter an appropriate file name in the **Name** box, the default file name in ***_Clip**. Click **OK**.
8. Remove ***Yr1_dom_type*** from the **Table of Contents** by right clicking on the layer and selecting **Remove**. Rename the ***Yr1_dom_type_clip.shp*** layer to ***Yr1_dom_type.shp***.
9. Right click on the ***Yr1_dom_type_clip*** theme, and select **Properties> Symbology**. Click **Categories>Unique values, many fields>Classes** in the first drop-down box. Under **Value fields**, select **Value_min** and **Value_max** in the second and third drop-down box, respectively. Click **Add All Values**. Note that the **Class** and maximum and minimum values are listed to the right of each symbol.
10. Uncheck the box next to **All other values**.
11. The **Symbology** editor allows the user to label the values. Under **Label**, click **Classes, Value_min, Value_max**. Enter ***Weed Type*** in the box that appears.
12. Click the label next to each symbol, and enter ***Broadleaf*** for the 0, −1, and −0.25 values, ***Weed free*** for the 1, −0.25, and 0.25 values, and **Grass** for 2, 0.25, and 1 values. Note that the numeric class values are now labeled with text. Use the same color for each weed type in the polygon feature for the weed types in the point features.
13. Repeat this procedure for ***Yr4_dom_type*** or other layers of interest that were created.

The newly created shapefiles, ***Yr1_dom_type.shp*** and ***Yr4_dom_type.shp***, will be used to calculate areas of each weed species type, define areas where weed species shifts occurred, and calculate areas of each shift. To calculate the area of weed species in Yr1,

1. In ArcToolbox, expand **Spatial Statistics Tools>Utilities>Calculate Areas.**
2. Select ***Yr1_dom_type*** in **Input Feature Class** box.
3. Click the folder next to **Output Feature Class** box, navigate to ***C:\Weed_Map\Shapefiles***, enter ***Yr1_dom_type_area*** in **Name** input box, and click **OK**. A new shapefile with the entered name will appear in the **Table of Contents**. Import the appropriate symbology, as directed above, in **Properties>Symbology** box.
4. The each polygon type area was calculated. To view results, right click on file name in the **Table of Contents**, and select **Open attribute table**. The **F_Area** field of the attribute table is the calculated area of each weed species in square meters. To convert the calculated areas to hectares, divide the **F_Area** by 10,000 (i.e., 1 ha = 10,000 m^2).
5. Repeat the above procedure for ***Yr4_dom_type***.

In Yr1, broadleaf species were dominant on about 31 ha of the 65 ha field (Table 16.6). Weed-free and grass species dominated on about 18 and 16 ha, respectively. In Yr4, broadleaf species dominance declined to 18 ha, total weed-free area was about 17 ha, whereas grass species dominance increased in coverage to about 30 ha.

These data indicate that species shifts occurred between Yr1 and Yr4. Where did these shifts occur? The areas will be defined using the **Intersect** tool in **Analysis Tools** of **ArcToolbox**, which computes the geometric intersections of the two input layers and creates a new layer with features and attributes from both input layers. Both ***Yr1_dom_type.shp*** and ***Yr4_dom_type.shp*** are the same feature type (polygon) and contain the same fields for each polygon, and the **Intersect** of these layers results in weed species type at the beginning and end of the 4 year evaluation period. To intersect layers,

TABLE 16.6
Areas of Dominant Weed Species Type in Yr1 and Yr4 Calculated from Shapefiles and Using Calculated Area in Utilities of Spatial Statistics Tools in ArcToolbox

	Area (ha)	
Species Type	**Yr1**	**Yr4**
Broadleaf	31.5	17.9
Weed free	15.7	17.2
Grass	17.8	29.9

1. In **ArcToolbox**, expand **Analysis Tool>Overlay>Intersect**.
2. In **Input Features** box, select ***Yr1_dom_type.shp*** and ***Yr4_dom_type.shp. Yr1_dom_type.shp*** should appear first in the list box.
3. Click the folder next to **Output Feature Class** box, navigate to an appropriate file for output, and enter a new name for the new file in the **Name** box, for example, ***Yr1_Yr4_intersect***. Click **OK**. A new layer with the name is added to the **Table of Contents**.
4. In **Properties>Symbology>**, Click **Categories>Unique values, many fields**. Select **Classes** in the first drop-down box and **Classes_1** in the next drop-down box. Click **Add All Values**. Note that the **Class** and maximum and minimum values are listed to the right of each symbol.
5. Uncheck the box next to **All other values**.
6. Under the **label** field, click the field next to **Heading**, and enter ***Weed Type Shift***. Note that there are nine classifications. Symbols are added to the layer with Yr4 grouping using arrows at the right of the legend editor. Move classes in order of "**0,0**," "**1,0**," "**2,0**," "**0,1**," "**1,1**," "**2,1**," "**0,2**," "**1,2**," and "**2,2**" where **0** = ***broadleaf***, **1** = ***weed free***, and **2** = ***grass***. Note that the values have changed from the original –1, 0, and 1 to 0, 1, and 2.
7. Click on the labels for each class, and rename appropriately (Table 16.7). For example, **0,0** is ***broadleaf to broadleaf***, and **1,2** is ***weed free to grass***. Click **Apply**.
8. Right click on one of the symbols in the **Symbology>Properties of all symbols**. Click on the icon next to **Outline color**, and select **No Color**. Click **Apply**. Change the color ramp, as desired. The map in this chapter used black and white shading to display changes (Figure 16.13), although a full color display is located on the accompanying CD.

TABLE 16.7
Estimated Area of Dominant Weed Species Shift from Yr1 to Yr4 Based on Shapefile of Intersected Yr1 and Yr4 Dominant Species Layers Created Using Inverse Weighting Distance Calculations in GeoAnalyst

Class Code		Dominant Weed Species			
Classes	Classes_1	Yr1	Yr4	Weed_Shift	Area (ha)
0	0	Broadleaf	Broadleaf	Broadleaf–broadleaf	11.8
1	0	Weed free	Broadleaf	Weed free–broadleaf	4.1
2	0	Grass	Broadleaf	Grass–broadleaf	2.0
0	1	Broadleaf	Weed free	Broadleaf–weed free	9.8
1	1	Weed free	Weed free	Weed free–weed free	4.6
2	1	Grass	Weed free	Grass–Weed free	2.8
0	2	Broadleaf	Grass	Broadleaf–grass	9.8
1	2	Weed free	Grass	Weed free–grass	6.9
2	2	Grass	Grass	Grass–grass	13.1

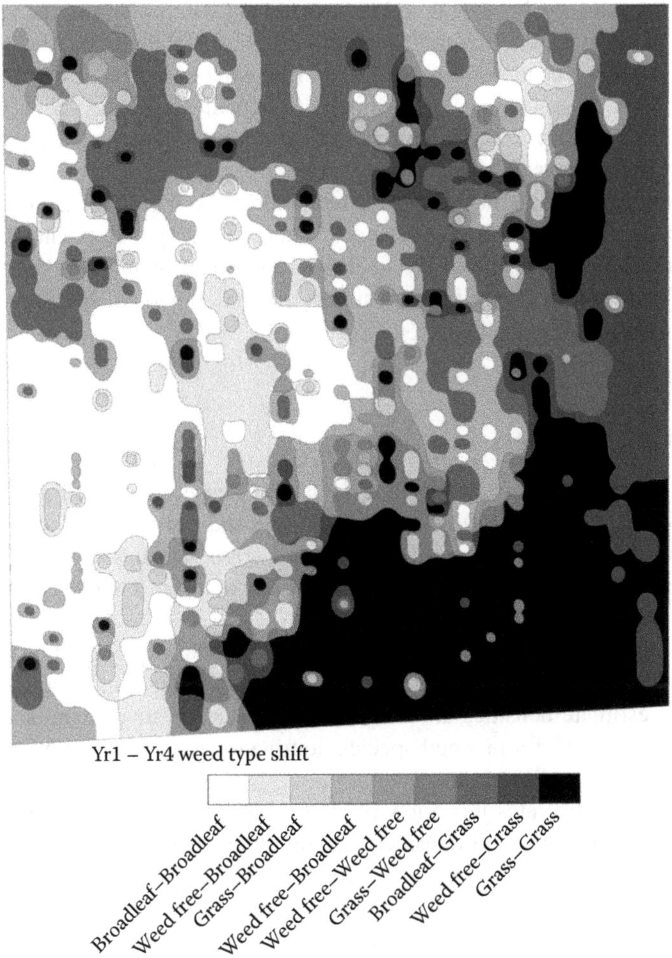

FIGURE 16.13 Results of intersected Y1 and Y4 data layers that display shifts in dominant weed type. The different shades are representative of the nine different conditions that may occur based on the three dominant types of conditions that are present each year.

Results of intersected Y1 and Y4 data layers display shifts in dominant weed type. The different shades are representative of the nine different conditions that may occur based on the three dominant types of conditions that are present in each year.

9. The area of each of the nine possible weed-shift combinations is calculated. In **ArcToolbox**, expand **Spatial Statistics Tools>Utilities>Calculate Areas**. Select ***Yr1_Yr4_intersect*** in **Input Feature Class** box and in **Output Feature Class** box, navigate to ***C:\Weed_Map\Shapefiles***, and enter ***Yr1_Yr4_intersect_areas*** in **Name** input box. Click **Save>OK**.
10. Right click ***Yr1_Yr4_intersect_areas***, and open the **Attribute Table** to see the calculated weed shift type area (Table 16.7).

The shift of grass to broadleaf only occurred on 2 ha, whereas the shift from broadleaf to grass was estimated to occur on 8.4 ha (Table 16.7). About 4.7 ha of the original weed-free areas remained weed free whereas 5.8 ha of weed-free areas shifted to grass species and 3.6 ha shifted to broadleaf species. New weed-free areas were more likely to come from broadleaf-dominated areas (9.2 ha) than from grass-dominated areas (3.1 ha).

This spatial analysis indicates that both broadleaf and weed-free areas were more apt to become infested with grass species. The greatest spatial extent of this shift occurred on the eastern side of the field where the field boundary abutted a grass pasture (Figure 16.13). On the western side of the field, broadleaf species dominance tended to remain unchanged.

16.5 CONCLUSIONS

This case study developed and used a dataset of densely collected georeferenced weed densities in a 65 ha (160 A field) to examine and improve field weed management. After 4 years of treatment, broadleaf and grass densities declined, but areas with grass species increased. The spatial extent of these changes could not be documented using these types of comparisons.

Collecting GPS data for each point allowed for spatial exploration of the dataset in ArcGIS™. ArcMap™ was used to create interpolated layers using kriging and IDW methods to estimate densities and locate weed species in unsampled areas. Using the **Intersect** tool, shifts in weed species dominance were mapped by location by combining two shapefiles.

This exercise illustrates just a few of many ways to examine this dataset. Other questions (e.g., locations and densities of individual weeds and shift in the occurrence of a specific broadleaf or grass species) could also be investigated. Including other data layers, including soil texture, yields, or topography, may reveal other spatial trends that could lead to alteration in weed control strategies.

There are many computer applications available that are appropriate for spatial analysis of this dataset. Software is often updated, and changes in keystrokes, utilities, or other significant updates are possible. However, this case study shows the power of using georeferenced data and GIS to work through scenarios. Becoming more proficient in the use of software of your preference may lead you to find other applications or data analysis methods that are comparable to or, better than, what is presented here. Happy Mapping!

ACKNOWLEDGMENTS

This research was partially funded by USDA-CSREES Grant 94-34214-1136, 99-36200-8702, and 2002-35108-11605; North Central IPM; EPS 0091948; South Dakota Soybean Research and Promotion Board; and South Dakota Corn Utilization Council. The authors would like to thank the producer, D. Deitrich, for the use of the field and technical assistance from K. Brix-Davis, T. DeSutter, C. Reese, S. Christopherson, G. Reicks, J. Lems, B. Kruetner, J. Kleinjan, F. Forcella,

C.G. Carlson, M.M. Ellsbury, and D. Clay. Thanks are also due to the many graduate and undergraduate students who experienced this field in an up-close and personal manner and learned seedling weeds as a result.

REFERENCES

1. Bigwood, D.W. and Inouye, D.W., Spatial distribution of the component species in an old field seed bank, and a comparison of sampling techniques, *Ecology*, 69, 497, 1988.
2. Van Groenendael, J.M., Patchy distribution of weeds and implications for modeling population dynamics: A short review, *Weed Res.*, 28, 437, 1988.
3. Halstead, S.J., Gross, K.L., and Renner, K.A., Geostatistical analysis of the weed seed bank, *Proc. N. Cent. Weed Sci. Soc.*, 45, 123, 1990.
4. Johnson, G.A., Mortensen, D., Young, L.J., and Martin, A., The stability of weed seedling population models and parameters in eastern Nebraska corn (*Zea mays*) and soybean (*Glycine max*) fields, *Weed Sci.*, 43, 604, 1995.
5. Marshall, E.J.P., Field-scale estimates of grass weed populations in arable land, *Weed Res.*, 28, 191, 1988.
6. Mortensen, D.A., Johnson, G.A., and Young, L.J., Weed distribution in agricultural fields. In: *Soil Specific Crop Management*, Robert, P.C., Rust, R.H., and Larson, W.E., Eds., ASA-CSSA-SSSA, Madison, WI, p. 113, 1993.
7. Nordmeyer, H. and Niemann, P., Möglichkeiten der gezielten Teilflächenbehandlung mit Herbiziden auf der Grundlage von Unkrautverteilung und Bodenvariabilität. *Z. Pflkrankh. Pflschutz Sonderheft*, 13, 539, 1992.
8. Thornton, P.K., Fawcett, R.H., Dent, J.B., and Perkins, T.J., Spatial weed distribution and economic thresholds for weed control, *Crop Prot.*, 9, 337, 1990.
9. Wiles, L.J., Oliver, G.W., York, A.C., Gold, H., and Wilkerson, G., Spatial distribution of broadleaf weed in North Carolina soybean (*Glycine max*) fields, *Weed Sci.*, 40, 554, 1992.
10. Clay, S.A., Lems, G.J., Clay, D.E., Forcella, F., Ellsbury, M.M., and Carlson, C.G., Sampling weed spatial variability on a fieldwide scale, *Weed Sci.*, 47, 674, 1999.
11. Scursoni, J., Forcella, F., and Gunsolus, J., Weed escapes and delayed weed emergence in glyphosate-resistant soybean, *Crop Prot.*, 26, 212, 2007.
12. Heisel, T., Andersen, C., and Ersbøll, A.K., Annual weed can be mapped with kriging, *Weed Res.*, 36, 325, 1996.
13. Trangmar, B.B., Yost, R.S., and Uehara, G., Application of geostatistics to spatial studies of soil properties, *Adv. Agron.*, 38, 45, 1985.

17 Creating and Using Weed Maps for Site-Specific Management

J. Anita Dille, Jeffrey W. Vogel, Tyler W. Rider, and Robert E. Wolf

CONTENTS

17.1 Executive Summary 405
17.2 Introduction 406
17.2.1 Obtaining Weed Spatial Distribution Information 406
17.2.2 Determining Economic Optimal Herbicide Rate Based on Weed Spatial Distribution 406
17.2.3 Developing the Prescription Map 408
17.3 Materials and Methods 408
17.3.1 General Procedures 408
17.3.2 Development of a Prescription Map 409
17.3.3 Collecting and Analyzing Data 411
17.4 Results and Discussion 411
17.4.1 Weed Species Composition and Herbicide Usage 411
17.4.2 Yield Results 413
17.5 Conclusion 417
Acknowledgments 417
References 417

17.1 EXECUTIVE SUMMARY

Information about weed spatial distribution and competitiveness, sprayer application technologies, and economics can be brought together within a geographic information system (GIS) in order to develop site-specific weed management (SSWM) approaches. Based on weed species, density, and size, potential crop yield loss can be determined, and this provides the basis to calculate the "economically optimal herbicide rate" to be applied. A prescription map can be created and applied using a variable rate sprayer. This approach was evaluated on one field over 2 years and then expanded to several crop fields. Results demonstrated that this method is a potentially practical approach to implementing SSWM.

17.2 INTRODUCTION

An integrated weed management (IWM) strategy maximizes producers' profits and reduces weed control inputs.[1,2] An integral part of an IWM strategy is the development of an economic threshold (ET), that is, the weed density at which the predicted yield loss equals the cost of control. An ET can assist in making decisions on whether or not to apply a management technique to a weed-infested field.[1,2] ET decisions are often made on a field-average-wide basis, except weeds are not typically spread evenly across a field but are distributed into patches.[3–5] An average weed population for a field may exceed an ET on a field-wide basis but may actually be below the ET in areas and warrant no weed management.[6,7] In other words, the ET value is too low because of overprediction of crop yield loss causing more weed management to be applied than needed.

Most crop fields have multiple weed species that need to be controlled. An ET can be converted into a value of competitive load that causes a given amount of yield loss. There is an opportunity to only apply weed control to a weed patch or to areas of a field that have weed densities greater than the ET. Uniform application involves spraying an herbicide at a constant rate across a field regardless of weed population densities. Conversely, variable rate application (VRA) technology can target weed populations by spatially adjusting herbicide rates as the sprayer travels through the field. Spatial distribution of weed species combined with yield loss predictions and economics can be used to develop prescription maps to optimize the use of chemical control practices.

17.2.1 Obtaining Weed Spatial Distribution Information

Many farmers and crop consultants recognize that weeds are not randomly distributed in agricultural fields. Producers often identify the areas of persistent weed problems, which then are used for selecting a preemergence (PRE) herbicide and are used to anticipate the year-to-year location of weeds. By intensively scouting fields for weed locations and densities, it has been documented that weeds were spatially distributed into patches.[3–5]

To obtain spatial distribution of multiple weed species in a field, we superimposed a uniform grid based on what our VRA equipment could accomplish. Thus, a 7.62 × 7.62 m grid (equal to 10 crop rows on 0.76 m centers × 7.62 m long) was developed based on sprayer boom width. A 1 × 1 m quadrat was located at the center of each grid cell and all weed species were identified, counted, and classified into size categories. These data were entered into a spreadsheet together with x and y coordinates of m east and m north. Several global positioning system (GPS) readings were taken throughout an area (at grid cell centers) in order to convert x and y coordinates into latitude and longitude coordinates to eventually guide the VRA equipment.

17.2.2 Determining Economic Optimal Herbicide Rate Based on Weed Spatial Distribution

An economical optimal herbicide rate (EOR) was determined and assigned to each grid cell. This was based on observed weed species, density, and size that were used to calculate the total competitive load (TCL) of that grid cell and the predicted yield loss (Y_L):[8,9]

$$\mathrm{TCL} \equiv \sum_{i=1}^{j} D_i \times \mathrm{ACI}_i \tag{17.1}$$

where

D is density of weed species (i)

ACI is adjusted competitive index value of weed species (i) with WeedSOFT® 2003 Kansas Version 8 (University of Nebraska, Lincoln, Nebraska) as the basis for adjusted competitive index values for each weed species

The TCL was then used to calculate the percent crop yield loss (Y_L) using a modified rectangular hyperbola model:

$$Y_L = (\mathrm{TCL} \times \mathrm{slope}) \times \mathrm{Adjustment} \tag{17.2}$$

where

the slope is the percent yield loss at low TCL

the Adjustment is the factor that changes the linear portion of the model to nonlinear at high TCL[8,9]

Actual crop yield loss (CY_L) is determined as a function of the expected crop yield goal (kg ha^{-1}):

$$CY_L = \left(\frac{Y_L}{100}\right) \times \text{yield goal} \tag{17.3}$$

The TCL predicted to remain after the postemergence (POST) herbicide application is calculated (PTCL):

$$\mathrm{PTCL} = \mathrm{TCL} \times \exp(-k \times H) \tag{17.4}$$

where

k is the coefficient of the herbicide efficacy

H is the herbicide application rate[10,11]

Once the PTCL is calculated, CY_L is recalculated using Equations 17.1 and 17.2. Profit is calculated using the following equation:

$$\mathrm{PROFIT} = CY_L \times \mathrm{CP} - H \times \mathrm{HP} \tag{17.5}$$

where

CP is the expected price of the crop at harvest (\$ kg^{-1})

HP is the price of the herbicide (\$ kg^{-1})

For each grid cell, the highest net profit was determined by optimizing the relationship between herbicide rate and actual yield. The herbicide rate generating the highest net profit was considered the EOR.

17.2.3 Developing the Prescription Map

Values generated by the spreadsheet for EOR are reviewed to ensure there are no recommendations to apply more than the labeled herbicide rate (1X). Given that the herbicide rate changes were based on volume, and that nozzle orifice size could not be changed during the variable rate POST application, the relative maximum range of herbicide rates was limited by the nonlinear increase in pressure as flow rates were increased. Considering all constraints, including minimum and maximum application volume quantity (L ha^{-1}) and recommended operating pressures of the nozzles, a relative maximum range of herbicide rates were established. EOR was converted into categories of 0X (i.e., no herbicide applied), and between 0.4X and 1.0X, and converted into values of L ha^{-1} spray volume.

Microsoft® Excel (Microsoft Corporation, Redmond, Washington) was used to calculate the EOR and ArcView GIS 3.2® (Environmental Systems Research Institute Inc., Redlands, California) was used to convert the spreadsheet recommendations to a two-dimensional shapefile that the herbicide application software could recognize. The *x* and *y* coordinates in the Excel spreadsheet (unique id numbers) were linked to the latitude and longitude coordinates (also unique id numbers) in the ArcView grid cell shapefile with the two tables (spreadsheet and ArcView) joined using these unique id numbers. The resulting prescription map based on spray volume changes was then uploaded into the appropriate VRA computer for application.

17.3 MATERIALS AND METHODS

17.3.1 General Procedures

These procedures were developed and evaluated on several fields in Kansas in 2003 and 2004. This particular example was a field study established at the Department of Agronomy Ashland Bottoms Research Farm positioned at 98° 38′ 07″ W and 39° 07′ 35″ N and located near Manhattan, KS. We proposed a general two-pass procedure to implement and evaluate SSWM. The first pass was a soil-applied (PRE) herbicide application followed by a map-based postemergence (POST) application. Soybean (cv "Asgrow AG 3302 RR/STS") were planted on June 10, 2003 at 296,000 plants ha^{-1} and on June 1, 2004 at 261,000 plants ha^{-1} in a 2.14 ha conventionally tilled field. A premix of flufenacet and metribuzin (206 + 302 g ha^{-1}) was applied PRE on June 11, 2003 and June 2, 2004 in 7.62 m strips for the entire length of the field at 0, 0.33, 0.67, and 1X of the labeled rate and replicated three to four times (Figure 17.1a). Application of reduced rates of PRE herbicide was based on changes in application volume, that is, a constant herbicide dilution was maintained in the tank and changes in L ha^{-1} of solution applied were achieved by altering spray nozzle, spray pressure, and applicator speed to increase or decrease the herbicide rate (Table 17.1).

After the PRE application, the 7.62 m strips were then further delineated into 7.62 m lengths and were treated as individual cells. Weed sampling points (a 1 × 1 m quadrat) were located in the center of each cell and weed density in each quadrat was assumed to represent the density of the 7.62 × 7.62 m grid cell. One day prior to the POST herbicide application when the soybeans were at the V3 growth stage in 2003 and the V2 growth

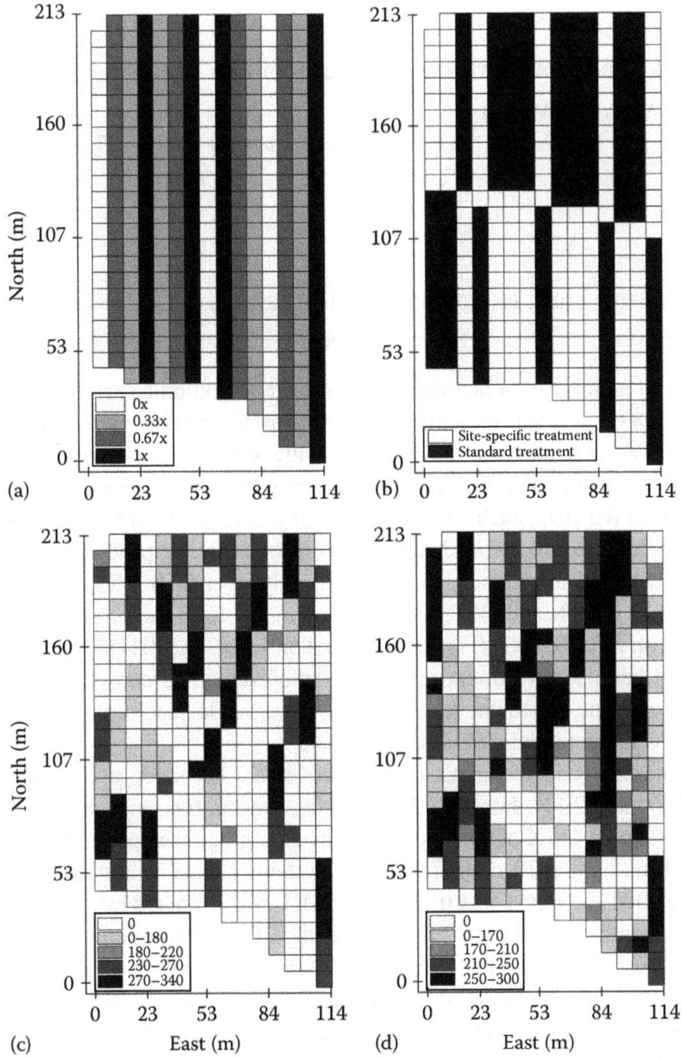

FIGURE 17.1 Treatment maps developed for (a) PRE herbicide strips applied in 2003 and 2004, (b) SSWM and standard postemergence treatments, and corresponding POST herbicide prescription map in L ha^{-1} for (c) 2003 and (d) 2004. Each box represents one 7.62 × 7.62 m grid cell.

stage in 2004, weed species were identified, classified into size categories, and counted at each weed sampling location on July 9, 2003 (PRE03) and June 23, 2004 (PRE04).

17.3.2 Development of a Prescription Map

Based on overall weed species and densities observed, the 1X labeled rate of a tank-mix of bentazon, sethoxydim, acifluorfen, and crop oil concentrate (227 g ha^{-1} + 85 g ha^{-1} + 57 g ha^{-1} + 1.2 L ha^{-1}) was recommended in 2003, whereas in 2004 the 1X

TABLE 17.1
Parameters Used to Apply Increasing Rates of PRE Herbicide

PRE Rate	Application Volume (L ha^{-1})	Pressure (kPa)	Spraying Nozzle	Travel Speed (km h^{-1})
0.33X	93	172	TT11002[a]	7.2
0.67X	197	172	TT11003	7.2
1.0X	80	241	TT11004	5.6

[a] Turbo TeeJet 11002, 11003, and 11004 nozzles, Spraying Systems Co., Wheaton, Illinois.

labeled rate of a tankmix of lactofen, imazamox, and nonionic surfactant (57 g ha^{-1} + 18 g ha^{-1} + .25% v v^{-1}) was recommended. For experimental purposes, each PRE herbicide strip was split into two ends (7.62 m wide × 12–14 cells long) and randomly assigned a site-specific or a standard herbicide treatment. Within the site-specific treatment end of a PRE strip, each 7.62 × 7.62 m cell received an individual recommendation. Within the standard herbicide treatment end of a PRE strip, at least four groups of three cells each (7.62 m wide × 23 m long) randomly received 0, 0.5, 0.75, or 1X labeled rate (Figure 17.1b). These POST herbicide treatments form the prescription map that was applied on July 10, 2003 (Figure 17.1c) and on June 24, 2004 (Figure 17.1d), where the standard herbicide treatment recommendation remained constant between years but the site-specific treatment received a new herbicide recommendation based on weed density and potential yield loss each year.

The equipment and the method of variable rate herbicide application differed between 2003 and 2004. For the 2003 growing season, the range was 0X (no herbicide applied) and 0.4–1X of the labeled rate by changing the application volume from 140 to 336 L ha^{-1} using TT11004 (Spraying Systems Co., Wheaton, Illinois) nozzles, whereas in 2004 the range was 0X and 0.45–1X of the labeled rate by changing the application volume from 140 to 299 L ha^{-1} using TT11004 spraying nozzles. For both years, speed was kept constant at 6.4 km h^{-1}. The large carrier volume was selected to keep recommended application volume, at any herbicide rate, above the minimum application volume on the herbicide label. In addition to following herbicide label directions, a large carrier volume made it easier for the sprayer controller to attain a particular target application rate. With conventional nozzles, a low application volume could enhance the application error and corresponding error in herbicide rate.

The herbicide tankmix was applied using the VRA sprayer constructed with commercially available equipment in the Department of Biological and Agricultural Engineering at Kansas State University during the spring of 2003. The sprayer was equipped with a 570 L tank and a 7.62 m three-section boom with Spraying Systems Turbo Teejet® (TT) 11002, 11003, and 11004 (Spraying Systems Co., Wheaton, Illinois) nozzles mounted on 15 three-position nozzle bodies spaced 51 cm apart. The sprayer controller was a Raven SCS 440 (Raven Industries Inc., Sioux Falls, South Dakota) with serial interface. To attain automatic product shutoff, the controller used a fast-close ball control valve on the 2.5 cm main product line placed after the flow meter. Sprayer-mounted radar was used to measure travel speed.

The variable rate sprayer was powered by a PTO-mounted Hypro belt-driven centrifugal pump (Hypro, New Brighton, Minnesota), and a Trimble® AgDGPS 132 (Trimble Navigation Limited, Sunnyvale, California) using Coast Guard differential correction with sub-meter accuracy communicated the sprayer's location to a Compaq™ Ipaq 3850 (Hewlett Packard Co., Palo Alto, California) running Farmworks® Farm Site Mate VRA (CTN Data Services Inc., Hamilton, Indiana) V.8.22 in 2003 and V.9.22 in 2004. With the creation and uploading of an appropriate prescription map to Farm Site Mate, this software program automatically transmitted recommended target herbicide rates from the prescription map to the sprayer controller and received actual application data that were logged and geo-referenced.

17.3.3 Collecting and Analyzing Data

Using the identical sampling points as the PRE03 and PRE04 weed surveys, weed populations were assessed 3 weeks after the POST herbicide treatment (3 WAT) on July 31, 2003 (POST03) and July 14, 2004 (POST04), and weed species were identified and placed into three categories: newly emerged, herbicide injured, and untreated weeds. To describe the weed species composition the following descriptive statistics were determined for each weed species: percent of quadrats occupied and mean and maximum density observed based on the initial weed survey at each location.

Soybean yield was obtained using a combine equipped with an Ag Leader© PF 3000 yield monitor (AgLeader Technology, Ames, Iowa) and a 6.7 m flex head that harvested the center eight crop rows of each 7.6 m PRE herbicide strip on October 23, 2003 and October 4, 2004. During harvest, the combine traveled at 4.8 km h^{-1} and yield was sensed at 1 s intervals so that each 7.62 × 7.62 m grid cell had four to five recorded yield data points. Unreasonable data points created by yield mapping errors (i.e., sudden starts or stops) were eliminated. The remaining yield data points were averaged within each grid cell and analyzed with Proc Mixed in SAS (Statistical Analysis Systems Inc., SAS Campus Drive, Cary, North Carolina) to evaluate yield differences due to PRE and POST herbicide treatments by using pairwise *t*-test comparisons.

17.4 RESULTS AND DISCUSSION

17.4.1 Weed Species Composition and Herbicide Usage

Nine weed species, including common waterhemp (*Amaranthus rudis* Sauer), honeyvine milkweed (*Cynanchum laeve* (Michx.) Pers.), ivyleaf morningglory (*Ipomoea hederacea* Jacq.), Palmer amaranth (*Amaranthus palmeri* S. Wats), prickly sida (*Sida spinosa* L.), smooth groundcherry (*Physalis longifolia* (Nutt.) var. *subglabrata* (Mackenzie & Bush) Cronq), velvetleaf (*Abutilon theophrasti*), shattercane (*Sorghum bicolor*), and yellow foxtail (*Setaria pumila* (Poir.) Roemer & J.A. Schultes) were present in the PRE03 and PRE04 weed surveys (Table 17.2). With the exception of Palmer amaranth, eight of the species were observed in the field on all four weed survey dates. A possible error in identification may have resulted in the absence of large crabgrass (*Digitaria sanguinalis* (L.) Scop.) in 2003 and its presence in 2004. In addition to individual weed species, composite totals of similar weed species groups

TABLE 17.2
Percent of Total Grid Cells (n = 362) with Weeds Present (%), Mean Density of Weeds in 1 m² Quadrat (Mean), and Maximum Density Observed for Each Weed Species (Maximum) Identified in the Ashland Bottoms Soybean Field for Each Weed Survey

	PRE03			POST03			PRE04			POST04		
		Number m^{-2}			Number m^{-2}			Number m^{-2}			Number m^{-2}	
Weed Species	%	Mean	Maximum	%	Mean	Maximum	%	Mean	Maximum	%	Mean	Maximum
Common waterhemp	9	0.15	6	5	0.06	3	38	3.21	90	6	0.27	32
Honeyvine milkweed	2	0.08	9	3	0.10	6	4	0.18	16	4	0.11	5
Ivyleaf morningglory	5	0.07	4	4	0.06	4	10	0.18	7	8	0.12	3
Palmer amaranth	5	0.07	4	1	0.01	1	2	1.76	186	1	0.05	12
Prickly sida	17	0.61	28	15	0.31	11	57	2.98	114	25	0.87	56
Smooth groundcherry	19	0.57	13	21	0.53	8	24	0.74	13	13	0.30	8
Velvetleaf	6	0.08	3	4	0.04	2	14	0.29	13	6	0.11	7
Large crabgrass	0	—	—	1	0.04	6	42	2.54	72	15	0.4	12
Shattercane	16	0.54	46	7	0.30	34	22	0.75	23	12	0.38	16
Yellow foxtail	11	0.27	7	17	0.33	9	16	0.63	16	13	0.38	16
Other spp.[a]	4	0.03	2	1	0.00	1	2	0.02	4	1	0.02	7
Total	61	7.24	92	57	1.05	49	92	11.4	288	65	5.46	112
Composite totals												
Pigweed spp.[b]	12	0.22	6	6	0.07	3	41	4.98	186	7	0.31	32
Grass spp.[c]	12	0.27	7	18	0.36	9	47	3.16	73	26	0.77	21

[a] Other spp. included common cocklebur, common sunflower, buffalobur, and common milkweed.
[b] Pigweed spp. included common waterhemp and Palmer amaranth.
[c] Grass spp. included yellow foxtail, large crabgrass, witchgrass, but excluded shattercane.

were calculated. Pigweed spp. included Palmer amaranth and common waterhemp, and grass spp. excluded shattercane but included yellow foxtail, large crabgrass, and witchgrass (*Panicum capillare* L.).

Overall, the percent of quadrats with weeds present and mean and maximum density observed were greater for almost every weed species with the PRE04 as compared to the PRE03 surveys. Total pigweed spp., prickly sida, and grass spp. percent of sampling quadrats occupied increased by 29%, 40%, and 35%, mean density increased by 4.76, 2.37, and 2.89 plants m^{-2}, and maximum observed density increased by 180, 86, and 66 plants m^{-2} between the PRE03 and PRE04 weed surveys, respectively. Velvetleaf and shattercane had the smallest increase between the PRE03 and PRE04 weed surveys. Both velvetleaf and shattercane increased in presence by 8%, both mean densities increased by 0.21 plants m^{-2}, and maximum weed density observed increased by 10 and 23 plants m^{-2}, respectively, between the PRE03 and PRE04 weed surveys. Overall, weeds were present in 61%–92% of the sampling quadrats, mean density increased from 7.24 to 11.4 plants m^{-2}, and maximum density observed increased from 92 to 288 plants m^{-2} between the PRE03 and PRE04 weed surveys, respectively. Based on the overall presence, mean density, maximum density observed, and propensity to result in yield loss, pigweed spp., prickly sida, grass spp., and shattercane were considered key weed species and were used to select the POST herbicide for both standard and site-specific treatments.

The increase in weed population densities between PRE03 and PRE04 weed surveys was not due to the increase in weed seedbank size from lack of weed control in 2003. Between 72% and 89% mortality was observed with SSWM in 2003 depending on PRE herbicide rate (data not shown). Although total rainfall prior to POST herbicide application was the same, the increase in weed population was due to the difference in total rainfall patterns between years (Figure 17.2) and an earlier planting date in 2004, which allowed more seeds to break dormancy resulting in more seedling recruitment.

In response to the weed populations, SSWM using the optimal rate based on weed species, density, and size, required treatment of 10%–56% of the area and used only 8%–40% of the herbicide as compared to a uniform 1X application across all PRE herbicide rates in 2003 and 2004, respectively (Table 17.3). In 2003, as PRE herbicide rate increased from 0 to 0.67X, area treated decreased from 17% to 4% and herbicide usage decreased from 10% to 2%. At the 1X PRE herbicide rate, area treated increased to 12% and herbicide usage increased to 8%. In 2004, area treated decreased from 93% to 28% and herbicide usage decreased from 84% to 16% between the 0X and 1X PRE herbicide rates.

17.4.2 Yield Results

Due to drought conditions in 2003 during the months of July and August, soybean yield averaged 1500 kg ha^{-1} across the field site, which was comparatively low to the 10 year Riley County nonirrigated soybean average yield of 2100 kg ha^{-1} (NASS 2005).[12] Between July 1, 2003 and August 27, 2003, which included soybean stages V-3 through R-5, only 5.3 cm of precipitation fell reducing soybean yield potential (Figure 17.2). Even with drought conditions, soybean yield differed

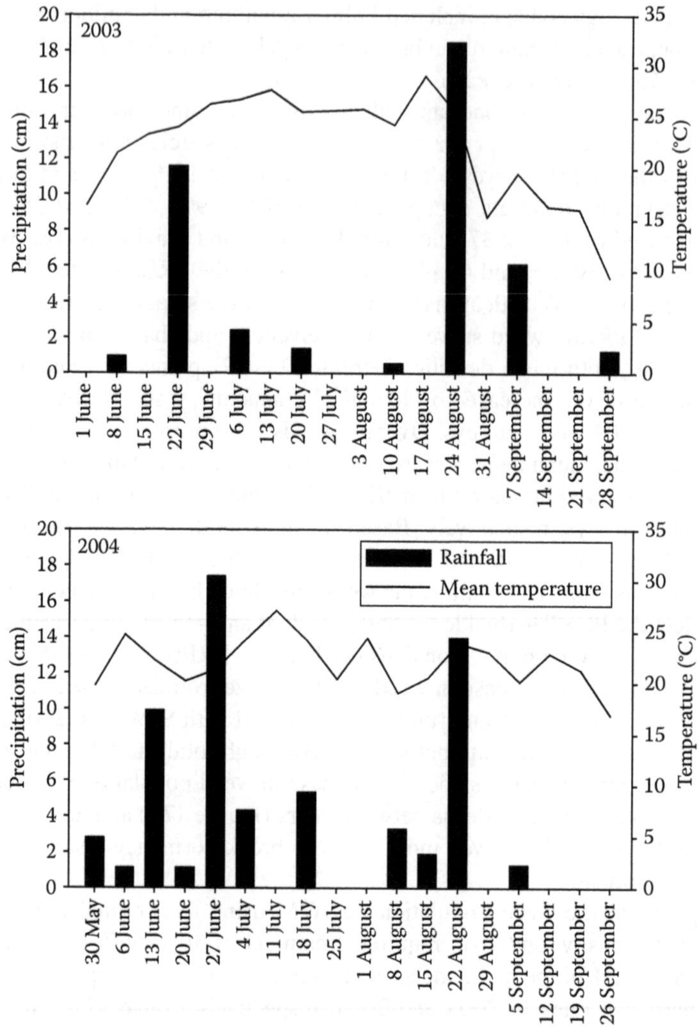

FIGURE 17.2 Weekly precipitation and mean temperature during the growing season at the Department of Agronomy Ashland Bottoms Research Farm, Manhattan, Kansas.

across PRE herbicide rates, SSWM and standard POST treatments, and POST herbicide rates (Table 17.4). Yield between the SSWM and standard POST herbicide treatment differed at the 0 and 0.5X POST herbicide rates within the 0.33X PRE rate (Table 17.4). In general, yield did not respond consistently to an increase in either PRE or POST herbicide rates. The differences in yield were due to other spatial factors that influenced the yield response more than the PRE herbicide rates or POST treatments. Consequently, it could not be determined in 2003 that SSWM increased or maintained soybean yield as compared to a uniform herbicide application.

TABLE 17.3
Percent Area Treated and Percent Herbicide Usage with POST SSWM and Total Herbicide Cost Reduction Associated with the PRE and POST SSWM Herbicide Applications as Compared to a Uniform 1X Herbicide Application for 2003 and 2004

		POST SSWM (%)		Total Cost Reduction
Year	PRE Rate	Area Treated	Herbicide Usage	(%)
2003	Across all PRE rates	10 (2.5)	8 (1.5)	
	0X	17 (5.0)	10 (3.3)	90
	0.33X	10 (5.3)	3 (3.5)	89
	0.67X	4 (2.3)	2 (1.3)	77
	1X	12 (6.2)	8 (4.0)	65
2004	Across all PRE rates	56 (7.0)	40 (3.0)	
	0X	93 (4.9)	84 (12)	46
	0.33X	78 (3.3)	50 (2.6)	54
	0.67X	43 (5.3)	23 (3.8)	67
	1X	28 (6.7)	16 (4.5)	64

Note: Number within parentheses indicates the standard error across replications.

TABLE 17.4
Soybean Yield for 2003 as Affected by PRE Herbicide Rate, POST Herbicide Treatment, and POST Herbicide Rate[a,b]

	POST	POST Rate (X Proportion of Label Rate)			
PRE Rate	Treatment	0X	0.5X	0.75X	1.0X
0X	SSWM	1230 (127) A[b]	1480 (213) A[a]	1330 (332) A[a]	—[c]
0.33X	SSWM	1470 (111) B[a,b,d]	1930 (241) A[a,d]	1750 (249) A[a]	—
0.67X	SSWM	1460 (109) A[a,b]	1160 (331) A[a]	1630 (327) A[a]	—
1X	SSWM	1710 (108) A[a]	1920 (240) A[a]	1450 (216) A[a]	1290 (328) A
0X	Standard	1400 (168) B[b]	1280 (155) B[a,b]	1420 (155) B[a]	1720 (147) A[a]
0.33X	Standard	1940 (141) A[a,d]	1210 (134) C[a,b,d]	1320 (131) BC[a]	1550 (129) B[a]
0.67X	Standard	1690 (137) A[a,b]	1070 (141) B[b]	1340 (134) B[a]	1540 (131) A[a]
1X	Standard	1480 (137) A[b]	1500 (134) AB[a]	1220 (134) C[a]	1580 (127) A[a]

Note: Number within parentheses is the standard error.

[a] Means within a row followed by a different uppercase letter are significantly different at P = 0.05.

[b] Means within a column separated by the dashed line with a different lowercase letter are significantly different at P = 0.05.

[c] Treatment combination was not applied.

[d] Means within PRE and POST herbicide rate and between SSWM and standard treatments are significantly different at P = 0.05.

Better environmental conditions in 2004 contributed to increased soybean yield, which averaged 3430 kg ha^{-1} across the field and was significantly higher than 2003. Between July 1, 2004, and August 27, 2004, the same drought interval as 2003, 38 cm of precipitation, fell, which increased the yield as compared to 2003 (Figure 17.2). In 2004, soybean yield differed between PRE and POST herbicide rate combinations but did not differ between the SSWM and standard POST herbicide treatments (Table 17.5). For the SSWM treatment with any PRE rate excluding 0X, soybean yield was the same except in one case. In general, for the standard herbicide treatment at any PRE rate excluding 0X, only two PRE and POST herbicide rate combinations produced lower yield.

With no PRE herbicide, yield was lower with no POST herbicide for both the SSWM and standard treatments as compared to the other POST herbicide rates. For the standard herbicide treatment, a reduction in yield was expected since weed density was not considered as part of choosing the rate applied. It was also expected that yield would remain consistent for all POST herbicide rates within the SSWM treatment since weed density and potential yield loss were considered. The reduction in yield with the SSWM treatment in areas of the field that received no PRE or POST herbicide treatment was due to limitations with our sampling strategy. Weed populations that reduced yield existed in areas of the cell that were not sampled with the quadrat. The yield results demonstrate that a reduced rate PRE herbicide application was necessary to minimize the impact that missed weed populations would cause.

TABLE 17.5
Soybean Yield for 2004 as Affected by PRE Herbicide Rate, POST Herbicide Treatment, and POST Herbicide Rate[a,b]

	POST	POST Rate (X Proportion of Label Rate)			
PRE Rate	Treatment	0X	0.5X	0.75X	1.0X
0X	SSWM	2360 (361) B[b]	3730 (361) A[a,b]	3690 (378) A[a]	3670 (176) A[a]
0.33X	SSWM	3260 (207) A[a]	3580 (164) A[a,b]	3450 (214) A[a]	3350 (260) A[a]
0.67X	SSWM	3450 (158) A[a]	3300 (183) A[b]	3400 (320) A[a]	—[c]
1X	SSWM	3480 (150) A[a]	3880 (228) A[a]	3330 (280) A[a]	—
0X	Standard	1790 (250) B[b]	3200 (231) A[a]	3300 (231) A[a]	3440 (223) A[a]
0.33X	Standard	3710 (214) A[a]	3660 (201) A[a]	3430 (201) A[a]	3700 (196) A[a]
0.67X	Standard	3690 (214) A[a]	3410 (201) A[a]	3590 (201) A[a]	3440 (196) A[a]
1X	Standard	3151 (214) B[a]	3700 (201) A[a]	3203 (201) B[a]	3534 (192) AB[a]

Note: Number within parentheses is the standard error.

[a] Means within a row followed by a different uppercase letter are significantly different at P = 0.05.

[b] Means within a column separated by the dashed line with a different lowercase letter are significantly different at P = 0.05.

[c] Treatment combination was not applied.

17.5 CONCLUSION

Weed populations expanded in the field from 2003 to 2004 causing a corresponding 32% increase in herbicide usage in 2004 for SSWM across all PRE herbicide rates but was still 60% less than a uniform 1X POST herbicide application. In 2003, total herbicide cost for SSWM was reduced by 90% at the 0X PRE herbicide rate, 89% at the 0.33X PRE herbicide rate, 77% at the 0.67X PRE herbicide rate, and 65% at the 1X PRE herbicide rate (Table 17.3). In 2004, total herbicide cost for SSWM was reduced by 46% at the 0X PRE herbicide rate, 54% at the 0.33X PRE herbicide rate, 67% at the 0.67X PRE herbicide rate, and 64% at the 1X PRE herbicide rate.

A comparison of total herbicide cost reduction and crop yield in 2003 was not useful due to the lack of relationship between herbicide treatments and crop yield. On the other hand, the yield results in 2004, when moisture did not limit soybean production, indicated that a PRE herbicide was necessary to reduce weed populations and maintain crop yield since the most significant reduction was in areas of the field that received no PRE or POST herbicide applications with either SSWM or standard treatments. Therefore, the 0.67X PRE herbicide rate in 2004 was the optimal PRE herbicide rate based on the highest reduction in total herbicide cost while maintaining crop yield (Table 17.3).

ACKNOWLEDGMENTS

This material is based upon work supported by the Cooperative State Research, Education, and Extension Service (CSREES), U.S. Department of Agriculture under Agreements 2003-34103-13192 and 2003-41530-01603 (North Central Regional IPM Research—Extension Grant). The authors would like to thank the Kansas State University Department of Biological and Agricultural Engineering and the Department of Agronomy. Additionally, we would like to thank our producer cooperators, Doug Keesling, John Oden, Harlan Ebright, Kenny Tucker, and Lee Scheufler, for permitting us to conduct research on their farms and helping implement SSWM across Kansas.

REFERENCES

1. Swanton, C.J. and Weise, S.F., Integrated weed management: The rationale and approach, *Weed Technology*, 5, 657, 1991.
2. Thornton, P.K., Fawcett, R.H., Dent, J.B., and Perkins, T.J., Spatial weed distribution and economic thresholds for weed control, *Crop Protection*, 9, 337, 1990.
3. Dieleman, J.A. and Mortensen, D.A., Characterizing the spatial pattern of *Abutilon theophrasti* seedling patches, *Weed Science*, 39, 455, 1999.
4. Johnson, G.A., Mortensen, D.A., Young, L.J., and Martin, A.R., The stability of weed seedling population models and parameters in eastern Nebraska corn (*Zea mays*) and soybean (*Glycine max*) fields, *Weed Science*, 43, 604, 1995.
5. Wiles, L.J., Oliver, G.W., York, A.C., Gold, H.J., and Wilkerson, G.G., Spatial distribution of broadleaf weeds in North Carolina soybean (*Glycine max*) fields, *Weed Science*, 40, 554, 1992.

6. Cardina, J., Sparrow, D.H., and McCoy, D.L., Analysis of spatial distribution of common lambsquarters (*Chenopodium album*) in no-till soybean (*Glycine max*), *Weed Science*, 43, 258, 1995.
7. Lindquist, J.L., Dieleman, J.A., Mortensen, D.A., Johnson, G.A., and Wyse-Pester, D.Y., Economic importance of managing spatially heterogeneous weed populations, *Weed Technology*, 12, 7, 1998.
8. Neeser, C., Dille, J.A., Krishnan, G., Mortensen, D.A., Rawlinson, J.T., Martin, A.R., and Bills, L.B., WeedSoft™: A weed management decision support system, *Weed Science*, 52, 115, 2004.
9. Rider, T.W., Vogel, J.W., Dille, J.A., Dhuyvetter, K.C., and Kastens, T.L., An economic evaluation of site-specific herbicide application, *Precision Agriculture*, 7, 379, 2006.
10. Dieleman, J.A., Hamill, A.S., Fox, G.C., and Swanton, C.J., Decision rules for postemergence control of pigweed (*Amaranthus* spp.) in soybean (*Glycine max*), *Weed Science*, 44, 126, 1996.
11. Pannell, D.J., An economic response model of herbicide application for weed control, *Australian Journal of Agricultural Economics*, 34, 223, 1990.
12. National Agriculture Statistics Service [NASS], U.S. Department of Agriculture, 2005 <http://www.nass.usda.gov:81/ipedbcnty/c_KScrops.htm> (verified May 2, 2005).

Index

A

Agent-based simulation and modeling (ABSM), 212–213
Animal Disease Research and Diagnostic Laboratory, 193
ArcGIS 9.3.1, 43
ArcGIS®
  ArcMap™
    base data, 236–237
    Create Graph Wizard dialog box, 250
    interpolation layer, 245
    Moran's *I* analysis, 242
    Spatial Analyst tools, 244, 246
  bar graph, 251
  coordinate systems and ESRI® shapefiles, 240
  latitude–longitude trap data importing, 237–238
  map layer symbology, 238–239
  tabulate area dialog box, 249
  Zonal statistics, 247
ArcMap™
  base data, 236–237
  Create Graph Wizard dialog box, 250
  data exploration
    Histogram option, 384–385
    Kurtosis value, 385, 387
    Leptokurtic distributions, 385
    mean and median values, 384
    Platykurtic distributions, 385
    skewness coefficient, 385
    standard deviation, 384–385
    Yr4 *vs.* Yrl data, 385–387
  data subset creation, 383–384
  interpolation layer, 245
  layer creation
    common coordinate system, 380
    event themes creation, 380–382
    layer conversion, 381–383
  model validation, 380
  Moran's *I* analysis, 242
  Spatial Analyst tools, 244, 246
  Tabulate Area tool, 248–251
ArcMap V.9.2
  data plotting, 179–181
  Myzus persicae population data, 175
  spatial autocorrelation, GS+, 178
ArcToolbox, 398–399
ArcView GIS 3.2®, 408
Australian Pest Animal Strategy, 94
Australian Weeds Strategy, 94

B

Bean pod mottle virus (BPMV), soyabean
  field-level information display, GIS software, 81
  Iowa soybean fields, 81–82
  logistic regression, 81
  Moran's Index, 81
  regression analysis, 83
  risk factors evaluation, 79
  seed infection reduction, 83
  winter temperature gradients, 84

C

Centaurea solstitialis, *see* Yellow starthistle
Chapter_15_data.xls file, 322–324, 326
Chilean needle grass (CNG), 256–257
  biodiversity, 258
  dispersal along roads, 268
  dispersal along watercourses, 267–268
  eradication evaluation, 270
  life history parameters, 268–269
  management, 273–274
  potential habitat, 267
  surveillance evaluation, 269–270
  wind dispersal, 267
CIMMYT
  GIS-based surveillance and monitoring systems, 136
  resistance mechanisms and virulence, Ug99, 134
  RustMapper, 148
  wheat areas, 144
  wheat cultivars, 145
Climate/environment
  above-average rainfall, 147–148
  CMORPH, 147
  dry and wet deposition, 146
  Sr24 variant, 147
  Ug99 assessment, 147
Climate Prediction Center Morphing Technique (CMORPH), 147
CMORPH, *see* Climate Prediction Center Morphing Technique
CNG, *see* Chilean needle grass
COKB3D.PAR file, 360–362

Corn rootworm (CRW)
ArcGIS®
ArcMap™ session, 236–237
coordinate systems and ESRI® shapefiles, 240
latitude–longitude trap data importing, 237–238
map layer symbology, 238–239
areawide management site, South Dakota, 234
field and insect trap locations, 235
IDW interpolation, 243–245
larval root feeding, 234
vs. soil texture
chi square analysis, 245
MAJORITY and MEDIAN values, 248
soil classifications, 245
Tabulate Area tool, 248–251
zonal statistical analysis, 245–248
spatial autocorrelation, Moran's *I*, 241–242
system requirements, 235–236

D

Decision-making process, 7
Diabrotica spp., Corn rootworm (CRW)
Digital Orthophoto Quadrangle (DOQ), 17–18, 21
Digital Orthophoto Quarter Quads (DOQQs), 18, 21
Digital raster graphic (DRG)
PackBits compression, 13
SDView Web site (*see* South Dakota View Web site)
UTM projection, 13
Disease management principles
chemical barriers, crops protection (y_0 and *r*), 64
disease risk avoidance (y_0 and/or *t*)
in space (*t*), 62
in time (*t*), 62–63
eradication
crop residues/debris removal and burial (y_0), 64
diseased plants roguing (y_0), 63
soil fumigation (y_0), 64
exclusion
quarantine (y_0), 62
seed/plant certification programs (y_0), 62
host resistance, initial inoculum reduction (y_0), 65
therapy (y_0 and sometimes r), 65–66
DOQ, *see* Digital Orthophoto Quadrangle
DOQQs, *see* Digital Orthophoto Quarter Quads
DRG, *see* Digital raster graphic

E

Earth Resources Observation and Science (EROS), 3
Economical optimal herbicide rate (EOR), 406–408
Economic threshold (ET), 406
ERDAS Imagine software
image uploading, 306
supervised image classification
accuracy assessment, 313–314
conduction procedure, 310–312
South Bend image, 313
test pixels, 309, 313
training pixels, 309
unsupervised image classification
accuracy assessment, 307–308
class distribution, 307
GPS-referenced ground scouting, 307
iterative process, 306
multispectral image, 307–308
producer's/reliability accuracy, 309
EROS, *see* Earth Resources Observation and Science
ESRI® shapefiles, 240

F

Fisher and Skellam models, 38

G

GAMV.par file, 335–337
GeoDA 0.9.5 software
attribute table addition, 203
empirical Bayes smoothing, 203–204
global Moran's *I* index, spatial autocorrelation, 204–205
local Moran's *I* index, spatial autocorrelation, 204–205
raw disease rates map creation, 203–204
shapefile opening, 202
spatial weights file computing, 203
Geographic information science (GIScience), 212
Geographic information systems (GIS)
data structures, 5–6
digital map, 5
early settlers and modern-day society comparison, 6–7
geocomputation, 212
geo-rectification process, 5
geospatial data, 4–5
GPS location information, 6
Internet-based tools, 5
software, 283
topographic knowledge, navigation, 4

Geo-rectification process, 5
Geostatistics, weed populations
analysis procedure, 321–322
cross-semivariograms and cokriging
biological and environmental variable, 356
COKB3D function, 360–362
GAMV function, 360
mean density, 359
PIXELPLT, 360, 363–364
primary and secondary variable, 356
data collection, 322–324
data transformation, 330–331
decision making, 369
detrending data
large-scale spatial trend removal, 330
median polishing, 330–331
trend estimation, linear regression, 331–334
empirical semivariograms
GAMV function, 335
GAMV.par file, 335–337
nugget, 335
PROC VARIOGRAM and **PROC GPLOT,** 335, 339–340
ranges, 335
Setaria viridis, 335, 340
setvi199394.dat file, 335–336
small-scale spatial trends, 334
WinGslib variogram, 335, 338
error analysis
prediction of weed location, 363, 365–369
prediction of weed means, 363
exploratory data analysis
Data1993 worksheet, 325, 329
frequency distribution, 323
mean, standard error, and skew functions, 323–324
mean-variance dependency, 324, 327
PROC GMAP program, 325, 328–329
PROC UNIVARIATE function, 325–326
Setaria viridis densities, 323, 325
GSLIB DOS version, 322
kriging
KB2D and **KB3D** functions, 356
PIXELPLT, 356
PROC KRIGE2D program, 356–358
seedling densities, 354
skewness and mean-variance correlation, 354
S. viridis, 357, 359
semivariogram model fitting
DUD *vs*. GAUSS, 345
iterative least-squares procedure, 340
linear and power models, 338
nested spherical model, 338–340, 345
PROC NLIN program, 340–344
spatial correlations, 335
white noise, 345
zonal anisotropy, 346
variogram parameter analysis
Glycine max, 354
linear model, 346, 354
pre- and postharvest seed production, 346
PROC GLM program, 346–353
seed dispersal and germination, 354–355
spatial variability, 354
GIS, *see* Geographic information systems
Global positioning systems (GPS), 6
Green Peach Aphid, *Myzus persicae (Sulzer)*
colorado potato beetle, *Leptinotarsa decemlineata*, 168
holocyclic life cycles, 169
HYSPLIT model
airborne particulate pollutants, 173
forward and back trajectory models, 143
implementation, 139
Lagrangian trajectory methods, 173
LLJ examination, Red River Valley, 183–186
meteorological forecasting models, 174
M. persicae prediction, 168
Peronospora tabacina, 174
results, Red River Valley, 186
RustMapper, 148
wind trajectories, 141–142
Lagrangian trajectory methods, 173
LLJ streams, 171–172
Peronospora tabacina, 174
PLRV, 169–170
site-specific pest management, 167–168
spatiotemporal colonization patterns, seed potato
aphid infestation, 174
data point map creation, 179–180
dataset, 175–176
interpolated map creation, 180–181
interpolated surface output, GS+, 178–179
observed colonization patterns, 182–183
spatial autocorrelation, 176–178
GS+ V.5.0
Myzus persicae population data, 175
spatial autocorrelation, semivariograms, 176–178

H

Hybrid Single-Particle Lagrangian Integrated Trajectory (HYSPLIT) model
airborne particulate pollutants, 173
forward and back trajectory models, 143
implementation, 139
Lagrangian trajectory methods, 173

LLJ examination, Red River Valley
backtrack plot, LLJ, 183–185
LLJ movement plot comparison, 185–186
meteorological forecasting models, 174
M. persicae prediction, 168
Peronospora tabacina, 174
results, Red River Valley, 186
RustMapper, 148
wind trajectories, 141–142

I

ICARDA, 134, 136
IDW method, *see* Inverse distance weighting method
Information collection, collation, and reporting
climate/habitat matching methods, 111–112
CLIMATE software, 112
data aggregation and scaling-up, 110–111
data consolidation, 110
geographic information systems tool
ArcView (version 3.3) ESRI, 105
ArcView screen capture, 106–108
GIS data capture routine, 106
land-use classification, 112
stepwise data collection and collation
abundant species trend, 109
data quality, 109–110
species classification, 108–109
species distribution, 108
species occurrence, 107–108
Insect damaged kernels (IDK), 209
Integrated weed management (IWM), 406
Invasive plants
ArcGIS, 256
eradication evaluation, 273
kernel definition, 257
methods
life history, 265–266
parameterization, CNG (*see* Chilean needle grass)
seed dispersal modeling (*see* Seed dispersal modeling)
surveillance simulation, 266–267
toolbox design elements, 258–260
wind dispersal kernel (*see* Wind dispersal kernel)
projected population growth, 270–272
Python programming language, 256
surveillance evaluation, 270, 272
wind-mediated seed dispersal, 257
Invasive species
definition, 92
detection, surveillance, and management
data synergy, 7
GIS (*see* Geographic Information Systems)
remote sensing (*see* Remote sensing)
distribution prediction
climate matching programs, 100
CLIMATE modeling, 100–102
potential distribution maps, 102–103
sleeper weeds, 103
monitoring and reporting, 121
monitoring efforts
data aggregation and scaling-up, 114
habitat and climate suitability, 114, 116
multiple attribute maps, 114, 116
multiple species data report, 114, 117
single attribute data report, 114–115
monitoring protocol
abundance, 122
data quality, 123
distribution, 122
occurrence, 122
trend, 122–123
spread estimation
Fisher and Skellam models, 38
gypsy moth, 39
radial rate of expansion, 40
Invasive weed species
atmospheric correction
DN haze value, 291–292
haze and atmospheric effects, 289
optical properties, 290–291
scene correction, 292
sensor parameter identification, 291
sunlight and visible haze, 291–292
environmental gradients, 278
georectification, 292–293
IDRISI software, 285
model development
linear/nonlinear parametric forms, 281
logit function, 280
on-the-ground presence–absence occurrence data, 280
plant community biomass, 281
productivity map, 279
network modeling, 296–297
plant community structure, 278
plant productivity and topography, 279
productivity model
components, 281–283
logit regression module, 295–296
spatial network models
anisotropic model, 283–284
cost assessment, 283
establishment and dispersal models, 284
GIS software, 283
isotropic model, 283–284
occurrence predictions, 285
topographic site correlation
preliminary steps, 286–287
slope, aspect, and sun angle differencing, 287–289

soil temperature and moisture, 285
USGS NED, 286
vegetation index calculation
NDVI layer development, 294
slope-/distance-based models, 293
TSAVI1 layer development, 294–295
Inverse distance weighting (IDW) method
CRW, 243–245
spatial data exploration
ArcToolbox, 398–399
broadleaf and weed-free areas, 402
Geostatistical Analyst Output Layers, 398
interpolation procedure, 397
layer intersection, 399–401
spatial distribution, 392
Yr1 weed species, 399
IWM, *see* Integrated weed management

L

Lesser grain borer (LGB)
behavior and ecology, 211–212
economic impact and management, 209–210
geocomputation
ABSM, 212–213
aspatial/spatial statistics, 208
computational 'black box' methods, 212
vs. GIS and GIScience, 212
NetLogo (*see* NetLogo)
IDK, 209
Triticum aestivum, 208
LGB ABS Demo A.nlogo, 229
LGB ABS Demo B.nlogo, 229
Likelihood ratio statistic, 200
Low level jet (LLJ) streams, 171–172

M

Mean prediction error, 369
Mean residual error, 363, 369
Mentha piperita, *see* Peppermint
Modifiable areal unit problem (MAUP), 193
Moko disease, banana
ArcGIS, 74
banana subsistence farms, 74–75
Chi-square analysis, 74
Ralstonia solanacearum race 2, 73
spatial analyses, *k*-function analysis, 74
yield-reducing factor, 73
yield-reducing risk factor, 75
Monte Carlo simulation, 200
Moran's Index (Moran's *I*), 177
ArcMap™, 242
BPMV, soyabean, 81
spatial autocorrelation
coefficient, 241–242, 251
degree of spatial dependency, 241
empirical Bayes methods, 201
GeoDA 0.9.5 software, 204–205
global statistics, 197, 204
local analysis, 198–199
scatterplot, 197–198
tool, 46, 48
spatial data exploration, 387
Z-score, 389

N

Nassella neesiana, *see* Chilean needle grass
National Aeronautics and Space Administration (NASA), 3
National Agricultural Imagery Program (NAIP), 21–22
NetLogo
"boom and bust" cycle, 228
bug-birth-energy slider, 222
bug movement, eating, reproduction, and death procedures, 220–222
Community Models library, 214
final view, 228–229
forest regrowth procedure and control slider, 222–223
gain-from-grain slider, 222
HubNet tool, 213
initial go procedure, 219–220
initial variables with sliders, 218–219
LGB ABS Demo A.nlogo, 229
LGB ABS Demo B.nlogo, 229
plot window
create and update, 226
monitors, 227–228
Plot Pens section, 227
revised go procedure, 226–227
setup and go buttons and energy switch, 215–216
show energy and display labels, 223–225
software tutorials, 214
turtle variables and setup procedure, 216–218
types of agents, 213–214
Nonnative species spread rate estimation, GIS
gypsy moth invasion, United States, 42–43
initial outbreak location, distance calculation, 43–44
OLS regression analysis (*see* Ordinary Least Squares regression analysis)
regional spread rate calculation, 49–50
residuals, diagnostic tool, 46
spatial autocorrelation (*see* Spatial autocorrelation)
temporal spread rate calculation, 47, 49

O

Online aerobiology process model
configuration, 162–163
model-based services, 160
online simulation
GIS tools, 164
main screen, 163–164
Web site, 163
wheat stem rust model output, 165–166
principles
deterministic model design, 160
stages, 161–162
Online simulation
aerobiology process model main screen, 163–164
GIS tools, 164
Web site, 163
wheat stem rust model output, 165–166
Ordinary kriging method, 393–396
Ordinary Least Squares (OLS) regression analysis
ArcMap, 43
OLS coefficient output table, 45–46
ordinary least squares dialogue box, 44–45
Spatial Statistics Tools, 44

P

Peppermint
crop health assessment, 314–315
GIS-based weed mapping, 315–316
grayscale infrared image, 304–305
herbicides, 302
hyperspectral remote sensing, 303
image uploading, 306
meadow mint, 302
multispectral remote sensing, 303–304
SSWM, 302–303
supervised image classification
accuracy assessment, 313–314
conduction procedure, 310–312
South Bend image, 313
test pixels, 309, 313
training pixels, 309
unsupervised image classification
accuracy assessment, 307–308
class distribution, 307
GPS-referenced ground scouting, 307
iterative process, 306
multispectral image, 307–308
producer's/reliability accuracy, 309
WAAS, 304
weed detection, 315
Persistently transmitted potato leafroll virus (PLRV)
aphid densities, 170
HYSPLIT results, 186
Pest animals and weeds distribution, Australia
agreed data attributes and standards, 104
consistent data collection methods/protocol, 104–105
current initiatives, 99–100
field manuals, monitoring, 104
GIS tools, 95
information collection, collation, and reporting (*see* Information collection, collation, and reporting)
information needs
Australian continent area, 97
consistent national datasets, 98
mapping classes definition, 95–96
invasive species (*see* Invasive species)
large-scale mapping and monitoring efforts, 113
limitations of methods
climate/habitat matching, 118–119
data collation and reporting, 116–118
habitat matching, land use data, 119–120
monitor and report information, 103–104
national level report, 120
occurrence, distribution, and density attributes, 123–125
previous mapping initiatives, 98–99
species impact reduction, 95
weed source, 93–94
zoonotic diseases, 94
PIXELPLT.EXE file, 360, 364
PIXELPLT.PAR file, 360, 364
Plant disease management
ash yellows disease, green ash
Geotracker GPS unit, 69
GPS maps, 69–70
phytoplasmas, 68
spatial dependence, 71
spatial patterns, 70
UTM coordinates, 69
Asian soybean rust
county scale maps, 68
ground-based reflectance measurements, 66–67
IKONOS satellite image, Cedara soybean field, 66–67
multispectral radiometer, 67
pathogen specific temporal and spatial signatures, 66
visual disease severity assessments, 66–67
corn gray leaf spot, 78–79
corn Stewart's disease, 76–77

disease management principles (*see* Disease management principles)
GIS tutorial, Moko disease
ArcMaP, 84
creating and printing map layouts, 85–87
map symbology, 84–85
Moko disease, banana
ArcGIS, 74
banana subsistence farms, 74–75
Chi-square analysis, 74
Ralstonia solanacearum race 2, 73
spatial analyses, *k*-function analysis, 74
yield-reducing factor, 73
yield-reducing risk factor, 75
plant disease epidemics, definition, 87
plum pox virus of *prunus* spp.
nearest neighbor type analysis, 72
Ripley's *k*-function analysis, 71
spatial dependence, 73
soybean pod mottle virus (*see* Bean pod mottle virus, soyabean)
PLRV, *see* Persistently transmitted potato leafroll virus
Population ecology, biological invasions
arrival
biological invaders, 33–34
invasion pathways, 32–33
establishment
factors, 34–35
space–time population persistence (*see* Space–time population persistence)
management, 40–42
nonnative species spread rate estimation, GIS
gypsy moth invasion, United States, 42–43
initial outbreak location, distance calculation, 43–44
OLS regression analysis (*see* Ordinary Least Squares regression analysis)
regional spread rate calculation, 49–50
residuals, diagnostic tool, 46
spatial autocorrelation (*see* Spatial autocorrelation)
temporal spread rate calculation, 47, 49
process, 30–31
spread
invasive species spread estimation (*see* Invasive species, spread estimation)
types, 37–38
Prediction standard error map, 395–396
Puccinia graminis f.sp. *tritici*, GIS applications, Wheat stem rust
Python functions, 256
terrain influences, wind dispersal, 263
toolbox design elements, 258
wind direction and strength, 262–263
Python programming language, 256

R

Raven SCS 440, 410
Relative operating characteristics (ROC), 295
Remote sensing
EDC, 3
electromagnetic wavelength spectrum, 2–3
standard remote sensing techniques, 4
visible and near-infrared spectral wavelengths, 2–3
Remote Sensing Institute (RSI), 3
Resistance mechanisms and virulence
Sr2 complex, 133
Sr24 gene breakdown, 135
Sr genes, 134
TTKST and TTTSK, 134
Rhyzopertha dominica, *see* Lesser grain borer
Ripley's *k*-function analysis, 69, 71
RSI, *see* Remote Sensing Institute
RustMapper, 148–149
RustMapper Web, 148–149

S

SAS program
descriptive statistics, 325–326
GSLIB DOS version, 322
large-scale trend, 331–333
mean-variance dependency, 324–327
prediction and decision errors, 363, 365–368
prn files, 321
PROC GCHART, 325
PROC GLM program, 346–353
PROC GMAP, 324–325, 328–329
PROC KRIGE2D and **PROC GMAP**, 356–359
PROC NLIN program, 340–345
PROC UNIVARIATE and **PROC GLM**, 325, 329
PROC VARIOGRAM and **PROC GPLOT**, 335, 339–340
UNIX environment, 321
Satellite and aerial imagery
DOQ image, Bruce quadrangle, 21
DOQQ, 18, 21
EarthExplorer and GloVis, 22
Landsat image, Brookings County, 23
NAIP image, Bruce quadrangle, 21–22
optical-sensor satellites, 22
SaTScan™ software package, 200
sd_equine_wnv.shp shapefile, 202
Seed dispersal modeling
anisotropic dispersal pattern, 262
Euclidian distance, 261
fat-tailed, exponential and Gaussian kernel, 260–261
functional form, 260

gamma function, 261
invasion simulation system, 259
stochastic/random event, 262
setvi199394.dat file, 335–336
Shuttle Radar Topographic Mission (SRTM), 150
Site-specific weed management (SSWM), 302–304
economic threshold, 406
materials and methods
data collection and analysis, 411
flufenacet and metribuzin, 408
prescription map, 409–411
treatment maps, 408–409
V3 and V2 growth, 408–409
weed spatial distribution
economic optimal herbicide rate, 406–407
information, 406
prescription map, 408
weed species composition and herbicide usage, 411–413
yield results
vs. PRE and POST herbicide rate, 414–416
weed density and potential yield loss, 416
weekly precipitation and mean temperature, 413–414
soil_crw_tarea file, 248
Soil taxonomy, 151
Soil textural class, 248–249
South Dakota View Web site
Bruce quadrangle, Brookings County, 17, 20
county boundaries, 16, 18
data services, 14–15
data services login, 14–15
digital base data, 14, 16
quadrangle download and unzip processes, 17, 19
quadrangle selection, 16, 19
USGS digital raster graphics metadata, 16–17
Space–time permutation statistic, 199
Space–time population persistence
population establishment, 35
space–time colony persistence, 36
space–time correspondence analysis, 36
space–time persistence patterns, 36–37
Spatial autocorrelation
clustered spatial pattern, 47
graphic output, 46, 48
GS+ measurement, 177–178
semivariogram, 176–177
Spatial Autocorrelation (Morans I) tool, 46, 48
Spatial data
data types, raster and vector, 13
definitions
data format, 10
dataset projection, 11–12
establishment, 10
geometric accuracy, 10–11
radiometric resolution, 10
spatial extent, 9
spatial resolution, 10
spectral resolution, 10
systematic and nonsystematic error, 11
factors, 12
raster data sources and examples
digital elevation data, 23–24
DRG (*see* Digital raster graphic)
satellite and aerial imagery (*see* Satellite and aerial imagery)
spatial data visualization and analysis software, 26–27
vector data sources and examples, 25–26
Spatial Statistics Tools
OLS regression analysis, 44
spatial autocorrelation, 46
Spatiotemporal colonization patterns, seed potato
aphid infestation, 174
data point map creation, 179–180
dataset, 175–176
interpolated map creation, 180–181
interpolated surface output, GS+, 178–179
observed colonization patterns, 182–183
spatial autocorrelation
GS+ measurement, 177–178
semivariogram, 176–177
Spearmint, 301–302
SRTM, *see* Shuttle Radar Topographic Mission
SSWM, *see* Site-specific weed management

T

Total competitive load (TCL), 406–407
Triangular irregular network, 180–181
Trimble® AgDGPS 132, 411
Turbo Teejet® (TT), 410

U

Ug99
lineage
biology of, 131–132
discovery and identification, 136–137
dispersal, 132–133
distribution map, 137–138
resistance mechanisms and virulence (*see* Resistance mechanisms and virulence)
movements
HYSPLIT model implementation, 139
migration routes prediction map, 139–140
rain scrubbing, 143

Sr24 and *Sr36* variants, 137
wind trajectories, 141–142
surveillance and monitoring systems, 135–136
UN Food and Agriculture Organization
challenges/future activities, 151
climate/environment, 147
GIS-based surveillance and monitoring systems, 136
United States Geological Survey (USGS), 3
Universal Transverse Mercator (UTM)
coordinates, 69
projection, 13, 18
U.S. Census of Agriculture, 194–195, 202

V

Variable rate application (VRA), 406, 410

W

Weed_data_key.xls file, 375–376
Weed_Data.xls file, 376
Weed map and management, *see* Peppermint
Weeds_Data.shp file, 381, 384
Weed shifts, corn/soybean rotation
analyses method, 375–376
ArcMap™
data exploration, 384–387
data subset creation, 383–384
layer creation, 380–383
model validation, 380
data aggregation, MS excel
direction distribution, 378–380
weed densities by year, 376–377
weed density and species change calculations, 377–378
field methods, 375–376
herbicide, 374
interpolation, 374
minimum recommended system requirements, 375
spatial data exploration
attribute tables, 391–392
autocorrelation, 388–389
directional distribution, 389–390
IDW method (*see* Inverse distance weighting method)
neighborhood count, 387–388
ordinary kriging method, 393–396
sample semivariogram, 392–393
Weeds of National Significance (WONS)
current mapping initiatives, 99
management, 94
previous mapping initiatives, 99
Weed_Sum cells, 376–377, 379
Weed training dataset, 289
Weed_Yr1Yr4.xls file, 375–376
weed_yr1yr4.xls file, 375
West Nile virus (WNv), South Dakota
arthropod-transmitted virus/arbovirus, 192–193
disease risk mapping
empirical Bayes smoothing, 195–196
GeoDa software package, 196
horse population mapping, 194–195
raw rates, 195–196
spatial empirical Bayes smoothing, 196–197
GeoDA 0.9.5 software
attribute table addition, 203
empirical Bayes smoothing, 203–204
global Moran's *I* index, spatial autocorrelation, 204–205
local Moran's *I* index, spatial autocorrelation, 204–205
raw disease rates map creation, 203–204
sd_equine_wnv.shp shapefile, 202
shapefile opening, 202
spatial weights file computing, 203
H5N1 avian influenza, 192
infectious disease emergence, 192
mapping, GIS, 193–194
spatial autocorrelation analysis
local Moran's *I* analysis, 198–199
Moran scatterplot, 197–198
Moran's *I* statistic, 197
permutation test, 198
statistical significance assessment, 197
spatiotemporal clustering
likelihood ratio statistic, 200
Monte Carlo simulation, 200
spatial analysis, 198
temporal distribution, 199–200
2002 WNv cases, 200–201
ZCTA level data, 199
symptoms, 193
Wheat stem rust, GIS applications
challenges/future activities
Bayesian statistics application, 150
circumstantial evidence, 149
digital elevation model, 150
race analysis, 152–153
remote sensing, farming systems, 151
soil taxonomy, 151
SRTM, 150
deposition/colonization factors
climate/environment (*see* Climate/environment)
crop calendars/crop growth stage, 145–146
wheat areas, 144–145
wheat cultivars susceptibility, 145
Green Revolution, 131
information tools, 148–149
spore deposition, 153–154
Ug99 (*see* Ug99)
yield losses, 130

Wide area augmentation system (WAAS), 304
Wind dispersal kernel
  multiple dispersal events, 265
  roads and rivers, 264–265
  terrain influences, 263–264
  wind direction and strength, 262–263
WNv, *see* West Nile virus, South Dakota
WONS, *see* Weeds of National Significance

Y

Yellow starthistle
  anisotropic model, 283–284
  control and management strategies, 279
  dispersal, 298
  evolution and location, 279
  isotropic model, 283–284
  Kamiah 1981 data, 296
  logit function, 280
  polar coordinate system, 281
  productivity, 281
  productivity model, 281–283
  sds germinataion, 286–287
  topographic trends, 280
Yr4_dom_type layers, 398–399
Yrl_dom_type.shp file, 399
Yr4_dom_type.shp file, 399
Yrl_gr_elp.shp file, 390

Z

Zip code tabulation areas (ZCTAs), 193–194